ISBN 978-1-334-25431-4
PIBN 10734568

PHILOSOPHICAL

TRANSACTIONS

OF THE

ROYAL SOCIETY OF LONDON.

SERIES A

CONTAINING PAPERS OF A MATHEMATICAL OR PHYSICAL CHARACTER.

VOL. 192.

LONDON:
PRINTED BY HARRISON AND SONS, ST. MARTIN'S LANE, W.C.,
Printers in Ordinary to Her Majesty.
1899.

CONTENTS.

(A)

VOL. 192.

LIST OF ILLUSTRATIONS.

ADVERTISEMENT.

The Committee appointed by the *Royal Society* to direct the publication of the *Philosophical Transactions* take this opportunity to acquaint the public that it fully appears, as well from the Council-books and Journals of the Society as from repeated declarations which have been made in several former *Transactions*, that the printing of them was always, from time to time, the single act of the respective Secretaries till the Forty-seventh Volume; the Society, as a Body, never interesting themselves any further in their publication than by occasionally recommending the revival of them to some of their Secretaries, when, from the particular circumstances of their affairs, the *Transactions* had happened for any length of time to be intermitted. And this seems principally to have been done with a view to satisfy the public that their usual meetings were then continued, for the improvement of knowledge and benefit of mankind: the great ends of their first institution by the Royal Charters, and which they have ever since steadily pursued.

But the Society being of late years greatly enlarged, and their communications more numerous, it was thought advisable that a Committee of their members should be appointed to reconsider the papers read before them, and select out of them such as they should judge most proper for publication in the future *Transactions;* which was accordingly done upon the 26th of March, 1752. And the grounds of their choice are, and will continue to be, the importance and singularity of the subjects, or the advantageous manner of treating them; without pretending to answer for the certainty of the facts, or propriety of the reasonings contained in the several papers so published, which must still rest on the credit or judgment of their respective authors.

It is likewise necessary on this occasion to remark, that it is an established rule of the Society, to which they will always adhere, never to give their opinion, as a Body, upon any subject, either of Nature or Art, that comes before them. And therefore the thanks, which are frequently proposed from the Chair, to be given to the authors of such papers as are read at their accustomed meetings, or to the persons through whose hands they received them, are to be considered in no other light than as a matter of civility, in return for the respect shown to the Society by those communications. The like also is to be said with regard to the several projects, inventions, and curiosities of various kinds, which are often exhibited to the Society; the authors whereof, or those who exhibit them, frequently take the liberty to report, and even to certify in the public newspapers, that they have met with the highest applause and approbation. And therefore it is hoped that no regard will hereafter be paid to such reports and public notices; which in some instances have been too lightly credited, to the dishonour of the Society.

1899.

List of Institutions entitled to receive the Philosophical Transactions or Proceedings of the Royal Society.

Institutions marked A are entitled to receive Philosophical Transactions, Series A, and Proceedings.
" " B " " " " Series B, and Proceedings.
" " AB " " " " Series A and B, and Proceedings.
" " p " " Proceedings only.

America (Central).

Mexico.

p. Sociedad Científica "Antonio Alzate."

America (North). (See United States and Canada.)

America (South).

Buenos Ayres.

AB. Museo Nacional.

Caracas.

B. University Library.

Cordova.

AB. Academia Nacional de Ciencias.

Demerara.

p. Royal Agricultural and Commercial Society, British Guiana.

La Plata.

B. Museo de La Plata.

Rio de Janeiro.

p. Observatorio.

Australia.

Adelaide.

p. Royal Society of South Australia.

Brisbane.

p. Royal Society of Queensland.

Melbourne.

p. Observatory.
p. Royal Society of Victoria.
AB. University Library.

Sydney.

p. Australian Museum.
p. Geological Survey.
p. Linnean Society of New South Wales.
AB. Royal Society of New South Wales.
AB. University Library.

Austria.

Agram.

p. Jugoslavenska Akademija Znanosti i Umjetnosti.
p. Societas Historico-Naturalis Croatica.

Austria (continued).

Brünn.

AB. Naturforschender Verein.

Gratz.

AB. Naturwissenschaftlicher Verein für Steiermark.

Innsbruck.

AB. Das Ferdinandeum.
p. Naturwissenschaftlich - Medicinischer Verein.

Prague.

AB. Königliche Böhmische Gesellschaft der Wissenschaften.

Trieste.

B. Museo di Storia Naturale.
p. Società Adriatica di Scienze Naturali.

Vienna.

p. Anthropologische Gesellschaft.
AB. Kaiserliche Akademie der Wissenschaften.
p. K.K. Geographische Gesellschaft.
AB. K.K. Geologische Reichsanstalt.
B. K.K. Naturhistorisches Hof-Museum.
B. K.K. Zoologisch-Botanische Gesellschaft.
p. Oesterreichische Gesellschaft für Meteorologie.
A. Von Kuffner'sche Sternwarte.

Belgium.

Brussels.

B. Académie Royale de Médecine.
AB. Académie Royale des Sciences.
B. Musée Royal d'Histoire Naturelle de Belgique.
p. Observatoire Royal.
p. Société Belge de Géologie, de Paléontologie, et d'Hydrologie.
p. Société Malacologique de Belgique.

Ghent.

AB. Université.

Belgium (continued).

Liége.

AB. Société des Sciences.

p. Société Géologique de Belgique.

Louvain.

B. Laboratoire de Microscopie et de Biologie Cellulaire.

AB. Université.

Canada.

Halifax, N.S.

p. Nova Scotian Institute of Science.

Hamilton.

p. Hamilton Association.

Montreal.

AB. McGill University.

p. Natural History Society.

Ottawa.

AB. Geological Survey of Canada.

AB. Royal Society of Canada.

St. John, N.B.

p. Natural History Society.

Toronto.

p. Astronomical and Physical Society.

p. Canadian Institute.

AB. University.

Windsor, N.S.

p. King's College Library.

Cape of Good Hope.

A. Observatory.

AB. South African Library.

Ceylon.

Colombo.

a. Museum.

China.

Shanghai.

p. China Branch of the Royal Asiatic Society.

Denmark.

Copenhagen.

AB. Kongelige Danske Videnskabernes Selskab.

Egypt.

Alexandria.

AB. Bibliothèque Municipale.

England and Wales.

Aberystwith.

AB. University College.

Bangor.

AB. University College of North Wales.

Birmingham.

AB. Free Central Library.

AB. Mason College.

p. Philosophical Society.

Bolton.

p. Public Library.

England and Wales (continued).

Bristol.

p. Merchant Venturers' School.

AB. University College.

Cambridge.

AB. Philosophical Society.

p. Union Society.

Cooper's Hill.

AB. Royal Indian Engineering College.

Dudley.

p. Dudley and Midland Geological and Scientific Society.

Essex.

p. Essex Field Club.

Falmouth.

p. Royal Cornwall Polytechnic Society.

Greenwich.

A. Royal Observatory.

Kew.

B. Royal Gardens.

Leeds.

p. Philosophical Society.

AB. Yorkshire College.

Liverpool.

AB. Free Public Library.

p. Literary and Philosophical Society.

A. Observatory.

AB. University College.

London.

AB. Admiralty.

p. Anthropological Institute.

AB. British Museum (Nat. Hist.).

AB. Chemical Society.

A. City and Guilds of London Institute.

p. "Electrician," Editor of the.

B. Entomological Society.

AB. Geological Society.

AB. Geological Survey of Great Britain.

p. Geologists' Association.

AB. Guildhall Library.

A. Institution of Civil Engineers.

p. Institution of Electrical Engineers.

A. Institution of Mechanical Engineers.

A. Institution of Naval Architects.

p. Iron and Steel Institute.

AB. King's College.

B. Linnean Society.

AB. London Institution.

p. London Library.

A. Mathematical Society.

p. Meteorological Office.

p. Odontological Society.

p. Pharmaceutical Society.

England and Wales (continued).

London (continued).

- *p.* Physical Society.
- *p.* Quekett Microscopical Club.
- *p.* Royal Agricultural Society.
- A. Royal Astronomical Society.
- B. Royal College of Physicians.
- B. Royal College of Surgeons.
- *p.* Royal Engineers (for Libraries abroad, six copies).
- AB. Royal Engineers. Head Quarters Library.
- *p.* Royal Geographical Society.
- *p.* Royal Horticultural Society.
- *p.* Royal Institute of British Architects.
- AB. Royal Institution of Great Britain.
- B. Royal Medical and Chirurgical Society.
- *p.* Royal Meteorological Society.
- *p.* Royal Microscopical Society.
- *p.* Royal Statistical Society.
- AB. Royal United Service Institution.
- AB. Society of Arts.
- *p.* Society of Biblical Archæology.
- *p.* Society of Chemical Industry (London Section).
- *p.* Standard Weights and Measures Department.
- AB. The Queen's Library.
- AB. The War Office.
- AB. University College.
- *p.* Victoria Institute.
- B. Zoological Society.

Manchester.

- AB. Free Library.
- AB. Literary and Philosophical Society.
- *p.* Geological Society.
- AB. Owens College.

Netley.

- *p.* Royal Victoria Hospital.

Newcastle.

- AB. Free Library.
- *p.* North of England Institute of Mining and Mechanical Engineers.
- *p.* Society of Chemical Industry (Newcastle Section).

Norwich.

- *p.* Norfolk and Norwich Literary Institution.

Nottingham.

- AB. Free Public Library.

Oxford.

- *p.* Ashmolean Society.
- AB. Radcliffe Library.
- A. Radcliffe Observatory.

England and Wales (continued).

Penzance.

- *p.* Geological Society of Cornwall.

Plymouth.

- B. Marine Biological Association.
- *p.* Plymouth Institution.

Richmond.

- A. "Kew" Observatory.

Salford.

- *p.* Royal Museum and Library.

Stonyhurst.

- *p.* The College.

Swansea.

- AB. Royal Institution.

Woolwich.

- AB. Royal Artillery Library.

Finland.

Helsingfors.

- *p.* Societas pro Fauna et Flora Fennica.
- AB. Société des Sciences.

France.

Bordeaux.

- *p.* Académie des Sciences.
- *p.* Faculté des Sciences.
- *p.* Société de Médecine et de Chirurgie.
- *p.* Société des Sciences Physiques et Naturelles.

Caen.

- *p.* Société Linnéenne de Normandie.

Cherbourg.

- *p.* Société des Sciences Naturelles.

Dijon.

- *p.* Académie des Sciences.

Lille.

- *p.* Faculté des Sciences.

Lyons.

- AB. Académie des Sciences, Belles-Lettres et Arts.
- AB. Université.

Marseilles.

- AB. Faculté des Sciences.

Montpellier.

- AB. Académie des Sciences et Lettres.
- B. Faculté de Médecine.

Nantes.

- *p.* Société des Sciences Naturelles de l'Ouest de la France.

Paris.

- AB. Académie des Sciences de l'Institut.
- *p.* Association Française pour l'Avancement des Sciences.
- *p.* Bureau des Longitudes.
- A. Bureau International des Poids et Mesures.
- *p.* Commission des Annales des Ponts et Chaussées.
- *p.* Conservatoire des Arts et Métiers.

France (continued).

Paris (continued).

- *p*. Cosmos (M. L'ABBÉ VALETTE).
- AB. Dépôt de la Marine.
- AB. École des Mines.
- AB. École Normale Supérieure.
- AB. École Polytechnique.
- AB. Faculté des Sciences de la Sorbonne.
- AB. Jardin des Plantes.
- *p*. L'Électricien.
- A. L'Observatoire.
- *p*. Revue Scientifique (Mons. H. DE VARIGNY).
- *p*. Société de Biologie.
- AB. Société d'Encouragement pour l'Industrie Nationale.
- AB. Société de Géographie.
- *p*. Société de Physique.
- B. Société Entomologique.
- AB. Société Géologique.
- *p*. Société Mathématique.
- *p*. Société Météorologique de France.

Toulouse.

- AB. Académie des Sciences.
- A. Faculté des Sciences.

Germany.

Berlin.

- A. Deutsche Chemische Gesellschaft.
- A. Die Sternwarte.
- *p*. Gesellschaft für Erdkunde.
- AB. Königliche Preussische Akademie der Wissenschaften.
- A. Physikalische Gesellschaft.

Bonn.

- AB. Universität.

Bremen.

- *p*. Naturwissenschaftlicher Verein.

Breslau.

- *p*. Schlesische Gesellschaft für Vaterländische Kultur.

Brunswick.

- *p*. Verein für Naturwissenschaft.

Carlsruhe. See Karlsruhe.

Charlottenburg.

- A. Physikalisch-Technische Reichsanstalt.

Danzig.

- AB. Naturforschende Gesellschaft.

Dresden.

- *p*. Verein für Erdkunde.

Emden.

- *p*. Naturforschende Gesellschaft.

Erlangen.

- AB. Physikalisch-Medicinische Societät.

Frankfurt-am-Main.

- AB. Senckenbergische Naturforschende Gesellschaft.

Germany (continued).

Frankfurt-am-Main (continued).

- *p*. Zoologische Gesellschaft.

Frankfurt-am-Oder.

- *p*. Naturwissenschaftlicher Verein.

Freiburg-im-Breisgau.

- AB. Universität.

Giessen.

- AB. Grossherzogliche Universität.

Görlitz.

- *p*. Naturforschende Gesellschaft.

Göttingen.

- AB. Königliche Gesellschaft der Wissenschaften.

Halle.

- AB. Kaiserliche Leopoldino - Carolinische Deutsche Akademie der Naturforscher.
- *p*. Naturwissenschaftlicher Verein für Sachsen und Thüringen.

Hamburg.

- *p*. Naturhistorisches Museum.
- AB. Naturwissenschaftlicher Verein.

Heidelberg.

- *p*. Naturhistorisch-Medizinischer Verein.
- AB. Universität.

Jena.

- AB. Medicinisch-Naturwissenschaftliche Gesellschaft.

Karlsruhe.

- A. Grossherzogliche Sternwarte.
- *p*. Technische Hochschule.

Kiel.

- *p*. Naturwissenschaftlicher Verein für Schleswig-Holstein.
- A. Sternwarte.
- AB. Universität.

Königsberg.

- AB. Königliche Physikalisch - Ökonomische Gesellschaft.

Leipsic.

- *p*. Annalen der Physik und Chemie.
- AB. Königliche Sächsische Gesellschaft der Wissenschaften.

Magdeburg.

- *p*. Naturwissenschaftlicher Verein.

Marburg.

- AB. Universität.

Munich.

- AB. Königliche Bayerische Akademie der Wissenschaften.
- *p*. Zeitschrift für Biologie.

Münster.

- AB. Königliche Theologische und Philosophische Akademie.

Germany (continued).
Potsdam.
A. Astrophysikalisches Observatorium.
Rostock.
AB. Universität.
Strasburg.
AB. Universität.
Tübingen.
AB. Universität.
Würzburg.
AB. Physikalisch-Medicinische Gesellschaft.

Greece.
Athens.
A. National Observatory.

Holland. (See NETHERLANDS.)

Hungary.
Buda-pest.
p. Königl. Ungarische Geologische Anstalt.
AB. Á Magyar Tudós Társaság. Die Ungarische Akademie der Wissenschaften.
Hermannstadt.
p. Siebenbürgischer Verein für die Naturwissenschaften.
Klausenburg.
AB. Az Erdélyi Muzeum. Das Siebenbürgische Museum.
Schemnitz.
p. K. Ungarische Berg- und Forst-Akademie.

India.
Bombay.
AB. Elphinstone College.
p. Royal Asiatic Society (Bombay Branch).
Calcutta.
AB. Asiatic Society of Bengal.
AB. Geological Museum.
p. Great Trigonometrical Survey of India.
AB. Indian Museum.
p. The Meteorological Reporter to the Government of India.
Madras.
B. Central Museum.
A. Observatory.
Roorkee.
p. Roorkee College.

Ireland.
Armagh.
A. Observatory.
Belfast.
AB. Queen's College.
Cork.
p. Philosophical Society.
AB. Queen's College.

Ireland (continued).
Dublin.
A. Observatory.
AB. National Library of Ireland.
B. Royal College of Surgeons in Ireland.
AB. Royal Dublin Society.
AB. Royal Irish Academy.
Galway.
AB. Queen's College.

Italy.
Acireale.
p. Accademia di Scienze, Lettere ed Arti.
Bologna.
AB. Accademia delle Scienze dell' Istituto.
Catania.
AB. Accademia Gioenia di Scienze Naturali.
Florence.
p. Biblioteca Nazionale Centrale
AB. Museo Botanico.
p. Reale Istituto di Studi Superiori.
Genoa.
p. Società Ligustica di Scienze Naturali e Geografiche.
Milan.
AB. Reale Istituto Lombardo di Scienze, Lettere ed Arti.
AB. Società Italiana di Scienze Naturali.
Modena.
p. Le Stazioni Sperimentali Agrarie Italiane.
Naples.
p. Società di Naturalisti.
AB. Società Reale, Accademia delle Scienze.
B. Stazione Zoologica (Dr. DOHRN).
Padua.
p. University.
Palermo.
A. Circolo Matematico.
AB. Consiglio di Perfezionamento (Società di Scienze Naturali ed Economiche).
A. Reale Osservatorio.
Pisa.
p. Il Nuovo Cimento.
p. Società Toscana di Scienze Naturali.
Rome.
p. Accademia Pontificia de' Nuovi Lincei.
p. Rassegna delle Scienze Geologiche in Italia.
A. Reale Ufficio Centrale di Meteorologia e di Geodinamica, Collegio Romano.
AB. Reale Accademia dei Lincei.
p. R. Comitato Geologico d' Italia.
A. Specola Vaticana.
AB. Società Italiana delle Scienze.
Siena.
p. Reale Accademia dei Fisiocritici.

Italy (continued).
Turin.
p. Laboratorio di Fisiologia.
AB. Reale Accademia delle Scienze.
Venice.
p. Ateneo Veneto.
AB. Reale Istituto Veneto di Scienze, Lettere ed Arti.

Japan.
Tokiô.
AB. Imperial University.
p. Asiatic Society of Japan.

Java.
Buitenzorg.
p. Jardin Botanique.

Luxembourg.
Luxembourg.
p. Société des Sciences Naturelles.

Malta.
p. Public Library.

Mauritius.
p. Royal Society of Arts and Sciences.

Netherlands.
Amsterdam.
AB. Koninklijke Akademie van Wetenschappen.
p. K. Zoologisch Genootschap 'Natura Artis Magistra.'
Delft.
p. École Polytechnique.
Haarlem.
AB. Hollandsche Maatschappij der Wetenschappen.
p. Musée Teyler.
Leyden.
AB. University.
Rotterdam.
AB. Bataafsch Genootschap der Proefondervindelijke Wijsbegeerte.
Utrecht.
AB. Provinciaal Genootschap van Kunsten en Wetenschappen.

New Zealand.
Wellington.
AB. New Zealand Institute.

Norway.
Bergen.
AB. Bergenske Museum.
Christiania.
AB. Kongelige Norske Frederiks Universitet.
Tromsoe.
p. Museum.
Trondhjem.
AB. Kongelige Norske Videnskabers Selskab.

Portugal.
Coimbra.
AB. Universidade.
Lisbon.
AB. Academia Real das Sciencias.
p. Secção dos Trabalhos Geologicos de Portugal.
Oporto.
p. Annaes de Sciencias Naturaes.

Russia.
Dorpat.
AB. Université.
Irkutsk.
p. Société Impériale Russe de Géographie (Section de la Sibérie Orientale).
Kazan.
AB. Imperatorsky Kazansky Universitet.
p. Société Physico-Mathématique.
Kharkoff.
p. Section Médicale de la Société des Sciences Expérimentales, Université de Kharkow.
Kieff.
p. Société des Naturalistes.
Moscow.
AB. Le Musée Public.
B. Société Impériale des Naturalistes.
Odessa.
p. Société des Naturalistes de la Nouvelle-Russie.
Pulkowa.
A. Nikolai Haupt-Sternwarte.
St. Petersburg.
AB. Académie Impériale des Sciences.
B. Archives des Sciences Biologiques.
AB. Comité Géologique.
p. Compass Observatory.
A. Observatoire Physique Central.

Scotland.
Aberdeen.
AB. University.
Edinburgh.
p. Geological Society.
p. Royal College of Physicians (Research Laboratory).
p. Royal Medical Society.
A. Royal Observatory.
p. Royal Physical Society.
p. Royal Scottish Society of Arts
AB Royal Society.
Glasgow.
AB. Mitchell Free Library.
p. Natural History Society.
p. Philosophical Society.

Servia.
Belgrade.
p. Académie Royale de Serbie.

Sicily. (See ITALY.)

Spain.

Cadiz.

A. Instituto y Observatorio de Marina de San Fernando.

Madrid.

p. Comisión del Mapa Geológico de Espāna.

AB. Real Academia de Ciencias.

Sweden.

Gottenburg.

AB. Kongl. Vetenskaps och Vitterhets Samhälle.

Lund.

AB. Universitet.

Stockholm.

A. Acta Mathematica.

AB. Kongliga Svenska Vetenskaps-Akademie.

AB. Sveriges Geologiska Undersökning.

Upsala.

AB. Universitet.

Switzerland.

Basel.

p. Naturforschende Gesellschaft.

Bern.

AB. Allg. Schweizerische Gesellschaft.

p. Naturforschende Gesellschaft.

Geneva.

AB. Société de Physique et d'Histoire Naturelle.

AB. Institut National Genevois.

Lausanne.

p. Société Vaudoise des Sciences Naturelles.

Neuchâtel.

p. Société des Sciences Naturelles.

Zürich.

AB. Das Schweizerische Polytechnikum.

p. Naturforschende Gesellschaft.

p. Sternwarte.

Tasmania.

Hobart.

p. Royal Society of Tasmania.

United States.

Albany.

AB. New York State Library.

Annapolis.

AB. Naval Academy.

Austin.

p. Texas Academy of Sciences.

Baltimore.

AB. Johns Hopkins University.

Berkeley.

p. University of California.

United States (continued).

Boston.

AB. American Academy of Sciences.

B. Boston Society of Natural History.

A. Technological Institute.

Brooklyn.

AB. Brooklyn Library.

Cambridge.

AB. Harvard University.

B. Museum of Comparative Zoology.

Chapel Hill (N.C.).

p. Elisha Mitchell Scientific Society.

Charleston.

p. Elliott Society of Science and Art of South Carolina.

Chicago.

AB. Academy of Sciences.

p. Field Columbian Museum.

p. Journal of Comparative Neurology.

Davenport (Iowa).

p. Academy of Natural Sciences.

Ithaca (N.Y.).

p. Physical Review (Cornell University).

Madison.

p. Wisconsin Academy of Sciences.

Mount Hamilton (California).

A. Lick Observatory.

New Haven (Conn.).

AB. American Journal of Science.

AB. Connecticut Academy of Arts and Sciences.

New York.

p. American Geographical Society.

p. American Mathematical Society.

p. American Museum of Natural History.

p. New York Academy of Sciences.

p. New York Medical Journal.

p. School of Mines, Columbia College.

Philadelphia.

AB. Academy of Natural Sciences.

AB. American Philosophical Society.

p. Franklin Institute.

p. Wagner Free Institute of Science.

Rochester (N.Y.).

p. Academy of Science.

St. Louis.

p. Academy of Science.

Salem (Mass.).

p. American Association for the Advancement of Science.

AB. Essex Institute.

San Francisco.

AB. California Academy of Sciences.

United States (continued).

Washington.

AB. Patent Office.

AB. Smithsonian Institution.

AB. United States Coast Survey.

B. United States Commission of Fish and Fisheries.

AB. United States Geological Survey.

United States (continued).

Washington (continued).

AB. United States Naval Observatory.

p. United States Department of Agriculture.

A. United States Department of Agriculture (Weather Bureau).

West Point (N.Y.)

AB. United States Military Academy.

Adjudication of the Medals of the Royal Society for the year 1898, by the President and Council.

The Copley Medal to Sir William Huggins, F.R.S., for his Researches in Spectrum Analysis applied to the Heavenly Bodies.

The Rumford Medal to Oliver Joseph Lodge, F.R.S., for his Researches in Radiation and in the relations between Matter and Ether.

A Royal Medal to Walter Gardiner, F.R.S., for his Researches on the Protoplasmic Connection of the Cells of Vegetable Tissues and on the Minute Histology of Plants.

A Royal Medal to the Rev. John Kerr, LL.D., F.R.S., for his Researches on the Optical Effect of Electrical Stress and on the Reflection of Light at the Surface of a Magnetized Body.

The Davy Medal to Johannes Wislicenus, For.Mem.R.S., for his Contributions to Organic Chemistry, especially in the Domain of Stereochemical Isomerism.

The Darwin Medal to Karl Pearson, F.R.S., for his work on the Quantitative Treatment of Biological Problems.

The Bakerian Lecture for 1899, "The Crystalline Structure of Metals," was delivered by Professor J. A. Ewing, F.R.S., and Mr. W. Rosenhain on May 18, 1899.

The Croonian Lecture for 1899, "On the Relation of Motion on Animals and Plants to the Electrical Phenomena which are associated with it," was delivered by Professor Burdon Sanderson, F.R.S., on March 16, 1899.

TIONS.

orphic Functions.

e, Cambridge.

)., F.R.S.

INDEX SLIP.

WHITTAKER, E. T.—On the Connexion of Algebraic Functions with Automorphic Functions.
Phil. Trans., A, 1898, vol. 192, pp. 1–32.

Algebraic Functions, expressed as Uniform Automorphic Functions.
WHITTAKER, E. T. Phil. Trans., A, 1898, vol. 192, pp. 1–32.

Automorphic Functions, corresponding to Algebraic Relations.
WHITTAKER, E. T. Phil. Trans., A, 1898, vol. 192, pp. 1–32.

Differential Equations satisfied by Automorphic Functions; Lamé's generalised.
WHITTAKER, E. T. Phil. Trans., A, 1898, vol. 192, pp. 1–32.

. . . . (1)

:ht) zero, then u and z
If, however, the genus
elliptic functions of a

31 by the discovery of
(1), u and z can be

functions. Instead of
of z on the Riemann
independent variable,
ie multiformity of the
of uniform functions.
inctions, however, has
forms ; in describing
nnexions which exist
hitherto which have
unity, are those given
ly uniformise special
ich the fundamental
i. These latter have

2.12.98.

PHILOSOPHICAL TRANSACTIONS.

I. *On the Connexion of Algebraic Functions with Automorphic Functions.*

By E. T. Whittaker, *B.A., Fellow of Trinity College, Cambridge.*

Communicated by Professor A. R. Forsyth, *Sc.D., F.R.S.*

Received April 23,—Read May 12, 1898.

§ 1. *Introduction.*

It is well known that if

$$f(u, z) = 0 \quad . \quad . \quad . \quad . \quad . \quad . \quad . \quad . \quad . \quad . \quad . \quad . \quad (1)$$

is the equation of an algebraic curve of genus (*genre, Geschlecht*) zero, then u and z can be expressed as rational functions of a single variable t. If, however, the genus of the curve (1) is unity, u and z can be expressed as uniform elliptic functions of a variable t.

The natural extension of these results was effected in 1881 by the discovery of automorphic functions; whatever be the genus of the curve (1), u and z can be expressed as uniform automorphic functions of a new variable.

This result is of great importance in the study of algebraic functions. Instead of taking z as the independent variable, and studying functions of z on the Riemann surface corresponding to the equation (1), we can take t as the independent variable, and consider the functions in the plane of t. We thus avoid the multiformity of the problem, and can apply the simpler and more developed theory of uniform functions.

Comparatively little of the published work on automorphic functions, however, has been written in connexion with the uniformisation of algebraic forms; in describing either groups applicable for the purpose, or the analytical connexions which exist between u, z, and t. The only automorphic functions known hitherto which have been applied to uniformise forms whose genus is greater than unity, are those given by certain sub-groups of the modular group (which will only uniformise special curves, containing no arbitrary constants), and those in which the fundamental polygon is the space outside a number of non-intersecting circles. These latter have been studied by SCHOTTKY,* WEBER,† and BURNSIDE,‡ and are capable of uniformising any algebraic form. As, however, the fundamental polygon is multiply-connected, the Abelian Integrals of the first kind, and the factorial functions associated with the algebraic form, are not uniform functions of the new variable.

With regard to the analytical connexion between the uniformising variable t and the variables u, z, of the algebraic form, POINCARÉ proved that if z is an automorphic function of t, then $\{t, z\}$ is another automorphic function of the same group, where $\{t, z\}$ is the Schwarzian derivative. t therefore satisfies a differential equation of the form

$$\{t, z\} = \phi(u, z),$$

where $\phi(u, z)$ is some rational function of u and z. SCHOTTKY and WEBER have determined $\phi(u, z)$, save for a number of undetermined constants, for the groups found by them, and KLEIN§ has obtained more general results, applying to any algebraic equation, but with a certain number of undetermined constants left in ϕ.

The problem has been formulated by KLEIN as one of conformal representation. The algebraic form which is given by

$$f(u, z) = 0$$

can be represented on a Riemann surface of class p, so that, corresponding to every pair of values (u, z) of the form, there is a place on the surface. By drawing $2p$ cuts we can make this surface simply-connected. Now let z be regarded as a function of a new variable t, having the following properties :—

1°. The dissected Riemann surface is to be conformally represented on a plane area in the t-plane, bounded by $4p$ curvilinear sides (namely, the conformal representations of the cuts, each cut giving two sides).

2°. Of the two sides of the t-area which correspond to any cut, one is to be derivable from the other by a projective substitution

$$\left(t, \frac{at + b}{ct + d}\right).$$

3°. The group formed by the combination and repetition of these $2p$ substitutions is to be discontinuous.

When a variable t has been found satisfying these conditions, u and z will be uniform automorphic functions of t; and we know by the existence-theorem of POINCARÉ and KLEIN that such a variable does exist, although the existence-theorem does not connect it analytically with z and u. The primary result of the present paper is, that the uniformisation of any algebraic form can be effected by automorphic func-

* 'Crelle,' vol. 101, 1897, p. 227.

† 'Göttinger Nachrichten,' 1886, p. 359.

‡ 'Proc. Lond. Math. Soc.,' vol. 23, 1891, p. 49.

§ 'Jahresbericht der Deutschen Mathematiker-Vereinigung,' 1894–5, p. 91.

tions of certain kinds of groups, which are described in § 3. These are either groups whose generating substitutions are of period two, or sub-groups of such groups. This theorem is made to depend on the well-known theorem that any algebraic form can, by birational transformation, be represented on a Riemann surface with only simple branch-points. A method is given for the division of the t-plane into polygons, corresponding to a group generated by real substitutions of period two, whose double points are not on the real axis; and the genus of the group is found. The group is of the kind called by POINCARÉ Fuchsian; the polygons into which the plane is divided are simply-connected, and cover completely the half of the t-plane which is above the real axis. Results are deduced relating to the possibility of uniformising any algebraic functions by automorphic functions of such groups, and the analytical connexion of the uniformising variable with the variables of the form.

In § 2, certain properties of substitutions of period two are found, which are of use later. These substitutions are for brevity termed "self-inverse" substitutions, owing to the fact that they are the same as their inverse substitutions.

In § 3, a method is given for carrying out the division of the plane into polygons, corresponding to a group generated by a given set of self-inverse substitutions. It is proved that the genus of the group is zero, although the group has sub-groups whose genus is greater than zero.

In § 4, the automorphic functions of the group are introduced. Since the group is of genus zero, these automorphic functions are all rational algebraic functions of one of them; the conformal representation of the polygons in the t-plane on the plane of this variable is considered. It is shown that the functions which have been obtained solve the following problem of conformal representation:—Draw from any point P, in the plane of a variable z, lines (not necessarily straight) to any other points A, B, C. . . . This set of rays is to be regarded as the boundary of the z-plane, and the problem is, to conformally represent the z-plane, thus bounded, on a simply-connected region in the plane of a variable t, in such a way that each of the lines PA, PB, PC, . . . gives rise to two distinct lines of the boundary of the t-region; and one of these lines is derivable from the other by a projective substitution

$$\left(t, \frac{at+b}{ct+d}\right).$$

The uniformisation of algebraic functions is afterwards made to depend on this problem of conformal representation.

In § 5, the analytical relation between the variables z and t is discussed. It is shown that they are connected by a differential equation which is a particular case of what has been named by KLEIN the "generalised LAMÉ's equation," and has been connected by BÔCHER with the differential equations of harmonic analysis.

In § 6, the functions which have been obtained are applied to the uniformising of algebraic forms. The differential equation in the hyperelliptic case is found to be the same as KLEIN's "unverzweigt" differential equation for hyperelliptic forms, save that a number of constants left arbitrary in KLEIN's equation are found to be zero. The conditions that $2p$ arbitrarily given substitutions may generate the group corresponding to a hyperelliptic equation of genus p are found.

In § 7, the consideration of the constants left undetermined in the differential equation of § 5 is resumed. If an algebraic form of genus p be given, the uniformising variable is one of ∞^{3p-3} variables, which are here termed "quasi-uniformising." Any quasi-uniformising variable affords a solution of the problem of conformally representing the Riemann surface of the form on a plane area whose sides are derived from each other in pairs by projective substitutions. The differential equations connecting the uniformising with the quasi-uniformising variables of a given algebraic form are obtained.

§ 2. *Properties of Self-inverse Substitutions.*

A projective substitution of a variable t is denoted by

$$\left(t, \frac{at+b}{ct+d}\right),$$

where we can always suppose that $ad - bc = 1$.

The substitutions, from which the groups considered in this paper are generated, are such that

$$a + d = 0.$$

Such a substitution is elliptic and of period two; its multiplier is -1, and it is its own inverse substitution. For brevity we shall call such substitutions "*self-inverse.*" Thus, if S denotes any self-inverse substitution, we have

$$S^2 = 1, \quad \text{and} \quad S = S^{-1}.$$

If T *be any substitution, and* S *be a self-inverse substitution, then* $T^{-1}ST$ *is a self-inverse substitution.* For the multiplier of a substitution is unaffected by the transformation which changes S into $T^{-1}ST$.

If there be any number of self-inverse substitutions, and a substitution be formed from them, then the substitution inverse to this is formed by taking the same substitutions in the reverse order. For if S_p, S_q, S_r, ..., are self-inverse substitutions, then obviously

$$S_pS_qS_r \ldots S_uS_vS_wS_wS_vS_u \ldots S_rS_qS_p = 1.$$

So if

$$T = S_pS_q \ldots S_uS_vS_w,$$

then

$$T^{-1} = S_wS_vS_u \ldots S_rS_qS_p.$$

The group formed by the combination and repetition of any two projective substitutions can be obtained as a self-conjugate sub-group of a group generated by three self-inverse substitutions.

For let

$$T_1 = \left(t, \frac{\alpha_1 t + \beta_1}{\gamma_1 t + \delta_1}\right) \quad \text{and} \quad T_2 = \left(t, \frac{\alpha_2 t + \beta_2}{\gamma_2 t + \delta_2}\right)$$

be the given substitutions; let

$$S_1 = \left(t, \frac{a_1 t + b_1}{c_1 t - a_1}\right), \quad S_2 = \left(t, \frac{a_2 t + b_2}{c_2 t - a_2}\right), \quad S_3 = \left(t, \frac{a_3 t + b_3}{c_3 t - a_3}\right)$$

be three self-inverse substitutions; then we have

$$S_3S_1(t) = \frac{(a_1a_3 + c_1b_3)\,t + (a_3b_1 - a_1b_3)}{(a_1c_3 - a_3c_1)\,t + (a_1a_3 + b_1c_3)}$$

and

$$S_3S_2(t) = \frac{(a_2a_3 + b_3c_2)\,t + (a_3b_2 - a_2b_3)}{(a_2c_3 - a_3c_2)\,t + (a_2a_3 + b_2c_3)}.$$

The equations to be satisfied by the coefficients of S_3, in order that we may have

$$S_3S_1 = T_1, \quad \text{and} \quad S_3S_2 = T_2,$$

reduce to

$$\left.\begin{aligned}(\alpha_1 - \delta_1)\,a_3 + \gamma_1 b_3 + \beta_1 c_3 = 0\\ (\alpha_2 - \delta_2)\,a_3 + \gamma_2 b_3 + \beta_2 c_3 = 0\end{aligned}\right\}.$$

These equations always admit of a solution for the ratios $a_3 : b_3 : c_3$, if the substitutions T_1 and T_2 are distinct. Thus, the substitution S_3 is determinate; and then S_1 and S_2 can be uniquely determined from the equations

$$S_1 = S_3T_1, \qquad S_2 = S_3T_2.$$

[Added June 2. In view of the subsequent limitations to substitutions for which $a^2 + bc$ is negative, it should be noticed that these equations may give either a positive or a negative value for $a^2 + bc$.]

Now let G denote the group formed from the generating substitutions S_1, S_2, S_3, and let H denote the group formed from the generating substitutions T_1 and T_2.

As T_1 and T_2 are themselves substitutions of the group G, the group H will be either the same as G, or a sub-group of it. We shall now show that H is a self-conjugate sub-group of G.

Since $S_r^2 = 1$, and $S_r = S_r^{-1}$, any substitution of G can be represented in the form

$$\Sigma = S_pS_qS_rS_s \ldots S_v, \quad \text{where} \quad p, q, r, s, \ldots v = 1, 2, 3.$$

But

$$S_1S_2 = T_1^{-1}T_2, \qquad S_1S_3 = T_1^{-1}, \qquad S_2S_1 = T_2^{-1}T_1,$$
$$S_2S_3 = T_2^{-1}, \qquad S_3S_1 = T_1, \qquad S_3S_2 = T_2.$$

Therefore every pair S_pS_q can be expressed in terms of T_1 and T_2.

So if the number of substitutions in Σ is even, the whole substitution can be expressed in the form

$$\Sigma = T_1^{\alpha}T_2^{\beta}T_1^{\gamma}T_2^{\delta} \dots$$

i.e., it is a substitution of the group H.

But if the number of substitutions in Σ is odd, there will be one substitution S_r left at the end unpaired. Now

$$S_1 = T_1^{-1}S_3, \qquad S_2 = T_2^{-1}S_3, \qquad S_3 = S_3,$$

so in any case

$$\Sigma = T_1^{\alpha}T_2^{\beta}T_1^{\gamma}T_2^{\delta} \dots T_r^{\nu}S_3.$$

So Σ is always either a substitution of H, or else the product of S_3 and a substitution of H.

Now let S_h be any substitution of H, and S_g any substitution of G.

Then $S_g^{-1}S_hS_g$ evidently contains, when decomposed into the substitutions S_1, S_2, S_3, an even number of them; for S_h contains an even number, and S_g^{-1} and S_g each contain the same number. Therefore $S_g^{-1}S_hS_g$ is a substitution of the group H; which establishes the required result, namely, that H is a self-conjugate sub-group.

As an example of this theorem, consider the modular group generated by the substitutions

$$(t, t + 1) \quad \text{and} \quad \left(t, -\frac{1}{t}\right).$$

This is a self-conjugate sub-group of the group formed from the three self-inverse substitutions

$$(t, -t-1), \qquad \left(t, \frac{1}{t}\right), \qquad (t, -t).$$

As another example, take the group which occurs in the theory of elliptic functions, which is formed from the generating substitutions

$$(t, t + 2w_1), \qquad (t, t + 2w_2).$$

This is a self-conjugate sub-group of the group formed from the three self-inverse substitutions

$$(t, c - 2w_1 - t), \qquad (t, c - 2w_2 - t), \qquad (t, c - t)$$

where c is an arbitrary constant.

In this exceptional case, an arbitrary constant, c, is introduced. The reason is, that the quantities $\alpha_1 - \delta_1$, $\alpha_2 - \delta_2$, γ_1, γ_2, all vanish, so the two equations for determining $a_3 : b_3 : c_3$ reduce to the single equation

$$c_3 = 0.$$

Any group of substitutions which is formed from $(k + 1)$ self-inverse substitutions as generating substitutions, always contains a self-conjugate sub-group which can be generated from k substitutions.

For let G be a group formed from $(k + 1)$ self-inverse substitutions $S_1, S_2, S_3, \ldots S_{k+1}$. Then, as before, any substitution of G can be written in the form

$$\Sigma = S_p S_q S_r S_s S_t \ldots S_v.$$

Now let

$$T_1 = S_{k+1} S_1,\ T_2 = S_{k+1} S_2,\ \ldots\ T_k = S_{k+1} S_k.$$

Then

$$S_p S_q = S_p S_{k+1} S_{k+1} S_q = T_p^{-1} T_q.$$

Therefore, if the number of substitutions in Σ is even, Σ can be expressed in the form

$$\Sigma = T_p^{-1} T_q T_r^{-1} T_s \ldots T_v,$$

so Σ is a substitution of the group generated from $T_1, T_2, \ldots T_k$.

If the number of substitutions in Σ is odd, we have, therefore,

$$\Sigma = T_p^{\alpha} T_q^{\beta} \ldots T_s^{\delta} S_r,$$

and as

$$S_r = T_r^{-1} S_{k+1},$$

we have, in this case,

$$\Sigma = T_p^{\alpha} T_q^{\beta} \ldots T_s^{\delta} T_r^{-1} S_{k+1}.$$

So any substitution of the group G can be expressed either in the form Σ_p, or in the form $\Sigma_p S_{k+1}$, where Σ_p is a substitution of the group H, which is formed from $T_1, T_2, \ldots T_k$. And as in the case $k = 2$, which has been already discussed, we see that H is a self-conjugate sub-group of G.

[Added June 2, 1898.—H may, of course, coincide with G; I am indebted to Professor Burnside for the example,

$$S_1^2 = 1,\ S_2^2 = 1,\ S_3^2 = 1,\ (S_1 S_2 S_3)^2 = 1,$$

in which this happens.]

To find the conditions that a group H, generated from any k arbitrary projective substitutions, $T_1, T_2, \ldots T_k$, may in this way be a self-conjugate sub-group of a group G formed from $(k + 1)$ self-inverse substitutions.

Let

$$T_1 = \left(t, \frac{\alpha_1 t + \beta_1}{\gamma_1 t + \delta_1}\right), \quad T_2 = \left(t, \frac{\alpha_2 t + \beta_2}{\gamma_2 t + \delta_2}\right), \ldots T_k = \left(t, \frac{\alpha_k t + \beta_k}{\gamma_k t + \delta_k}\right).$$

Let

$$S_{k+1} = \left(t, \frac{at + b}{ct - a}\right), \quad \text{and let} \quad S_r = S_{k+1} T_r.$$

Then

$$S_r = \left(t, \frac{(a\alpha_r + b\gamma_r)t + (a\beta_r + b\delta_r)}{(c\alpha_r - a\gamma_r)t + (c\beta_r - a\delta_r)}\right).$$

If this is a self-inverse substitution, we have

$$a(\alpha_r - \delta_r) + b\gamma_r + c\beta_r = 0.$$

Thus the coefficients of the substitution S_{k+1} must satisfy the conditions

$$\left.\begin{array}{l} (\alpha_1 - \delta_1)a + \gamma_1 b + \beta_1 c = 0 \\ (\alpha_2 - \delta_2)a + \gamma_2 b + \beta_2 c = 0 \\ \cdot\ \cdot\ \cdot\ \cdot\ \cdot\ \cdot\ \cdot\ \cdot\ \cdot\ \cdot\ \cdot \\ \cdot\ \cdot\ \cdot\ \cdot\ \cdot\ \cdot\ \cdot\ \cdot\ \cdot\ \cdot\ \cdot \\ (\alpha_k - \delta_k)a + \gamma_k b + \beta_k c = 0 \end{array}\right\}.$$

The elimination of $a : b : c$, from these equations gives $(k - 2)$ conditions between the coefficients of the substitutions T.

[Added June 2, 1898.—These conditions are sufficient, but are not actually necessary, as it may be possible to generate the group from a different set of substitutions, for which these conditions are satisfied, although they may not be satisfied by $T_1, T_2, \ldots T_k$.]

We shall, later, take $k = 2p$, and show that these $(2p - 2)$ conditions must be satisfied by the coefficients of $2p$ substitutions, whose group gives rise to automorphic functions which uniformise a hyperelliptic form of genus p.

§ 3. *The Division of the t-plane, corresponding to a group formed of Self-inverse Substitutions with Real Coefficients.*

A method will now be given for dividing the t-plane into regions, corresponding to a group generated from a given set of self-inverse substitutions. These regions are to be derivable from each other by applying the substitutions of the group.

Let

$$S = \left(t, \frac{at + b}{ct - a}\right)$$

be a self-inverse substitution with real coefficients a, b, c. Then the substitution transforms real values of t into other real values, so the real axis in the t-plane is unaffected by the substitution. If (a^2+bc) is negative, it is easily seen that the part of the t-plane above the real axis transforms into itself; if (a^2+bc) is positive, the part of the t-plane above the real axis transforms into the part below the real axis. We shall suppose that our groups are generated only from the former kind of substitutions, so we need only consider the half of the t-plane above the real axis.

Assuming then throughout that (a^2+bc) is negative for the substitution considered, it is obvious that the double points of the substitution are conjugate complex quantities; for the double points are the roots of the equation

$$ct^2 - 2at - b = 0.$$

Now draw any circle through the double points of the substitution. This circle cuts the real axis orthogonally.

Then *the substitution transforms the parts of the t-plane outside and inside this circle into each other.*

For, let the double points be

$$t = \gamma + i\delta, \qquad \text{and} \qquad t = \gamma - i\delta,$$

and let t' be the point into which any point t is transformed. Then the substitution may be written

$$\frac{t' - \gamma + i\delta}{t' - \gamma - i\delta} = - \frac{t - \gamma + i\delta}{t' - \gamma - i\delta}.$$

This shows that the angle subtended by t at the double points is changed into its supplement by the transformation; and therefore the circumferences of all circles through the double points transform into themselves, the part on one side of the double points transforming into the part on the other side of them. By considering the whole plane as made up of the circumferences of circles through the double points, we obtain the theorem.

Now consider the infinite group generated from a number $(n+2)$ of these self-inverse substitutions,

$$S_1 = \left(t, \frac{a_1 t + b_1}{c_1 t - a_1}\right), \qquad S_2 = \left(t, \frac{a_2 t + b_2}{c_2 t - a_2}\right), \ldots S_{n+2} = \left(t, \frac{a_{n+2} t + b_{n+2}}{c_{n+2} t - a_{n+2}}\right),$$

which satisfy the relation

$$S_1 S_2 S_3 \ldots S_{n+2} = 1.$$

If $n = 1$, we find that it is impossible to satisfy this relation by self-inverse substitutions with conjugate complex double points; and if $n = 2$, it will be seen later that the method about to be given for the division of the plane into regions breaks down; but if $n > 2$, the relation can be satisfied, in an infinite number of ways, by substitutions of the required kind. A worked-out example is given below.

[Added June 2, 1898.—The possibility of the construction given below depends on the satisfying of certain inequalities among the constants of the substitutions; as in general, when the construction described is carried out, the sides of the polygon may cross each other.]

Now let $D_1, D_2, \ldots D_{n+2}$ be those double points, of the substitutions $S_1, S_2, S_3, \ldots S_{n+2}$ respectively, which are above the real axis.

Let C_1 be the point derived from D_{n+2} by applying the substitution S_1; or, as we can write it, let

$$C_1 = S_1(D_{n+2}).$$

Similarly, let

$$C_2 = S_2(C_1), \qquad C_3 = S_3(C_2), \ldots, C_{n+1} = S_{n+1}(C_n).$$

Then

$$\begin{aligned} C_{n+1} &= S_{n+1}S_n \ldots S_2S_1(D_{n+2}) \\ &= S_{n+2}(D_{n+2}), \qquad \text{since } S_1S_2 \ldots S_{n+2} = 1, \\ &= D_{n+2}. \end{aligned}$$

Now, by the last theorem, any point, and the point which is derived from it by a self-inverse substitution, lie on a circle through the double points of the substitution.

Therefore $D_{n+2}D_1C_1$ lie on a circle orthogonal to the real axis.

Similarly $C_1D_2C_2$, $C_2D_3C_3, \ldots$, $C_nD_{n+1}C_{n+1}$, all lie on circles orthogonal to the real axis.

Therefore *a curvilinear polygon can be formed, whose* $(n + 1)$ *sides are arcs of circles orthogonal to the real axis and pass through the points* $D_1, D_2, D_3, \ldots D_{n+1}$, *respectively, and whose corners are the points* $D_{n+2}, C_1, C_2, \ldots C_n$.

Now suppose we transform the polygon by the substitution S_r, where $r = 1, 2, \ldots (n + 1)$. We obtain another polygon, likewise formed of arcs of circles orthogonal to the real axis, and having contact with the original polygon along the side $C_{r-1}D_rC_r$. The side of this new polygon which is the conformal representation of $C_{p-1}D_pC_p$ passes through the double points of the self-inverse substitution $S_rS_pS_r$; and on applying this substitution to the new polygon, we obtain a third polygon, having contact with the second along the side which is the conformal representation of $C_{p-1}D_pC_p$. In this way we can, as every new polygon is formed, surround it with other polygons, each having one side in common with it.

Now consider what happens at any angular point of the polygon, say D_{n+2}, when we derive polygons in this way. If we derive a fresh polygon by applying the substitution S_1, the derived polygon adjoins the original one along the side $D_{n+2}C_1$. If now we derive a fresh polygon from the original one by applying the substitution S_1S_2, this second derived polygon adjoins the first along its free side through D_{n+2}. If now again we derive a fresh polygon from the original one by applying the substitution $S_1S_2S_3$, this third derived polygon adjoins the second along its free side through D_{n+2}. Proceeding round D_{n+2} in this way, we obtain at last a polygon which is derived from the original one by the substitution $S_{n+1}S_n \ldots S_2S_1S_{n+1}S_n \ldots S_2S_1$.

But since

$$S_{n+1}S_n \ldots S_2S_1 = S_{n+2}, \quad \text{and} \quad S_{n+2}^2 = 1,$$

this is the identical substitution; in other words, *the* $2(n+1)^{\text{th}}$ *polygon as we go round* A *is the original polygon.*

In the same way we can prove, that at every corner $2(n+1)$ polygons meet. The sides of the polygons are all portions of circles orthogonal to the real axis. As we approach the real axis, the polygons become smaller and more crowded together.

If from the original polygon we derive others, by transforming it with all the substitutions of the group generated by $S_1, S_2, \ldots S_{n+2}$, *we cover the half-plane once and only once.* So the original polygon is a "fundamental region" for the group of substitutions

In the annexed figure, the polygons in a portion of the plane are drawn to scale for the group formed from the substitutions

$$S_1 = \left(t, \frac{5t-74}{t-5}\right), \qquad S_2 = \left(t, \frac{2t-5}{t-2}\right), \qquad S_3 = \left(t, \frac{5t-29}{t-5}\right),$$
$$S_4 = \left(t, \frac{253t-2061}{33t-253}\right), \qquad S_5 = \left(t, \frac{132t-1675}{11t-132}\right), \qquad S_6 = \left(t, \frac{281t-4786}{17t-281}\right),$$

which are self-inverse substitutions satisfying the required relation

$$S_1S_2S_3S_4S_5S_6 = 1.$$

Here $n = 4$; the double points are given by

$$D_1 = 5 + 7i, \; D_2 = 2 + i, \; D_3 = 5 + 2i, \; D_4 = 7\tfrac{2}{3} + \sqrt{\tfrac{364}{99}}\,i, \; D_5 = 12 + \sqrt{\tfrac{91}{11}}\,i.$$

The vertex at the intersection of the S_1 and S_2 circles is at the point $t = 1 + i$.

Since the polygons are conformal representations of each other, they are equiangular to each other.

From the construction of the polygon, all the angular points are equivalent in respect of the group.

The sum of the angles round any vertex is 2π; but these angles are the conformal representations of the angles of a polygon, taken twice. Hence *the sum of the angles of any polygon is* π.

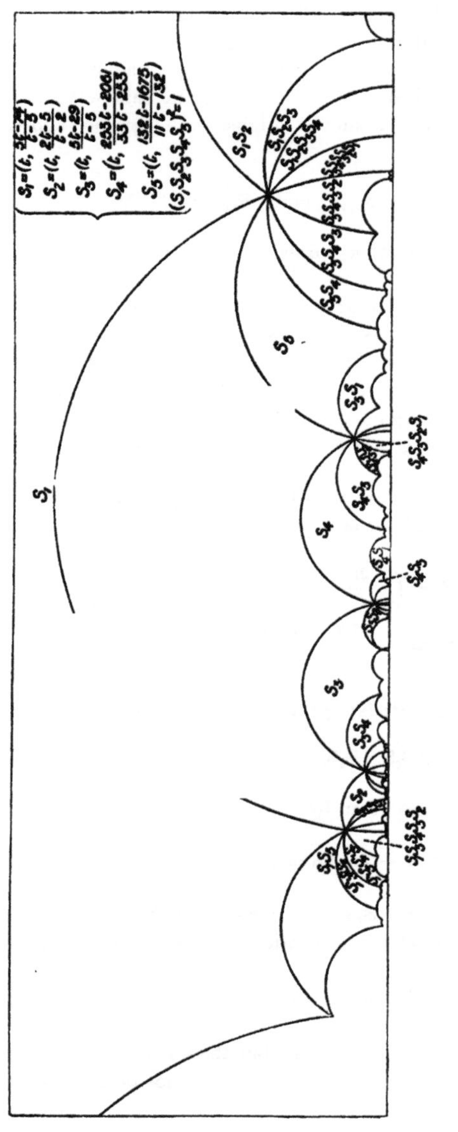
$S_2 = \left(t, \frac{2t-5}{t-2}\right)$
$S_3 = \left(t, \frac{5t-29}{t-5}\right)$
$S_4 = \left(t, \frac{253t-2001}{33t-253}\right)$
$S_5 = \left(t, \frac{132t-1675}{11t-132}\right)$
$(S_1 S_2 S_3 S_4 S_5)^2 = 1$
S_1
$S_1 S_2$
S_3
S_4

Let $t = u + iv$; if we measure the distance between two points, in the non-Euclidian sense, by $\int \frac{|dt|}{v}$ taken along the circle orthogonal to the real axis and joining the points, then we can easily prove that the lengths of corresponding sides of the polygons are in this sense all equal; if we measure the area of any region by $\iint \frac{du\,dv}{v^2}$ taken over that region, we can show that the areas of all the polygons are also in this sense equal; and the areas and lengths of corresponding regions and lines in the polygons are all equal. *The substitutions by which the polygons are derived from each other are, in this non-Euclidian sense, simple displacements, which leave their dimensions unchanged.* All the theorems of LOBATCHEWSKI's geometry hold if, where LOBATCHEWSKI uses the word "straight line," we understand "circle orthogonal to the real axis."

Thus, in non-Euclidian phraseology, we can say that the network of polygons has been obtained by drawing a rectilinear polygon of $(n + 1)$ sides, deriving new polygons from it by turning the polygon through an angle π round the middle points of its sides, and deriving fresh polygons from these by the same process, until the whole non-Euclidian plane is covered. This enables us to see that *our figure is the natural extension of the division of a whole plane into parallelograms, so familiar in the theory of elliptic functions.* For that division can be obtained by drawing any rectilinear triangle in the Euclidian plane, deriving fresh triangles by turning it through an angle π round the middle points of its sides, and deriving new triangles from these by the same process, until the whole Euclidian plane is covered. The groups for which the elliptic functions are automorphic are sub-groups of the groups so obtained; and similarly the groups, whose automorphic functions are required in the uniformisation of algebraic forms of genus higher than unity, are sub-groups of the group we have found. The reason why we have to pass from Euclidian to non-Euclidian geometry is, that in the Euclidian plane it is impossible to obtain a rectilinear figure with more sides than three, the sum of whose angles is π.

If to the original polygon we apply the substitution S_{n+1}, the point D_{n+1} is unchanged, and the arcs $D_{n+2}D_{n+1}$ and $D_{n+1}C_n$ are transformed into each other. So the parts of the boundary of the polygon which correspond to each other in the transformations of the group are $D_{n+2}D_{n+1}$ to C_nD_{n+1}, C_nD_n to $C_{n-1}D_n$, . . ., C_1D_1 to $D_{n+2}D_1$, respectively. If now we suppose the polygon lifted up from the plane, and these corresponding arcs pieced together, we obtain a simple closed surface, without multiple connectivity.

Therefore *the genus* (genre, Geschlecht) *of the group* (as defined by POINCARÉ) *is zero.* The group however may have, and will in fact be proved to have, sub-groups whose genus is greater than zero.

§ 4. *The Automorphic Functions of the Group.*

From the fact which has just been proved, that the genus of all groups of the kind we have found is zero, we know that the algebraic relation between any two automorphic functions of the group is of genus zero; therefore *all the automorphic functions of the group can be expressed as rational functions of a certain one of them.* We shall denote this one by z.

First, let us see what degree of arbitrariness there is in the choice of the function z.

If a, b, c, d, are any four constants (which can without loss of generality be taken to satisfy the relation $ad - bc = 1$), then

$$\frac{az + b}{cz + d}$$

is another such function as z. Hence the function z contains three distinct arbitrary constants.

z takes every value once, and only once, in each polygon of the figure. The three arbitrary constants may be taken to be the place of its zero, the place of its infinity, and a multiplicative constant.

Now consider *the conformal representation of a t-polygon on the z-plane.*

The function z takes every value once in the polygon; therefore the conformal representation of the polygon will cover the whole z-plane. Also, z takes the same value, say e_{n+2}, at each of the corners of the polygon; suppose that z takes the values $e_1, e_2, e_3, \ldots e_{n+1}$, at the points $D_1, D_2, D_3, \ldots D_{n+1}$, respectively.

As t describes the boundary of the polygon, beginning at D_{n+2}, z begins with the value e_{n+2} and varies until, at D_1, the value e_1 is reached; then, retracing the same series of values, z returns to the value e_{n+2} at C_1. Then at D_2 the value e_2 is reached, and at C_2 z takes the value e_{n+2} again; and so on round the polygon.

Thus the conformal representation of the boundary of the polygon is a series of lines (not necessarily straight), radiating from the point e_{n+2} to the points $e_1, e_2, e_3, \ldots e_{n+1}$, in succession. *The polygon corresponds to the whole z-plane, with this regarded as boundary.* Small arbitrary variations in the form of the lines radiating from e_{n+2} to $e_1, e_2, \ldots e_{n+1}$, merely correspond to small arbitrary variations in the boundary of the polygon.

Thus *we see the nature of the solution of the problem: To conformally represent the whole plane of a variable z, bounded by a set of finite lines radiating from a point, on a curvilinear polygon in the plane of a variable t; this polygon being the fundamental region of an infinite discontinuous group of real projective substitutions of the variable t, and z being an automorphic function of the group.*

We may note that *dz/dt is zero at each of the double points.* For if t and t' are two points very near a double point, which are transformed into each other by the substitution corresponding to the point, we have approximately

$$dt' = - dt.$$

Thus dz/dt has values equal in magnitude, but opposite in sign, at the points t and t'; and therefore, making t and t' to coalesce in the double point, dz/dt is zero at the double point.

Let us now enumerate the constants at our disposal, in order to see the correspondence between the arrangement in the z-plane and the group of substitutions.

The t-figure is determined by $n+2$ self-inverse substitutions, $S_1, S_2, \ldots S_{n+2}$, satisfying the relation

$$S_1S_2S_3 \ldots S_{n+2} = 1. \quad . \quad . \quad . \quad . \quad . \quad . \quad . \quad . \quad . \quad (1).$$

There are three real constants, a, b, c, in each substitution. But by reason of the relation

$$a^2 + bc = -1,$$

these three are only equivalent to two. Thus from the $(n+2)$ substitutions we get $(2n+4)$ real constants.

The relation (1) defines three of these constants in terms of the rest. Also, this group is not essentially different from one which is obtained by transforming it with any real substitution, which shows that three more of the constants are non-essential. So there are altogether $(2n-2)$ essential real constants involved in the t-figure.

Now considering the z-plane, there are $n+2$ points $e_1, e_2, \ldots e_{n+2}$; and each of these is defined by two real co-ordinates, giving $2n+4$ as the number of real constants. But we can make a homographic transformation of the plane, so as to transform any three of the points into three arbitrary points. This shows that 6 of the constants can be disregarded as non-essential. So we have $(2n-2)$ essential constants in the z-figure.

Hence *the number of essential constants is the same in the z-figure as in the t-figure.*

[ADDED June 2, 1898.—This does not in itself prove that for every z-figure there exists a corresponding t-figure; but the general existence-theorem of POINCARÉ and KLEIN can be applied to complete the proof.]

Hitherto we have derived the z-figure from the t-figure. The next section is chiefly concerned with the converse problem of deriving the t-figure from the z-figure.

§ 5. *The Analytical Relations between z and t.*

The analytical relations between z and t are of two kinds; (α) those which express z in terms of t and the constants of the substitutions, and (β) those which express t in terms of z and the quantities $e_1, e_2, \ldots e_{n+2}$.

The Thetafuchsian series of POINCARÉ solve the first problem for all classes of automorphic functions. We shall therefore only discuss relations of the kind (β).

As any quantity of the form $(at+b)/(ct+d)$, where a, b, c, d, are arbitrary real constants, is a solution of the problem (β) equally with t, we shall expect t to be given by a differential equation of which the general integral is $(at + b)/(ct + d)$; in other words, by a differential equation of the form

$$\tfrac{1}{2}\{t, z\} = \mathrm{R}(z),$$

where $\mathrm{R}(z)$ is some function of z, and

$$\{t, z\} = -\frac{d^3z/dt^3}{(dz/dt)^3} + \frac{3}{2}\frac{(d^2z/dt^2)}{(dz/dt)^4}$$

is a Schwarzian derivative.

As $\{t, z\}$ is unaltered by a change of t into $(at + b)/(ct + d)$, $\mathrm{R}(z)$ is an automorphic function of the group, and therefore $\mathrm{R}(z)$ *is a rational function of* z. We have to find $\mathrm{R}(z)$.

Considering the conformal representation, we see that z and t are regular functions of each other, except near the points $z = e_1, e_2, \ldots e_{n+2}, \infty$. Hence, except at these special points, $\frac{1}{2}\{t, z\}$ is a regular function of z, and we shall not get an infinity of $\mathrm{R}(z)$. As z is a uniform automorphic function of t, $\frac{dz}{dt}$ is infinite only at $z = \infty$.

Near $z = \infty$ (supposing for the present that no one of the quantities $e_1, e_2, \ldots e_{n+2}$, is infinite), z and t are uniform functions of each other, so

$$z = \frac{a}{t - t_0} + b + c(t - t_0) + \ldots, \text{ where } a \text{ is not zero.}$$

This gives

$$\tfrac{1}{2}\{t, z\} = \frac{3c}{a^3}(t - t_0)^4 + \ldots$$

Hence at $z = \infty$, $\frac{1}{2}\{t, z\}$ must be zero to at least the order $\frac{1}{z^4}$.

Near $z = e_r$, z is a uniform function of t, but dz/dt is zero. So near this point,

$$z - e_r = c(t - t_0)^2 + d(t - t_0)^3 + \ldots,$$

where c is not zero, since t has at the point a simple branch-point, considered as a function of z.

This gives

$$\tfrac{1}{2}\{t, z\} = \frac{3}{16c^2(t - t_0)^4} + \ldots \qquad = \frac{3}{16(z - e_r)^2} + \ldots$$

Thus the only infinities of the rational function $\mathrm{R}(z)$ are at the points $e_1, e_2, \ldots e_{n+2}$; and these points are poles of the kind just found.

Hence

$$\mathrm{R}(z) = \tfrac{3}{16}\sum_{r=1}^{n+2}\frac{1}{(z - e_r)^2} + \sum_{r=1}^{n+2}\frac{a_r}{z - e_r} + \mathrm{P}(z),$$

where P (z) is a polynomial in z, and a's are constants. Now at $z = \infty$ we must have a zero of at least the order $\frac{1}{z^4}$. Hence P $(z) = 0$; and since near $z = \infty$, R (z) can be expanded in the form

$$\mathrm{R}(z) = \tfrac{3}{16}\sum_{r=1}^{n+2}\left(\frac{1}{z^2} + \frac{2e_r}{z^3} + \ldots\right) + \sum_{r=1}^{n+2}\left(\frac{a_r}{z} + \frac{a_r e_r}{z^2} + \frac{a_r e_r^2}{z^3} + \ldots\right),$$

by equating to zero the coefficients of $\frac{1}{z}$, $\frac{1}{z^2}$, and $\frac{1}{z^3}$, respectively, we obtain

$$\sum_{r=1}^{n+2} a_r = 0,$$

$$\sum_{r=1}^{n+2} a_r e_r = -\frac{3(n+2)}{16},$$

$$\sum_{r=1}^{n+2} a_r e_r^2 = -\tfrac{3}{8}\sum_{r=1}^{n+2} e_r.$$

These conditions enable us to write R (z) in the form

$$\mathrm{R}(z) = \tfrac{3}{16}\sum_{r=1}^{n+2}\frac{1}{(z-e_r)^2} + \tfrac{3}{16}\,\frac{-(n+2)z^n + n\,.\,\Sigma e_r\,.\,z^{n-1} + c_1 z^{n-2} + \ldots + c_{n-1}}{(z-e_1)(z-e_2)\ldots(z-e_{n+2})},$$

where $c_1, c_2, \ldots c_{n-1}$ are constants as yet undetermined.

Hence *the required analytical relation between t and z is*

$$\tfrac{1}{2}\{t, z\} = \tfrac{3}{16}\sum_{r=1}^{n+2}\frac{1}{(z-e_r)^2} + \tfrac{3}{16}\,\frac{-(n+2)z^n + n\,.\,\Sigma e_r\,.\,z^{n-1} + c_1 z^{n-2} + \ldots + c_{n-1}}{(z-e_1)(z-e_2)\ldots(z-e_{n+2})}.$$

It will be seen that this is the differential equation for the quotient of two solutions of a linear differential equation of the second order with $(n + 2)$ singularities, at each of which the exponent-difference is $\frac{1}{2}$. Such linear differential equations have been studied by KLEIN,* as being the generalisation of LAMÉ's equation; and BÔCHER's book, 'Ueber die Reihenentwickelungen der Potential-Theorie' (Leipsic, TEUBNER, 1894), is chiefly concerned with them. BÔCHER proves that the differential equations of harmonic analysis are limiting cases of them.

We can *transform this equation to a simpler form.*

Put

$$w = \int\frac{dz}{\sqrt{(z-e_1)(z-e_2)\ldots(z-e_{n+2})}},$$

so w is a known function of z.

* 'Göttinger Nachrichten,' 1890, pp. 85–95.

Then the differential equation becomes

$$\tfrac{1}{2}\{t, w\} = \frac{n^2-4}{16}z^n - \frac{n(n-2)}{16}\sum_{r=1}^{n+2} e_r \,.\, z^{n-1} + d_1 z^{n-2} + d_2 z^{n-3} + \ldots + d_{n-1},$$

where $d_1, d_2, \ldots d_{n-1}$, are new undetermined constants replacing the c's.

This can be written

$$\tfrac{1}{2}\{t, w\} = \frac{n-2}{8(n+1)} \frac{d^2z/dw^2}{dz/dw} + k_1 z^{n-2} + k_2 z^{n-3} + \ldots + k_{n-1},$$

or

$$\tfrac{1}{2}\{t, w\} = \frac{n-2}{8(n+1)} \frac{1}{u} \frac{d^2u}{dw^2} + k_1 z^{n-2} + k_2 z^{n-3} + \ldots + k_{n-1} \quad . \quad . \quad . \quad (1),$$

where

$$u^2 = (z-e_1)(z-e_2)\ldots(z-e_{n+2}),$$

and where $k_1, k_2, \ldots k_{n-1}$ are new undetermined constants, replacing $d_1, d_2, \ldots d_{n-1}$.

If z has its infinity at a double point of one of the substitutions, we get a slightly different form of the equation.

In this case, one of the e's is infinite. Let $e_{n+2} = \infty$. Then, near $z = \infty$, the expansions are of the form

$$z = \frac{c}{(t-t_0)^2} + \ldots \quad \text{and} \quad \mathrm{R}(z) = \frac{3}{16z^2} + \ldots,$$

whence, by the same reasoning as before, we find that

$$\tfrac{1}{2}\{t, z\} = \tfrac{3}{16}\sum_{r=1}^{n+1} \frac{1}{(z-e_r)^2} + \tfrac{3}{16}\frac{-nz^{n-1} + c_1 z^{n-2} + \ldots + c_{n-1}}{(z-e_1)(z-e_2)\ldots(z-e_{n+1})}.$$

Put

$$w = \int \frac{dz}{\sqrt{(z-e_1)(z-e_2)\ldots(z-e_{n+1})}}.$$

Then the equation becomes

$$\tfrac{1}{2}\{t, w\} = \frac{n(n-2)}{16} z^{n-1} + d_1 z^{n-2} + d_2 z^{n-3} + \ldots + d_{n-1},$$

where again the quantities $d_1, d_2, \ldots d_{n-1}$, are undetermined constants.

This can be written

$$\tfrac{1}{2}\{t, w\} = \frac{n-2}{8(n+1)} \frac{d^2z/dw^2}{dz/dw} + k_1 z^{n-2} + k_2 z^{n-3} + \ldots + k_{n-1},$$

or,

$$\tfrac{1}{2}\{t, w\} = \frac{n-2}{8(n+1)} \frac{1}{u} \frac{d^2u}{dw^2} + k_1 z^{n-2} + k_2 z^{n-3} + \ldots + k_{n-1} \quad . \quad . \quad . \quad (2),$$

where

$$u^2 = (z-e_1)(z-e_2)\ldots(z-e_{n+1}).$$

The differential equations (1) *and* (2) *determine* t *in terms of* z *in the two cases respectively.*

The constants $k_1, k_2, \ldots k_{n-1}$, are as yet undetermined. The reason is, that we have not yet made any use of the condition which in fact does determine them; namely, that all the projective substitutions, which t undergoes when the independent variable z of the differential equation describes a circuit round one of the singularities, are such as to leave unchanged a certain circle. This circle is, in the figure we have drawn, the real axis of the variable t, which is unchanged by all the substitutions of the group; but it may more generally be any circle in the t-plane. This condition will be shown in § 7 to be equivalent to the determination of $(n-1)$ complex quantities, which are the constants $k_1, k_2, \ldots k_{n-1}$. But a further consideration of this is deferred to § 7. For the present we shall suppose $k_1, k_2, \ldots k_{n-1}$ determined in such a way as to give the required representation.

§ 6. *Application of the Preceding Theory to the Uniformising of Algebraic Forms.*

We have proved that the genus of groups of the kind we have found is zero, and hence the automorphic functions of the group as it stands will not uniformise algebraic forms whose genus is greater than zero. But we can find sub-groups of the original group, and these will be found to be of genus greater than zero.

The process of deriving these sub-groups is analogous to the method of building up a Riemann surface of any genus by superposing a number of plane sheets and connecting them along branch lines. We join together a certain number of the polygons in the figure, and regard them as forming one new polygon. This will, in certain cases, be the fundamental polygon of a sub-group of the original group, and may have a genus greater than zero.

Consider a double polygon, made up by taking together the original polygon, and the polygon derived from it by transforming with the substitution S_{n+1}, and erasing the boundary which separates them. The new polygon has $2n$-sides. By erasing all the lines corresponding to the line already erased, we obtain a division of the half-plane into $2n$-gons. The opposite sides of the $2n$-gon are easily seen to be transformed into each other by the n substitutions

$$T_1 = S_{n+1}S_1,\ T_2 = S_{n+1}S_2, \ldots T_n = S_{n+1}S_n,$$

respectively.

This $2n$*-gon is a "fundamental region" for the group generated from the substitutions* $T_1, T_2, \ldots T_n$. We proved in § 2 that the group generated by $T_1, T_2, \ldots T_n$, is a self-conjugate sub-group of the group formed by $S_1, S_2, \ldots S_{n+2}$; and that any substitution of the latter group is equivalent to a substitution of the former group acting on either the identical substitution or on S_{n+1}. This corresponds to the fact that a point in any of the derived $2n$-gons can be obtained by transformation with the substitutions T from a point in either the original $(n+1)$-gon or the $(n+1)$-gon derived from this by the substitution S_{n+1}.

We have, therefore, obtained a new division of the half-plane into $2n$-gons, and found the group of substitutions corresponding to it. We can now find the genus p of this group.

The opposite sides of the $2n$-gon are transformed into each other by substitutions of the group. If we suppose the $2n$-gon lifted up from the plane, and opposite sides pieced together, we obtain a surface of connectivity $(n+1)$. If n is even, this surface is of genus p where $n = 2p$. In what follows we shall suppose n even.

Hence, *the algebraic relation between any two automorphic functions of this group is, in general, of genus* $p = \frac{1}{2}n$.

The function z, which has been obtained, takes every value once in each $(n+1)$-gon; and therefore it takes every value twice in each $2n$-gon. But this is the condition that the algebraic form, made up of the automorphic functions of the group, should be hyperelliptic.

Hence, *the algebraic form, which is made up of the automorphic functions of the group, is hyperelliptic, and of genus* $\frac{1}{2}n$; and, as z is a variable which takes every value twice in each polygon, the form consists of rational functions of z and u, where u is a function of z defined by an equation

$$u^2 = (z - a_1)(z - a_2) \ldots (z - a_{n+2}),$$

where $a_1, a_2, \ldots a_{n+2}$ are constants to be determined. But the function

$$\sqrt{(z - e_1)(z - e_2) \ldots (z - e_{n+2})}$$

is an automorphic function of the group, for it has the same value, save for a change of sign, at corresponding points in adjacent $(n+1)$-gons, and therefore the same value at corresponding points in different $2n$-gons.

Hence

$$a_1 = e_1,\ a_2 = e_2, \ldots a_{n+2} = e_{n+2},$$

and we see that the automorphic functions of the group generated from the substitutions $T_1, T_2, \ldots T_n$ are the algebraic functions of the form defined by the equation

$$u^2 = (z - e_1)(z - e_2) \ldots (z - e_{n+2}).$$

Thus we have the solution of the problem, "*To find a variable of which the functions rational on the Riemann surface of the equation*

$$u^2 = (z - e_1)(z - e_2) \ldots (z - e_{n+2})$$

are uniform functions."

We could have foreseen this by regarding the problem as one of conformal representation. The algebraic functions can be regarded as uniform functions on a Riemann surface which covers the z-plane twice, the branch-points being at the points $e_1, e_2, \ldots e_{n+2}$. Now join the point e_{n+2} to each of the points $e_1, e_2, e_3, \ldots e_{n+1}$. Then each of the sheets, regarded as an infinite plane bounded by these lines, is represented conformally on one of the $(n+1)$-gons in the t-plane; by taking two adjacent $(n+1)$-gons, we obtain a $2n$-gon, which corresponds to the fact that by taking the two z-planes, and connecting them along the line $e_{n+2}e_{n+1}$, we obtain the Riemann surface as dissected by n cross-cuts.

The analytical connexion between the variables in a hyperelleptic form and the uniformising variable t is therefore given by the equations of § 5. It can be shown that the differential equation found there is, as might be expected, one of KLEIN'S* "unverzweigt" differential equations for hyperelliptic forms. It can be obtained by equating $(p-2)$ of the arbitrary constants in KLEIN'S equation to zero.

There are p integrals of the first kind connected with the form. It is easily proved that if v is one of them, then v undergoes a projective substitution of the form

$$(v, c - v),$$

where c is a constant, when t is transformed by one of the generating substitutions of the group.

The theory of Abelian integrals of the form can be developed with t as independent variable; but developments of this kind are outside the scope of this paper.

One consequence of the results just obtained is that we can find *the conditions that $2p$ arbitrarily given projective substitutions may generate the group corresponding to a hyperelliptic equation of genus p.*

Let the substitutions be $T_1, T_2 \ldots T_{2p}$, where

$$T_r = \left(t, \frac{a_r t + b_r}{c_r t + d_r}\right).$$

On comparing the results of this section with those of § 2, we see that *the conditions may be expressed in the form*

$$\begin{vmatrix} a_r - d_r & b_r & c_r \\ a_s - d_s & b_s & c_s \\ a_t - d_t & b_t & c_t \end{vmatrix} = 0, \quad (r, s, t, = 1, 2, 3, \ldots 2p).$$

[Added June 3, 1898.—These conditions are not, however, proved to be strictly necessary, since the group may be generated by another set of substitutions to which these conditions apply, although they do not apply to $T_1, T_2, \ldots T_{2p}$. And the

* 'Göttinger Nachrichten,' 1890, p. 85.

inequalities expressing the conditions that the sides of the generating polygon do not cross must also be satisfied.]

In all our work hitherto it has been assumed that $p > 1$. *The case $p = 1$ is exceptional;* algebraic forms of genus unity cannot be uniformised by groups of the kind we have found. For if the construction which has been given were possible for $p = 1$, we should have, as the fundamental polygon of the group, a triangle whose sides are, in the non-Euclidian sense, straight lines, and the sum of whose angles is π. But this is impossible, for in LOBATCHEWSKI's geometry the sum of the angles of a triangle is always less than π. When the sum is equal to π we arrive at the limiting case of Euclidian geometry. Therefore the construction fails, and we have to devise instead a construction in which Euclidian geometry replaces non-Euclidian. We take four substitutions, S_1, S_2, S_3, S_4, satisfying the relation

$$S_1 S_2 S_3 S_4 = 1,$$

which are self-inverse and leave the Euclidian absolute unchanged, *i.e.*, which are all of the type

$$(t, \quad c - t),$$

where c is a complex constant. By reasoning exactly analogous to that in § 3, we see that these substitutions generate a group, to which corresponds a division of the plane into rectilinear triangles. The sub-group which is got by taking adjacent triangles in pairs gives a division of the plane into parallelograms; and this is the well-known group of the doubly-periodic functions, which uniformise algebraic curves of genus unity.

The following shows how the former construction breaks down in this case.

If possible, let S_1, S_2, S_3, S_4, be four self-inverse substitutions with real coefficients satisfying the relation

$$S_1 S_2 S_3 S_4 = 1.$$

Then if

$$S_r = \left(t, \ \frac{a_r t + b_r}{c_r t - a_r}\right),$$

we have

$$S_1 S_2 S_3 (t) = \frac{(a_1 a_2 a_3 + a_1 b_2 c_3 + a_3 b_1 c_2 - a_2 b_1 c_3)\, t + (a_1 a_2 b_3 - a_1 b_2 a_3 - b_1 c_2 b_3 + b_1 a_2 a_3)}{(c_1 a_2 a_3 + c_1 b_2 c_3 - a_1 c_2 a_3 + a_1 a_2 c_3)\, t + (c_1 a_2 b_3 - c_1 b_2 a_3 - a_1 c_2 b_3 - a_1 a_2 a_3)}.$$

This has to be a self-inverse substitution, since S_4 is self-inverse.

So

$$a_1 b_2 c_3 + b_1 a_3 c_2 - a_2 b_1 c_3 + c_1 a_2 b_3 - c_1 b_2 a_3 - a_1 c_2 b_3 = 0,$$

or

$$\begin{vmatrix} a_1 & a_2 & a_3 \\ b_1 & b_2 & b_3 \\ c_1 & c_2 & c_3 \end{vmatrix} = 0.$$

Let $\gamma_r + i\delta_r$ and $\gamma_r - i\delta_r$ be the double points of S_r, then

$$a_r = \gamma_r, \quad b_r = -(\gamma_r^2 + \delta_r^2), \quad c_r = 1.$$

Therefore

$$\begin{vmatrix} \gamma_1^2 + \delta_1^2 & \gamma_2^2 + \delta_2^2 & \gamma_3^2 + \delta_3^2 \\ \gamma_1 & \gamma_2 & \gamma_3 \\ 1 & 1 & 1 \end{vmatrix} = 0.$$

This shows that the double points of all three substitutions lie on a circle orthogonal to the real axis. Since $S_2S_3S_4$ is a self-inverse substitution, the double points of S_4 lie on the same circle.

Hence, if we attempt to construct the fundamental polygon, we find that all its angular points lie on the same circle orthogonal to the real axis, and therefore all its sides coalesce, and its area is zero. This explains why the method fails in this case.

We now proceed to *the uniformisation of algebraic forms which are not hyper-elliptic*. These only occur when the genus is greater than two.

If we are given any algebraic form of genus p, it is known that it can by birational transformation be represented on a Riemann surface of which all the branch-points are simple, *i.e.*, only two sheets interchange at any branch-point.

Let $f(u, z) = 0$ be an algebraic equation corresponding to this surface. Suppose the branch-points are at the values of z for which $z = e_1, e_2, e_3, \ldots e_{n+2}$, respectively. It may of course happen that for some of these values of z there are several branch-points superposed on each other on the Riemann surface.

Now in the z-plane, join the point e_{n+2} to each of the points $e_1, e_2, \ldots e_{n+1}$, and conformally represent this, in the plane of a variable t, on the fundamental polygon of a group from $(n + 2)$ self-inverse substitutions, as before explained.

Then, as before, z is a uniform function of t. At each of the points $z = e_1, e_2, \ldots e_{n+2}$, say e_r, u is expansible in a series of ascending powers of either $(z - e_r)^{\frac{1}{2}}$ or $(z - e_r)$, according as the point $z = e_r$ happens to be a branch-point or not in the sheet in which the point is situated. But near this point $(z - e_r)^{\frac{1}{2}}$ is expansible in a power-series in terms of $(t - t_0)$, where t_0 is the value of t at the point; so in either case, u is expansible as a series of ascending powers of $(t - t_0)$; that is, u has no branch-point, considered as a function of t, at this point.

But since z is a uniform function of t, the only points where u can have branch-points, considered as a function of t, are the points where u has branch-points considered as a function of z; that is, the points $e_1, e_2, \ldots e_{n+2}$. Hence, u is a uniform function of t.

Thus, *any algebraic curve can be uniformised by means of groups of substitutions formed from self-inverse substitutions.*

It will be seen that a great similarity exists between the place occupied by self-inverse substitutions, in the theory of groups of projective substitutions, and the place occupied by branch-points at which only two branches interchange, in the theory of Riemann surfaces; the usefulness of the method of self-inverse substitutions depends on the fact that algebraic forms can be represented on Riemann surfaces with only simple branch-points.

Algebraic functions are not, however, the only ones which can be uniformised. POINCARÉ* has proved a general existence-theorem that, if $u_1, u_2, \ldots u_m$, are any multiform analytical functions of a variable z, a variable t always exists, such that $z, u_1, u_2, \ldots u_m$, are uniform functions of t. The existence-theorem, however, does not connect t analytically with the other variables. If $u_1, u_2, \ldots u_m$, are transcendental functions of z, their multiformity will not in general be capable of being expressed by simple branch-points, and so the groups generated by self-inverse substitutions cannot be used.

§ 7. *The Undetermined Constants in the Differential Equation connecting z and t.*

In § 5, certain constants $k_1, k_2, \ldots k_{n-1}$, in the differential equation connecting z and t, were left undetermined. It was there explained that they are to be determined by the consideration that the group of substitutions of t leaves unchanged a fundamental circle. In general, however, arbitrary constants occurring in similar differential equations cannot be determined by this consideration, as the group may be "Kleinian," *i.e.*, it may not conserve a fundamental circle. The following discussion approaches the subject from this more general point of view.

The Riemann surface, corresponding to the algebraic form $f(u, z) = 0$, can be made simply-connected by drawing $2p$ cuts, and the problem of finding the uniformising variable t can be divided into two parts, as follows :—

1. Finding all the variables τ, which are such that the dissected Riemann surface is represented on the τ-plane by a curvilinear polygon, whose $4p$ sides can be derived from each other in pairs by projective substitutions of τ.

2. Selecting from among these variables τ, a variable t, which is such that the group generated from these projective substitutions is a discontinuous group.

We shall call the variables τ *quasi-uniformising* variables, to distinguish them from the true uniformising variable t.

In the case of the groups we have found, the differential equation of § 5 *gives the quasi-uniformising variables; the determination of $k_1, k_2, \ldots k_{n-1}$ is equivalent to selecting the uniformising variable from among them.*

In this section the connexion between the uniformising and quasi-uniformising variables is considered for more general groups.

As an example of the nature of quasi-uniformising variables, take the algebraic equation

$$u^2 = 4z^3 - g_2 z - g_3.$$

* 'Bulletin de la Société Math. de France,' 1883, vol. 11, p. 112.

To this corresponds a Riemann surface of two sheets, which can be resolved by two cuts into a simply-connected surface.

Let P be the Weierstarssian elliptic function associated with this curve; and w_1, w_2, its periods.

Consider u and z as functions of τ, where

$$u = \mathrm{P}'(\log \tau), \qquad z = \mathrm{P}(\log \tau).$$

In the τ-plane, form a curvilinear parallelogram ABCD, of which the side CB is derived from AD by the projective substitution

$$(\tau, e^{w_1}\tau),$$

and the side CD is derived from AB by the projective substitution

$$(\tau, e^{w_2}\tau).$$

Then within this parallelogram ABCD, the dissected Riemann surface corresponding to the curve

$$u^2 = 4z^3 - g_2 z - g_3$$

is conformally represented; the sides AD, CB of the parallelogram correspond to the two edges of one cross-cut, and the sides AB, CD to the other; and, as we have seen, the opposite sides of the parallelogram are derived from each other by projective substitutions. But in spite of this, u and z are not uniform functions of τ. The reason is, that τ is only a quasi-uniformising variable; when we derive all possible polygons from ABCD by applying the group of substitutions generated from

$$(\tau, e^{w_1}\tau) \text{ and } (\tau, e^{w_2}\tau),$$

the polygons so derived cover the plane more than once.

The connexion between the uniformising and quasi-uniformising variables for any algebraic form is given by the following theorems.

If t is a uniformising variable of an equation

$$f(u, z) = 0,$$

and T *is any holomorphic Thetafuchsian function of t of order two, then the quotient of any two solutions of the differential equation*

$$\frac{d^2v}{dt^2} + \mathrm{T}v = 0 \quad . \quad . \quad . \quad . \quad . \quad . \quad . \quad . \quad . \quad . \quad (1)$$

is a quasi-uniformising variable.

The term "holomorphic Thetafuchsian function of order two" may require some explanation.

Let $\left(t, \frac{at+b}{ct+d}\right)$ be any one of the substitutions of the group associated with the given uniformising variable t. Then a Thetafuchsian function T of order m is such that

$$\mathrm{T}\left(\frac{at+b}{ct+d}\right) = (ct+d)^{2m}\mathrm{T}(t).$$

We have said that T is to be holomorphic (except at the singularities of the group). Such functions exist; for instance, if w be an Abelian integral of the first kind associated with the curve, then dw/dt is a holomorphic Thetafuchsian function of order one, and $(dw/dt)^2$ is a holomorphic Thetafuchsian function of order two.

To prove the theorem, let

$$\tau = v_1/v_2,$$

where v_1 and v_2 are any two solutions of (1). Then v_1 and v_2 have singularities, considered as functions of t, only where T has singularities. But in any one of the polygons in the t-plane, T has no singularities. Therefore, v_1 and v_2 are holomorphic functions of t (except at the essential singularities of the group, which for the present we do not consider).

Also, v_1 and dv_1/dt cannot be zero together at any point; for if they were, by equation (1), v_1 would be permanently zero. Similarly for v_2.

Therefore, at all points p within any one of the polygons in the t-plane, we have expansions beginning with

$$v_1 = c + d(t - t_0) + \ldots,$$

where c and d are not both zero, and

$$v_2 - e + f(t - t_0) + \ldots,$$

where e and f are not both zero.

And we may not have d and f zero together, as v_1 and v_2 are independent solutions of the differential equation.

So, at all points except the singularities of the group,

$$\tau = \frac{c + d(t - t_0) + \ldots}{e + f(t - t_0) + \ldots}.$$

gives either

$$\tau = \mathrm{A} + \mathrm{B}(t - t_0) + \ldots,$$

or,

$$\tau = \mathrm{A}(t - t_0) + \mathrm{B}(t - t_0)^2 + \ldots,$$

or,

$$\tau = \frac{\mathrm{A}}{t - t_0} + \mathrm{B} + \mathrm{C}(t - t_0) + \ldots$$

In all these cases t and τ are uniform functions of each other, near the point considered. So u and z are, near the point, uniform functions of τ. This is easily seen to be true also of $t = \infty$.

Now, let accented letters denote the effect of operating on t with a substitution

$$\left(t, \frac{at+b}{ct+d}\right)$$

of the group.

We have

$$\frac{d^2v'}{dt'^2} + T'v' = 0.$$

Now $T' = (ct+d)^4 T$. Write $v' = \frac{\xi}{ct+d}$.

Then

$$\frac{d^2v'}{dt'^2} = (ct+d)^2 \frac{d}{dt}\left\{(ct+d)^2 \frac{d}{dt}\left(\frac{\xi}{ct+d}\right)\right\} = (ct+d)^3 \frac{d^2\xi}{dt^2}.$$

Therefore

$$(ct+d)^3 \frac{d^2\xi}{dt^2} + (ct+d)^3 T\xi = 0,$$

or

$$\frac{d^2\xi}{dt^2} + T\xi = 0.$$

So $\xi = Av_1 + Bv_2$, where A and B are constants, and

$$v' = \frac{Av_1 + Bv_2}{ct+d}.$$

Therefore

$$\tau' = \frac{v_1'}{v_2'} = \frac{A_1v_1 + B_1v_2}{A_2v_1 + B_2v_2},$$

or

$$\tau' = \frac{A_1\tau + B_1}{A_2\tau + B_2}.$$

This shows that, when t is transformed by a projective substitution of the group, τ is transformed by a corresponding projective substitution

$$\left(\tau, \frac{A_1\tau + B_1}{A_2\tau + B_2}\right).$$

Thus the theorem is proved, namely, that the dissected Riemann surface can be conformally represented on a polygon in the τ-plane, and the sides of this polygon can be derived from each other in pairs by certain projective substitutions ; in other words, τ is a quasi-uniformising variable. An infinite number of variables τ can be got in this way, for T depends linearly on several arbitrary constants.

In the above theorem, for the sake of simplicity, we have made a restriction which is really unnecessary, namely, we have supposed that t is a uniformising variable. *t can, however, be any quasi-uniformising variable* if we make the corresponding extension in the meaning of T. T will now have to be a function of t, which is holomorphic in any of the polygons, and which obeys the law

$$\mathrm{T}\left(\frac{at+b}{ct+d}\right) = (ct+d)^4\, \mathrm{T}(t)$$

for substitutions of the group generated from the substitutions which change the sides of the t-polygon into each other. Such functions exist; for, as before, if w is an Abelian integral of the first kind connected with the curve, $(dw/dt)^2$ is such a function. T is, of course, really a multiform function of t, if t is a quasi-uniformising variable; but as it is not possible to pass from one of its values to another by any paths contained within one of the polygons, we can regard it as uniform within that polygon. The proof in this extended case is just as before. Thus we have the more general theorem:

If t is any uniformising or quasi-uniformising variable of an algebraic form

$$f(u, z) = 0,$$

and T *is any holomorphic Thetafuchsian function of t of order two, then the quotient of any two solutions of the differential equation*

$$\frac{d^2v}{dt^2} + \mathrm{T}v = 0$$

is another uniformising or quasi-uniformising variable.

To complete the theorem, we must prove that the converse is also true. Suppose, then, that τ and t both belong to the set of uniformising and quasi-uniformising variables, so that a polygon in the τ-plane corresponds to a polygon in the t-plane, point for point, and to each of the substitutions of the group $\left(\tau, \frac{\alpha\tau+\beta}{\gamma\tau+\delta}\right)$ corresponds a substitution $\left(t, \frac{at+b}{ct+d}\right)$.

Now τ is the quotient of two integrals of the equation

$$\frac{d^2v}{dt^2} + \mathrm{T}v = 0$$

if

$$\mathrm{T} = -\tfrac{1}{2}\frac{d^3t/d\tau^3}{(dt/d\tau)^3} + \tfrac{3}{4}\frac{(d^2t/d\tau^2)^2}{(dt/d\tau)^4}.$$

Now τ has no branch-point, considered as a function of t, and t has no branch-point, considered as a function of τ, except at the limiting points of the groups. So, if we consider any point in the t-plane, which is not one of the singularities of the group, $dt/d\tau$ and $d\tau/dt$ are, in its vicinity, regular functions of t.

So T is holomorphic at all points except the singularities of the group.

Now, denoting as before the effect of a substitution of the group by accents, we have

$$\begin{aligned} T' &= -\tfrac{1}{2}\frac{d^3t'/d\tau'^3}{(dt'/d\tau')^2} + \tfrac{3}{4}\frac{(d^2t'/d\tau'^2)^2}{(dt'/d\tau')^4} \\ &= \left(\frac{dt}{dt'}\right)^2\left[-\tfrac{1}{2}\frac{d^3t/d\tau^3}{(dt/d\tau)^2} + \tfrac{3}{4}\frac{(d^2t/d\tau^2)^2}{(dt/d\tau)^4}\right] \\ &= (\gamma t + \delta)^4\, T. \end{aligned}$$

So, T is a function of t of the kind already specified.

So, *the converse of the theorem is true.*

Thus, *if we can find any one quasi-uniformising variable of an algebraic form, we can find the totality of all uniformising and quasi-uniformising variables by this equation.*

We can now *find the functions* T.

If

$$t' = \frac{at + b}{ct + d},$$

we have

$$\frac{dt'}{dt} = \frac{1}{(ct + d)^2},$$

and so

$$(dz/dt')^2 = (ct + d)^4\,(dz/dt)^2.$$

Thus $(dz/dt)^2$ is a Thetafuchsian function of order two; any other Thetafuchsian function of order two can be written in the form

$$T = R(z, u)\,.\,(dz/dt)^2,$$

where $R(z, u)$ is an automorphic function of the group, *i.e.*, a rational function of the algebraic form.

If the algebraic form is of genus p, it is known* that any function $R(z, u)$ for which T is holomorphic is a linear function of $(3p - 3)$ special functions. These we can write

$$R_1(z, u),\quad R_2(z, u), \ldots R_{3p-3}(z, u).$$

The case $p = 1$ is exceptional; here there is one such function, T, namely, a constant.

* Humbert, 'Liouville's Journal,' (4), vol. 2, p. 239, 1886.

In general, therefore, we have

$$T = [a_1 R_1(z, u) + a_2 R_2(z, u) + \ldots + a_{3p-3} R_{3p-3}(z, u)] (dz/dt)^2,$$

where $R_1(z, u)$, $R_2(z, u)$, ... $R_3(z, u)$ are functions which can be found, and $a_1, a_2, \ldots a_{3p-3}$, are arbitrary constants.

We can now *find the form of the differential equation which gives all the quasi-uniformising variables.* Take any quasi-uniformising variable τ of the algebraic equation

$$f(u, z) = 0.$$

For it, we have

$$\tfrac{1}{2}\{\tau, z\} = \phi(z, u),$$

where ϕ is some rational function of z and u.

If t is the most general quasi-uniformising variable, we have seen that t is given as the quotient of two solutions of the differential equation

$$d^2v/dt^2 + Tv = 0,$$

where

$$T = [a_1 R_1(z, u) + a_2 R_2(z, u) + \ldots + a_{3p-3} R_{3p-3}(z, u)] (dz/dt)^2.$$

Hence

$$\tfrac{1}{2}\{t, \tau\} = T.$$

But

$$\{t, z\} = \{\tau, z\} + (d\tau/dz)^2 \{t, \tau\}.$$

Therefore

$$\tfrac{1}{2}\{t, z\} = \phi(z, u) + T(d\tau/dz)^2,$$

or

$$\tfrac{1}{2}\{t, z\} = \phi(z, u) + a_1 R_1(z, u) + a_2 R_2(z, u) + \ldots + a_{3p-3} R_{3p-3}(z, u).$$

Thus, *the solution of the problem of finding all the variables t, which will conformally represent the Riemann surface of a given algebraic form on a curvilinear polygon, whose sides are derived from each other in pairs by projective substitutions, is given by a differential equation containing $(3p - 3)$ arbitrary parameters linearly, and the problem of finding the uniformising variable is equivalent to that of determining these parameters in order that that group generated by these substitutions may be discontinuous.*

Now let us return to the differential equation of § 5, which we can write

$$\tfrac{1}{2}\{\tau, w\} = \frac{n-2}{8(n+1)} \frac{1}{u} \frac{d^2u}{dw^2} + k_1 z^{n-2} + k_2 z^{n-3} + \ldots + k_{n-1}.$$

If we take any set of values $k_1, k_2 \ldots k_{n-1}$ for the undetermined constants, this differential equation will give a variable τ in terms of z, which will not in general be the variable t of §§ 3 and 4. But the variables τ so found will solve the problem of conformally representing the z-plane, regarded as bounded by a number of finite lines radiating from a point, on a curvilinear polygon in the τ-plane, such that the sides of the boundary can be transformed into each other in pairs by certain projective substitutions. The variable t is one of these variables, characterised by the condition that the infinite group generated from these substitutions is a discontinuous group.

We can, in fact, find the functions T in this case. We must have

$$\mathrm{T} = \mathrm{R}(z)\left(\frac{dz}{dt}\right)^2,$$

and R (z) must be such that T is holomorphic. So the only possible poles of R (z) are the places where dz/dt is zero, *i.e.*, the places $z = e_1, e_2, \ldots e_{n+2}$. At these places dz/dt is zero of the first order: so $(dz/dt)^2$ is zero of the second order, and R (z) may have a pole of the second order.

Therefore

$$\mathrm{R}(z) = \frac{\mathrm{I}(z)}{u^2},$$

where

$$u^2 = (z - e_1)(z - e_2)\ldots(z - e_{n+2}),$$

and I (z) is an integral function of z. At $z = \infty$, dz/dt has a pole of the second order, and u^2 a pole of the $(n + 2)^{\text{th}}$ order. So I (z) may have a pole of the $(n - 2)^{\text{th}}$ order.

Therefore

$$\mathrm{I}(z) = k_1' z^{n-2} + k_2' z^{n-3} + \ldots + k_{n-1}'$$

and

$$\mathrm{T} = \frac{k_1' z^{n-2} + k_2' z^{n-3} + \ldots + k_{n-1}'}{u^2}\left(\frac{dz}{dt}\right)^2.$$

Thus if τ is the quotient of two solutions of the equation

$$d^2v/dt^2 + \mathrm{T}v = 0$$

and t is defined by the equation

$$\tfrac{1}{2}\{t, z\} = \mathrm{R}(z),$$

then τ is defined by the equation

$$\tfrac{1}{2}\{\tau, z\} = \mathrm{R}(z) + \frac{k_1' z^{n-2} + k_2' z^{n-3} + \ldots k_{n-1}'}{u^2}.$$

Comparing this with the equation of § 5, we see that *the variables t given by it, when the constants $k_1, k_2, \ldots k_{n-1}$, are arbitrary, are the quasi-uniformising variables.*

We can now prove that the number of conditions which have to be satisfied in order that the group of substitutions of t may be discontinuous, *i.e.*, in this case may conserve a fundamental circle, is equal to the number of the constants k.

In order that a self-inverse substitution with complex coefficients,

$$\left(t, \frac{at + b}{ct - a}\right),$$

may leave unchanged a given circle, two of the four real constants contained in the substitution must be determinate in terms of the others.

Now there are $(n + 2)$ fundamental self-inverse complex substitutions, containing $4(n + 2)$ real constants; of these, the relation

$$S_1 S_2 S_3 \ldots S_{n+2} = 1$$

accounts for six. So $(2n + 1)$ of the real constants are determined in terms of the other $(2n + 1)$ by the condition that the group is to conserve a fundamental circle; but as the fundamental circle may be any whatever, and so involves three constants, we must deduct three from the number of equations, giving $(2n - 2)$. Thus, $2n - 2$ real, or $n - 1$ complex, constants can be determined from the condition that the substitutions of t conserve a fundamental circle. *This accords with the fact, otherwise arrived at, that the constants* $k_1, k_2, \ldots k_{n-1}$, *in the differential equation have to be determined from this consideration.*

Among the quasi-uniformising variables of any algebraic form there are several distinct uniformising variables. The groups we have found in § 3 have simply-connected fundamental polygons. But automorphic functions exist, for which the fundamental polygons are multiply-connected.

The simplest example of such a function is

$$z = P\left(\frac{iw_1}{\pi}\log t\right),$$

where P is Weierstrass' elliptic function with periods $2w_1$ and $2w_2$; the fundamental polygon is the space between two circles in the t-plane.

The automorphic functions studied by SCHOTTKY, WEBER, and BURNSIDE may be regarded as generalisations of this. As these uniformising variables with multiply-connected fundamental polygons are included in the general set of quasi-uniformising variables, they are defined by the same differential equations as the uniformising variables with simply-connected polygons, except that the constants k will have different values.

Gyrostatic Vortex

University College,

INDEX SLIP.

I.)

10.12.98

THE chief part of the following investigation (Sects. i. and iii.) was undertaken with the view of discovering whether it was possible to imagine a kind of vortex motion which would impress a gyrostatic quality which the forms of vortex aggregates hitherto known do not possess. The other part (Sect. ii.) deals with the non-gyrostatic vortex aggregates, the discovery of which we owe to HILL,* and investigates the conditions under which two or more aggregates may be combined into one. It is shown that it is allowable to suppose one or more concentric shells of vortex aggregates to be applied over a central spherical nucleus, subject to one relation between the radii and the vorticities. In all cases the vorticities must be in opposite directions in alternate shells. The special case when the aggregates are built up of the same vortical matter is considered, and the magnitudes of the radii and the positions of the equatorial axes determined. The cases of motion in a rigid spheroidal shell and of dyad spheroidal aggregates are also considered.

The chief part of the paper refers to gyrostatic aggregates. The investigation has brought to light an entirely new system of spiral vortices. The general conditions for the existence of such systems, when the motion is symmetrical about an axis, are determined in Sect. i., and are worked out in more detail for a particular case of spherical aggregate in Sect. iii. It is found that the motion in meridian planes is determined from a certain function ψ in the usual manner. The velocity along a parallel of latitude is given by $v = f(\psi)/\rho$ where ρ is the distance of the point from the axis. The function ψ, however, does not depend on the differential equation of the ordinary non-spiral type, but is a solution of the equation

$$\frac{d^2\psi}{dr^2} + \frac{1}{r^2}\frac{d^2\psi}{d\theta^2} - \frac{\cot\theta}{r^2}\frac{d\psi}{d\theta} = \rho^2\mathrm{F} - f\frac{df}{d\psi},$$

where F and f are both functions of ψ. The case F and $fdf/d\psi$ both uniform is briefly treated. It refers to a spiral aggregate with a central solid nucleus, and is not of great interest. The case F uniform and $f \propto \psi$ is treated more fully. If $f \equiv \lambda\psi/a$ where a is the radius of the aggregate

$$\psi = \mathrm{A}\left\{\mathrm{J}_2\left(\frac{\lambda r}{a}\right) - \frac{r^2}{a^2}\mathrm{J}_2\lambda\right\}\sin^2\theta.$$

The most striking and remarkable fact brought out is that with increasing parameter λ, we get a periodic system of families of aggregates. The members of each family differ from one another in the number of layers and equatorial axes they possess. I have ventured to call them singlets, doublets, triplets, &c., in contradistinction to the more or less fortuitous and arbitrary compounds dealt with later, and which I have named monads, dyads, triads, &c. Of these families two are investigated more in detail than the others. In one family (the λ_2 family) all the members remain at rest in the surrounding fluid. In the other (the λ_1 family) the

* "On a Spherical Vortex," 'Phil. Trans.,' A, vol. 185, 1894.

distinguishing feature common to all the members is that the stream lines and the vortex lines are coincident.

The parameter λ defines the total angular pitch of the stream lines, on the outer current-sheet, viz., up the polar axis and down the outside; although in the aggregates with more than one axis these lines are not one continuous stream line. The first aggregates—with $\lambda < 5 \cdot 7637$ (the first λ_2 parameter)—behave abnormally. Beyond these we get successive series, in one set of which the velocity of translation is in the same direction as the polar motion of the central nucleus, in the alternate set the velocity is opposite, and the aggregate regredes in the fluid as compared with its central aggregate (see fig. 3, Plate 1). The physical analogue of these aggregates is obvious. It is specially enlarged upon in the abstract.*

Suppose we set ourselves the problem of making a set of aggregates with greater and greater angular pitch. As we do so we shall find that as the pitch increases the equatorial axis contracts, and the surface velocity diminishes. On the outer layers (ring shaped) the spiral is chiefly produced on the inner side facing the polar axis, until on the boundary itself the stream lines flow in meridians, and the twist is altogether on the polar axis. The pitch can be increased up to a certain degree. As this is done, the stream lines and vortex lines fold up towards one another, coincide at a certain pitch, and exchange sides. When an external angular pitch of about 330° is attained it is impossible to go further if a simple aggregate is desired. If a higher pitch is desired it is attained by taking it in two parts. First, a central spherical nucleus of the same nature as the former, in which a portion of the twist is produced, and outside this a spherical shell, in which the spirals have the same direction of twist, and complete the pitch to the desired amount but in which the spirals are traversed in the opposite direction. With increasing pitch this layer becomes thicker, and its equatorial axis contracts relatively to the mid-point of the shell until another limit is reached; the stream and vortex lines again fold together, cross, and expand as this second limit is reached. If a larger pitch still is desired there must be a third layer, and so on. The first coincidence of vortex and stream lines takes place for an aggregate whose pitch is 257°·27′. Whenever a maximum pitch is attained the aggregate is at rest in the fluid. This is first attained for an external pitch of 330°·14′. Beyond this there are two equatorial axes. For an external pitch of 442°·37′ the stream and vortex lines again coincide, the internal nucleus gives 257°·27′ of the pitch and the outer shell the remainder, and so on.

At the end a theory of compound aggregates is developed similar to that in Sect. ii. for non-gyrostatic vortices. It is not worked out in detail in the present communication, but the conditions are determined for dyad compounds, whilst a similar theory holds for triad and higher ones. Each element of a poly-ad may consist of singlets, doublets, &c. The equations of condition leave three quantities arbitrary—

* 'Roy. Soc. Proc.,' vol. 62, p. 332.

as, for instance, ratio of volumes, ratio of primary cyclic constants, ratio of secondary cyclic constants. The full development of this theory is, however, left for a future communication. It is clear that spiral or gyrostatic vortex aggregates are not confined to forms symmetrical about an axis. Their theory is however much more complicated.

If we take any particular spherical aggregate with given λ and primary cyclic constant (μ), the energy is determinate. We may, however, alter the energy. If it be increased, the spherical form begins to open out into a ring form, whose shape and properties have not yet been investigated. If the energy be increased sufficiently the aperture becomes large compared with the thickness of the rotational core, and approximate calculation can be applied. The differential equation for ψ is given in Sect. i., but its development is left for a future occasion. After that I hope to deal with the question of stability, and then more fully with that of the conditions of combination. The new field opens up so many questions of interest that other workers in it are welcomed.

Section i.—*General Theorems.*

1. To give an idea of the nature of the motions considered in the present investigation, consider the case of motion of an infinitely long cylindrical vortex of sectional radius a. The velocity perpendicular to the axis inside the vortex will be of the form $v = f(r)$, where $f(0) = 0$. Outside it will be given by $v = \mathrm{V}a/r$, where $\mathrm{V} = f(a)$.

We may, however, have a motion in which the fluid moves parallel to the axis inside the cylinder with rest outside. The velocity will be of the form $u = \mathrm{F}(r)$ inside, where $\mathrm{F}(a) = 0$, and zero outside. Both $f(r)$ and $\mathrm{F}(r)$ are arbitrary functions subject only to the conditions $f(0) = 0$ and $\mathrm{F}(a) = 0$.

Putting aside for the present the question of the stability of these simple motions or of their resultant, it is clear that if we superpose the two we get another state of motion in which we have vortex-filaments in the shape of helices lying on concentric cylindric surfaces. The problem to be considered is whether it is possible to conceive a similar superposition of two motions in the case of any vortex aggregate whose motions are symmetric about an axis.

There are an infinite number of either ring-shaped vortices, or singly connected aggregates (of which HILL's vortex may serve as a type), differing from one another in the law of vorticity of the different parts—the most important being those in which the vorticity is uniform. The motions in all these are known in terms of the stream function ψ. The value of ψ is however at present only actually known for an infinitely thin ring-filament or for a spherical aggregate.

2. We are to consider two superposed motions. The one component is in meridian planes through an axis and can be defined in terms of the stream-function ψ.* The

* Throughout ψ is taken as the total flow *up* through the circle whose radius is ρ. In other words the velocity perpendicular to ds is $\dfrac{1}{2\pi\rho}\dfrac{d\psi}{ds}$.

other component is everywhere perpendicular to these meridian planes. The vortex aggregates will be moving with rectilinear translation through the fluid with a velocity calculable, when the distribution of vortex motion is known, by HELMHOLTZ's method. Bring the aggregate to rest by impressing everywhere a velocity equal and opposite to the velocity of translation. The motion then consists of a flow up through the centre in the direction of previous translation, the fluid then streaming (in this most general case) in spirals round a certain circle. The circle may conveniently be called the *equatorial axis* of the aggregate. The line of symmetry through the centre in the direction of translation may then be termed the *polar axis*. Whether we deal with ring-shaped or singly connected aggregates, the surfaces ψ will always be ring-shaped inside. In fact they are so also at the boundary, for the surface value of ψ really consists in the latter case of the outer boundary together with the polar axis.

3. Conceive now the aggregate divided up into a large number of ring-surfaces given by values of a parameter ψ differing by $d\psi$, and confine attention to what is going on between the two surfaces ψ and $\psi + d\psi$. We shall suppose ψ to increase as we pass from the outside inwards. Let dn denote the distance at a point between the surfaces ψ and $\psi + d\psi$, dn to be measured also inwards. In the shell considered the lines of flow will be spiral, and the vortex-filaments also spirals, as indicated in the figure, the thin line Pf representing a line of flow, the thick Pv a vortex-filament,

Fig. 1.

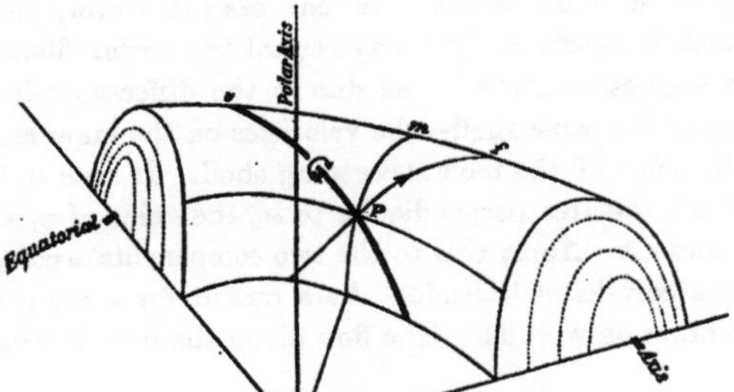

and the line Pm a meridian section. Denote the velocity at P by v and the angle it makes with the meridian by ϕ. Also let ω denote the molecular rotation at P, and χ the angle the filament makes with the meridian—estimated positive when on the opposite side of the meridian to v.

Consider the flow between the two surfaces ψ and $\psi + d\psi$ across the "parallel of latitude" through P. The total flow must be the same for every parallel. The area through which the flow takes place is $2\pi\rho dn$, where ρ is the distance of P from the polar axis. Hence $2\pi\rho v \cos\phi dn$ is constant over the surface ψ. It must therefore be of the form $f(\psi)\, d\psi$. So far ψ is only defined as the parameter which determines the particular surface. Choose the parameter so that $f(\psi) = 1$. ψ is then analogous to the stream-function in the simple case. It acts in fact as the stream-function for the component of velocity $v \cos\phi$. Similar reasoning leads to the conclusion that $\omega\rho \cos\chi dn$ is also of the form $f(\psi)\, d\psi$, say $f_1 d\psi$. Hence

$$2\pi\rho v \cos\phi dn = d\psi \; . \; . \; . \; . \; . \; . \; . \; . \; . \; . \; (1)$$

$$2\pi\rho\omega \cos\chi dn = f_1\, d\psi \; . \; . \; . \; . \; . \; . \; . \; . \; . \; (2).$$

We started with the supposition that the stream-lines and vortex-lines must lie on the same surfaces ψ. In other words, there must be no component rotation perpendicular to ψ. This may be expressed in other words by the statement that the circulation round any circuit drawn wholly on ψ must vanish. Take for this circuit any two parallels of latitude. The condition gives that the flow along one must equal the flow along the other. In other words, the flow round a parallel of latitude must be the same for all parallels on the same surface ψ. Hence

$$2\pi\rho v \sin\phi = f \; . \; . \; . \; . \; . \; . \; . \; . \; . \; . \; (3)$$

where f is a function of ψ.

Equations 1, 2, 3 give conditions which any motion possible between any two given surfaces ψ and $\psi + d\psi$ must satisfy. In our case, however, the motions in the separate shells must fit together. We may regard the vortex-filaments as due to the velocities in two successive shells, *or* as due to the different velocities on the inner and outer surfaces of the same shell—the velocities on the inner surface of one being the same as on the outer of the next succeeding shell. If now ω_1 be any component of a filament, and dA the area perpendicular to ω_1, the value of $\omega_1 dA$ is given by half the circulation round dA. Apply this to the two components $\omega \cos\chi$ along a meridian and $\omega \sin\chi$ along a parallel of latitude. As a circuit for $\omega \cos\chi$ take two parallels one on ψ and the other on $\psi + d\psi$. The flow along the first is $2\pi\rho v \sin\phi$ and along the latter

$$2\pi\rho v \sin\phi + 2\pi \frac{d}{dn}(\rho v \sin\phi)\, dn.$$

Hence

$$2\omega \cos\chi \,.\, 2\pi\rho dn = -2\pi \frac{d}{dn}(v\rho \sin\phi)\, dn.$$

But by (3),

$$2\pi\rho v \sin\phi = f.$$

Hence

$$4\pi\rho\omega \cos\chi = -\frac{df}{dn} \; . \; . \; . \; . \; . \; . \; . \; . \; . \; . \; (4).$$

Comparing with (2) it follows that

$$f_1 \frac{d\psi}{dn} = -\tfrac{1}{2}\frac{df}{dn},$$

or

$$f_1 = -\tfrac{1}{2}\frac{df}{d\psi}.$$

We may regard then Eq. (2) as replaced by (4), which includes it as the greater does the less.

For the circuit for $\omega \sin \chi$ take a small circuit formed by a small arc ds of a meridian PP′ on ψ, the normals (dn) at P, P′ and the portion of the meridian arc on $\psi + d\psi$ cut off by these normals. The flow along the normals dn is zero. Along ds it is $v \cos\phi\, ds$; along ds' it is

$$v\cos\phi\, ds + \frac{d}{dn}(v\cos\phi\, ds)\, dn.$$

The area of the cross-section of $\omega \sin\chi$ is $dn\, ds$.

Hence

$$2\omega \sin\chi\, dn\, ds = -\frac{d}{dn}(v\cos\phi\, ds)\, dn.$$

But by (1),

$$v\cos\phi = \frac{1}{2\pi\rho}\frac{d\psi}{dn};$$

therefore

$$4\pi\omega \sin\chi\, ds = -\frac{d}{dn}\left(\frac{1}{\rho}\frac{d\psi}{dn}ds\right).$$

Since $d\psi/ds = 0$ ψ will give any component of velocity *in the meridian plane* in the same way as the ordinary stream-function.

4. It will often be found advantageous to express ψ in terms of curvilinear co-ordinates. Denote these by u, v. Displacements perpendicular to the u will be denoted by dn, and to v by dn', to be estimated positive in the directions in which u, v respectively increase.

The differential equation satisfied by ψ is found by expressing the circulation round a small area bounded by the curves $u, u + du, v, v + dv$. Let ω_1 ($= \omega \sin\chi$) denote the rotation at a point of the area. We shall regard this as positive when it goes clockwise. The circulation is then $2\omega_1 \times \text{area} = 2\omega_1\, dn, dn'$.

The velocities along PQ, PP′ (see fig. 2) are respectively

$$\frac{1}{2\pi\rho}\frac{d\psi}{dn}, \qquad -\frac{1}{2\pi\rho}\frac{d\psi}{dn'},$$

The flows along them are therefore (clockwise)

$$\frac{1}{2\pi\rho}\frac{d\psi}{dn}dn' \text{ and } +\frac{1}{2\pi\rho}\frac{d\psi}{dn'}dn.$$

Fig. 2.

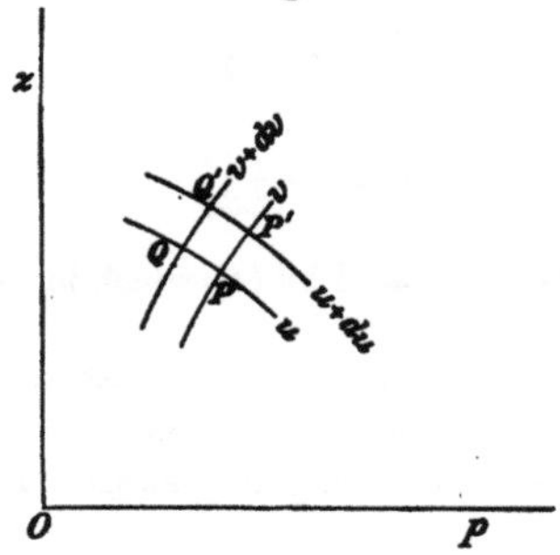

Hence the total flow round PQQ'P' is

$$-\frac{d}{du}\left(\frac{1}{2\pi\rho}\frac{d\psi}{dn}dn'\right)du-\frac{d}{dv}\left(\frac{1}{2\pi\rho}\frac{d\psi}{dn'}dn\right)dv,$$

or

$$-\frac{d}{du}\left(\frac{1}{2\pi\rho}\frac{d\psi}{du}\cdot\frac{du}{dn}\cdot\frac{dn'}{dv}\right)du\,dv-\frac{d}{dv}\left(\frac{1}{2\pi\rho}\frac{d\psi}{dv}\cdot\frac{dv}{dn'}\cdot\frac{dn}{du}\right)du\,dv.$$

But this is $2\omega_1\,dn\,dn'$. Hence

$$\frac{d}{du}\left(\frac{1}{\rho}\frac{d\psi}{du}\cdot\frac{du}{dn}\cdot\frac{dn'}{dv}\right)+\frac{d}{dv}\left(\frac{1}{\rho}\frac{d\psi}{dv}\frac{dv}{dn'}\cdot\frac{dn}{du}\right)=-4\pi\omega_1\frac{dn}{du}\frac{dn'}{dv}$$

$$=-4\pi\omega\sin\chi\frac{dn}{du}\frac{dn'}{dv}\quad . \quad (5).$$

In many cases $\rho+z\iota=f(u+v\iota)$, giving $du/dn=dv/dn'$, and the equation simplifies to

$$\frac{d}{du}\left(\frac{1}{\rho}\frac{d\psi}{du}\right)+\frac{d}{dv}\left(\frac{1}{\rho}\frac{d\psi}{dv}\right)=-4\pi\omega_1\left(\frac{dn}{du}\right)^2.$$

The following cases will be required :—

(1) *Cylindrical co-ordinates.* (ρ, z),

$$du=d\rho=dn \qquad dv=dz=dn',$$

and

$$\frac{d}{d\rho}\left(\frac{1}{\rho}\frac{d\psi}{d\rho}\right)+\frac{1}{\rho}\frac{d^2\psi}{dz^2}=-4\pi\omega\sin\chi,$$

or

$$\frac{d^2\psi}{d\rho^2}-\frac{1}{\rho}\frac{d\psi}{d\rho}+\frac{d^2\psi}{dz^2}=-4\pi\rho\omega\sin\chi\quad . \; . \; . \; . \; . \; . \quad (6).$$

(2) *Polar Co-ordinates.* (r, θ),

$$\rho = r\sin\theta \qquad du = dr = dn \qquad dv = d\theta \qquad dn' = r\,d\theta,$$

and

$$\frac{d}{dr}\left(\frac{r}{\rho}\frac{d\psi}{dr}\right) + \frac{d}{d\theta}\left(\frac{1}{r\rho}\frac{d\psi}{d\theta}\right) = -4\pi r\omega\sin\chi,$$

or

$$\frac{d^2\psi}{dr^2} + \frac{1}{r^2}\frac{d^2\psi}{d\theta^2} - \frac{\cot\theta}{r^2}\frac{d\psi}{d\theta} = -4\pi\rho\omega\sin\chi \quad . \quad . \quad . \quad . \quad . \quad . \quad (7).$$

(3) *Spheroids.*

(α) Prolate. Here $\rho + z\iota = \lambda\sinh(u + v\iota)$,
whence

$$\rho = \lambda\sinh u\cos v, \qquad z = \lambda\cosh u\sin v.$$

The surfaces u, v are respectively the ellipses and hyperbolas

$$\frac{\rho^2}{\sinh^2 u} + \frac{z^2}{\cosh^2 u} = \lambda^2 \quad \text{and} \quad \frac{z^2}{\sin^2 v} - \frac{\rho^2}{\cos^2 v} = \lambda^2,$$

u increases from 0 at the origin to ∞ at an infinite distance ; v increases from $-\frac{1}{2}\pi$ at points on the negative part of the axis of z, through 0 for points on the equatorial plane to $\frac{1}{2}\pi$ at points on the positive part of the axis of z.

Again

$$\left(\frac{dn}{du}\right)^2 = \left(\frac{d\rho}{du}\right)^2 + \left(\frac{dz}{du}\right)^2 = \frac{d}{du}(\rho + z\iota)\frac{d}{du}(\rho - z\iota)$$
$$= \lambda^2\cosh(u + v\iota)\cosh(u - v\iota)$$
$$= \lambda^2(\cosh^2 u - \sin^2 v).$$

Hence the differential equation is (writing C and S for $\cosh u$, $\sinh u$),

$$\frac{1}{\cos v}\frac{d}{du}\left(\frac{1}{\mathrm{S}}\frac{d\psi}{du}\right) + \frac{1}{\mathrm{S}}\frac{d}{dv}\left(\frac{1}{\cos v}\frac{d\psi}{dv}\right) = -4\pi\lambda^3\omega\sin\chi\,(\mathrm{C}^2 - \sin^2 v) \quad . \quad . \quad (8).$$

(β) Oblate. Here $\rho + z\iota = \lambda\cosh(u + v\iota)$

$$\rho = \lambda\cosh u\cos v, \qquad z = \lambda\sinh u\sin v,$$

$$\left(\frac{dn}{du}\right)^2 = \lambda^2\sinh(u + v\iota)\sinh(u - v\iota)$$
$$= \lambda^2(\cosh^2 u - \cos^2 v),$$

and the differential equation is

$$\frac{1}{\cos v}\frac{d}{du}\left(\frac{1}{\mathrm{C}}\frac{d\psi}{du}\right) + \frac{1}{\mathrm{C}}\frac{d}{dv}\left(\frac{1}{\cos v}\frac{d\psi}{dv}\right) = -4\pi\lambda^3\omega\sin\chi\,(\mathrm{C}^2 - \cos^2 v) \quad . \quad . \quad (9).$$

(4) *Toroidal Functions.*—Here ['Phil. Trans.,' 1881, Part III., p. 614]

$$u + v\iota = \log\frac{\rho + a + z\iota}{\rho + a + z\iota}, \qquad \rho = a\frac{\sinh u}{\cosh u - \cos v}, \qquad \frac{du}{dn} = \frac{\sinh u}{\rho},$$

whence

$$\frac{d}{dv}\left(\frac{C - \cos v}{S}\frac{d\psi}{du}\right) + \frac{1}{S}\frac{d}{dv}\left((C - \cos v)\frac{d\psi}{dv}\right) = -4\pi a\frac{\rho^2}{S^2}\omega\sin\chi,$$

$$= -\frac{4\pi a^3}{(C - \cos v)^2}\omega\sin\chi \quad . \quad (10).$$

5. Equations 1, 3, 4, 5 or 6 give the conditions for a possible motion. It is open to us to choose ψ arbitrarily. In this case the equations give v, ω, χ, ϕ. The motion is instantaneously possible, but in general it will at once proceed to change the configuration—the motion will not be steady. The application of this theory to values of ψ which are already known (HILL's vortex for example) leads to interesting results, but the absence of steadiness robs the theory of importance. If we impose the condition of steady motion, it is no longer open to us to choose ψ at will. Let us then impose this condition. The condition that the motion shall be steady involves:—

(1) ψ must be a surface containing both vortex-lines and stream-lines.

This is already the case.

(2) $v\omega\sin(\phi + \chi)\,dn$ must be constant over the surface.

It must therefore be of the form $\mathrm{F}d\psi$, where F is a function of ψ. Hence

$$v\omega\sin(\phi + \chi) = \mathrm{F}\frac{d\psi}{dn} \quad . \quad . \quad . \quad . \quad . \quad . \quad . \quad . \quad . \quad (11).$$

Expanding this, and substituting from 1, 3, 4, 7,

$$-f\frac{df}{d\psi} - \left\{\frac{d^2\psi}{dr^2} + \frac{1}{r^2}\frac{d^2\psi}{d\theta^2} - \frac{\cot\theta}{r^2}\frac{d\psi}{d\theta}\right\} = 8\pi^2\rho^2\mathrm{F},$$

or

$$\frac{d^2\psi}{dr^2} + \frac{1}{r^2}\frac{d^2\psi}{d\theta^2} - \frac{\cot\theta}{r^2}\frac{d\psi}{d\theta} = -8\pi^2\rho^2\mathrm{F} - f\frac{df}{d\psi} \quad . \quad . \quad . \quad . \quad (12),$$

where f and F are arbitrary functions of ψ. Choosing these, equation 12 will give the type of ψ.*

We proceed to apply these general theorems to certain special cases of spherical aggregates. In order to exemplify the method employed we will take first the case in which there is no secondary spin, the type in which HILL's spherical vortex is the simplest case.

* For another proof of this equation, due to one of the referees, see end of present paper.

Section ii.—*Aggregates with no Secondary Spin* ($f = 0$) *and with Uniform Vorticity.*

6. We begin with the spherical aggregate, the simplest type of which is the HILL'S vortex. The equation for ψ is that given by equation 7, in which ω is put $k\rho$ where k is uniform and $\chi = \frac{1}{2}\pi$. It is

$$\frac{d^2\psi}{dr^2} + \frac{1}{r^2}\frac{d^2\psi}{d\theta^2} - \frac{\cot\theta}{r^2}\frac{d\psi}{d\theta} = -4\pi k\rho^2 = -4\pi k r^2 \sin^2\theta,$$

in which θ is measured from the pole to the equator. A particular solution of this is $-\frac{1}{2}\pi k r^4 \sin^4\theta$.

In

$$\frac{d^2\psi}{dr^2} + \frac{1}{r^2}\frac{d^2\psi}{d\theta^2} - \frac{\cot\theta}{r^2}\frac{d\psi}{d\theta} = 0,$$

put $\psi = r^n Z_n$, Z_n being a function of θ only. Then

$$\frac{d^2 Z_n}{d\theta^2} - \cot\theta\frac{dZ_n}{d\theta} + n(n-1)Z_n = 0.$$

The integral of this is

$$Z_n = -\sin\theta\frac{dP_{n-1}}{d\theta},$$

where P_{n-1} is a zonal harmonic of degree $n-1$.

Hence the general solution of the equation in ψ is

$$\psi = -\tfrac{1}{2}\pi k r^4 \sin^4\theta + \Sigma\left(A_n r^n + \frac{B}{r^{n-1}}\right)Z_n.$$

Since

$$P_n = \frac{1.3.5\ldots(2n-1)}{n!}\left\{\cos^n\theta - \frac{n(n-1)}{2(2n-1)}\cos^{n-2}\theta + \ldots\right\}$$

the values of Z_n are easily found, except for Z_1 or Z_0. It is easily found from the direct equation in this case that $Z_1 = Z_0 = \cos\theta$. The following results are easily deduced:—

$$Z_2 = \sin^2\theta, \qquad Z_3 = 3\sin^2\theta\cos\theta,$$

$$Z_4 = \tfrac{3}{2}(4\sin^2\theta - 5\sin^4\theta), \qquad \sin^4\theta = \tfrac{4}{5}Z_2 - \tfrac{2}{15}Z_4.$$

Consider now first the case of a homogeneous spherical aggregate. In this case the functions $\frac{B}{r^{n-1}}Z_n$ apply only to the space outside, and Ar^nZ_n to the space inside. Let ψ_1 denote the value of ψ inside and ψ_2 outside. Hence

$$\psi_1 = -\tfrac{1}{2}\pi k r^4 \sin^n\theta + \Sigma A_n r^n Z_n$$

$$\psi_2 = \Sigma \frac{B_n}{r^{n-1}} Z_n.$$

Let a denote the radius of the sphere. Along the boundary of the sphere $\psi_1 = \psi_2$, and also $d\psi_1/dr = d\psi_2/dr$. Expressing $\sin^4\theta$ in terms of Z_2 and Z_4,

$$\psi_1 = -\tfrac{2}{5}\pi k r^4 Z_2 + \tfrac{1}{15}\pi k r^4 Z_4 + \Sigma A_n r^n Z_n.$$

The term in $\tfrac{1}{15}\pi k r^4 Z_4$ may be supposed merged in $A_4 r^4 Z_4$, and may therefore be treated as absent. The conditions

$$\left.\begin{aligned}\psi_1 &= \psi_2\\ \frac{d\psi_1}{dr} &= \frac{d\psi_2}{dr}\end{aligned}\right\}\text{ when } r = a$$

give

$$A_1 = 0, \qquad A_n = 0 \text{ when } n > 2,$$
$$B_1 = 0, \qquad B_n = 0 \text{ when } n > 2,$$

and for $n = 2$

$$\left.\begin{aligned}A_2 a^2 - \tfrac{2}{5}\pi k a^4 &= \frac{B_2}{a}\\ 2\,A_2\,a - \tfrac{8}{5}\pi k a^3 &= -\frac{B_2}{a^2}\end{aligned}\right\}$$

Hence

$$A_2 = \tfrac{2}{3}\pi k a^2, \qquad B_2 = \tfrac{4}{15}\pi k a^5,$$

and

$$\psi_1 = 2\pi k\left(\tfrac{1}{3}a^2 r^2 - \tfrac{1}{5}r^4\right)\sin^2\theta$$

$$\psi_2 = \tfrac{4}{15}\pi k \frac{a^5}{r}\sin^2\theta.$$

The velocity along the normal to the aggregate is

$$\frac{1}{2\pi\rho}\frac{d\psi_2}{r d\theta} = \tfrac{4}{15}k a^2 \cos\theta.$$

Hence the aggregate moves forward through the surrounding fluid with a velocity

$$V = \tfrac{4}{15}k a^2.$$

Referred to the aggregate at rest therefore

$$\psi_1 = \tfrac{2}{5}\pi k r^2 (a^2 - r^2)\sin^2\theta.$$

The cyclic constant (μ) is the circulation taken round a meridian section, up the polar axis and down outside. It is the sum of the circulation round the elementary areas of which the section is composed. Hence

$$\mu = \Sigma \text{ (elementary circulations)} = \Sigma 2\omega\, d\text{A} = k\Sigma 2\rho\, d\text{A}$$

$$= \frac{k}{\pi}\,\Sigma 2\pi\rho\, d\text{A} = \frac{k}{\pi} \times \text{volume of aggregate} = \frac{mk}{\pi}.$$

Thus

$$\omega = k\rho = \frac{\pi\mu}{m}\rho, \qquad \text{V} = \frac{\mu}{5a},$$

which are HILL's results obtained by direct methods.

7. *Heterogeneous Aggregates.*—We may, however, superpose on an aggregate such as the foregoing other spherical layers of different vorticities. It will be advisable to consider first the case where there is one such layer of vorticity determined by (say) k'. We may call them dyads. In this outer portion both terms in $\text{A}r^n$ and B/r^{n-1} can appear. Let ψ_1, ψ_2, ψ denote the stream functions for each part and for the surrounding fluid. Then

$$\psi_1 = -\tfrac{2}{5}\pi k r^4 \text{Z}_2 + \Sigma \text{A}_n r^n \text{Z}_n,$$

$$\psi_2 = -\tfrac{2}{5}\pi k' r^4 \text{Z}_2 + \Sigma\left(\text{A}_n' r^n + \frac{\text{B}_n'}{r^{n-1}}\right)\text{Z}_n,$$

$$\psi = \Sigma \frac{\text{B}_n}{r^{n-1}}\,\text{Z}_n.$$

Let a, b denote the radii of the two spherical surfaces ($a > b$), and apply the same conditions as before to the two surfaces.

Again all the co-efficients vanish except for $n = 2$, and there results

$$\left.\begin{aligned} \text{A}_2 b^2 - \tfrac{2}{5}\pi k b^4 &= \text{A}_2' b^2 + \frac{\text{B}_2'}{b} - \tfrac{2}{5}\pi k' b^4 \\ 2\text{A}_2 b - \tfrac{8}{5}\pi k b^3 &= 2\text{A}_2' b - \frac{\text{B}_2'}{b^2} - \tfrac{8}{5}\pi k' b^3 \end{aligned}\right\}$$

and

$$\left.\begin{aligned} \frac{\text{B}_2}{a} &= \text{A}_2' a^2 + \frac{\text{B}_2'}{a} - \tfrac{2}{5}\pi k' a^4 \\ -\frac{\text{B}_2}{a^2} &= 2\text{A}_2' a - \frac{\text{B}_2'}{a^2} - \tfrac{8}{5}\pi k' a^3 \end{aligned}\right\}$$

The first two give at once

$$\text{B}_2' = \tfrac{4}{15}\pi\,(k - k')\, b^5,$$

the last two

$$\text{A}_2' = \tfrac{2}{3}\pi a^2 k';$$

also

$$\text{B}_2 = \tfrac{4}{15}\pi\,\{(k - k')\, b^5 + k' a^5\}$$

$$\text{A}_2 = \tfrac{2}{3}\pi\,\{k' a^2 + (k - k')\, b^2\}.$$

Whence

$$\psi_1 = 2\pi \left\{\frac{k'a^2 + (k - k') b^2}{3} r^2 - \tfrac{1}{5}kr^4\right\} \sin^2\theta,$$

$$\psi_2 = 2\pi \left\{\tfrac{1}{3}k'a^2r^2 + \tfrac{2}{15}(k - k')\frac{b^5}{r} - \tfrac{1}{5}k'r^4\right\} \sin^2\theta,$$

$$\psi = \frac{4\pi}{15}\frac{(k - k') b^5 + k'a^5}{r} \sin^2\theta.$$

The normal velocity at the outer boundary is

$$\frac{1}{2\pi\rho}\frac{d\psi}{rd\theta} \text{ (when } r = a) = \tfrac{4}{15}\frac{(k - k') b^5 + k'a^5}{a^3} \cos\theta.$$

The outer boundary therefore progresses unchanged with velocity of translation

$$V = \tfrac{4}{15}\frac{(k - k') b^5 + k'a^5}{a^3}.$$

Bring the outer boundary to rest by impressing on every part of the fluid a velocity equal and opposite to this, *i.e.*, adding to the stream-functions a term

$$-\tfrac{4}{15}\pi\frac{(k - k') b^5 + k'a^5}{a^3} r^2 \sin^2\theta.$$

The relative motions are then given by

$$\psi_1 = \frac{2\pi}{5}\left\{\tfrac{1}{3}\left(5k'a^2 + 5(k - k') b^2 - 2k'a^2 - 2(k - k')\frac{b^5}{a^3}\right) - kr^2\right\} r^2 \sin^2\theta$$

$$= \frac{2\pi}{5}\left\{k'a^2 - kr^2 + \tfrac{1}{3}(k - k')\left(5 - \frac{2b^3}{a^3}\right) b^2\right\} r^2 \sin^2\theta$$

$$\psi_2 = \frac{2\pi}{5}\left\{k'(a^2 - r^2) r^2 + \tfrac{2}{3}(k - k')\frac{b^5}{a^3 r}(a^3 - r^3)\right\} \sin^2\theta.$$

If, however, the motion is to be steady, the inner sphere must now be at rest, that is $\psi_1 = 0$ when $r = b$. We get, therefore, the following necessary relation between k, k', a, b,

$$k'a^2 - kb^2 + \tfrac{1}{3}(k - k')\left(5 - \frac{2b^3}{a^3}\right) b^2 = 0.$$

This may be written

$$2b^2k(a^3 - b^3) + k'\{3a^3(a^2 - b^2) - 2b^2(a^3 - b^3)\} = 0.$$

Both the expressions in the brackets are positive, hence k/k' must be negative or the rotations in opposite directions in the two portions.

Denote the cyclic constants of the inner and outer portions by μ_1, μ_2. As before, we see that they are respectively

$$\frac{k}{\pi} \times \text{vol.}$$

That is

$$\mu_1 = \frac{mk}{\pi} = \tfrac{4}{3}\, b^3 k,$$

$$\mu_2 = \frac{m'k'}{\pi} = \tfrac{4}{3}\,(a^3 - b^3)\, k'.$$

Substituting for k, k' in terms of μ_1, μ_2

$$V = \tfrac{1}{5}\left\{\mu_1 \frac{b^2}{a^3} + \mu_2 \frac{a^5 - b^5}{a^3 (a^3 - b^3)}\right\}.$$

The result is that a double aggregate is possible. If, however, the size is given the ratio of the vorticities must have a special value, and *vice versâ*. In terms of the radii it may be shown that

$$V = -\frac{4 (k - k')\, b^2 (a - b)\, \{2a^3 + 3b^3 + 4a^2 b + 6ab^2\}}{45 a^3 (a + b)}.$$

Three cases specially invite attention, (1) equal volumes, (2) both parts made of similar matter, *i.e.*, vorticities equal, and (3) equal cyclic constants.

Case i.—Here $a^3 = 2b^3$.

$$\frac{k'}{k} = -\frac{2b^5}{6b^3 (a^2 - b^2) - 2b^5} = -\frac{1}{3 \times 2^{2/3} - 4},$$

$$\frac{k}{k'} = -\cdot 76220 = -\tfrac{3}{4} \text{ nearly}.$$

Case ii.—$k' = -k$.

$$3a^3 (a^2 - b^2) - 4b^2 (a^3 - b^3) = 0.$$

Put $a/b \equiv x$, we get

$$3x^4 + 3x^3 - 4x^2 - 4x - 4 = 0.$$

This has three negative roots; the positive one is

$$x = 1\cdot 3283 \quad \text{or} \quad \frac{a}{b} = \tfrac{4}{3} \text{ nearly}.$$

Case iii.—$\mu = -\mu'$ or

$$\frac{k}{a^3 - b^3} = -\frac{k'}{b^3} = \frac{k - k'}{a^3},$$

whence

$$-a^2 b^3 - b^2 (a^3 - b^3) + \tfrac{1}{3} a^3 \left(5 - \frac{2b^3}{a^3}\right) b^2 = 0,$$

$$2a^3 + b^3 - 3a^2 b = (a - b)(2a^2 - ab - b^2) = (a - b)^2 (2a + b) = 0.$$

Equal circulations are therefore impossible.

8. *Polyads.*—Passing on now to the consideration of any number of layers, let the radii of the spherical boundaries from the inside outwards be denoted by $a_1, a_2 \ldots a_n$; the vorticities by $k_1, k_2 \ldots k_n$, and the stream-functions by $\psi_1, \psi_2 \ldots \psi_n$ and ψ. Then

$$\left.\begin{aligned} \psi_1 &= 2\pi \{ A_1 r^2 - \tfrac{1}{5} k_1 r^4 \} \sin^2\theta \\ \psi_p &= 2\pi \left\{ A_p r^2 + \frac{B_p}{r} - \tfrac{1}{5} k_p r^4 \right\} \sin^2\theta \\ \psi_{n+1} &= 2\pi \frac{B_{n+1}}{r} \sin^2\theta \end{aligned}\right\}.$$

Applying the conditions of continuity at the pth boundary, there results

$$A_p a_p^2 + \frac{B_p}{a_p} - \tfrac{1}{5} k_p a_p^4 = A_{p+1} a_p^2 + \frac{B_{p+1}}{a_p} - \tfrac{1}{5} k_{p+1} a_p^4,$$

$$2A_p a_p^2 - \frac{B_p}{a_p} - \tfrac{4}{5} k_p a_p^4 = 2A_{p+1} a_p^2 - \frac{B_{p+1}}{a_p} - \tfrac{4}{5} k_{p+1} a_p^4,$$

with

$$B_1 = 0, \qquad A_{n+1} = 0.$$

Adding

$$A_{p+1} - A_p = \tfrac{1}{3}(k_{p+1} - k_p) a_p^2 \quad \text{with} \quad A_{n+1} = 0.$$

Similarly

$$B_{p+1} - B_p = -\tfrac{2}{15}(k_{p+1} - k_p) a_n^5 \quad \text{with} \quad B_1 = 0.$$

Clearly the A's evolve from the outside, the B's from inside.

Write

$$\tfrac{1}{3}(k_p - k_{p+1}) = \lambda_p.$$

Then

$$A_p - A_{p+1} = \lambda_p a_p^2 \quad \text{with} \quad A_{n+1} = 0,$$

$$B_{p+1} - B_p = \tfrac{2}{5} \lambda_p a_p^5 \quad \text{with} \quad B_1 = 0.$$

Hence

$$A_p = \Sigma_p^n \lambda_p a_p^2, \qquad B_p = \tfrac{2}{5} \Sigma_1^{p-1} \lambda_p a_p^5.$$

Thus the ψ are completely determined.

For steadiness of motion it is necessary that the translatory velocity of the different boundaries be the same. This is obtained if the velocities of the inner and outer boundaries of each layer are equal.

Hence we get $n - 1$ equations ($p = 2$ to n)

$$\tfrac{1}{2} V = A_p + \frac{B_p}{a_p^3} - \tfrac{1}{5} k_p a_p^2 = A_p + \frac{B_p}{a_{p-1}^3} - \tfrac{1}{5} k_p a_{p-1}^2,$$

or

$$B_p \left(\frac{1}{a_p^3} - \frac{1}{a_{p-1}^3} \right) = -\tfrac{1}{5} k_p (a_{p-1}^2 - a_p^2),$$

or

$$B_p = -\tfrac{1}{5} k_p a_p^3 a_{p-1}^3 \frac{a_p^2 - a_{p-1}^2}{a_p^3 - a_{p-}^3}.$$

If the volumes of all the layers are equal,

$$a_p^3 - a_{p-1}^3 = a_1^3 \quad \text{and} \quad a_p^3 = pa_1^3.$$

Hence

$$B_p = -\tfrac{1}{5} k_p p (p-1) \{p^{2/3} - (p-1)^{2/3}\} a_1^5$$

or

$$p(p-1)\{p^{2/3} - (p-1)^{2/3}\} k_p = -2\{(p-1)^{5/3}\lambda_{p-1} + \ldots + \lambda_1\}.$$

Now

$$\lambda_{p-1} = \tfrac{1}{3}(k_{p-1} - k_p).$$

Hence

$$\{(p-1)p^{5/3} - (p+\tfrac{2}{3})(p-1)^{5/3}\} k_p = -\tfrac{2}{3}(p-1)^{5/3}k_{p-1} - 2\{(p-2)^{5/3}\lambda_{p-2} + \ldots + \lambda_1\}$$
$$= -\tfrac{2}{3}\{((p-1)^{5/3} - (p-2)^{5/3})\, k_{p-1} + \ldots + (2^{5/3} - 1) k_2 + k_1\},$$

or subtracting two consecutive equations

$$\{p^{5/3} - (p+\tfrac{2}{3})(p-1)^{2/3}\} k_p = -\{(p-2)^{5/3} - (p-\tfrac{2}{3})(p-1)^{2/3}\} k_{p-1}.$$

Thus the k can be determined in order from the inside. The peculiarity is that the process can stop at any point. That is that if we have two poly-ads, with m and n layers respectively ($m > n$) then the first n layers in the first will be precisely similar to those in the second. The values are

$$k_2 = -\frac{1}{3 \times 2^{2/3} - 4} k_1 = -1 \cdot 3120 k_1$$
$$k_3 = +1 \cdot 4717 k_1$$
$$k_4 = -1 \cdot 5866 k_1$$

and when p is large

$$k_p = -k_{p-1}.$$

As another example, take the case where the layers are formed of the same material, *i.e.*, the vorticities alternately equal and opposite. Then $k_p = (-)^{p-1} k_1$

$$\lambda_p = \tfrac{2}{3} k_p = \tfrac{2}{3} k_1 (-)^{p-1} \quad \text{but} \quad \lambda_n = \tfrac{1}{3} k_n = (-)^{n-1} \tfrac{1}{3} k_1$$

$$\tfrac{4}{15}(a_1^5 - a_2^5 + \ldots + \overline{-}|^{p-2} a_{p-1}^5) = \tfrac{1}{5}(-)^{p-2} a_p^3 a_{p-1}^3 \frac{a_p^2 - a_{p-1}^2}{a_p^3 - a_{p-1}^3}.$$

Let x_p denote the ratio a_{p+1}/a_p.
These values are then given by

$$x_p^3 \frac{x_p^2 - 1}{x_p^3 - 1} = \tfrac{4}{3}\left\{1 - \frac{1}{x_{p-1}^5} + \frac{1}{x_{p-1}^5 x_{p-2}^5} - \ldots\right\}$$

and may be found in succession. The equations are, if b_p denote $1 - \frac{1}{x_{p-1}^5} + \ldots$

$$x_p^3(x_p + 1) = \tfrac{4}{3} b_p (x_p^2 + x_p + 1).$$

In which it is clear that

$$b_p = 1 - \frac{b_{p-1}}{x_{p-1}^5}.$$

If $b_p = \frac{1}{2}$, the equation is

$$x^3(x+1) = \tfrac{3}{2}(x^2+x+1),$$

the positive root of which is $x = 1$. In this case

$$b_{p+1} = 1 - \frac{\frac{1}{2}}{1} = \tfrac{1}{2}.$$

If ever b_p is nearly $\frac{1}{2} = \frac{1}{2} + \alpha$ (say), x_p is nearly $1 = 1 + \xi$ (say). Then, regarding α and ξ of same order

$$(1 + 3\xi + 3\xi^2)(2 + \xi) = \tfrac{3}{2}(3 + 3\xi + \xi^2)(1 + 2\alpha)$$

$$\xi = \tfrac{4}{5}\alpha + \tfrac{4}{5}\alpha\xi - \tfrac{5}{3}\xi^2 = \tfrac{4}{5}\alpha - \tfrac{32}{75}\alpha^2 = \tfrac{4}{5}\alpha(1 - \tfrac{8}{15}\alpha)$$

and

$$b_{p+1} = 1 - \frac{\frac{1}{2} + \alpha}{\{1 + \frac{4}{5}\alpha - \frac{32}{75}\alpha^2\}^5} = 1 - (\tfrac{1}{2} + \alpha)(1 - 4\alpha + \tfrac{176}{15}\alpha^2) = \tfrac{1}{2} + \alpha - \tfrac{88}{15}\alpha^2.$$

Hence b_p continually converges to $\frac{1}{2}$ and the value of x_p to 1 as p increases.

The first seven values are

$x_1 = 1\cdot3283,$	$b_2 = \cdot7582.$
$x_2 = 1\cdot1840,$	$b_3 = \cdot6741.$
$x_3 = 1\cdot1284,$	$b_4 = \cdot6315.$
$x_4 = 1\cdot0987,$	$b_5 = \cdot6056.$
$x_5 = 1\cdot0802,$	$b_6 = \cdot5882.$
$x_6 = 1\cdot0674,$	$b_7 = \cdot5753.$
$x_7 = 1\cdot0580,$	$b_8 = \cdot5660.$

The succeeding values will be given to four figures by the foregoing approximations.

The velocities of translation of the series of aggregates are

Monad	$V_1 = \frac{4}{15}k_1a_1^2$	$= V_1.$
Dyad	$V_2 = \frac{1}{2}(7 - 5x_1^2)V_1$	$= -\cdot9110\,V_1.$
Triad	$V_3 = \frac{1}{2}(7 - 10x_1^2 + 5x_1^2x_2^2)V_1$	$= +\cdot8615\,V_1.$
4-ad	$V_4 = \frac{1}{2}(7 - 10x_1^2 + 10x_1^2x_2^2 - 5x_1^2x_2^2x_3^2)V_1$	$= -\cdot8282\,V_1.$
5-ad	$V_5 = \frac{1}{2}(7 - 10x_1^2 + 10x_1^2x_2^2 - 10x_1^2x_2^2x_3^2 + 5x_1^2x_2^2x_3^2x_4^2)V_1$	$= \cdot8023\,V_1.$
&c.	V_6	$= -\cdot7833\,V_1.$
	V_7	$= \cdot7618\,V_1.$
	V_8	$= -\cdot7462\,V_1.$

9. The form of the stream lines for a monad aggregate have been delineated by HILL. The *general* form of the stream lines for a poly-ad is obvious, and there is no special reason for drawing them accurately at present. It will be well, however, to determine the position of the equatorial axes, for the particular case of homogeneous poly-ads, that is in which $k_p = (-)^p k_1$.

The condition at an equatorial axis is that

$$\frac{1}{2\pi\rho}\frac{d\psi}{dr} = 0 \quad \text{when} \quad \theta = \frac{\pi}{2},$$

in which ψ denotes the stream-function referred to the boundary at rest. Applying this to the p-th layer in an n-ad

$$\psi_p = 2\pi\left\{A_p r^2 + \frac{B_p}{r} - \tfrac{1}{5}k_p r^4 - A_p r^2 - \frac{B_p r^2}{a_p^3} + \tfrac{1}{5}k_p a_p^2 r^2\right\}\sin^2\theta.$$

The equation for the equatorial axis is therefore

$$-B_p\left(\frac{1}{r^3} + \frac{2}{a_p^3}\right) - \tfrac{4}{5}k_p r^2 + \tfrac{2}{5}k_p a_p^2 = 0,$$

or

$$r^5(a_p^3 - a_{p-1}^3) - \tfrac{1}{2}(a_p^5 - a_{p-1}^5)r^3 - \tfrac{1}{4}a_p^3 a_{p-1}^3(a_p^2 - a_{p-1}^2) = 0.$$

This may be written

$$r^5 - \tfrac{1}{2}\cdot\frac{x_{p-1}^5 - 1}{x_{p-1}^3 - 1}a_{p-1}^2 r^3 - \tfrac{1}{4}\frac{x_{p-1}^2 - 1}{x_{p-1}^3 - 1}x_{p-1}^3 a_{p-1}^5 = 0.$$

Now

$$x_{p-1}^3(x_{p-1}^2 - 1) = \tfrac{4}{3}b_{p-1}(x_{p-1}^3 - 1),$$

therefore

$$\frac{x_{p-1}^5 - 1}{x_{p-1}^3 - 1} - 1 = \tfrac{4}{3}b_{p-1},$$

and

$$r^5 - \tfrac{1}{2}(\tfrac{4}{3}b_{p-1} + 1)a_{p-1}^2 r^3 - \tfrac{1}{3}b_{p-1}a_{p-1}^5 = 0.$$

For a monad

$$b = 0 \qquad r^2 = \frac{a^2}{2} \qquad r = \frac{a}{\sqrt{2}},$$

for a dyad

$$b_1 = 1 \qquad r^5 - \tfrac{7}{6}a_1^2 r^3 - \tfrac{1}{3}a_1^5 = 0 \qquad r = 1\cdot1720a_1.$$

Beyond dyads

$$r = a_{p-1} \text{ nearly } = (1 + \xi)a_{p-1}.$$

Then

$$a_{p-1}^5\{1 - \tfrac{1}{2}(\tfrac{4}{3}b_{p-1} + 1) - \tfrac{1}{3}b_{p-1}\} + a_{p-1}^5\xi\{5 - \tfrac{3}{2}(\tfrac{4}{3}b_{p-1} + 1)\} = 0.$$

Now b_{p-1} is nearly $\tfrac{1}{2}$

$$= \tfrac{1}{2} + f_{p-1} \qquad (\tfrac{5}{2} - 2f_{p-1})\xi = f_{p-1} \qquad \xi = \frac{2f_{p-1}}{5} \qquad r_p = \left(1 + \frac{2f_{p-1}}{5}\right)a_{p-1}.$$

The distance from the inner layer is therefore

$$r_p - a_{p-1} = \tfrac{2}{5} f_{p-1} a_{p-1}.$$

From the outer it is

$$a_p - r = a_p - a_{p-1} - \tfrac{2}{5} f_{p-1} a_{p-1}$$

$$\text{Ratio} = \frac{\tfrac{2}{5} f_{p-1}}{x_{p-1} - 1 - \tfrac{2}{5} f_{p-1}}$$

But (p. 50)

$$x_{p-1} = 1 + \tfrac{4}{5} f_{p-1},$$

therefore

$$\text{Ratio} = 1,$$

or the equatorial axis, with increasing number of layers, tends to bisect the distance between the two boundaries of the layer.

10. *Energy.*—The energy within any region is

$$\mathrm{E} = \tfrac{1}{2} \iint \frac{1}{(2\pi\rho)^2} \left\{ \left(\frac{d\psi}{d\rho}\right)^2 + \left(\frac{d\psi}{dz}\right)^2 \right\} 2\pi\rho \, d\rho \, dz,$$

the integral extending within the boundary of the region. By the ordinary method this is reduced to the form

$$\mathrm{E} = -\frac{1}{4\pi} \int \frac{\psi}{\rho} \frac{d\psi}{dn} ds + \iint \omega\psi \, d\rho \, dz.$$

Since

$$\frac{d}{d\rho}\left(\frac{1}{\rho}\frac{d\psi}{d\rho}\right) + \frac{d}{dz}\left(\frac{1}{\rho}\frac{d\psi}{dz}\right) = -4\pi\omega.$$

If the boundary be infinite and the fluid at rest then the first integral is zero, and

$$\mathrm{E} = \iint \omega\psi \, d\rho \, dz.$$

The integral extending only to spaces which contain rotational motion. If the motion is of uniform vorticity $\omega = k\rho$, and

$$\mathrm{E} = k \iint \rho\psi \, d\rho \, dz.$$

In the cases here considered ψ is of the form $f(r) \sin^2 \theta$, and

$$\mathrm{E} = 2k \iint_0^{\frac{\pi}{2}} r^2 f(r) \, dr \sin^3 \theta \, d\theta = \tfrac{4}{3} k \int r^2 f(r) \, dr.$$

In the case of a poly-ad $f(r)$ is different for the various layers, and

$$\mathrm{E} = \tfrac{4}{3} \left\{ k_1 \int_0^{a_1} r^2 f_1(r) \, dr + k_2 \int_{a_1}^{a_2} r^2 f_2(r) \, dr + \ldots \right\}.$$

We work out the case for a dyad aggregate, in which $k_2 = -k_1$,

$$f_1(r) \equiv 2\pi \{(\tfrac{2}{3} a_1^2 - \tfrac{1}{3} a_2^2) k_1 r^2 - \tfrac{1}{5} k_1 r^4\},$$
$$f_2(r) \equiv 2\pi \{-\tfrac{1}{3} k_1 a_2^2 r^2 + \tfrac{4}{15} k_1 a_1^5/r + \tfrac{1}{5} k_1 r^4\},$$

and

$$\begin{aligned} \mathrm{E} &= \frac{8\pi k_1^2}{3} \{\tfrac{1}{15}(2a_1^2 - a_2^2) a_1^5 - \tfrac{1}{5 \cdot 7} a_1^7 + \tfrac{1}{15} a_2^2 (a_2^5 - a_1^5) \\ &\qquad - \tfrac{2}{15} a_1^5 (a_2^2 - a_1^2) - \tfrac{1}{5 \cdot 7} (a_2^7 - a_1^7)\} \\ &= \frac{8\pi k_1^2}{15} \{\tfrac{4}{3} a_1^7 - \tfrac{4}{3} a_2^2 a_1^5 + \tfrac{4}{5 \cdot 7} a_2^7\} \\ &= \frac{32\pi k_1^2}{45} (a_1^7 - a_2^2 a_1^5 + \tfrac{1}{7} a_2^7) \\ &= 1{\cdot}945 \times \frac{32\pi k_1^2}{45 \times 7} a_1^7. \end{aligned}$$

If the two parts had been single monads their combined energy (when far apart) would have been

$$\mathrm{E} = 1{\cdot}534 \times \frac{32\pi k_1^2}{45 \times 7} a_1^7.$$

The energy when combined is therefore greater than when they are separate.

11. It may not be out of place to make a short digression here as to the relation of a HILL's vortex to the vortex rings which have been investigated in previous parts of these researches. As is known the translation velocity of an ordinary ring decreases as the energy increases, and formulæ are given in a former paper* whereby those quantities can be calculated for comparatively thick rings up to $\mathrm{R}/r = 4$ with considerable accuracy, and possibly further. Here R is the radius of the equatorial axis and r the mean radius of the section of the ring. Refer all measurements to the spherical form, and let c denote its radius, V_0 its velocity of translation, and E_0 its energy. Take now a ring of the same volume and circulation as the sphere, and let V and E denote its translation velocity and energy. We get the following value of E/E_0, V/V_0 for different apertures.

$\frac{\mathrm{R}}{r}$.	$\frac{\mathrm{R}}{c}$.	$\frac{r}{c}$.	$\frac{\mathrm{V}}{\mathrm{V}_0}$.	$\frac{\mathrm{E}}{\mathrm{E}_0}$.
100	..	..	·199	176
50	8·09	·162	·282	95
10	2·77	·277	·593	20·8
5	1·745	·349	·784	10·25
4	1·500	·375	·856	8
3	1·239	·413	*·946*	*6*

* "Researches in the Theory of Vortex Rings," Part II., p. 757, 'Phil. Trans.,' 1885, Part II.

These numbers are graphically represented in fig. 1, Plate 1, where the abscissæ give E/E_0 and the ordinates V/V_0. Dotted lines refer to points where calculation cannot be applied. On the same figure are placed outlines of the aggregates drawn to scale. Two things at once strike the eye. First, that the spherical aggregate evidently lies on the E.V curve of the rings, belongs, in fact, to the same family; and, secondly, that the variation of V with the energy is small over a very large range. The shape and nature of the aggregate when the energy is nearly that of the spherical form have not yet been determined. It is probable that as the energy diminishes the form lengthens along the polar axis, until when the energy is very small it becomes a long, thin, cylindrical aggregate. When this is so long that the end portions form only a small portion of the whole, it is possible to obtain an approximation to the energy, for when very long the fluid outside will be very nearly at rest (as in case of force outside a long helix). The velocity of propagation will then be the velocity at the axis. Let a be the radius of the cylinder, l its length. Then

$$la^2 = \tfrac{4}{3}c^3.$$

Again, if V denote the velocity along the axis, the velocity outside is zero, and the variation at the ends only a small part of the whole. Hence the circulation is given by

$$\mu = Vl.$$

Again let v denote the velocity at a distance r from the axis. Take a small rectangular circuit, b parallel to the axis, one inside distant r from the axis, the other outside. The circulation round this is bv. But it is also the value $\Sigma\omega dA$ taken over the area of the rectangle.

Therefore

$$bv = k\Sigma 2r\,dA = \frac{k}{\pi}\,(\text{volume}) = \frac{k}{\pi}\,.\,b\pi(a^2 - r^2),$$

$$v = k(a^2 - r^2), \qquad \mu = lka^2;$$

therefore

$$v = \frac{\mu}{l}\left(1 - \frac{r^2}{a^2}\right).$$

$$\text{Energy in E} = \int_0^r 2\pi r\,.\,l\,dr\,.\,\tfrac{1}{2}v^2$$

$$= \frac{\pi\mu^2}{2l}\int_0^r\left(1 - \frac{r^2}{a^2}\right)^2 d(r^2)$$

$$= \frac{\pi\mu^2 a^2}{6l} = \frac{\mu^2}{6l^2}\,.\,m = \tfrac{2}{9}\pi\mu^2\frac{c}{l}$$

$$= \tfrac{1}{6}\,mV^2 = \tfrac{2}{9}\,\pi c^3 V^2.$$

It is thus the same as a mass of one-third its own mass moving with its velocity of translation. Now

$$E_0 = \tfrac{2}{35}\pi\mu^2 c, \qquad V_0 = \frac{\mu}{5c},$$

therefore

$$\frac{E}{E_0} = \tfrac{35}{9}\frac{c^2}{l^2}, \qquad \frac{V}{V_0} = \frac{5c}{l},$$

therefore

$$\frac{E}{E_0} = \tfrac{7}{45}\left(\frac{V}{V_0}\right)^2.$$

This only holds, however, when V/V_0 is small. It is a small part of a parabola in the figure touching the axis of E/E_0.

12. *Spheroidal Aggregates.*—As is known from HILL's investigations, the spheroid, although an instantaneously possible form, is not steady. It proceeds at once to change its shape into a non-spheroidal one. It seems, however, advisable to give the general outline of the method as adopted in this paper and as applied to the spheroids, in order to investigate whether by superposing a second or third layer it may be possible to obtain a steady form.

The functions involved and the differential equation for ψ are given in Eqs. 8, 9. Writing C for $\cosh u$ and S for $\sinh u$, the differential equation in ψ is

$$\frac{1}{\cos v}\frac{d}{du}\left(\frac{1}{S}\frac{d\psi}{du}\right) + \frac{1}{S}\frac{d}{dv}\left(\frac{1}{\cos v}\frac{d\psi}{dv}\right) = -4\pi k\lambda^4 S\cos v\,(C^2 - \sin^2 v),$$

since

$$\omega = k\rho = k\lambda S\cos v.$$

As in the former case, a particular integral is

$$\psi = -\frac{\pi}{2}k\rho^4 = -\tfrac{1}{2}\pi k\lambda^4 S^4\cos^4 v.$$

It remains to integrate

$$\frac{1}{\cos v}\frac{d}{du}\left(\frac{1}{S}\frac{d\psi}{du}\right) + \frac{1}{S}\frac{d}{dv}\left(\frac{1}{\cos v}\frac{d\psi}{dv}\right) = 0.$$

This can be satisfied by writing $\psi = \Sigma X_m Z_m$ where X and Z are functions respectively of u and v only, and

$$\left.\begin{aligned} \frac{d}{dv}\left(\frac{1}{\cos v}\frac{dZ}{dv}\right) &= -\frac{mZ}{\cos v} \\ \frac{d}{du}\left(\frac{1}{S}\frac{dX}{du}\right) &= \frac{mX}{S} \end{aligned}\right\},$$

m being any constant. These equations are

$$\left.\begin{aligned}\frac{d^2Z}{dv} + \tan v\frac{dZ}{dv} + mZ = 0\\ \frac{d^2X}{du^2} - \coth u\frac{dX}{du} - mX = 0\end{aligned}\right\}.$$

As will be seen later m must be of the form $n(n-1)$, n being any integer. Writing for a moment $v = \frac{\pi}{2} - \theta$, the equation in Z becomes

$$\frac{d^2Z}{d\theta^2} - \cot\theta\frac{dZ}{d\theta} + n(n-1)Z = 0,$$

whence

$$Z_n = -\sin\theta\frac{dP_{n-1}}{d\theta}.$$

Therefore

$$\begin{aligned}Z_2 &= \sin^2\theta = \cos^2 v\\ Z_4 &= 6\sin^2\theta - \tfrac{15}{2}\sin^4\theta\\ &= 6\cos^2 v - \tfrac{15}{2}\cos^4 v\\ \cos^4 v &= \tfrac{4}{5}Z_2 - \tfrac{2}{15}Z_4.\end{aligned}$$

To determine X, we proceed by the same analogy to put

$$X = S\frac{dP}{du}.$$

Then

$$\frac{d^2X}{du^2} - \coth u\frac{dX}{du} - n(n-1)X = S\frac{d}{du}\left\{\frac{d^2P}{du^2} + \coth u\frac{dP}{du} - n(n-1)P\right\}.$$

If then P denote a zonal harmonic with imaginary argument and of order $n-1$, the right hand of the above vanishes, and the value of X is a solution. That is

$$X_n = S\frac{dP_{n-1}}{du}.$$

Now we have

$$P_n = \frac{1.3\ldots(2n-1)}{n!}\left\{C^n - \frac{n(n-1)}{2(2n-1)}C^{n-2} + \ldots\right\}.$$

Hence

$$X_2 = S^2. \qquad X_4 = 6S^2 + \tfrac{15}{2}S^4. \qquad S^4 = \tfrac{2}{15}X_4 - \tfrac{4}{5}X_2.$$

This set of solutions gives values finite and continuous at all points inside a given ellipse of the family, but infinitely large at an infinite distance. Let Y denote the second integral of the equation. Then, in the usual way, it may be shown that

$$Y = X\int\frac{S}{X^2}du,$$

whence it is easy to prove that

$$Y_2 = S^2 \int \frac{du}{S^3} = \tfrac{1}{4} S^2 \log \frac{C+1}{C-1} - \tfrac{1}{2} C,$$

$$Y_4 = \tfrac{1}{24} X_4 \log \frac{C+1}{C-1} - \tfrac{1}{24} C \,(15C^2 - 13).$$

With these values the particular integral is

$$\tfrac{2}{25} \pi k \lambda^4 (2X_2 - \tfrac{1}{8} X_4)(2Z_2 - \tfrac{1}{3} Z_4).$$

The terms in X_2Z_2, X_4Z_4 may be supposed merged in the general solution. We may then write

$$\psi_1 = (A_2X_2 - \tfrac{4}{75} \pi k \lambda^4 X_4) Z_2 + (A_4X_4 - \tfrac{4}{75} \pi k \lambda^4 X_2) Z_4,$$

$$\psi_2 = B_2Y_2Z_2 + B_4Y_4Z_4.$$

From these it is easy to deduce the values of A_2, etc., for a single free aggregate, by applying the conditions $\psi_1 = \psi_2$ and $d\psi_1/du = d\psi_2/du$ at the surface. It is unnecessary to do this, as from HILL's work we know that it is not steady.

The case of motion inside a rigid spheroidal boundary is also given by HILL.* The solution follows immediately by impressing the condition $\psi_1 = 0$ when $u = \mathbf{u}$.

Hence

$$A_2 = \tfrac{4}{75} \pi k \lambda^4 \frac{\mathbf{X}_4}{\mathbf{X}_2},$$

$$A_4 = \tfrac{4}{75} \pi k \lambda^4 \frac{\mathbf{X}_2}{\mathbf{X}_4},$$

where thick type denotes values at the surface, and

$$\psi_1 = \tfrac{4}{75} \pi k \lambda^4 \frac{X_2\mathbf{X}_4 - \mathbf{X}_2X_4}{\mathbf{X}_2} Z_2 - \tfrac{4}{75} \pi k \lambda^4 . \frac{X_2\mathbf{X}_4 - \mathbf{X}_2X_4}{\mathbf{X}_4} Z_4,$$

which easily reduces to

$$\psi_1 = \frac{2\pi k \lambda^4}{4 + 5\mathbf{S}^2} (\mathbf{S}^2 - S^2) S^2(\mathbf{S}^2 + \cos^2 v) \cos^2 v.$$

The total circulation is $\tfrac{4}{3} k \lambda^3 \mathbf{C}\mathbf{S}^2$.

The equatorial axis is given by

$$\frac{d\psi}{du} = 0, \quad \text{when } v = 0.$$

That is by the equation

$$2\mathbf{S}^2 S - 4S^3 = 0, \quad \text{or } S = \frac{1}{\sqrt{2}} \mathbf{S}.$$

* 'Phil. Trans.' Part II., 1884, p. 403.

The equatorial axis therefore lies in the equatorial section in a similar position to that for a sphere.

13. *Dyad Spheroids.*—Poly-ad spheroids clearly occur in the same way as for spheres; they are, however, also unsteady. It will be sufficient merely to indicate the steps of the proof.

Let $u = u'$ and $u = u''$ denote the two boundaries. ψ will involve terms in Z_2 and Z_4. By applying the surface conditions in the same way as for the spheres to both sets of terms independently, the coefficients are determined, whilst the condition that the internal interface has the same translational velocity as the outer gives for Z_4 an equation which u' and u'' must satisfy. This is

$$\frac{C'S'^4}{S''}M'' - \frac{C''S''^4}{S'}M' + \frac{C''S''^4 - C'S'^4}{6 - \frac{15}{2}S'^2} = 0,$$

where

$$M = X_2\frac{dY_4}{du} - Y_4\frac{dX_2}{du},$$

and the dashed letters refer to values at the outer and inner boundaries u'', u'.

The same applied to the Z_2 terms give

$$\frac{C'S'^4}{S''}N'' - \frac{C''S''^4}{S'}N' + (C''S''^4 - C'S'^4)\frac{X_4'}{X_2'} = 0,$$

where

$$N = X_4\frac{dY_2}{du} - Y_2\frac{dX_4}{du}.$$

The existence of steadily-moving spheroids depends on the possibility of finding values of u', u'' to satisfy these two equations.

It is easy to show that

$$6M + N = \tfrac{1}{2}S(25S^2 + 14).$$

Hence, adding 6 times the first equation to the second, there results an equation free of logarithmic terms and which can easily be reduced to

$$\frac{S'^2(S''^2 - S'^2)}{C''S''^4 - C'S'^4} = \frac{2C'}{5C'^2 - 1}.$$

Putting $C' = y$, $C'' = x$, the factor $(x - y)^2$ divides out, and the equation may be put in the form

$$2y(x^3 + 2x^2y + 3xy^2 - 2x) + (y^2 - 1)(3y^2 + 1) = 0.$$

Now $x > y > 1$. Hence $3xy^2 - 2x \equiv xy^2 + 2x(y^2 - 1)$ is positive. The expression on the left is therefore always positive and no suitable values of x, y satisfy the equation. A prolate spheroidal dyad is therefore not steady.

The condition for the oblate spheroid can be found by writing $S\sqrt{-1}$ for C. It can be shown that this also has no suitable root.

Section iii.—Gyrostatic Aggregates.

14. Passing on now to the consideration of the more general problem where a secondary spin exists, the simplest case is that in which in equation (12) both F and $fdf/d\psi$ are uniform.

Suppose

$$f\frac{df}{d\psi} = \text{A}, \text{ or } f = \sqrt{(2\text{A}\psi)}.$$

The differential equation in ψ is now

$$\frac{d^2\psi}{dr^2} + \frac{1}{r^2}\frac{d^2\psi}{d\theta^2} - \frac{\cot\theta}{r^2}\frac{d\psi}{d\theta} = -8\pi^2\rho^2\text{F} - \text{A},$$

a particular integral of which is

$$\psi = -\pi^2\rho^4\text{F} - \tfrac{1}{2}\text{A}r^2,$$

and the general integral is the same as that considered in the previous section, viz. :

$$\left(\text{A}r^n + \frac{\text{B}}{r^{n-1}}\right)\text{Z}_n.$$

It will however not be found possible to satisfy the boundary conditions unless the term $\psi = \text{A}_1 r$ be introduced. This term, as well as that in $\tfrac{1}{2}\text{A}r^2$, makes the motion discontinuous at the polar axis. However, we will suppose for the moment this portion of space excluded, and see later if it is possible to do so. The stream-functions are then,—inside

$$\psi_1 = -\pi^2\rho^4\text{F} - \tfrac{1}{2}\text{A}r^2 + \text{A}_1 r + \Sigma_2\text{A}_n r^n\text{Z}_n,$$

outside

$$\psi_2 = \Sigma\frac{\text{B}_n}{r^{n-1}}\text{Z}_n,$$

and ρ^4 can be replaced as before by $\tfrac{4}{5}r^4\text{Z}_2$.

Applying the conditions $\psi_1 = \psi_2$ and $d\psi_1/dr = d\psi_2/dr$, when $r = a$ it is easy to deduce that

$$\psi_1 = -\tfrac{1}{2}\text{A}(a-r)^2 - \tfrac{4}{5}\pi^2\text{F}r^4\text{Z}_2 + \tfrac{4}{3}\pi^2a^2\text{F}r^2\text{Z}_2,$$

$$\psi_2 = \tfrac{8}{15}\pi^2\text{F}\frac{a^5}{r}\text{Z}_2.$$

The velocity normal to the sphere is

$$\left[\frac{1}{2\pi\rho}\frac{d\psi}{rd\theta}\right]_{r=a} = \tfrac{8}{15}\pi\text{F}a^2\cos\theta.$$

That is, the sphere progresses bodily with a velocity given by

$$\text{V} = \tfrac{8}{15}\pi a^2\text{F}.$$

Impress $-$ V on every point, that is, deduct $\frac{8}{15}\pi^2 a^2 F r^2 Z_2$. Then the stream-function referred to the boundary is

$$\psi_1 = -\tfrac{1}{2}A(a-r)^2 + \tfrac{4}{5}\pi^2 F r^2 (a^2 - r^2) Z_2.$$

At the outer boundary $\psi = 0$. If we trace the stream-line $\psi = 0$, it is seen that it consists of the circle $r = a$ and the curve

$$\tfrac{1}{2}A(a-r) = \tfrac{4}{5}\pi^2 F r^2 (a+r)\sin^2\theta.$$

This passes through the poles ($r = a$, $\theta = 0$) and touches the circle there. Hence the space between this and the outer boundary does not contain the polar axis. The motion given by ψ is therefore finite and continuous there. The space inside it must be excluded as giving a motion not possible—or rather, a motion due to sources and sinks on the polar axis. We shall suppose it excluded by replacing the fluid by a solid nucleus of the shape required.

The radius of an equatorial axis is given by $d\psi/dr = 0$ when $\theta = \pi/2$, or by

$$A(a-r) + \tfrac{8}{5}\pi^2 F r(a^2 - 2r^2) = 0.$$

In this write $r/a = x$ and $\frac{5A}{16\pi^2 F a^2} = b$. Then

$$x^3 + (b - \tfrac{1}{2})x - b = 0 \quad . \quad . \quad . \quad . \quad . \quad . \quad . \quad . \quad . \quad (13).$$

This has one root between 0 and 1. The other roots must either be both imaginary, or, if real, one at least must be negative, since the coefficient of x^2 is zero. As, further, $x = -\infty$ and $x = 0$ both make the expression on the left of the same sign, both these roots must be negative. Hence there is one and only one root between 0 and 1. That is, there is only one equatorial axis.

In the special case $b = \frac{1}{2}$, the radius of the equatorial axis is $a \,.\, 2^{-\frac{1}{3}} = \cdot 7937a$. For this curve

$$\psi_1 = \tfrac{1}{2}A\left\{\frac{r^2}{a^2}(a^2 - r^2) Z_2 - (a-r)^2\right\}.$$

The curves are drawn in fig. 1, Plate 2, for values of $2\psi/Aa^2 = -\cdot 1, 0, +\cdot 1$. The value at the equatorial axis is $\cdot 397$. The value ($-\cdot 1$) is drawn to show how the discontinuity enters.

The velocity along a parallel of latitude is given by the equation

$$2\pi\rho v \sin\phi = f = \sqrt{(2A\psi)}.$$

This is zero at the surface and on the spindle-shaped nucleus, and increases to a maximum at the equatorial axis. The secondary cyclic constant is the circulation round the two circles (1) the equator of the sphere, and (2) the equatorial axis. It is therefore given by

$$\nu = \sqrt{(2A\psi')},$$

or

$$\nu = Aa\sqrt{(2x - 1 - x^4)}$$

where x is the root of equation (13).

On account of the artificial nature of the internal nucleus the further discussion of this case is scarcely called for. We pass on, therefore, to the more important case—the next simplest one—in which F is uniform, but the second terms varies as ψ.

15. *Case* $f\dfrac{df}{d\psi} \propto \psi$.—Here also f varies as ψ.

Write $f = \dfrac{\lambda}{a}\psi$ where a is a length, which may be taken to be the radius of the sphere, and λ is a pure number. Also write $F = \dfrac{8\pi^2a^2}{V}$ where V is a velocity. Then the equation in ψ is

$$\frac{d^2\psi}{dr^2} + \frac{1}{r^2}\frac{d^2\psi}{d\theta^2} - \frac{\cot\theta}{r^2}\frac{d\psi}{d\theta} = -\frac{\rho^2}{a^2}V - \frac{\lambda^2}{a^2}\psi.$$

A particular integral is $-\dfrac{V}{\lambda^2}\rho^2$ and the general integral depends on

$$\frac{d^2\psi}{dr^2} + \frac{1}{r^2}\frac{d\psi}{d\theta^2} - \frac{\cot\theta}{r^2}\frac{d\psi}{d\theta} + \frac{\lambda^2}{a^2}\psi = 0.$$

In this put $\psi = J_nZ_n$ where Z_n is the function of θ already discussed (§ 6) and J_n is a function of r only. Then

$$\frac{d^2J_n}{dr^2} - \left\{\frac{n(n-1)}{r^2} - \frac{\lambda^2}{a^2}\right\}J_n = 0.$$

$J_n/\sqrt{r}$ is therefore a BESSEL's function of order $n - \frac{1}{2}$, which can, as is known, be expressed in finite form involving circular functions. In what immediately follows, the values of J_2 will alone be required. The equation is, writing x for r/a, and dropping the subscript 2,

$$\frac{d^2J}{dx^2} - \left(\frac{2}{x^2} - \lambda^2\right)J = 0.$$

If J and Y denote the two integrals

$$J = \frac{\sin\lambda x}{\lambda x} - \cos\lambda x,$$

$$Y = \frac{\cos\lambda x}{\lambda x} + \sin\lambda x,$$

or, in more general terms,

$$\text{Integral} = \mathrm{C}\left\{\frac{\sin(\alpha+\lambda x)}{\lambda x} - \cos(\alpha+\lambda x)\right\}$$

where C and α are arbitrary constants.

J and Y may be expressed in infinite convergent series. Thus

$$\mathrm{J}(y) \equiv \frac{\sin y}{y} - \cos y = \tfrac{1}{3}y^2 - \frac{1}{2.3.5}y^4 + \ldots (-)^{n+1}\frac{2n}{(2n+1)!}y^{2n}$$

$$= \tfrac{1}{3}y^2\left\{1 - \frac{y^2}{10} + \ldots + (-)^m \frac{3}{2m+3}\frac{y^{2m}}{(2m+1)!} + \ldots\right\} \quad . \quad (14)$$

$$\mathrm{Y}(y) \equiv \frac{\cos y}{y} + \sin y = \frac{1}{y} + \tfrac{1}{2}y + \ldots + (-)^{n+1}\frac{2n-1}{(2n)!}y^{2n-1} + \ldots$$

$$= \frac{1}{y}\left\{1 + \tfrac{1}{2}y^2 + \ldots + (-)^{n+1}\frac{2n-1}{(2n)!}y^{2n} + \ldots\right\} \quad . \quad . \quad (15),$$

also,

$$\left.\begin{aligned} \frac{d\mathrm{J}(y)}{dy} &= \sin y - \frac{\mathrm{J}}{y} = \mathrm{Y} - \frac{\sin y}{y^2} \\ \frac{d\mathrm{Y}(y)}{dy} &= \cos y - \frac{\mathrm{Y}}{y} = -\mathrm{J} - \frac{\cos y}{y^2} \\ \text{and} \quad \mathrm{Y}\frac{d\mathrm{J}}{dy} &- \mathrm{J}\frac{d\mathrm{Y}}{dy} = 1 \end{aligned}\right\} \quad . \quad . \quad . \quad . \quad . \quad (16).$$

Clearly the functions J refer only to space excluding infinity; Y to space excluding the origin.

16. For the problem in question the stream-functions are, therefore,

inside,

$$\psi_1 = -\frac{\mathrm{V}}{\lambda^2}r^2\sin^2\theta + \Sigma \mathrm{A}_n \mathrm{J}_n \mathrm{Z}_n,$$

outside,

$$\psi_2 = \Sigma \frac{\mathrm{B}_n}{r^{n-1}}\mathrm{Z}_n.$$

Applying the surface conditions that when $x = 1$, $\psi_1 = \psi_2$, and $d\psi_1/dx = d\psi_2/dx$, it follows that when

$$n > 2, \qquad \mathrm{A}_n = \mathrm{B}_n = 0,$$

when

$$n = 2,$$

$$-\frac{\mathrm{V}}{\lambda^2}a^2 + \mathrm{A}_2\mathrm{J}' = \frac{\mathrm{B}_2}{a},$$

$$-\frac{2\mathrm{V}}{\lambda^2}a^2 + \mathrm{A}_2\frac{d\mathrm{J}'}{dx} = -\frac{\mathrm{B}_2}{a},$$

where J′ and $d\mathrm{J}'/dx$ mean the values of J and $d\mathrm{J}/dx$ when $x = 1$, that is

$$J' = \frac{\sin \lambda}{\lambda} - \cos \lambda,$$

$$\frac{dJ'}{dx} = \lambda \sin \lambda - \frac{\sin \lambda}{\lambda} + \cos \lambda = \lambda \sin \lambda - J',$$

the two equations for A_2, B_2 give

$$A_2 = \frac{3V}{\lambda^3 \sin \lambda} a^2, \qquad B_2 = \frac{Va^3}{\lambda^2}\left(\frac{3J'}{\lambda \sin \lambda} - 1\right).$$

The aggregate moves through the fluid with a velocity of translation given by

$$U = \frac{2B_2}{2\pi a^3} = \frac{V}{\pi \lambda^2}\left(\frac{3J'}{\lambda \sin \lambda} - 1\right).$$

By its formation the above value of ψ satisfies all the equations of condition except that in those equations ψ is the velocity-function referred to fixed axes. Here it is not—it represents the motion referred to the instantaneous position of the sphere. It is, therefore, not directly applicable unless the velocity of translation given by it vanishes, that is, unless

$$J' - \tfrac{1}{3}\lambda \sin \lambda = 0.$$

If λ be a root of this equation we get a steady motion of a vortex aggregate, at rest in the surrounding fluid.

If we, however, take the above general function, it gives a velocity of translation

$$U = \frac{V}{\pi \lambda^2}\left(\frac{3J'}{\lambda \sin \lambda} - 1\right) \quad . \quad . \quad . \quad . \quad . \quad . \quad . \quad . \quad (17).$$

Bring the aggregate to rest by impressing a velocity—U on the whole fluid—that is, add to the stream-function a term $- \pi U \rho^2 = - \pi U a^2 x^2 \sin^2 \theta$.

We get a new value of ψ, referred to axes remaining fixed, viz.,

$$\psi = \frac{3Va^2}{\lambda^3 \sin \lambda}(J - x^2 J') \sin^2 \theta.$$

Take this value of ψ, and put $f = \frac{\lambda}{a}\psi$. Then equations (1, 3, 4, 7) become

$$v\rho \cos \phi = \frac{1}{2\pi}\frac{d\psi}{dn}$$

$$v\rho \sin \phi = \frac{\lambda}{2\pi a}\psi$$

$$\omega\rho \cos \chi = -\frac{\lambda}{4\pi a}\frac{d\psi}{dn}$$

$$\omega\rho \sin \chi = \frac{3V}{4\pi\lambda \sin \lambda} J \sin^2 \theta.$$

These give v, ω, ϕ, χ.

Now substitute in $v\omega \sin \overline{\phi + \chi}$. The result is that

$$v\omega \sin (\phi + \chi)\, dn = \frac{3\mathrm{V}}{8\pi^2 a^2 \lambda \sin \lambda} \mathrm{J}'.d\psi$$

so that the motion given by the new ψ is a steady one. There exist, therefore, systems travelling through the fluid with velocities given by (17) and with a steady motion. The system given by $\mathrm{J}' = \frac{1}{3}\lambda \sin \lambda$ is contained as a special case.

17. There are two circulations to be considered. That along a circuit up the polar axis and down over the surface of the sphere, and that due to the motion round the polar axis. Call them respectively the primary and secondary cyclic constants, and denote them by μ, ν.

$$\mu = 2\int_0^a \left\{\frac{1}{2\pi\rho}\frac{d\psi}{r\,d\theta}\right\}_{\theta=0} dr + 2\int_0^{\pi/2} \left\{-\frac{1}{2\pi\rho}\frac{d\psi}{dr}\right\}_{r=a} a\, d\theta.$$

In finding this the term $x^2\mathrm{J}' \sin^2\theta$ may be omitted as giving no circulation, and we may take

$$\psi = \frac{3\mathrm{V}a^2}{\lambda^3 \sin \lambda} \mathrm{J} \sin^2\theta$$

$$\mu = \frac{3\mathrm{V}a^2}{\pi\lambda^3 \sin \lambda}\left\{2\int_0^a \frac{\mathrm{J}}{r^2}\, dr - \frac{d\mathrm{J}'}{dr}\int_0^{\pi/2} \sin\theta\, d\theta\right\}$$

$$= \frac{3\mathrm{V}a}{\pi\lambda^2 \sin \lambda}\left\{2\int_0^\lambda \frac{\mathrm{J}}{y^2}\, dy - \frac{d\mathrm{J}'}{dy}\right\}$$

where

$$y \equiv \frac{\lambda r}{a}.$$

Now

$$\int \frac{\mathrm{J}}{y^2}\, dy = -\frac{\mathrm{J}}{y} + \int \frac{1}{y}\frac{d\mathrm{J}}{dy}\, dy = -\frac{\mathrm{J}}{y} + \int\left(\frac{\sin y}{y} - \frac{\mathrm{J}}{y^2}\right) dy,$$

therefore,

$$2\int_0^\lambda \frac{\mathrm{J}}{y^2}\, dy = -\left[\frac{\mathrm{J}}{y}\right]_0^\lambda + \int_0^\lambda \frac{\sin y}{y}\, dy.$$

Also, $\mathrm{J}(y)$ is of the order y^2 when y is small, therefore,

$$2\int_0^\lambda \frac{\mathrm{J}}{y^2}\, dy = -\frac{\mathrm{J}'}{\lambda} + \mathrm{Si}\lambda,$$

and

$$\mu = \frac{3\mathrm{V}a}{\pi\lambda^3 \sin \lambda}(\mathrm{Si}\lambda - \sin\lambda).$$

If we replace V as a constant of the motion by μ,

$$\psi = \frac{\pi\mu a}{\lambda(\mathrm{Si}\lambda - \sin\lambda)}(\mathrm{J} - x^2\mathrm{J}') \sin^2\theta.$$

Before discussing the value of ν it will be well to get some general idea of the nature of the motions. One of the most striking peculiarities of these aggregates is the quasi-periodicity of type as λ increases from 0 to infinity. The best way to illustrate this is to use a graphical construction. Now

$$\psi \propto \{J(\lambda x) - x^2 J(\lambda)\}.$$

In fig. 2, Plate 1, the curve $y = J(\lambda)$ is drawn. P_1 corresponds to a given type (λ) of aggregate. A parabola is drawn with vertex at O and passing through P_1. Represent any abscissa to the left of λ (or of P_1) by λx, where $x < 1$. Then the differences of ordinates between the curve and the parabola up to P represent

$$J(\lambda x) - x^2 J(\lambda).$$

It is clear from the figure that, in the position P_1, this function never vanishes for $x < 1$. In the second position, P_2, however, the parabola intersects the curve at another point p. For this point (suppose $x = x_0$) ψ vanishes for all values of θ, and the corresponding current sheet is a sphere internal to the boundary. The aggregate consists of two portions with independent motions. The primary circulations are in opposite directions, and there will be *two* equatorial axes. So, as P moves on along the curve, *i.e.*, as λ increases, we get families of aggregates with three, four, &c., layers, and a corresponding number of equatorial axes. We shall denote any transition value of λ by λ_2. Each layer will have its own secondary circulation, given by the circulation round the double circuit formed by its equatorial axis, and an equator on its boundary.

Now the secondary spin velocity is given by

$$v\rho \sin\phi = \frac{\lambda}{2\pi a}\psi.$$

And since $\psi = 0$ on the boundary, it follows that

$$\nu_n = 2\pi\rho v \sin\phi, \text{ along the equatorial axis only, } = \frac{\lambda}{a}\psi_n,$$

where ψ_n is the value of ψ at the nth equatorial axis, or

$$\nu_n = \frac{\pi\mu}{Si\lambda - \sin\lambda}\{J_n - x_n^2 J'\},$$

where $J_n \equiv J(\lambda x_n)$ and $J' \equiv J(\lambda)$.

18. The moment of angular momentum is

$$M = 2\int_0^a\int_0^{\pi/2} 2\pi\rho r\, dr\, d\theta v\rho \sin\phi$$

$$= \frac{2\lambda}{a}\int_0^a\int_0^{\pi/2}\psi r^2 \sin\theta\, dr\, d\theta$$

$$= \frac{2\pi\mu a^3}{Si\lambda - \sin\lambda}\int_0^1 (J - x^2 J')\, x^2\, dx \int_0^{\pi/2} \sin^3\theta\, d\theta$$

$$= \tfrac{4}{3}\pi a^3 \frac{\mu}{Si\lambda - \sin\lambda}\int_0^1\left(\frac{x}{\lambda}\sin\lambda x - x^2\cos\lambda x - x^4 J'\right)dx$$

$$= m\frac{\mu}{Si\lambda - \sin\lambda}\left\{\left(\tfrac{1}{5} - \frac{3}{\lambda^2}\right)\left(\cos\lambda - \frac{\sin\lambda}{\lambda}\right) - \frac{\sin\lambda}{\lambda}\right\} \quad . \quad . \quad . \quad . \quad (18)$$

where m denotes the volume of the aggregate.

19. The internal energy of the aggregate, supposed without translation is

$$E = \tfrac{1}{2}\iint 2\pi\rho\, d\rho\, dz\, (v^2\cos^2\phi + v^2\sin^2\phi)$$

$$= \tfrac{1}{2}\iint \frac{1}{2\pi\rho}\left\{\left(\frac{d\psi}{d\rho}\right)^2 + \left(\frac{d\psi}{dz}\right)^2 + \frac{\lambda^2}{a^2}\psi^2\right\} d\rho\, dz$$

where

$$\psi = A\,(J - x^2 J')\sin^2\theta$$

and

$$A = \frac{\pi\mu a}{Si\,\lambda - \sin\lambda}.$$

Hence, as in the usual way,

$$E = -\frac{1}{4\pi}\int\frac{\psi}{\rho}\frac{d\psi}{dn}\, ds - \frac{1}{4\pi}\iint\psi\left\{\frac{d}{d\rho}\left(\frac{1}{\rho}\frac{d\psi}{d\rho}\right) + \frac{d}{dz}\left(\frac{1}{\rho}\frac{d\psi}{dz}\right) - \frac{\lambda^2}{a^2}\psi\right\} d\rho\, dz.$$

Now along the boundary $\psi = 0$. Also

$$\frac{d}{d\rho}\left(\frac{1}{\rho}\frac{d\psi}{d\rho}\right) + \frac{d}{dz}\left(\frac{1}{\rho}\frac{d\psi}{dz}\right) = -\frac{\lambda^2}{a^2}A\frac{J\sin^2\theta}{\rho} = -\frac{\lambda^2\psi}{a^2\rho} - \frac{\lambda^2 A}{a^4}\rho J',$$

therefore

$$E = \frac{\lambda^2}{2\pi a^2}\iint\frac{\psi^2 r\, dr\, d\theta}{\rho} + \frac{\lambda^2 AJ'}{4\pi a^4}\iint\psi\rho r\, dr\, d\theta$$

$$= \frac{\lambda^2 A^2}{2\pi a}\left\{2\int_0^1\int_0^{\pi/2}(J - x^2 J')^2\sin^3\theta\, d\theta\, dx + J'\int_0^1\int_0^{\pi/2}(J - x^2 J')\, x^2\sin^3\theta\, d\theta\, dx\right\}$$

$$= \frac{\lambda^2 A^2}{3\pi a}\int_0^1\left\{x^4 J'^2 - 3J'\left(\frac{x\sin\lambda x}{\lambda} - x^2\cos\lambda x\right) + \frac{2\sin^2\lambda x}{\lambda^2 x^2} - 2\frac{\sin 2\lambda x}{\lambda x} + 2\cos^2\lambda x\right\}dx$$

$$= \frac{\lambda^2 A^2}{3\pi a}\left\{\tfrac{1}{5}J'^2 - 3J'\left(\frac{3\sin\lambda}{\lambda^3} - \frac{3\cos\lambda}{\lambda^2} - \frac{\sin\lambda}{\lambda}\right) + 1 + \frac{\sin 2\lambda}{2\lambda} - \frac{2\sin^2\lambda}{\lambda^2}\right\}$$

$$= \frac{\lambda^2 A^2}{3\pi a}\left\{\left(\tfrac{1}{5} - \frac{9}{\lambda^2}\right)J'^2 + \frac{\sin^2\lambda}{\lambda^2} - \frac{2\sin\lambda\cos\lambda}{\lambda} + 1\right\}$$

$$= \frac{\lambda^2 A^2}{3\pi a}\left\{\left(\tfrac{6}{5} - \frac{9}{\lambda^2}\right)J'^2 + \sin^2\lambda\right\},$$

or

$$E = \frac{\pi\mu^2 a}{3} \frac{\left(\frac{6}{5} - \frac{9}{\lambda^2}\right) J'^2 + \sin^2\lambda}{(\mathrm{Si}\lambda - \sin\lambda)^2} \quad . \quad . \quad . \quad . \quad . \quad . \quad . \quad . \quad (19).$$

The energy due to translation is that due to the bodily translation of the sphere + ½ the same.

The velocity of translation is

$$U = \frac{\mu}{a} \frac{J' - \frac{1}{3}\lambda \sin\lambda}{\lambda(\mathrm{Si}\lambda - \sin\lambda)}.$$

Hence this part of the energy is

$$\tfrac{3}{2} \cdot \tfrac{1}{2} \cdot \tfrac{4}{3} \pi a^3 U^2.$$

Therefore total energy is

$$= \frac{\pi\mu^2 a}{(\mathrm{Si}\lambda - \sin\lambda)^2} \left\{ \left(\tfrac{2}{5} - \frac{3}{\lambda^2}\right) J'^2 + \tfrac{1}{3}\sin^2\lambda + \left(\frac{1}{\lambda} J' - \tfrac{1}{3}\sin\lambda\right)^2 \right\}$$

$$= \frac{\pi\mu^2 a}{(\mathrm{Si}\lambda - \sin\lambda)^2} \left\{ 2\left(\tfrac{1}{5} - \frac{1}{\lambda^2}\right) J'^2 - \frac{2}{3\lambda} J' \sin\lambda + \tfrac{4}{9}\sin^2\lambda \right\} \quad . \quad . \quad . \quad . \quad (20).$$

A verification is afforded by putting $\lambda = 0$ (HILL's vortex). Then

$$(\mathrm{Si}\lambda - \sin\lambda)^2 = \tfrac{1}{81}\lambda^6.$$

$$\text{Large bracket} = \tfrac{2}{5 \cdot 7 \cdot 9 \cdot 9}\lambda^6.$$

$$E = \tfrac{2}{35}\pi\mu^2 a,$$

which is correct.

The preceding formulæ refer to the whole aggregate. When, however, $\lambda >$ the lowest λ_2, there are more than one component, and it will be well to give the requisite formulæ for each of these separately. Denote $\lambda r_n/a$ by y_n, where r_n is the radius of the nth interface from the centre. Also for shortness let $S(x)$ denote the function $\mathrm{Si}x - \sin x$. Then

$$\frac{\mu_n}{\mu} = \frac{S(y_n) - S(y_{n-1})}{S(\lambda)} \quad . \quad . \quad . \quad . \quad . \quad . \quad . \quad . \quad . \quad . \quad (21).$$

$$\frac{\nu_n}{\mu_n} = \frac{\pi}{S(y_n) - S(y_{n-1})} \{J_n - x_n^2 J'\} \quad . \quad . \quad . \quad . \quad . \quad . \quad . \quad (22),$$

J_n denoting the value of J at the equatorial axis.

$$M_n = \frac{m\mu}{\lambda^3(\mathrm{Si}\lambda - \sin\lambda)} \int_{y_{n-1}}^{y_n} \left(y \sin y - y^2 \cos y - y^4 \frac{J'}{\lambda^2}\right) dy$$

$$= \frac{1}{\lambda^3} \frac{m\mu_n}{S(y_n) - S(y_{n-1})} \left\{ 3y_n J(y_n) - 3y_{n-1} J(y_{n-1}) - y_n^2 \sin y_n + y_{n-1}^2 \sin y_{n-1} - \frac{y_n^5 - y_{n-1}^5}{5} \cdot \frac{J(\lambda)}{\lambda^2} \right\}.$$

But

$$Jy - y^2 \frac{J\lambda}{\lambda^2} = 0.$$

Hence

$$M_n = \frac{1}{\lambda^3} \frac{m\mu_n}{S(y_n) - S(y_{n-1})} \left[\left\{ 3(y_n^3 - y_{n-1}^3) - \frac{y_n^5 - y_{n-1}^5}{5} \right\} \frac{J(\lambda)}{\lambda^2} - y_n^2 \sin y_n + y_{n-1}^2 \sin y_{n-1} \right]. \quad (23).$$

20. The velocity of translation is given by

$$U = -\frac{\mu}{3a} \frac{\lambda \sin \lambda - 3(\sin \lambda / \lambda - \cos \lambda)}{\lambda (Si\lambda - \sin \lambda)}$$

$$= -\frac{\mu}{3a(Si\lambda - \sin \lambda)} \frac{d}{d(\lambda x)} \{J(\lambda x) - x^2 J(\lambda)\}_{x=1}.$$

To see how this varies with the parameter λ, refer to the graphical construction in fig. 2, Plate 1. The curve J and the parabola intersect in P. If A be a point on

Fig. 3.

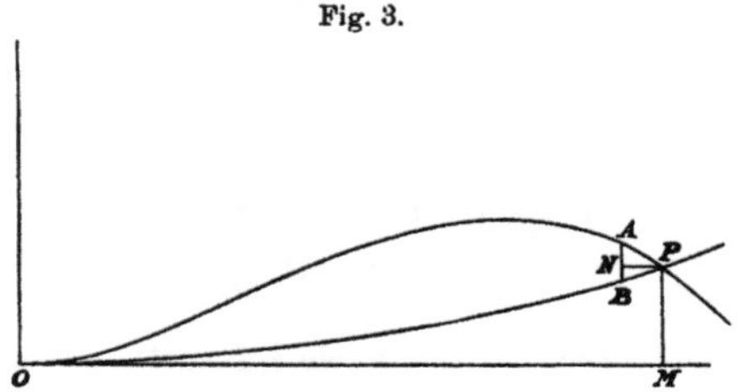

the curve (fig. 3), and B on the parabola with the same abscissa near P, and PN be the perpendicular on AB,

$$U = \frac{\mu}{3a(Si\lambda - \sin \lambda)} \frac{AB}{PN}$$

$$= -\frac{\mu}{3a(Si\lambda - \sin \lambda)} \frac{\sin(\alpha - \beta)}{\cos \alpha \cos \beta}$$

$$= -\frac{\mu}{3aA} \cdot \frac{\sin(\alpha - \beta)}{\cos \alpha \cos \beta}.$$

Where α, β are the angles which the tangents to the curve and the parabola at P make with the axis of x, and A denotes the area of the curve OAPMO.

The factor $\frac{\mu}{3a(Si\lambda - \sin \lambda)}$ is always finite, except for $\lambda = 0$, and positive. It is then easy to see in general how the velocity alters as the parameter λ increases.

As P (fig. 2, Plate 1), travels along the curve, U is positive. Leaving out of sight for the present its value for λ small, it later on diminishes to zero when P reaches a certain point Q where the parabola touches the curve. It then changes sign and remains negative until P reaches another Q point where the parabola again touches the curve, and so on.

We shall call the values of λ corresponding to the Q points the λ_2 values, and denote them in order by $\lambda_2^{(1)}, \lambda_2^{(2)}, \ldots \lambda_2^{(n)}$. Thus for values of $\lambda < \lambda_2^{(1)}$ the aggregate moves in the direction of the rotational flow up the axis. At $\lambda = \lambda_2^{(1)}$ the aggregate is at rest, the velocity of the fluid on the boundary is zero; as λ increases beyond this, the aggregate takes on another layer with primary rotation in the opposite direction, and it moves in the fluid in a direction opposed to the rotational motion of the innermost layer. It regredes relatively to this. The velocity at first increases and then diminishes until P reaches the second λ_2 point, when the corresponding aggregate is at rest in the fluid, and so on.

The periodic nature of the aggregates is thus evident. We get for example a whole periodic family of aggregates whose peculiar property is that they remain at rest in the fluid. The members of the family differ, amongst other things, in the number of independent layers each possesses.

So we get another family formed by values of λ, corresponding to points where the J-curve cuts the axis of x. We will call values of λ, corresponding to these the λ_1 parameters, and denote the orders in the same way as for the λ_2 parameters. As we shall see shortly, the distinguishing property of this family is that in each of them the vortex lines and the stream lines coincide.

For small values of λ it is preferable to express the value of U in terms of the lowest powers of λ.

It is easy to show that

$$\mathrm{S}i\lambda - \sin\lambda = \frac{2}{3.3!}\lambda^3 - \ldots - (-)^n \frac{2n}{2n+1}\frac{\lambda^{2n+1}}{(2n+1)!},$$

whence

$$\mathrm{U} = \frac{\mu}{5a}\left(1 - \tfrac{2}{175}\lambda^2\right).$$

This gives for $\lambda = 0$ the value of U already known for HILL'S vortex.

The curve $y = \mathrm{U}/\mathrm{U}_0$, where U_0 is the velocity of the non-gyrostatic aggregate of same cyclic constant and volume, is drawn in fig. 3, Plate 1, up to $\lambda_2^{(6)}$. The periodic quality is evident.

21. The directions of the lines of flow and of the vortex lines are given by

$$\tan\phi = \frac{v\rho\sin\phi}{v\rho\cos\phi} = \frac{\lambda}{a}\frac{\psi}{\frac{d\psi}{dn}} \quad \ldots \ldots \ldots \quad (24),$$

$$\tan\chi = \frac{\omega\rho\sin\chi}{\omega\rho\cos\chi} = -\frac{\lambda}{a}\frac{\mathrm{J}\sin^2\theta}{\frac{d}{dn}\left\{(\mathrm{J} - x^2\mathrm{J}')\sin^2\theta\right\}}.$$

Hence

$$\tan\chi = -\tan\phi - \frac{\lambda}{a}\frac{J'x^2\sin\theta}{\frac{d}{dn}\left\{(J - x^2J')\sin^2\theta\right\}} \quad . \quad . \quad . \quad . \quad (25),$$

also

$$\frac{\tan\phi}{\tan\chi} = -1 + x^2\frac{J'}{J}.$$

Equation (25) shows that when $J' = 0$, *i.e.*, for the λ_1 parameters, the stream lines and vortex lines coincide. (It is to be remembered that we have supposed in the foregoing that ϕ and χ lie on opposite sides of meridian lines, and therefore $\tan\phi = -\tan\chi$ means that they lie on the same side and coincide.)

I. From $\lambda = 0$ up to $\lambda = \lambda_1^{(1)}$, $J > x^2J'$ and $J - x^2J' < J$. Hence between these limits, the stream lines and vortex line are on the same side of the meridians, and $\chi > \phi$, *i.e.*, the stream lines lie between the vortex lines and meridians. At $\lambda = \lambda_1^{(1)}$ they coincide.

II. Between $\lambda_1^{(1)}$ and $\lambda_2^{(1)}$ $J > x^2J'$, but $J - x^2J' > J$. For any given λ, J changes from + to − as x passes through the value $\lambda x = \lambda_1^{(1)}$. For this value of x, or $r = \frac{\lambda_1}{\lambda}a$, $\chi = 0$. Thus, for an aggregate whose parameter λ lies between the first λ_1 and λ_2 roots, the vortex lines lie between the stream lines and the meridians for all points at a less distance from the centre than $r = \frac{\lambda_1}{\lambda}a$. At this distance $\chi = 0$, or the vortex lines coincide with the meridian planes, and beyond this distance up to the boundary the vortex lines and stream lines are on opposite sides of the meridians.

For values of λ between the first and second λ_2 parameters we have to deal with two layers. In the outer $J - x^2J'$ is negative, whilst J is negative between $\lambda_1^{(1)}$ and $\lambda_1^{(2)}$, positive between $\lambda_1^{(2)}$ and $\lambda_2^{(2)}$. Referring to fig. 2, Plate 1, let the point p where the parabola cuts the J curve be given by λ', corresponding in the aggregate to a distance from the centre $\lambda x = \lambda'$ or $r = \frac{\lambda'}{\lambda}a$. It is clear that J ($\lambda$) and J ($\lambda'$) are of the same sign. Hence, if λ lies between $\lambda_2^{(1)}$ and $\lambda_1^{(2)}$ (corresponding to P between Q_1 and R_2), λ' lies between $\lambda_1^{(1)}$ and $\lambda_2^{(1)}$, whereas if λ lies between $\lambda_1^{(2)}$ and $\lambda_2^{(2)}$, λ' lies between 0 and $\lambda_1^{(1)}$—or, taking closer limits still, between π and $\lambda_1^{(1)}$. We find, therefore, the following results.

III. P between Q_1 and R_2. In the inner spherical nucleus the vortex lines lie on the same side of the stream lines as the meridians—they are, in fact, exactly similar to the second category. At the boundary between the central nucleus and the outer layer $\phi = 0$, the stream lines coincide with the meridians. In the outer layer the stream lines lie on the other side of the meridian, with the vortex lines beyond. When P coincides with R_2 or λ is the second λ_1 parameter the stream lines coincide with the vortex lines again, but on the opposite side of the meridians.

IV. For P between R_2 and Q_2, we get still two layers, the boundary being given by (say) λ' (P at p), where $J\lambda'$ and $J\lambda$ are both positive. $J - x^2J'$ is positive between 0 and $x = \lambda'/\lambda$ and negative between $x = \lambda'/\lambda$ and 1. In the inner spherical nucleus ($r = 0$ to $r = \lambda' a/\lambda$) the stream lines lie between the vortex lines and the meridians (similar to the first category). At the interface the stream-lines coincide with the meridian. In the outer layer the stream lines and vortex lines lie on opposite sides of the meridian for points whose distance from the centre are less than $\frac{\lambda_1^{(1)}}{\lambda} a$, or greater than $\frac{\lambda_1^{(2)}}{\lambda} a$. For points at a distance $\frac{\lambda'_1}{\lambda} a$ and $\frac{\lambda_1^{(2)}}{\lambda} a$, the vortex lines coincide with the meridians, and between them the two lines lie on the same side of the meridian. In the same way the behaviour for aggregates whose parameter is greater than $\lambda_2^{(2)}$, may be determined. The periodic nature of the aggregate is again very clearly seen.

It is perhaps easier to describe the nature of the changes above indicated by supposing our eyes placed in a prolongation of the polar axis. Call the vortex lines blue lines and the stream lines red lines, and suppose for λ small that the stream or red lines lie on the right of the meridians. For $\lambda = 0$, or HILL'S vortex, the red lines lie along meridians and the blue lines perpendicular to these, along parallels of latitude. As λ increases the red and blue lines swing round towards each other, the reds to the right and the blues to the left, and this goes on with increasing values of λ up to $\lambda_1^{(1)}$, when they coincide. Beyond $\lambda = \lambda_1^{(1)}$ and up to $\lambda = \lambda_2^{(1)}$ the red and blue lines interchange their relative positions. In any given aggregate the blue lines move more and more towards meridians as we pass from the centre outwards. At a distance $\frac{\lambda_1^{(1)}}{\lambda} a$ from the centre the blue lines all coincide with the meridians, both red and blue lines are swinging round to the left. Beyond the distance $\frac{\lambda_1^{(1)}}{\lambda} a$ the blue lines cross to the left of the meridians and the red lines close up towards the meridians until at the surface of the aggregate they coincide with them.

Between $\lambda_2^{(1)}$ and $\lambda_2^{(2)}$ we have doublets. The aggregates lying between $\lambda_2^{(1)}$ and $\lambda_1^{(2)}$ and between $\lambda_1^{(2)}$ and $\lambda_2^{(2)}$ are however essentially different.

In the first set in the central nucleus the blue lines lie to the left of the red, and both to the right of the meridians for points near the centre. As we pass outwards from the centre they swing round to the left, the blue lines swing past the meridians whilst at the surface of the nucleus the red lines just reach it. Beyond, in the outer layer as we pass out, the blue and red swing further to the left, and later at least the red swing back again towards the meridian, coinciding with it at the surface. When $\lambda = \lambda_1^{(2)}$ red and blue coincide everywhere. They lie to the right in the inner nucleus and to the left in the outer layer.

Between $\lambda_1^{(2)}$ and $\lambda_2^{(2)}$ we get aggregates in which red and blue lines again change sides. In the inner nucleus both lie to the right of the meridian, blue furthest out. They close up to the meridian as we pass out from the centre to the nucleus surface. In the outer layer the red lines swing further to the left and back again, the blue lines follow after in the same way, crossing the meridian twice; once in each direction.

Beyond $\lambda_2^{(2)}$ we get triplets.

In general, between $\lambda_1^{(n)}$ and $\lambda_1^{(n+1)}$ the blue lines lie to the right of the red or the opposite according as n is even or odd. They coincide for the λ_1 parameter. Also, if n is even, both lie to the right of the meridian for the inner nucleus, the reds to the left for the second layer, to the right for the third, and so on. Whilst the opposite takes place if n is odd.

The forms of the spirals may be obtained by finding the polar equations to their projections on the equatorial plane. Let (ρ, η) be the polar co-ordinates of a point on the projection of a flow; (ρ, ς) of a vortex line lying on a given sheet ψ. Then

$$\rho \frac{d\eta}{ds} = \tan \phi, \qquad \rho \frac{d\varsigma}{ds} = \tan \chi,$$

where ds is an element of a meridian curve. Hence

$$d\eta = \frac{\lambda}{a} \frac{\psi\, ds}{\rho \dfrac{d\psi}{dn}} = \frac{\lambda}{a} \frac{\psi\, dr}{\rho \dfrac{d\psi}{r\, d\theta}}.$$

Provided dr is not perpendicular to ds, *i.e.*, on the outer boundary, but then $\psi = 0$ and $\eta = 0$.

$$d\eta = \frac{\lambda}{a} \frac{dr}{2 \cos \theta}, \qquad \eta = \frac{\lambda}{2} \int_{x_1}^{x} \frac{dx}{\cos \theta},$$

where x_1 corresponds to the inner circle of the two in which the current sheet ψ cuts the equatorial plane. The total angular pitch of the spiral is

$$\lambda \int_{x_1}^{x_2} \frac{dx}{\cos \theta} \quad . \quad . \quad . \quad . \quad . \quad . \quad . \quad . \quad . \quad . \quad . \quad (26),$$

where x_1, x_2 are the two roots of

$$\mathrm{J}(\lambda x) - x^2 \mathrm{J}\lambda = \frac{\lambda \psi (\mathrm{S}i\lambda - \sin \lambda)}{\pi \mu a} = b, \text{ say.}$$

The above may also be written

$$\eta = \tfrac{1}{2}\lambda \int_{x_1}^{x} \left\{ \frac{\mathrm{J} - x^2 \mathrm{J}'}{\mathrm{J} - x^2 \mathrm{J}' - b} \right\}^{\frac{1}{2}} dx \quad . \quad . \quad . \quad . \quad . \quad . \quad . \quad (27).$$

Equation (26) enables us easily to determine the form graphically when the surfaces ψ are drawn. So

$$\varsigma = \eta + \frac{\lambda}{2} \mathrm{J}' \int_{x_1}^{x} \{(\mathrm{J} - x^2 \mathrm{J}')(\mathrm{J} - x^2 \mathrm{J}' - b)\}^{-\frac{1}{2}} x^2\, dx \quad . \quad . \quad . \quad (28),$$

the case of a spherical boundary being excepted as before.

For the outside stream-lines the pitch is

$$\eta = 2 \times \frac{\lambda}{2}\int_0^1 dx = \lambda.$$

For values of λ, however, lying beyond $\lambda_2^{(1)}$ there are several layers in which the stream-lines are distinct. If $x_1, x_2, x_3 \ldots$ denote the values of x corresponding to the interfaces of the layers, the pitches of the stream-lines on those surfaces as we pass outwards are

$$(x_1 - 0)\,\lambda, \qquad (x_2 - x_1)\,\lambda, \qquad (x_3 - x_2)\,\lambda, \text{ \&c.}$$

We have seen that on these surfaces the stream-lines coincide with the meridian. These parts therefore produce no part of the pitch. The twist must be supposed as taking place in the part of the stream-line along the polar axes. It is easy to see that this is so by considering current sheets near the interfaces.

We may therefore regard the physical meaning of λ to be the criterion of the total external pitch of the stream-lines. We will return to the consideration of the pitch, and the shape of these lines later.

The total angular pitch of a stream spiral on any stream sheet ψ can easily be expressed in terms of the volume of the fluid inside that sheet. For

$$d\varsigma = d\eta - \frac{ds}{\rho}\cdot\frac{\lambda J'}{a^3}\cdot\frac{\rho^2}{\dfrac{d\psi}{dn}} \times \frac{\pi\mu a}{\lambda(\mathrm{Si}\lambda - \sin\lambda)}$$

$$= d\eta - \frac{\lambda J'}{2\pi a^3}\frac{2\pi\rho\, ds\, dn}{d\psi'}\ \text{ if } \psi = \frac{\pi\mu a}{\lambda(\mathrm{Si}\lambda - \sin\lambda)}\psi'.$$

Integrate round the stream surface

$$\varsigma = \eta - \frac{\lambda J'}{2\pi a^3}\frac{d}{d\psi'}\int 2\pi\rho\, ds\, dn = \eta - \frac{\lambda J'}{2\pi a^3}\frac{dm}{d\psi'} \quad . \quad . \quad . \quad . \quad . \quad (29),$$

where m denotes the volume inside ψ.

22. The discriminating properties of the λ_1 and λ_2 parameters make it important to determine their values. The λ_1 parameters are the roots of the equation

$$J(\lambda) \equiv \frac{\sin\lambda}{\lambda} - \cos\lambda = 0, \quad \text{or} \quad \tan\lambda = \lambda$$

The large roots are clearly nearly $(2n+1)\frac{\pi}{2}$.

Put

$$\lambda = (2n+1)\frac{\pi}{2} - y = a - y \text{ say}.$$

Then

$$\frac{\cos y}{a - y} - \sin y = 0.$$

Expanding this in powers of y, it is easily proved by successive approximation that

$$y = \frac{1}{a} + \frac{2}{3a^3} + \frac{13}{15a^5}$$

or

$$\lambda_1^{(n)} = (2n+1)\frac{\pi}{2} - \frac{2}{(2n+1)\pi} - \frac{16}{3(2n+1)^3\pi^3} - \frac{13 \times 32}{15(2n+1)^5\pi^5}$$

$$= 1{\cdot}57079\,(2n+1) - \frac{{\cdot}63662}{2n+1} - \frac{{\cdot}17201}{(2n+1)^3} - \frac{{\cdot}03558}{(2n+1)^5} \quad . \quad . \quad . \quad (30).$$

The first root is by numerical calculation

$$\lambda = 4{\cdot}49341 = 257^\circ\ 27'\ 10''$$

The foregoing formula gives for this case ($n = 1$)

$$\lambda = 4{\cdot}49366.$$

For higher values the formula is correct to five places at least
The first three roots are

$$\left.\begin{aligned} 4{\cdot}49341 &= 270^\circ - 12^\circ\ 32'\ 50' \\ 7{\cdot}72528 &= 450^\circ - 7^\circ\ 22'\ 27'' \\ 10{\cdot}90408 &= 630^\circ - 5^\circ\ 14'\ 23'' \end{aligned}\right\} \quad . \quad . \quad . \quad . \quad . \quad . \quad . \quad . \quad (31).$$

The λ_2 parameters are roots of the equation

$$\cot\lambda = \frac{1}{\lambda} - \frac{\lambda}{3}.$$

The large roots are clearly nearly $n\pi = n\pi - y$ say, where

$$\cot y = \frac{n\pi - y}{3} - \frac{1}{n\pi - y},$$

or

$$\cos y = \left(\frac{n\pi - y}{3} - \frac{1}{n\pi - y}\right)\sin y.$$

Writing $n\pi \equiv \beta$, and expanding in terms of y it is easy to prove, as in the former case, that

$$y = \frac{3}{\beta} + \tfrac{1}{3}\left(\frac{3}{\beta}\right)^3 + \tfrac{1}{5}\left(\frac{3}{\beta}\right)^5$$

$$\lambda = n\pi - \frac{3}{n\pi} - \tfrac{1}{3}\cdot\left(\frac{3}{n\pi}\right)^3 - \tfrac{1}{5}\left(\frac{3}{n\pi}\right)^5$$

$$= 3{\cdot}14159n - \frac{{\cdot}95493}{n} - \frac{{\cdot}29026}{n^3} - \frac{{\cdot}15881}{n^5} \quad . \quad . \quad . \quad . \quad . \quad (32).$$

There is no root corresponding to $n = 1$. The first root is

$$\lambda_2 = 5{\cdot}76346 = 360^\circ - 29^\circ\ 46'\ 41''.$$

The formula gives for this root

$$\lambda = 5{\cdot}76448.$$

For $n > 2$ it is exact to five places.

The first three roots are

$$\left.\begin{array}{l} 5{\cdot}76346 = 360^\circ - 29^\circ\ 46'\ 41'' \\ 9{\cdot}09506 = 540^\circ - 18^\circ\ 53'\ 29'' \\ 12{\cdot}32296 = 720^\circ - 13^\circ\ 56'\ 48'' \end{array}\right\} \quad \dots\dots \quad (33).$$

23. *Equatorial Axes.*—An equatorial axis is the line of particles which remains at rest. It is given by the equation

$$\frac{d\psi}{dr} = 0, \qquad \text{when } \theta = 0,$$

or by

$$\frac{dJ}{dx} - \frac{d}{dx}(x^2 J') = 0.$$

The positions of the axes are, therefore, readily observed by means of the graphical construction in fig. 2, Plate 1. They depend on the abscissæ of points for which the tangents to the J curve and the parabola are parallel. For values of $\lambda > \lambda_1^{(1)}$, the inclination of the parabola to the axis of x is always small. Hence the equatorial axes must always be near the crests (or bottoms) of the J curve, *i.e.*, near values $(2m + 1)\frac{1}{2}\pi$.

The equation for the axes becomes, if y be put for λx,

$$\cos y + \left(y - \frac{1}{y}\right)\sin y - 2y^2 \frac{J'}{\lambda^2} = 0 \quad \dots\dots \quad (34),$$

in which the roots $< \lambda$ are required.

As the values of the secondary cyclic constants and other important properties depend on the position of the equatorial axes, it will be necessary to determine their values. We shall do this (1) for the case of λ small, and (2) for the case of λ large. As, however, the case of the λ_1 values is special, we shall treat these separately. In the case of values other than λ_1, say, *e.g.*, λ_2 parameters, all the axes of any aggregate depend on the particular λ value. In the case of λ_1, however, they are independent of the particular λ_1. In fact, the successive λ_1 aggregates may be built up by taking any one and putting outside of this a suitable vortex shell. Moreover, the values of the axes for the λ_1 roots are the crests, and bottoms, of the J curve, and so are important for their own sakes.

Case of λ *small.*—Here y is also small. If equation (34) be expanded in powers of y and λ, there results

$$2\lambda^2 y^2 \Sigma_1 (-)^n \frac{\lambda^{2n-2}}{(2n+3)(2n+1)!} + 4\Sigma_2 (-)^n \frac{n^2}{(2n+1)!} y^{2n} = 0.$$

Dividing by $2y^2/15$, this may be written

$$y^2 = \frac{\lambda^2}{2} + 30\Sigma_2 (-)^n \frac{n+1}{(2n+3)!} \{(n+1) y^{2n} - \lambda^{2n}\},$$

whence y can be expressed in terms of λ by successive approximation. To λ^6 it will be found that

$$y^2 = \frac{\lambda^2}{2}\left\{1 - \frac{\lambda^2}{112} - \tfrac{50}{27}\left(\frac{\lambda^2}{112}\right)^2\right\},$$

$$y = \frac{\lambda}{\sqrt{2}}\left\{1 - \frac{\lambda^2}{224} - \tfrac{227}{54}\left(\frac{\lambda^2}{224}\right)^2\right\} \quad . \quad . \quad . \quad . \quad . \quad . \quad . \quad (35).$$

This gives the equatorial axis at

$$r = \frac{a}{\sqrt{2}}\left\{1 - \frac{\lambda^2}{224} - \tfrac{227}{54}\left(\frac{\lambda^2}{224}\right)^2\right\}.$$

When $\lambda = 0$, this agrees with HILL's vortex.

Case of λ_1.—The equation in y for this case becomes

$$\cos y + \left(y - \frac{1}{y}\right)\sin y = 0,$$

y is always nearly $n\pi = n\pi - z$ say, where z is small. Then

$$\cos z - \left(n\pi - z - \frac{1}{n\pi - z}\right)\sin z = 0.$$

Whence

$$y = n\pi - \frac{1}{n\pi} - \frac{5}{3(n\pi)^3} - \frac{7^3}{15(n\pi)^5}$$

$$= n\pi - \frac{\cdot 31831}{n} - \frac{\cdot 05375}{n^3} - \frac{\cdot 01590}{n^5}.$$

This formula gives for the two first roots

2·75363, 6·11682.

The values obtained by numerical calculation are

2·74371, 6·11676.

The roots beyond this are therefore given by the formula correct to five places. The radii of the equatorial axes are $r = ya/\lambda$. Hence using the values of λ_1 given in (31), the first three are. For $\lambda_1^{(1)}$,

$$r = \frac{2\cdot74371}{4\cdot49341} a = \cdot61062\, a.$$

For $\lambda_1^{(2)}$,

$$\left.\begin{aligned} r_2 &= \frac{6\cdot11676}{7\cdot72528} a = \cdot79179\, a \\ r_1 &= \frac{2\cdot74371}{7\cdot72528} a = \cdot35516\, a \end{aligned}\right\}.$$

For $\lambda_1^{(3)}$

$$\left.\begin{aligned} r_3 &= \frac{9\cdot31663}{10\cdot90408} a = \cdot85442\, a \\ r_2 &= \frac{6\cdot11676}{10\cdot90408} a = \cdot56096\, a \\ r_1 &= \frac{2\cdot74371}{10\cdot90408} a = \cdot25162\, a \end{aligned}\right\}.$$

Case of λ *large.*—The number of equatorial axes depends on the order of the λ_2 parameter next greater than λ. If λ lie between $\lambda_2^{(n-1)}$ and $\lambda_2^{(n)}$, there are n such axes. It seems then natural to refer the magnitude of λ to $\lambda_2^{(n)}$. Suppose then

$$\lambda = \lambda_2^{(n)} - \mathrm{X},$$

where the maximum value of X is about π,—*or* we may write $\lambda = \lambda_2^{(n-1)} + \mathrm{X}$, and if both be allowed X will have a maximum of the order $\frac{1}{2}\pi$.

$$\lambda_2^{(n)} = n\pi - \frac{3}{n\pi} - \tfrac{1}{3}\left(\frac{3}{n\pi}\right)^3 - \tfrac{1}{5}\left(\frac{3}{\pi}\right)^5.$$

The equation in y is

$$\cos y + \left(y - \frac{1}{y}\right)\sin y - 2y^2 \frac{\mathrm{J}\lambda}{\lambda^2} = 0,$$

in which the first n roots are to be determined. For small roots the parabola of fig. 3 is almost coincident with the axis of x, and consequently the small y roots are very nearly equal to the corresponding values for λ_1. It will be best to obtain an expression for the large roots and then see how far back it holds for the smaller roots. Clearly y is always near $m\pi$ where m is an integer $< n$.

Put

$$y = m\pi + z = \alpha + z \text{ say}.$$

Then

$$\frac{\cos z}{y} + \left(1 - \frac{1}{y^2}\right)\sin z - (-)^m\, 2y \frac{\mathrm{J}\lambda}{\lambda^2} = 0.$$

J (λ) may be either + or −, it is of order of magnitude 1 at most.

Since z is not large (it is of order $1/\alpha$), we get

$$\frac{1}{\alpha}\left(1-\frac{z}{\alpha}+\frac{z^2}{\alpha^2}\right)\left(1-\frac{z^2}{2}+\frac{z^4}{4!}\right)+\left(1-\frac{1}{\alpha^2}+\frac{2z}{\alpha^3}\right)$$
$$\times\left(z-\frac{z^3}{6}+\frac{z^5}{5!}\right)-(-)^m 2\frac{\alpha}{\lambda}\left(1+\frac{z}{\alpha}\right)\frac{J\lambda}{\lambda}=0.$$

Write

$$2(-)^m\frac{\alpha}{\lambda}\cdot\frac{J\lambda}{\lambda}\equiv\frac{b}{\alpha}.$$

The greatest value of α/λ is <1. $J'\lambda/\lambda$ is of order $1/\lambda$, therefore at least of order $1/\alpha$. Hence in the most unfavourable cases b is $\lesseqgtr 2$. The above equation can be written

$$-z=\frac{1}{\alpha}-\frac{b}{\alpha}-\frac{bz}{\alpha^2}-\frac{2z}{\alpha^2}-\frac{z^2}{2\alpha}-\frac{z^3}{6}+\frac{2}{3}\frac{z^3}{\alpha^2}+\frac{3z^2}{\alpha^3}+\frac{z^4}{\alpha.4!}+\frac{z^5}{5!},$$

$$z=\frac{b}{\alpha}-\frac{1}{\alpha}=\frac{b-1}{\alpha}, \qquad \text{(1st approx.)}$$

$$z=\frac{b-1}{\alpha}+\frac{(b+2)(b-1)}{\alpha^3}+\frac{(b-1)^2}{2\alpha^3}+\frac{(b-1)^3}{6\alpha^3}, \quad \text{(2nd approx.)}$$

$$=\frac{b-1}{\alpha}+\frac{(b-1)(b+2)(b+5)}{6\alpha^3} \qquad \text{(36).}$$

It will be convenient to put $b-1\equiv c$. Then

$$z=\frac{c}{\alpha}+\frac{c(c+3)(c+6)}{6\alpha^3}+\frac{1}{\alpha^5}\left\{\frac{c(c+3)(c+6)(c^2+4c+6)}{12}-3c^2-\tfrac{2}{3}c^3-\frac{c^4}{4!}-\frac{c^5}{5!}\right\}.$$

If λ is a λ_1 root, $c=-1$ and

$$z=-\frac{1}{\alpha}-\frac{5}{3\alpha^3}-\frac{7^3}{15\alpha^5},$$

which agrees with the result already found.

24. *The Spiral Forms taken by the Lines of Flow and Vortex Filaments.*—The equations determining these are given in § (21). Unfortunately, however, they are not integrable in finite forms.

We give a graphical method for the stream-lines later. At present it is proposed to determine (1) the forms of the stream and vortex lines when λ is small, (2) the pitch of the spirals near the equatorial axes, and (3) the pitch of the same on the outer surface.

Let the stream surface ψ, the streams and filaments on which we have to investigate, cut the equatorial plane in circles given by $r/a=x_1$ and x_2. Then

$$\eta=\frac{\lambda}{2}\int_{x_1}^{r}\sqrt{\left\{\frac{J-x^2J'}{J-x^2J'-b}\right\}}\,dx$$

where

$$b = \frac{\lambda}{\pi\mu a}(Si\lambda - \sin\lambda)\,\psi.$$

In determining the vortex filaments we will take s to be measured in the same direction as η : that is, to the right of meridians as looked at from the polar axis.

In this case

$$s = \lambda\int_{x_1}^{x}\frac{J\sin^2\theta\,dx}{\sin\theta\,(J - x^2J')\,2\sin\theta\cos\theta} = \frac{\lambda}{2}\int_{x_1}^{x}\frac{J\,dx}{\sqrt{\{(J - x^2J')(J - x^2J' - b)\}}}$$

or

$$s = \eta + \frac{\lambda}{2}J'\int_{x_1}^{x}\frac{x^2\,dx}{\sqrt{(J - x^2J')(J - x^2J' - b)}}.$$

(1.) Case of λ small.

$$J - x^2J' = \tfrac{1}{3}\lambda^4x^2\left\{-\frac{x^2-1}{10} + \frac{3\lambda^2}{7.5\,!}(x^4 - 1)\right\}$$

also b is of order λ^4. Put

$$b = \tfrac{1}{30}\lambda^4 c.$$

Hence

$$\eta = \frac{\lambda}{2}\int_{x_1}^{x}\sqrt{\left\{\frac{(1-x^2)\left(1 - \frac{\lambda^2}{28}\overline{x^2+1}\right)}{x^2(1-x^2) - \frac{\lambda^2x^2}{28}(1-x^4) - c}\right\}}\,x\,dx$$

x_1, x_2 are the roots of the denominator equated to 0, viz., of

$$x^4 - x^2 + c = \frac{\lambda^2}{28}(x^6 - x^2).$$

A first approximation is

$$x^2 = \frac{1 \pm \sqrt{1-4c}}{2} = r_1 \text{ or } r_2 \text{ (say), where } r_1 \text{ denotes the smaller root.}$$

Let for a second approximation

$$x^2 = r_1 + \xi \text{ where } \xi \text{ is of order } \lambda^2.$$

$$x^4 - x^2 + c = r_1^2 + 2r_1\xi - r_1 - \xi + c = (2r_1 - 1)\,\xi.$$

$$x^6 - x^2 = r_1^3 + 3r_1\xi - r_1 - \xi.$$

Therefore,

$$(2r_1 - 1)\xi = \frac{\lambda^2}{28}(r_1^3 - r_1).$$

$$\xi = \frac{\lambda^2}{28}\,\frac{r_1^3 - r_1}{2r_1 - 1} = -\frac{\lambda^2}{28}\,\frac{r_1r_2(r_1+1)}{r_1 - r_2}.$$

Since $r_1 + r_2 = 1$, also $r_1 r_2 = c$,

$$\xi = -\frac{\lambda^2 c}{28} \cdot \frac{r_1+1}{r_1-r_2}.$$

Hence the roots are

$$x_1^2 = r_1 + \frac{\lambda^2 c}{28} \cdot \frac{r_1+1}{r_2-r_1}$$

and

$$x_2^2 = r_2 - \frac{\lambda^2 c}{28} \cdot \frac{r_2+1}{r_2-r_1},$$

and the denominator becomes

$$(x^2 - x_1^2)(x_2^2 - x^2)\left(1 - \frac{\lambda^2}{28} - \frac{\lambda^2}{28} x^2\right).$$

Whence

$$\eta = \frac{\lambda}{2} \int_{x_1}^{x} \sqrt{\frac{(1 - x^2)\left(1 - \frac{\lambda^2}{28} - \frac{\lambda^2}{28} x^2\right)}{(x^2 - x_1^2)(x_2^2 - x^2)\left(1 - \frac{\lambda^2}{28} - \frac{\lambda^2}{28} x^2\right)}}\, x\, dx,$$

$$= \frac{\lambda}{4} \int_{y_1}^{y} \sqrt{\left\{\frac{1-y}{(y-y_1)(y_2-y)}\right\}} dy, \text{ where } y = x^2.$$

In this put

$$y = \frac{y_2+y_1}{2} - \frac{y_2-y_1}{2} \cos\theta.$$

So that

$$y - y_1 = \frac{y_2-y_1}{2}(1 - \cos\theta) \qquad y_2 - y = \frac{y_2-y_1}{2}(1 + \cos\theta).$$

Then

$$\eta = \frac{\lambda}{4} \int_0^{\theta} \sqrt{1 - y}\, d\theta,$$

$$= \frac{\lambda}{4} \int_0^{\theta} \sqrt{\left\{1 - y_1 - (y_2 - y_1)\sin^2\frac{\theta}{2}\right\}} d\theta,$$

$$= \frac{\lambda}{2} \sqrt{1 - y_1} \int_0^{\phi} \sqrt{1 - k^2 \sin^2\phi}\, d\phi,$$

$$= \frac{\lambda}{2} \sqrt{1 - y_1}\, \mathrm{E}(k, \phi), \text{ where } \sin^2\phi = \frac{x^2-x_1^2}{x_2^2-x_1^2} \quad . \quad . \quad . \quad . \quad . \quad (37),$$

and

$$k^2 = \frac{y_2-y_1}{1-y_1} = \frac{x_2^2-x_1^2}{1-x_1^2},$$

$$= \frac{r_2 - r_1 - \frac{\lambda^2 c}{28}\frac{r_2+r_1+2}{r_2-r_1}}{r_2 + r_1 - r_1 - \frac{\lambda^2 c}{28}\frac{r_1+1}{r_2-r_1}}$$

$$= 2\frac{1 - 4c - \frac{3\lambda^2 c}{28}}{1-4c+\sqrt{1-4c} - \frac{\lambda^2 c}{28}(3 - \sqrt{1 - 4c})}.$$

At the equatorial axis $k = 0$, on the surface $k = 1$. Thus k increases from 0 to 1 for the various current sheets in order from the axis to the surface. The pitch of the helix on any sheet is

$$\text{Pitch} = \lambda \sqrt{1 - y_1}\, \text{E}.$$

At the surface this is λ, at the axis it is

$$= \lambda \sqrt{1 - y_0} \cdot \frac{\pi}{2} = \frac{\pi\lambda}{2} \sqrt{\left\{1 - \tfrac{1}{2}\left(1 - \frac{\lambda^2}{112}\right)\right\}} = \frac{\pi\lambda}{2\sqrt{2}}\left(1 + \frac{\lambda^2}{112}\right).$$

Since $\pi/(2\sqrt{2}) = 1{\cdot}11$, the pitch at the axis is about 11 per cent. larger than on the surface when λ is small.

The corresponding quantity for the vortex filaments is given by

$$s = \eta + \tfrac{1}{2}\lambda \text{J}' \int_{x_1}^{x} \frac{x^2\, dx}{\sqrt{(\text{J} - x^2\text{J}')(\text{J} - x^2\text{J}' - b)}}.$$

By what has immediately gone before

$$s - \eta = \tfrac{1}{2}\lambda\text{J}' \int_{x_1}^{x} \frac{30}{\lambda^4} \frac{x\, dx}{\left\{1 - \frac{\lambda^2}{28}(x^2 + 1)\right\} \sqrt{\{(1 - x^2)(x^2 - x_1^2)(x_2^2 - x^2)\}}}$$

$$= \tfrac{15}{2} \frac{\text{J}'}{\lambda^3} \int_{y_1}^{y} \frac{dy}{\left(1 - \frac{\lambda^2}{28} - \frac{\lambda^2}{28} y\right) \sqrt{\{(1 - y)(y - y_1)(y_2 - y)\}}}$$

$$= \tfrac{15}{2} \frac{\text{J}'}{\lambda^3} \int_{0}^{\theta} \frac{d\theta}{\left(1 - \frac{\lambda^2}{28} - \frac{\lambda^2}{28} y\right)\sqrt{1 - y}}$$

$$= \frac{15\text{J}'}{2\lambda^3\sqrt{1 - y_1}} \int_0^{\phi} \frac{d\phi}{\left\{1 - \frac{\lambda^2}{28}(1 + y_1) - \frac{\lambda^2}{28}(y_2 - y_1)\sin^2\phi\right\} \sqrt{(1 - k^2 \sin^2\phi)}},$$

and

$$\text{J}' = \tfrac{1}{3}\lambda^2 \left(1 - \tfrac{1}{10}\lambda^2\right).$$

Therefore

$$s - \eta = \frac{5\left(1 - \frac{\lambda^2}{10}\right)}{2\lambda\sqrt{1 - y_1}\left\{1 - \frac{\lambda^2}{28}(1 + y_1)\right\}} \int_0^{\phi} \frac{d\phi}{(1 - n \sin^2\phi)\sqrt{(1 - k^2 \sin^2\phi)}},$$

where

$$n = \frac{\lambda^2}{28}(y_2 - y_1) = \frac{\lambda^2}{28}\sqrt{(1 - 4c)}.$$

Thus

$$s = \frac{\lambda}{2}\sqrt{1 - y_1}\,\text{E}(k \,.\, \phi) + \frac{5 - \frac{9}{28}\lambda^2 + \frac{5\lambda^2 y_1}{28}}{2\lambda\sqrt{1 - y_1}}\, \Pi(-n, k, \phi). \quad . \quad . \quad (38).$$

At the equatorial axis $n = 0$, $k = 0$; $\Pi = \pi/2$ for a half turn.

Thus the pitch at the equatorial axis is

$$= \frac{\pi\lambda}{2}\sqrt{1-y_0} + \frac{5 - \frac{9}{28}\lambda^2 + \dfrac{5\lambda^2 y_0}{28}}{2\lambda\sqrt{1-y_0}}\,\pi,$$

and

$$y_0 = \tfrac{1}{2}\left(1 - \frac{\lambda^2}{112}\right).$$

Therefore

$$\text{Pitch} = \frac{\pi}{2}\left\{\frac{\lambda}{\sqrt{2}}\left(1+\frac{\lambda^2}{112}\right) + \frac{\sqrt{2}}{\lambda}\left(5 - \tfrac{13}{56}\lambda^2\right)\left(1-\frac{\lambda^2}{112}\right)\right\}$$

$$= \frac{\pi}{2\lambda\sqrt{2}}\{\lambda^2 + 2\,(5 - \tfrac{31}{112}\lambda^2)\}$$

$$= \frac{5\pi}{\lambda\sqrt{2}}\,(1 + \tfrac{5}{112}\lambda^2).$$

If $\lambda = 0$, the pitch is ∞, as it clearly ought to be, since all the vortex filaments then lie along parallels.

The Form of the Spirals near an Equatorial Axis.

The meridian sections of a current sheet near an axis will evidently in general be elliptic. To find η it is therefore necessary to determine for an ellipse the value of

$$\int\frac{dr}{\cos\theta}.$$

The following general theorem enables us easily to do this. Transfer the origin to any point O′ in the equatorial plane, at a distance c; and let the new polar co-ordinates of a point P be $r'.\theta'$, corresponding to $r.\theta$. Also let $x.y$ denote the Cartesian co-ordinates referred to O′. Then

$$r^2 = r'^2 + c^2 + 2cx,$$

$$rdr = r'dr' + cdx,$$

$$\int\frac{dr}{\cos\theta} = \int\frac{rdr}{r\cos\theta} = \int\frac{r'dr' + cdx}{r'\cos\theta'} = \int\frac{dr'}{\cos\theta'} + c\int\frac{dx}{y}.$$

For the spirals near the axis the point of interest is to determine the angular pitch. Now clearly for a complete ellipse, whose axes are parallel and perpendicular to the equatorial plane, and whose centre is at O′

$$\int\frac{dr'}{\cos\theta'} = 0.$$

Further, if the axes are α, β, respectively in the equatorial plane and perpendicular to it,

$$x = \alpha \sin \theta, \qquad y = \beta \cos \theta,$$

where θ is the excentric angle of a point on it. Hence,

$$\int \frac{dx}{y} = \int_{-\frac{\pi}{2}}^{+\frac{\pi}{2}} \frac{\alpha \cos \theta \, d\theta}{\beta \cos \theta} = \pi \frac{\alpha}{\beta}.$$

Therefore,

$$\eta \text{ (for half-turn)} = \frac{\pi \alpha c}{\beta} \cdot \frac{\lambda}{2a},$$

or,

$$\text{angular pitch} = \frac{\pi \alpha c}{\beta} \frac{\lambda}{a}.$$

To apply this, it is necessary to determine the form of the current sheets near the axis.

Let the co-ordinates of the equatorial axis be c, o.

The equation to a current sheet is

$$\left\{ \mathrm{J}\left(\frac{\lambda r}{a}\right) - \frac{r^2}{a^2} \mathrm{J}(\lambda) \right\} \sin^2 \theta = \text{constant},$$

or,

$$\frac{\rho^2}{r^2} \left\{ \mathrm{J}\left(\frac{\lambda r}{a}\right) - k\left(\frac{\lambda r}{a^2}\right)^2 \right\} = \text{constant}.$$

$k = \mathrm{J}(\lambda)/\lambda^2$, and r, ρ are nearly $= c$.

Denote $\mathrm{J}\frac{\lambda r}{a} - k\left(\frac{\lambda r}{a}\right)^2$ by f, and suppose it expressed in terms of x, y co-ordinates. Refer to O′.

Then $x = c + \xi$, $y = o + \eta$, where ξ, η are small. Hence, if f now denote the value at O′,

$$\frac{(c+\xi)^2}{(c+\xi)^2 + \eta^2} \left[f + \frac{df}{dx} \cdot \xi + \frac{df}{dy} \eta + \tfrac{1}{2} \left\{ \xi^2 \frac{d^2 f}{dx^2} + 2\xi\eta \frac{d^2 f}{dx\,dy} + \eta^2 \frac{d^2 f}{dy^2} \right\} \right] = \text{constant},$$

$$\frac{df}{dx} = \frac{df}{dr} \frac{x}{r}, \qquad \frac{df}{dy} = \frac{df}{dr} \frac{y}{r},$$

and $df/dr = o$, for c is given by this equation.

Denote df/dr by f', d^2f/dr^2 by f''. Then

$$\frac{d^2 f}{dx^2} = \left(\frac{1}{r} - \frac{x^2}{r^3}\right) f' + \frac{x^2}{r^2} f'' = \frac{x^2}{r^2} f'' = f'', \text{ since } x = r \text{ to 1st order},$$

$$\frac{d^2 f}{dx\,dy} = -\frac{xy}{r^3} f' + \frac{xy}{r^2} f'' = 0, \text{ since } y = 0 \text{ to 1st order},$$

$$\frac{d^2 f}{dy^2} = \left(\frac{1}{r} - \frac{y^2}{r^3}\right) f' + \frac{y^2}{r^2} f'' = 0.$$

Hence

$$\left(1 - \frac{\eta^2}{c^2}\right)(f + \tfrac{1}{2}\xi^2 f'') = \text{constant}.$$

The constant is nearly $f = f - \alpha$ say; then

$$-\frac{f}{c^2}\eta^2 + \tfrac{1}{2}f''\xi^2 = -\alpha,$$

or the curve is the ellipse

$$\frac{\xi^2}{2\alpha / -f''} + \frac{\eta^2}{\alpha c^2/f} = 1.$$

Hence the angular pitch $= \frac{\pi\lambda c}{a}\sqrt{\frac{2f}{-c^2 f''}} = \frac{\pi\lambda}{a}\sqrt{-\frac{2f}{f''}}$.

Now

$$f'' = \frac{d^2}{dr^2}\left\{\frac{J\lambda r}{a} - \frac{r^2}{a^2}J\lambda\right\} = \frac{\lambda^2}{a^2}\frac{d^2}{dy^2}\left\{Jy - y^2\frac{J(\lambda)}{\lambda^2}\right\}$$

$$= \frac{\lambda^2}{a^2}\left\{\frac{d^2 J}{dy^2} - 2\frac{J\lambda}{\lambda^2}\right\} = \frac{\lambda^2}{a^2}\left\{\left(\frac{2}{y^2} - 1\right)J - \frac{2J}{\lambda^2}\right\} = \frac{\lambda^2}{a^2}\left(\frac{2f}{y^2} - Jy\right).$$

The angular pitch of the stream-lines is therefore

$$\pi\sqrt{\left\{\frac{2f}{J(y) - \frac{2f}{y^2}}\right\}} = \frac{\pi}{p},$$

where

$$p^2 = \tfrac{1}{2}\frac{J(y)}{J(y) - y^2\frac{J(\lambda)}{\lambda^2}} - \frac{1}{y^2}.$$

Now y is determined by $J(y) - y\sin y + 2y^2\frac{J(\lambda)}{\lambda^2} = 0$, therefore

$$p^2 = \tfrac{1}{2}\frac{\sin y - 2y\frac{J(\lambda)}{\lambda^2}}{\sin y - 3y\frac{J(\lambda)}{\lambda^2}} - \frac{1}{y^2}$$

$$= \tfrac{1}{2} - \frac{1}{y^2} + \tfrac{1}{2}\frac{1}{\frac{\sin y}{y\frac{J\lambda}{\lambda^2}} - 3}$$

$\left(\text{Note for } \lambda_1 \text{ aggregates } p^2 = \tfrac{1}{2} - \frac{1}{y^2}\right)$.

When the value of λ is fairly large we substitute for y from equation (36)

$$y = m\pi + z,$$

where

$$z = \frac{b-1}{\alpha} + \frac{(b-1)(b+2)(b+5)}{6\alpha^3},$$

$$\alpha \equiv m\pi, \qquad b \equiv 2(-)^m \alpha^2 \frac{J(\lambda)}{\lambda^2}.$$

Therefore

$$p^2 = \tfrac{1}{2} - \frac{1}{(\alpha+z)^2} + \tfrac{1}{2} \frac{1}{(-1)^m \dfrac{\sin z}{(\alpha+z)\dfrac{b}{2(-)^m\alpha^2}} - 3}$$

$$= \tfrac{1}{2} - \frac{1}{\alpha^2} + \tfrac{1}{2} \frac{1}{\dfrac{2\alpha^2}{\alpha+z} \cdot \dfrac{z - \frac{1}{6}z^3}{b} - 3}$$

$$= \tfrac{1}{2} - \frac{1}{\alpha^2} + \tfrac{1}{2} \frac{1}{\dfrac{2\alpha z}{b}\left(1 - \dfrac{z}{\alpha} + \dfrac{z^2}{\alpha^2}\right)\left(1 - \dfrac{z^2}{6}\right) - 3}$$

$$= \tfrac{1}{2} - \frac{1}{\alpha^2} + \tfrac{1}{2} \frac{1}{\dfrac{2(b-1)}{b}\left(1 + \dfrac{\overline{b+2}\,\overline{b+5}}{6\alpha^2}\right)\left(1 - \dfrac{(b-1)^2}{6\alpha^2}\right)\left(1 - \dfrac{b-1}{\alpha^2}\right) - 3}$$

$$= \tfrac{1}{2} - \frac{1}{\alpha^2} + \tfrac{1}{2} \frac{1}{\dfrac{2(b-1)}{b}\left\{1 + \dfrac{1}{6\alpha^2}(\overline{b+2}\,\overline{b+5} - \overline{b-1}|^2 - 6\,\overline{b-1})\right\} - 3}$$

$$= \tfrac{1}{2} - \frac{1}{\alpha^2} + \tfrac{1}{2} \frac{1}{\dfrac{b-1}{b}\left(2 + \dfrac{b+5}{\alpha^2}\right) - 3}$$

$$= \tfrac{1}{2} - \frac{1}{\alpha^2} + \tfrac{1}{2} \frac{b}{-(b+2) + \dfrac{(b-1)(b+5)}{\alpha^2}}.$$

$$p^2 = \frac{1 - \dfrac{(b-1)(b+5)}{2\alpha^2}}{b + 2 - \dfrac{(b-1)(b+5)}{\alpha^2}} - \frac{1}{\alpha^2} = \frac{1 - \dfrac{b^2+6b-1}{2\alpha^2}}{b + 2 - \dfrac{(b-1)(b+5)}{\alpha^2}},$$

and the pitch is

$$\pi \sqrt{\left\{\frac{b + 2 - \dfrac{(b-1)(b+5)}{\alpha^2}}{1 - \dfrac{b^2+6b-1}{2\alpha^2}}\right\}} = \pi \sqrt{\left\{b + 2 + \frac{(b+2)^2 - 9b}{2\alpha^2}\right\}}.$$

Now

$$b = 2(-)^m \left(\frac{\alpha}{\lambda}\right)^2 J\lambda,$$

where

$$\lambda = \lambda_2^{(n)} - X$$

and

$$-\lambda_2^{(n)} = n\pi - \left(\frac{3}{n\pi}\right) - \tfrac{1}{3}\left(\frac{3}{n\pi}\right)^3 - \ldots$$

and

$$X \overline{\underset{<}{=}} \pi = q\pi \text{ say},$$

denote $n\pi$ by β; then

$$\frac{a^2 J\lambda}{\lambda^2} = (-)^n \left(\frac{m}{n}\right)^2 \left(1 - \frac{3}{\beta^2} - \frac{X}{\beta}\right)^{-2} \left\{ - \frac{\sin\left(X + \frac{3}{\beta} + \frac{9}{\beta^3}\right)}{\beta\left(1 - \frac{3}{\beta^2} - \frac{X}{\beta}\right)} - \cos\left(X + \frac{3}{\beta} + \frac{9}{\beta^3}\right)\right\}$$

$$= -(-)^n \left(\frac{m}{n}\right)^2 \left(1 + \frac{2X}{\beta} + \frac{6 + 3X^2}{\beta^2}\right) \left\{\frac{1}{\beta}\left(1 + \frac{X}{\beta}\right)\left(\sin X + \frac{3}{\beta}\cos X\right) + \cos X\left(1 - \frac{9}{2\beta^2}\right) - \sin X\left(\frac{3}{\beta}\right)\right\}$$

$$= -(-)^n \left(\frac{m}{n}\right)^2 \left(1 + \frac{2X}{\beta} + \frac{6 + 3X^2}{\beta^2}\right)\left\{\cos X\left(1 - \frac{3}{2\beta^2}\right) + \sin X\left(-\frac{2}{\beta} + \frac{X}{\beta^2}\right)\right\},$$

therefore

$$b + 2 = 2\left[1 - (-)^{m+n}\left(\frac{m}{n}\right)^2\left\{\left(1 + \frac{2X}{\beta} + \frac{9 + 6X^2}{2\beta^2}\right)\cos X - \left(\frac{2}{\beta} + \frac{3X}{\beta^2}\right)\sin X\right\}\right].$$

For very large values of λ we may neglect powers of $\frac{1}{\beta}$, and then

$$\text{pitch} = \pi\sqrt{2}\left\{1 - (-)^{m+n}\left(\frac{m}{n}\right)^2 \cos X\right\}^{\frac{1}{2}}.$$

For the outside shell $m = n$,

$$\text{pitch} = 2\pi \sin \tfrac{1}{2}X.$$

Thus in the case of the λ_2 aggregates the pitch of the outer layer is very small.

If we number the shells backward from the outside, we write $n - p + 1$ for m, and the pitch is

$$\pi\sqrt{2}\left\{1 + (-)^p\left(\frac{n - p + 1}{n}\right)^2 \cos X\right\}^{\frac{1}{2}}.$$

It is seen, therefore, that there are two series of shells in aggregates of large λ, one in which the pitches increase as we pass inwards, and an alternate series in which it decreases. If λ lies between a λ_1 and a λ_2 parameter ($\lambda_2 > \lambda_1$), the outer series belongs to the first category. If λ lies between a λ_2 and a λ_1 value ($\lambda_1 > \lambda_2$), the opposite is the case. In other words, if the parametral point P in fig. 2, Plate 1, lie above the line of abscissæ, the outside layer has a very small pitch, and those of alternate shells increase as we go to the centre. If P lie below the opposite is the case.

The vortex spirals are given by

$$\varsigma - \eta = \frac{\lambda}{2} J(\lambda) \int_{x_1}^{x} \frac{x^2 \, dx}{\{(J - x^2 J')(J - x^2 J' - b)\}^{\frac{1}{2}}}.$$

Write $J - x^2 J' = f$.

Let x_0 be the value of x at the axis, so that $f' = 0$ when $x = x_0$.

For points near the axis,

$$x = x_0 + \xi,$$

$$f(x) = f + af'.\xi + \tfrac{1}{2}a^2 f''.\xi^2 = f + \tfrac{1}{2}a^2 f''.\xi^2,$$

and

$$\varsigma - \eta = \tfrac{1}{2}\frac{\lambda J'}{a^2}\int_{-\xi_1}^{\xi_1}\frac{(x_0+\xi)^2\,d\xi}{(\frac{1}{2}f'')\left\{\left(\frac{2f}{a^2f''}+\xi^2\right)\left(2\frac{f-b}{a^2f''}+\xi^2\right)\right\}^{\frac{1}{2}}},$$

where $(\frac{1}{2}f'')$ means the positive value of $\frac{1}{2}f''$.

To the first order,

$$\xi_1^2 = \xi_2^2 = -2\frac{f-b}{a^2 f''},$$

$$\varsigma - \eta = \frac{\lambda J'}{a^2(f'')}\int_{-\xi_1}^{\xi_1}\frac{(x_0+\xi)^2\,d\xi}{\left\{(\xi_1^2-\xi^2)\left(-\frac{2f}{a^2f''}-\xi^2\right)\right\}^{\frac{1}{2}}}.$$

Hence for the total pitch,

$$\varsigma - \eta = \frac{2\lambda J'}{a^2(f'')}\int_{-\xi_1}^{+\xi_1}\frac{(x_0^2+\xi^2)\,d\xi}{\left\{(\xi_1^2-\xi^2)\left(-\frac{2f}{a^2f''}-\xi^2\right)\right\}^{\frac{1}{2}}}.$$

Put

$$\xi = \xi_1 \sin\theta.$$

$$\varsigma - \eta = \frac{4\lambda J'}{a^2(f'')}\int_0^{\frac{1}{2}\pi}\frac{(x_0^2+\xi_1^2\sin^2\theta)\,d\theta}{\sqrt{\left(-\frac{2f}{a^2f''}-\xi_1^2\sin^2\theta\right)}}$$

or writing

$$k^2 = -\frac{a^2f''\xi_1^2}{2f}$$

$$\varsigma - \eta = \frac{4\lambda J'}{a\sqrt{(-2ff'')}}\int_0^{\frac{1}{2}\pi}\frac{x_0^2-\frac{2f}{a^2f''}+\frac{2f}{a^2f''}+\xi_1^2\sin^2\theta}{\sqrt{(1-k^2\sin^2\theta)}}\,d\theta$$

$$= \frac{4\lambda J'}{a\sqrt{(-2ff'')}}\left\{\left(x_0^2-\frac{2f}{a^2f''}\right)F+\frac{2f}{a^2f''}E\right\}.$$

At the axis itself $k^2 = 0$.

$$\varsigma - \eta = \frac{4\lambda J' x_0^2}{a\sqrt{(-2ff'')}}.\frac{\pi}{2}$$

$$\varsigma = \pi\lambda\sqrt{-\frac{2f}{a^2f''}} + 2\pi\lambda\frac{J'x_0^2}{a\sqrt{(-2ff'')}}$$

$$= \frac{2\pi\lambda}{a\sqrt{(-2ff'')}}\{(f)+x_0^2J'\}$$

$$= \frac{2\pi\lambda J(y)}{a\sqrt{(-2ff'')}} \quad \text{if} \quad J(y) > y^2\frac{J\lambda}{\lambda^2}$$

$$= \frac{2\pi Jy}{\sqrt{2f\left(Jy-\frac{2f}{y^2}\right)}} \quad \text{if} \quad J(y) > y^2\frac{J(\lambda)}{\lambda}$$

or

$$\varsigma - \eta = \frac{2\pi}{\sqrt{2f\left(Jy - \frac{2f}{y^2}\right)}}\left(\frac{2y^2J\lambda}{\lambda^2} - Jy\right), \text{ if } Jy < \frac{y^2J\lambda}{\lambda^2},$$

where $y = \lambda x_0/a$.

Spirals on the bounding surface, or interface between two shells. This is the case where the transformation (Eq. 26) fails. Taking first the stream spirals

$$d\eta = \frac{\lambda}{a} \cdot \frac{\psi}{\rho} \frac{ds}{\frac{d\psi}{dn}}.$$

On a spherical boundary this is zero, except for the λ_2 aggregates, in which, however, there is no flow at all. The other part of the stream surface is the portion up the polar axis. Here $ds = dr$ and $dn = rd\theta$. Therefore

$$\text{Twist on axis alone} = \frac{2\lambda}{a}\int_0^a \frac{\psi dr}{\rho \frac{d\psi}{rd\theta}}(\theta = 0)$$
$$= \frac{\lambda}{a}\int_0^a dr = \lambda.$$

There is no twist on the spherical boundary. Hence

Angular pitch of stream spiral $= \lambda$.

Next for the vortex spirals. Here there are two portions as in the former case—the polar axis, and the spherical boundary.

$$d\varsigma = \frac{\lambda}{a} \cdot \frac{J \sin^2\theta}{\rho \frac{d}{dn}\{(J - x^2J')\sin^2\theta\}} ds.$$

Hence, supposing at present we are dealing with a singlet only

$$\varsigma = \frac{2\lambda}{a}\int_0^a \frac{1}{2}\frac{J}{J - x^2J'} \cdot dr + \frac{2\lambda}{a}\int_0^{\frac{1}{2}\pi} \frac{J\lambda . a d\theta}{a \sin\theta \frac{d}{dr}(J - x^2J')_{r=a}}$$
$$= \lambda\int_0^1 \frac{J}{J - x^2J'} dx + \frac{2\lambda J\lambda}{\frac{d}{dx}(J - x^2J')_{x=1}}\int_0^{\frac{1}{2}\pi}\frac{d\theta}{\sin\theta}.$$

Both these integrals become infinite at the poles. We must therefore treat this part separately.

$$\varsigma = \lambda\int_0^{1-\epsilon}\frac{J}{J - x^2J'} dx + \frac{2\lambda J(\lambda)}{\lambda \sin\lambda - 3J(\lambda)}\int_\epsilon^{\frac{1}{2}\pi}\frac{d\theta}{\sin\theta} + 2\frac{\lambda}{a}\int_{\xi_1 a'}^{\xi_1' a}\frac{J \sin^2\theta \, ds}{\rho\left\{\left(\frac{d\psi'}{dr}\right)^2 + \left(\frac{d\psi'}{rd\theta}\right)^2\right\}^{\frac{1}{2}}},$$

in which ξ, α are small, and in the third integral, x and θ are nearly 1, 0 respectively. Let $x = 1 - \xi$, $\sin\theta = \eta$. Then near the pole the rectangular co-ordinates of a point referred to the pole are connected by (if $f \equiv \mathrm{J} - x^2\mathrm{J}'$)

$$f\sin^2\theta = \text{small constant} = \beta \text{ (say)},$$

$$-\frac{df}{dx}\cdot\xi\sin^2\theta = \beta, \qquad \xi\eta^2 = -\frac{\beta}{\frac{df}{dx}} = \gamma^3 \text{ say},$$

so that when $\xi = \eta$ each $= \gamma$.

The third integral is

$$2\lambda\int\frac{\mathrm{J}\sin^2\theta\sqrt{(d\xi^2+d\eta^2)}}{r\sin\theta\{(df/dr)^2\sin^4\theta + 4\sin^2\theta\cos^2\theta(f/r)^2\}^{\frac{1}{2}}}$$

$$= 2\lambda\int\frac{\mathrm{J}\sqrt{(d\xi^2+d\eta^2)}}{(1-\xi)\left\{(df/dx)^2\sin^2\theta + 4\cos^2\theta\left(\frac{df}{xdx}\cdot\xi\right)^2\right\}^{\frac{1}{2}}}$$

$$= \frac{2\lambda\mathrm{J}'}{df/dx}\left[\int_{\xi_1}^{\gamma}\sqrt{\left(\frac{1+(d\eta/d\xi)^2}{\eta^2+4\xi^2}\right)}d\xi + \int_{\gamma}^{\eta_1}\sqrt{\left(\frac{1+(d\xi/d\eta)^2}{\eta^2+4\xi^2}\right)}d\eta\right].$$

The curve is given by

$$\xi\eta^2 = \gamma^3.$$

Therefore

$$\frac{d\eta}{d\xi} = -\frac{\eta^2}{2\xi\eta} = -\frac{1}{2}\frac{\eta}{\xi} = -\frac{1}{2}\left(\frac{\gamma}{\xi}\right)^{3/2} = -\frac{1}{2}\left(\frac{\eta}{\gamma}\right)^3.$$

Therefore

$$\text{Integral} = \frac{2\lambda\mathrm{J}'}{df/dx}\left[\int_{\xi_1}^{\gamma}\sqrt{\left(\frac{1+\frac{1}{4}(\gamma/\xi)^3}{\gamma^3/\xi+4\xi^2}\right)}d\xi + \int_{\gamma}^{\eta_1}\sqrt{\left(\frac{1+4\gamma^6/\eta^6}{\eta^2+4\gamma^6/\eta^4}\right)}d\eta\right]$$

$$= \frac{2\lambda\mathrm{J}'}{df/dx}\left[\frac{1}{2}\int_{\xi_1}^{\gamma}\sqrt{\left(\frac{4\xi^3+\gamma^3}{4\xi^3+\gamma^3}\right)}\frac{d\xi}{\xi} + \int_{\gamma}^{\eta_1}\sqrt{\left(\frac{\eta^6+4\gamma^6}{\eta^6+4\gamma^6}\right)}\cdot\frac{d\eta}{\eta}\right]$$

$$= \frac{2\lambda\mathrm{J}'}{df/dx}\left[\frac{1}{2}\log\frac{\gamma}{\xi_1} + \log\frac{\eta_1}{\gamma}\right] = \frac{\lambda\mathrm{J}'}{df/dx}\log\frac{\eta_1^2}{\gamma\xi_1}.$$

The second integral is

$$\frac{2\lambda\mathrm{J}'}{df/dx}\int_{\theta}^{\frac{1}{2}\pi}\frac{d\theta}{\sin\theta} = \frac{2\lambda\mathrm{J}'}{df/dx}\left[\log\tan\frac{\theta}{2}\right]_{\theta}^{\frac{1}{2}\pi}$$

$$= \frac{\lambda\mathrm{J}'}{df/dx}\log\frac{1+\cos\theta}{1-\cos\theta} = \frac{\lambda\mathrm{J}'}{df/dx}\log\frac{(1+\cos\theta)^2}{\sin^2\theta},$$

and θ is nearly $= 0$. Therefore

$$\text{Second Integral} = \frac{\lambda\mathrm{J}'}{df/dx}\log\frac{4}{\eta_1^2},$$

and the second and third together

$$= \frac{\lambda\mathrm{J}'}{df/dx}\log\frac{4}{\gamma\xi_1}.$$

The first integral is

$$\lambda \int_0^{1-\xi_1} \frac{J}{J - x^2 J'} dx = \lambda + \lambda J' \int_0^{1-\xi_1} \frac{x^2\, dx}{J - x^2 J'}.$$

Now $J - x^2J' = x^2(1 - x^2)F(x)$, where $F(x)$ is finite for x between 0 and 1 and does not vanish. Hence

$$\text{Int.} = \lambda + \lambda J' \int_0^{1-\xi_1} \frac{dx}{(1-x)(1+x)Fx}$$

$$= \lambda + \lambda J' \int_0^{1-\xi_1} \left\{ - \frac{1}{df/dx} \frac{dx}{1-x} + \ldots \right\}$$

$$= \lambda + \frac{\lambda J'}{df/dx} \log \xi_1 + \text{finite quantity},$$

Therefore

$$s = \lambda + \frac{\lambda J'}{df/dx} \log \frac{4}{\gamma} + \text{finite}.$$

$$s = \lambda + \frac{\lambda J'}{\lambda \sin \lambda - 3J'} \log \frac{4a\sqrt{2}}{s} + \lambda J' \int_0^1 \left\{ \frac{x^2}{J - x^2 J'} + \frac{1}{\frac{df}{dx}(1-x)} \right\} dx,$$

where s is the distance from the pole of the point at which the stream sheet ψ cuts a line joining the pole to a point on the equator. The angular pitch is therefore infinite at the surface owing to the filaments being parallel to the equator at points close to the pole.

25. *Graphical Methods.*—The graphical construction indicated in § 17 affords a very convenient method of obtaining a general qualitative view of the properties of these aggregates. It serves also for a rough quantitative one, and at least gives for many determinations the rough starting point which is always the most troublesome obstacle in numerical approximations. It may be well, therefore, here, to collect and enlarge on what has gone before in this respect.

The first thing is to trace on a large scale the curve $y = J(\lambda)$ where λ is the abscissa. This is very easily done, since J is expressed in simple functions which are tabulated. The curve is drawn for the first three undulations in fig. (2), Plate (1). Now λ determines completely the *nature* of the aggregate (except its volume and its intensity). The point P on the J curve, corresponding to λ, we will call the parametral point. Draw through P a parabola touching the axis at the origin. For all points beyond the first few undulations a circle will suffice, or the curve drawn by a thin lath bent to touch the axis at O and to pass through P. If x denote r/a, λx will correspond to a point on the J curve between O and P. If P_1, P_2 denote the corresponding points on the J curve and the parabola, the value of ψ in the aggregate at the point $(r = xa, \theta)$ is given by $P_1P_2 \sin^2\theta$ (note P_2P_1 will be negative). The velocity of propagation will depend on the angle at which the parabola and curve intersect at P (see fig. 3, § 20). If they touch, the angle is zero, and the translation velocity zero. In fact the parameters of the points are the λ_2 values. We will call them the Q points. They are easily formed by fixing a lath at O and bending it to touch successive loops of the J curve. It is easy to do this correct to two decimal places, when numerical calculation will carry it to any degree of approximation desired.

The points where the J curve cuts the axis of x correspond to the λ_1 parameters. We will call them the R points.

Denote the points where the parabola through P cuts the J curve again by the letters p. These points give the sizes of the shells into which the aggregate divides. If ON be the abscissa of any such point, $\lambda x = \text{ON}$, and $r = \frac{\text{ON}}{\lambda} \,.\, a$ gives the radius of the corresponding interface between two shells. It is evident at once from the construction that the thicknesses of the shells, as we pass in or out, are alternately greater and less—that there are two categories, in one of which the thickness increases as we pass in, and an alternate series in which it decreases. There will be, however, some irregularity in the two inner components.

The position of the equatorial axes is determined by those abscissæ, for which the tangents to the J curve and the parabola are parallel. They are easily recognized by the eye, and thus a starting point for calculation is readily obtained. The difference of ordinates of these points (P_1P_2) is proportional to the secondary circulations of the corresponding shells. In fact, when multiplied by $\pi\mu/(\text{S}i\lambda - \sin\lambda)$, the products give the values of those constants. It is therefore clear from the figure that these circulations are in opposite directions alternately, and that we get two alternate series of ascending and descending values.

The function $\text{S}(\lambda) \equiv \text{S}i\lambda - \sin\lambda$ denotes the area between the J curve and the axis of x up to the point λ. It is clear, therefore, that it has its maximum values at the odd λ_1 points, and its minimum at the even ones.

The tracing of the current sheets is particularly easy from the fact that they are given by functions of the form

$$\psi = \text{F}(r) \,.\, \sin^2\theta.$$

Let

$$f(r) \equiv \text{J}\left(\frac{\lambda r}{a}\right) - \frac{r^2}{a^2}\,\text{J}(\lambda),$$

and let ψ_0 and r_0 denote values at the equatorial axis (*i.e.*, ψ_0 a numerical maximum). Then

$$\frac{\psi}{\psi_0} = \frac{f(r)}{f(r_0)}\sin^2\theta, \qquad \sin\theta = \sqrt{\frac{\psi}{\psi_0} \cdot \frac{f(r_0)}{f(r)}}.$$

On squared paper, draw a series of circles, radii sub-multiples of a, say at intervals of $\cdot 05a$ or $\cdot 1a$, also the circle $r = r_0$. This last circle has the property that all the current sheets cut it at right angles.

Let us trace first one sheet (say $\psi = \cdot 1\psi_0$). We do this by tabulating the values of $\sin\theta$ for values of r, corresponding to the series of circles drawn. Now mark on the bounding circle $(r = a)$ points whose abscissæ are those tabulated values (which is done at once on the squared paper). Mark the points where the radii vectores to these points cut the corresponding circles. Join these points by a continuous curve, and the shape of the particular ψ curve is obtained; call it the ψ_1 sheet. This first curve should be obtained with care and as much accuracy as possible. We may now proceed to draw from this as many of the other sheets as we please. Suppose we want to draw the curve $\psi = k \,.\, \psi_0$. We set a pair of proportional compasses (or any similar method) to the ratio $\sqrt{10k}$. Suppose the ψ_1 cuts any particular circle at P, set the short legs of the compasses to its abscissa. Turn it round and find the point on the *same* circle whose abscissa is the new value. Proceed thus with the other circles and the sheet is rapidly traced. Although this may appear cumbrous in stating, it is very expeditious in practice, and with a moderate amount of care very accurate.

Having traced the ψ curves, we may now easily trace the projections of the stream lines, for these are given by

$$\eta = \int \frac{dr}{\cos\theta} = \Sigma \text{PN} \quad \text{(see fig. 4).}$$

Fig. 4.

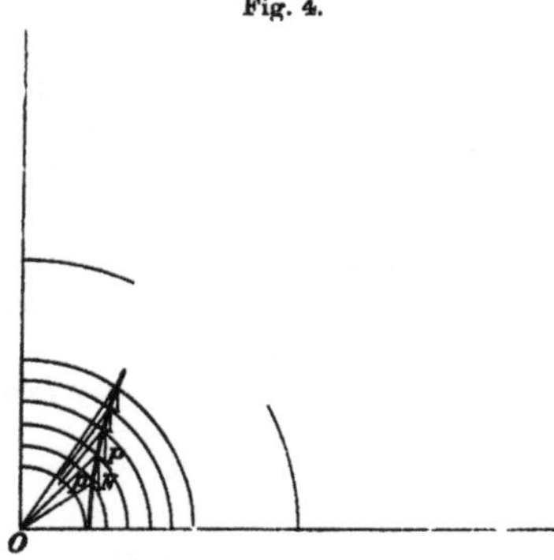

26. It will be interesting to go into further details for a few cases, and for this purpose we take the first two aggregates of the λ_2 and λ_1 families.

The distinguishing feature of the λ_2 types, is that the aggregates are at rest in the surrounding fluid. The distinguishing feature of the λ_1 types is that the vortex and stream lines are coincident.

λ_2 aggregates. Here

$$\text{U} = 0, \quad \text{M} = \frac{m\mu}{15} \frac{-\lambda \sin\lambda}{Si\lambda - \sin\lambda},$$

$$\text{E} = \tfrac{2}{45}\pi\mu^2 a \frac{\lambda^2 \sin^2\lambda}{(Si\lambda - \sin\lambda)^2} = \tfrac{7}{9} \cdot \frac{\lambda^2 \sin^2\lambda}{(Si\lambda - \sin\lambda)^2} \text{E}_0,$$

where E_0 is the energy of a Hill's aggregate of equal volume and circulation.

The first parameter is $\lambda_2^{(1)} = 5{\cdot}7637 = 330^\circ\ 14'$.
The equatorial axis has a radius $= {\cdot}5130a$,

$$M = {\cdot}0985m\mu$$
$$E = 1{\cdot}6979E_0$$
$$\nu = 2{\cdot}1020\mu.$$

Angular pitch of stream lines at surface $= 330^\circ\ 14'$.
" " " at axis $= 334^\circ\ 58'$.
" vortex lines at axis $= 267^\circ$.

The forms of the current sheets (ψ) are drawn in fig. 2, Plate 2. The projections of the stream lines in fig. 3, Plate 2. These latter were determined by the graphical method described above.

The second λ_2 parameter is

$$\lambda_2^{(2)} = 9{\cdot}0950 = 3\pi - 18^\circ\ 53'\ 40''.$$

The equatorial axes are given by

$$\lambda_2 x = 2{\cdot}6616 \text{ and } 6{\cdot}2718,$$
$$\text{or } r = {\cdot}2926a \text{ and } {\cdot}6896a.$$

Radius of internal nucleus $= {\cdot}4694a$.

$M = -\ {\cdot}1459m\mu$	$M_1 = {\cdot}08904m_1\mu_1$
$E = 3{\cdot}727E_0$	$M_2 = {\cdot}1890m_2\mu_2$
$\mu_1 = 1{\cdot}9403\mu$	$\nu_1 = 1{\cdot}1747\mu_1$
$\mu_2 = -\ {\cdot}9403\mu$	$\nu_2 = 3{\cdot}6415\mu_2$.
$\mu_1/\mu_2 = -\ 2{\cdot}063$	

Total angular pitch of stream lines outside $= 521^\circ\ 6'$.
" " " inside nucleus $= 244^\circ\ 37'$.
" " " outer shell $= 276^\circ\ 29'$.
Angular pitch of stream lines at 1st axis $= 284^\circ\ 21'$.
" " " 2nd axis $= 320^\circ\ 9'$.
" " vortex lines at 1st axis $= 308^\circ\ 48'$.
" " " 2nd axis $= 422^\circ\ 11'$.

The λ_1 aggregates. Here

$$U = \frac{\mu}{3a}\frac{-\sin\lambda}{Si\lambda - \sin\lambda} = \tfrac{5}{3}\frac{-\sin\lambda}{Si\lambda - \sin\lambda}U_0,$$

$$M = m\mu\frac{-\sin\lambda}{\lambda\,(Si\lambda - \sin\lambda)},$$

$$E = \tfrac{4}{9}\pi\mu^2 a\frac{\sin^2\lambda}{(Si\lambda - \sin\lambda)^2} = \tfrac{70}{9}\frac{\sin^2\lambda}{(Si\lambda - \sin\lambda)^2}E_0,$$

where U_0 is the velocity of translation of a HILL's aggregate of the same volume and circulation.

The stream pitch of these aggregates at the axis takes a very simple form, viz.,

$$\eta = \frac{\pi}{\sqrt{\left(\frac{1}{2} - \frac{1}{y^2}\right)}}.$$

The first λ_1 root is $\lambda_1^{(1)} = 4 \cdot 4935 = 257° \; 27' \; 30.''$

The equatorial axis has a radius $= \cdot 6106a$.

$$U = \cdot 6189 U_0, \qquad M = \cdot 0826 m\mu,$$
$$E = 10 \cdot 724 E_0, \qquad \nu = \cdot 8016 \mu.$$

Angular pitch of stream lines at surface $= 257° \; 27' \; 30''$.
" " " axis $= 297° \; 4'$.

The forms for the stream sheets ψ are shown in fig. 4, Plate 2. It is to be noticed that there is a considerable difference between the angular pitches outside and on the axis, whereas in the $\lambda_2^{(1)}$ aggregate they were very nearly the same.

The second λ_1 parameter is $\lambda_1^{(2)} = 7 \cdot 7253 = 450° - 7° \; 22' \; 27''$.

The equatorial axes are given by

$$y = \lambda_1 x = 2 \cdot 7437 \quad \text{and} \quad 6 \cdot 1168,$$

or,

$$r = \cdot 3552a \quad \text{and} \quad \cdot 7918a.$$

Radius of internal nucleus $= \cdot 5816a$.

$$\begin{array}{ll} U = -3 \cdot 094 U_0 & \mu_1/\mu_2 = -1 \cdot 2500 \\ M = -\ \cdot 2403 m\mu & M_1 = \cdot 0826 m_1\mu_1 \\ E = 26 \cdot 803 E_0 & M_2 = \cdot 1005 m_2\mu_2 \\ \mu_1 = 4 \cdot 9740 \mu & \nu_1 = 1 \cdot 2707 \mu_1 \\ \mu_2 = -3 \cdot 9740 \mu & \nu_2 = 1 \cdot 5135 \mu_2 \end{array}$$

Total angular pitch of outside $= 442° \; 37' \; 33''$.
" " on inside nucleus $= 257° \; 27' \; 30''$.
" " " outer shell $= 185° \; 10'$.
Angular pitch at inner axis $= 297° \; 4'$.
" " outer " $= 261° \; 39'$.

In all the λ_1 aggregates the expression for the angular pitch at an axis is

$$\frac{\pi}{\sqrt{\left(\frac{1}{2} - \frac{1}{y^2}\right)}}.$$

Hence, when λ_1 is large, the outer layers have their pitches at the axes about $\pi\sqrt{2} = 254° 31'$.

Fig. 5, Plate 2, shows the relative positions of the shells and axes for the λ_2^2 and λ_1^2 aggregates. The thin lines belong to the λ_2, the dotted to the λ_1. A, A are the position of the λ_2 equatorial axes. B, B those of the λ_1.

27. In the preceding investigation we find doublets, triplets, &c., naturally arising. We may have also built-up systems consisting of monads, dyads, &c., as in the cases developed in the previous section. Each element of a poly-ad may consist again of singlets, doublets, &c. I do not propose now to develop this theory of multiple combination to any length, but merely to draw attention to it, and to determine the necessary conditions for the case of a dyad only.

Referring to § 15, the general solution of the differential equation contains not only J functions, but also the functions $Y_2 = \frac{\cos y}{y} + \sin y$, which are suitable only for space not containing the origin. They are therefore suitable for any shell embracing an interior aggregate. In the shell the functions will be of the form AJ + BY, or as it may be written

$$\frac{\sin(\alpha + y)}{y} - \cos(\alpha + y).$$

It will be convenient to denote this by $f(\alpha, y)$.

Let now the radius of the interior aggregate be a, that of the exterior b. Let also λ, λ' denote the corresponding parameters.

Then we may write

$$\text{Inside} \quad \psi_1 = \mathrm{L}\left\{\mathrm{J}\left(\frac{\lambda r}{a}\right) - \frac{r^2}{a^2}\mathrm{J}\lambda\right\}\sin^2\theta \quad . \quad . \quad . \quad . \quad . \quad . \quad (39),$$

$$\text{Shell} \quad \psi_2 = q\mathrm{L}\left\{f\left(\alpha, \frac{\lambda' r}{b}\right) - \frac{r^2}{b^2}f(\alpha, \lambda')\right\}\sin^2\theta \quad . \quad . \quad . \quad (40),$$

$$\text{Outside} \quad \psi = -\pi\mathrm{V}\left(r^2 - \frac{b^3}{r}\right)\sin^2\theta.$$

At the interface $\psi_1 = \psi_2$ and $\psi_1 = 0$, therefore

$$f\left(\alpha, \frac{\lambda' a}{b}\right) - \frac{a^2}{b^2}f(\alpha, \lambda') = 0.$$

Write $a/b \equiv p$. This equation, when developed, gives

$$\tan\alpha = -\frac{\mathrm{J}(\lambda' p) - p^2\mathrm{J}(\lambda')}{\mathrm{Y}(\lambda' p) - p^2\mathrm{Y}(\lambda')} \quad . \quad . \quad . \quad . \quad . \quad . \quad . \quad . \quad (41).$$

Moreover, the tangential velocities must be the same. Hence, when $r = a$,

$$d\psi_1/dr = d\psi_2/dr.$$

Therefore

$$\frac{1}{a}\{\lambda \sin \lambda - 3J(\lambda)\} = \frac{q}{bp}\{-f(\alpha, \lambda'p) + \lambda'p \sin(\alpha + \lambda'p) - 2p^2 f(\alpha, \lambda')\}.$$

But

$$f(\alpha, \lambda'p) = p^2 f(\alpha, \lambda').$$

Hence

$$\lambda \sin \lambda - 3J(\lambda) = q\{\lambda'p \sin(\alpha + \lambda'p) - 3f(\alpha, \lambda'p)\} \quad . \quad . \quad . \quad (42).$$

So, also, making $d\psi_2/dr = d\psi/dr$ when $r = b$ we get

$$V = \frac{qL}{\pi b^2}\{f(\alpha\lambda') - \tfrac{1}{3}\lambda' \sin \lambda'\} \quad . \quad . \quad . \quad . \quad . \quad . \quad . \quad . \quad (43).$$

Equation (41) determines α; Equation (42) gives a relation between λ, λ', p, and q. We can therefore impress in general three further conditions. For instance, ratio of volumes, ratio of primary circulations, and ratio of secondary circulations.

There is a natural connection of the various singlets which go to make up an aggregate of the kind first discussed. At any interface all the differential co-efficients are continuous. In the polyad aggregates this is not so. Differential co-efficients beyond the first are not continuous. Monads, &c., which go to form them, are artificially combined. It is possible we may, on this basis, develop a theory of special aggregates which will unite with one another, or split up and be capable of uniting again in another manner. Some progress has been made with such a theory, but before an attempt is made to carry such a theory out it will be necessary to investigate the stability of the various systems. I hope soon to be able to take up this question.

[May 6, 1898.—By the permission of Professor HILL, to whose careful reading of the MS. I owe a great debt, I append an independent and very suggestive proof by him of the general theorem of gyrostatic vortices, based on the equations of motion.]

Take as co-ordinates r, θ, z.

$$\begin{pmatrix} x = r\cos\theta \\ y = r\sin\theta \end{pmatrix}$$

Let p be the pressure,
ρ the density,
V the potential of the impressed forces.

Let τ be the velocity increasing r,
σ be the velocity increasing θ,
w be the velocity increasing z.

Then the equations of motions are

$$\left(\frac{d}{dt} + \tau\frac{d}{dr} + \sigma\frac{d}{r\,d\theta} + w\frac{d}{dz}\right)\tau - \frac{\sigma^2}{r} = -\frac{d}{dr}\left(\frac{p}{\rho} + \mathrm{V}\right)$$

$$\left(\frac{d}{dt} + \tau\frac{d}{dr} + \sigma\frac{d}{r\,d\theta} + w\frac{d}{dz}\right)\sigma + \frac{\sigma\tau}{r} = -\frac{d}{r\,d\theta}\left(\frac{p}{\rho} + \mathrm{V}\right)$$

$$\left(\frac{d}{dt} + \tau\frac{d}{dr} + \sigma\frac{d}{r\,d\theta} + w\frac{d}{dz}\right)w \quad = -\frac{d}{dz}\left(\frac{p}{\rho} + \mathrm{V}\right)$$

$$\frac{d}{dr}(r\tau) + \frac{d\sigma}{d\theta} + \frac{d}{dz}(rw) = 0.$$

It is desired to find a solution in which all the quantities are independent of θ.

Therefore

$$\frac{d\tau}{d\theta} = 0, \quad \frac{d\sigma}{d\theta} = 0, \quad \frac{dw}{d\theta} = 0, \quad \frac{d}{d\theta}\left(\frac{p}{\rho} + \mathrm{V}\right) = 0.$$

The last gives

$$\left(\frac{d}{dt} + \tau\frac{d}{dr} + w\frac{d}{dz}\right)(r\sigma) = 0.$$

If therefore ψ be the equation of a surface always containing the same particles of fluid, it is possible to take

$$r\sigma = \frac{1}{2\pi}f(\psi).$$

Also

$$\frac{d}{dr}(r\tau) + \frac{d}{dz}(rw) = 0.$$

Let κ be the current function (which I distinguish throughout from ψ).

Therefore

$$\tau = -\frac{1}{2\pi r}\frac{d\kappa}{dz}, \qquad w = \frac{1}{2\pi r}\frac{d\kappa}{dr}.$$

Substituting in

$$\left(\frac{d}{dt} + \tau\frac{d}{dr} + w\frac{d}{dz}\right)\psi = 0,$$

it follows that

$$\frac{d\psi}{dt} - \frac{1}{2\pi r}\frac{d\kappa}{dz}\frac{d\psi}{dr} + \frac{1}{2\pi r}\frac{d\kappa}{dr}\frac{d\psi}{dz} = 0.$$

Now make the further supposition that the surfaces $\psi =$ const. move without alteration parallel to the axis of z with velocity $\dot{\mathrm{Z}}$.

Therefore

$$\frac{d\psi}{dt} = -\dot{\mathrm{Z}}\frac{d\psi}{dz}.$$

Therefore

$$\frac{d\psi}{dr}\frac{d}{dz}(\kappa - \pi r^2\dot{\mathrm{Z}}) - \frac{d\psi}{dz}\frac{d}{dr}(\kappa - \pi r^2\dot{\mathrm{Z}}) = 0.$$

Hence we can take

$$\kappa - \pi r^2 \dot{Z} = \psi.$$

Therefore

$$\tau = -\frac{1}{2\pi r}\frac{d\psi}{dz}, \qquad w = \frac{1}{2\pi r}\frac{d\psi}{dr} + \dot{Z}.$$

We have now

$$-\frac{d}{dr}\left(\frac{p}{\rho} + V + \frac{\tau^2 + w^2}{2}\right) = \frac{d\tau}{dt} + w\left(\frac{d\tau}{dz} - \frac{dw}{dr}\right) - \frac{\sigma^2}{r},$$

$$-\frac{d}{dz}\left(\frac{p}{\rho} + V + \frac{\tau^2 + w^2}{2}\right) = \frac{dw}{dt} - \tau\left(\frac{d\tau}{dz} - \frac{dw}{dr}\right).$$

Now

$$\frac{d\tau}{dt} = -\dot{Z}\frac{d\tau}{dz} = -\dot{Z}\frac{dw}{dr} - \dot{Z}\left(\frac{d\tau}{dz} - \frac{dw}{dr}\right),$$

$$\frac{dw}{dt} = -\dot{Z}\frac{dw}{dz}.$$

Therefore

$$-\frac{d}{dr}\left(\frac{p}{\rho} + V + \frac{\tau^2 + w^2}{2} - \dot{Z}w\right) = (w - \dot{Z})\left(\frac{d\tau}{dz} - \frac{dw}{dr}\right) - \frac{1}{4\pi^2 r^3}[f(\psi)]^2$$

$$= \frac{1}{2\pi r}\frac{d\psi}{dr}\left(\frac{d\tau}{dz} - \frac{dw}{dr}\right) - \frac{1}{4\pi^2 r^3}[f(\psi)]^2,$$

$$-\frac{d}{dz}\left(\frac{p}{\rho} + V + \frac{\tau^2 + w^2}{2} - \dot{Z}w\right) = -\tau\left(\frac{d\tau}{dz} - \frac{dw}{dr}\right)$$

$$= \frac{1}{2\pi r}\frac{d\psi}{dz}\left(\frac{d\tau}{dz} - \frac{dw}{dr}\right).$$

Hence

$$\frac{d}{dz}\left[\frac{d\psi}{dr}\left(\frac{\frac{d\tau}{dz} - \frac{dw}{dr}}{r}\right) - \frac{[f(\psi)]^2}{2\pi r^3}\right] = \frac{d}{dr}\left[\frac{d\psi}{dz}\left(\frac{\frac{d\tau}{dz} - \frac{dw}{dr}}{r}\right)\right].$$

Therefore

$$\frac{d\psi}{dr}\frac{d}{dz}\left(\frac{\frac{d\tau}{dz} - \frac{dw}{dr}}{r}\right) - \frac{d\psi}{dz}\left[\frac{d}{dr}\left(\frac{\frac{d\tau}{dz} - \frac{dw}{dr}}{r}\right) + \frac{f(\psi)f'(\psi)}{\pi r^3}\right] = 0.$$

Therefore

$$\frac{d\psi}{dr}\frac{d}{dz}\left[\frac{1}{r}\left(\frac{d\tau}{dz} - \frac{dw}{dr}\right) - \frac{f(\psi)f'(\psi)}{2\pi r^2}\right] - \frac{d\psi}{dz}\frac{d}{dr}\left[\frac{1}{r}\left(\frac{d\tau}{dz} - \frac{dw}{dr}\right) - \frac{f(\psi)f'(\psi)}{2\pi r^2}\right] = 0.$$

Therefore

$$\frac{1}{r}\left(\frac{d\tau}{dz} - \frac{dw}{dr}\right) = \frac{f(\psi)f'(\psi)}{2\pi r^2} + F'(\psi),$$

$$\frac{d^2\psi}{dr^2} - \frac{1}{r}\frac{d\psi}{dr} + \frac{d^2\psi}{dz^2} = -2\pi r^2 F'(\psi) - f(\psi)f'(\psi).$$

Therefore

$$-\frac{d}{dr}\left(\frac{p}{\rho}+\mathrm{V}+\frac{\tau^2+w^2}{2}-\dot{\mathrm{Z}}w\right)=\frac{1}{2\pi r}\frac{d\psi}{dr}\left[\frac{f(\psi)f'(\psi)}{2\pi r}+r\mathrm{F}'(\psi)\right]-\frac{[f(\psi)]^2}{4\pi^2r^3}$$

$$=\frac{d}{dr}\left[\frac{[f(\psi)]^2}{8\pi^2r^2}+\frac{\mathrm{F}(\psi)}{2\pi}\right],$$

and

$$-\frac{d}{dz}\left(\frac{p}{\rho}+\mathrm{V}+\frac{\tau^2+w^2}{2}-\dot{\mathrm{Z}}w\right)=\frac{1}{2\pi r}\frac{d\psi}{dz}\left[\frac{f(\psi)f'(\psi)}{2\pi r}+r\mathrm{F}'(\psi)\right]$$

$$=\frac{d}{dz}\left[\frac{[f(\psi)]^2}{8\pi^2r^2}+\frac{\mathrm{F}(\psi)}{2\pi}\right].$$

Therefore

$$\frac{p}{\rho}+\mathrm{V}+\frac{\tau^2+w^2}{2}-\dot{\mathrm{Z}}w+\frac{[f(\psi)]^2}{8\pi^2r^2}+\frac{\mathrm{F}(\psi)}{2\pi}=\text{arbitrary function of } t.$$

Therefore

$$\frac{p}{\rho}+\mathrm{V}+\tfrac{1}{2}[\tau^2+\sigma^2+(w-\dot{\mathrm{Z}})^2]+\frac{1}{2\pi}\mathrm{F}(\psi)=\text{arbitrary function of } t.$$

This arbitrary function of t is in this paper always a constant.

The last equation, together with the following, are the important equations :

$$\kappa=\psi+\pi\dot{\mathrm{Z}}r^2, \qquad \sigma=\frac{1}{2\pi r}f(\psi),$$

$$\tau=-\frac{1}{2\pi r}\frac{d\kappa}{dz}=-\frac{1}{2\pi r}\frac{d\psi}{dz},$$

$$w=\frac{1}{2\pi r}\frac{d\kappa}{dr}=\frac{1}{2\pi r}\frac{d\psi}{dr}+\dot{\mathrm{Z}},$$

$$\frac{d\tau}{dz}-\frac{dw}{dr}=\frac{1}{2\pi r}f(\psi)f'(\psi)+r\mathrm{F}'(\psi),$$

$$\frac{d^2\psi}{dr^2}-\frac{1}{r}\frac{d\psi}{dr}+\frac{d^2\psi}{dz^2}=-f(\psi)f'(\psi)-2\pi r^2\mathrm{F}'(\psi).$$

Whenever the conditions for the continuity of the τ and w components of the velocity have been satisfied at a separating surface whose equation is $\psi=$ const., then if the irrotational motion outside the surface have $\sigma=0$, we must have $\sigma=0$ when ψ is equal to the parameter of separating surface, if there is to be no slip there.

Therefore $f(\psi)=0$, when ψ is equal to the parameter of separating surface.

This is the case in the Third Section of the Paper.]

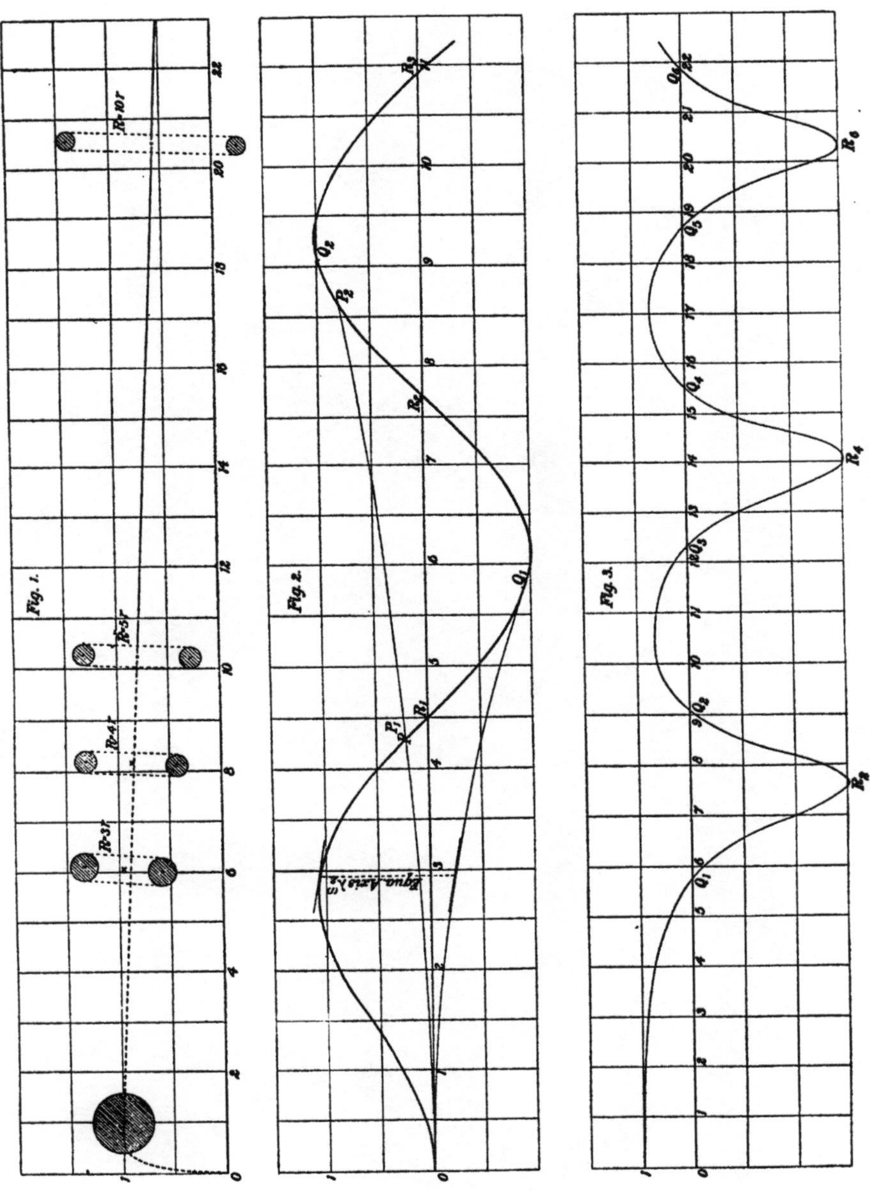
Fig. 1.
Fig. 2.
Fig. 3.

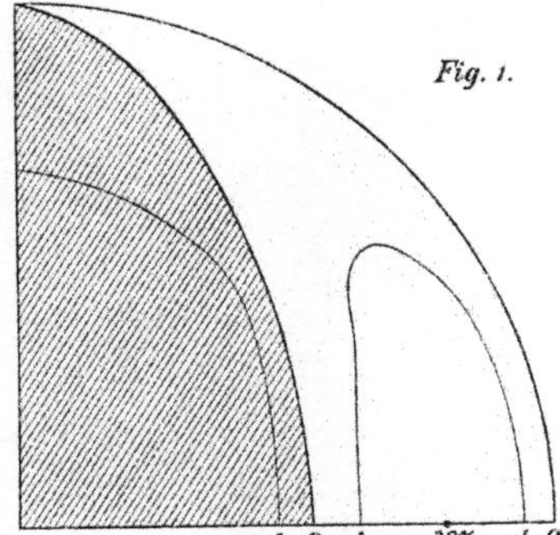
Fig. 1.
-·1 0 ·1 ·337 ·1 0

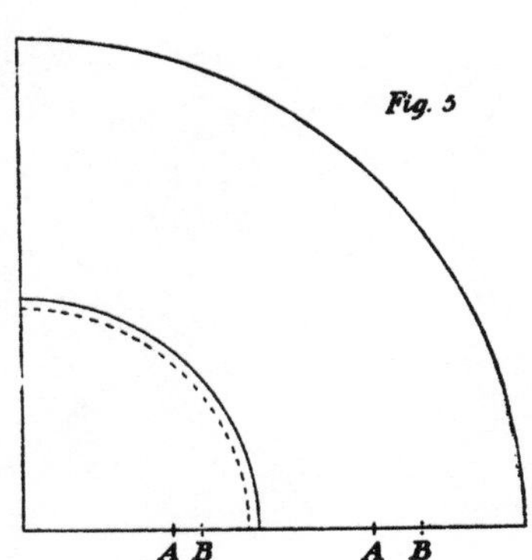
Fig. 5
A B A B

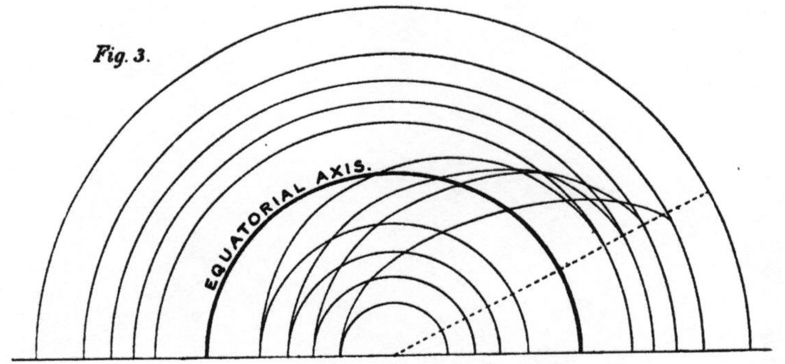
Fig. 3.
EQUATORIAL AXIS.

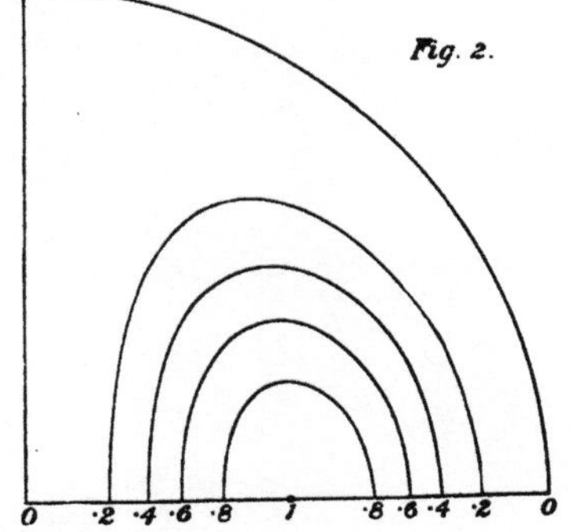
Fig. 2.
0 ·2 ·4 ·6 ·8 1 ·8 ·6 ·4 ·2 0

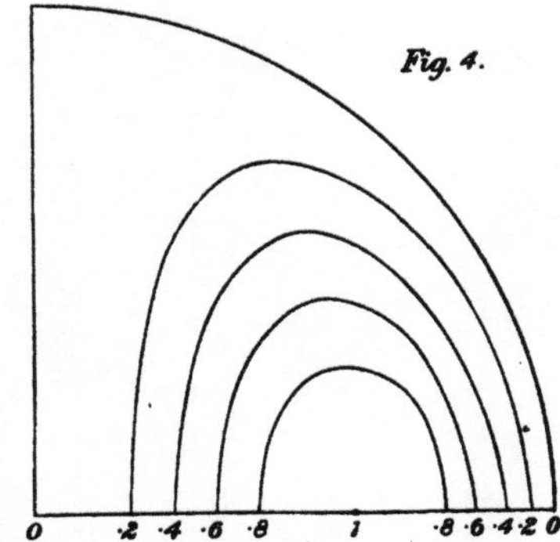
Fig. 4.
0 ·2 ·4 ·6 ·8 1 ·8 ·6 ·4 ·2 0

INDEX SLIP.

Cases of Normal Distribution and
ion.

llow of Trinity College, Cambridge.

R. FORSYTH, *F.R.S.*

of the 'Normal Curve' of Statistics," &c.)
7,—Revised July 15, 1898.

PAGE.

15.12.98

III. *On the Application of the Theory of Error to Cases of Normal Distribution and Normal Correlation.*

By W. F. Sheppard, *M.A., LL.M., Formerly Fellow of Trinity College, Cambridge.*

Communicated by Professor A. R. Forsyth, *F.R.S.*

Received (under the title "On the Geometrical Treatment of the 'Normal Curve' of Statistics," &c.) October 9, 1897,—Read November 25, 1897,—Revised July 15, 1898.

Contents.

15.12.98

Introductory.

In his 'Lettres sur la Théorie des Probabilités' (1846), QUETELET has shown that in certain anthropometrical statistics, *e.g.*, in statistics of height or of chest-measurement, the curve of frequency is approximately of the same form as the curve known to mathematicians as the "curve of error," but better described for statistical purposes as the *normal curve.* A similar conclusion has been arrived at by later observers with regard to a large number of biological measurements. The general similarity thus established has been extended, primarily by Mr. FRANCIS GALTON, to certain cases of statistical correlation of two or more attributes. It has been found in these cases that not only are the curves of frequency of the separate attributes approximately normal curves, but the frequencies of joint occurrence of different measures of these attributes follow (approximately) a simple law, corresponding to the law of correlation of errors of observation.

Since we can never observe more than a finite number of individuals, it is impossible to decide with absolute certainty as to the existence, in any particular case, of this (or any other) law of distribution or correlation. But if the number of observed individuals is large, and if they are obtained by random selection from a "community" comprising (practically) an indefinitely great number of individuals, the theory of error provides us with a test for deciding whether any particular law, suggested by the given observations, may be regarded as holding for the original community.

The main object of the present memoir is to obtain formulæ for testing the existence, in any particular case, of the *normal distribution* and *normal correlation* described above. As the treatment of multiple correlation presents some difficulty, I have restricted myself to the cases of one attribute, supposed to be normally distributed, and of two attributes, supposed to be normally correlated. Where the hypothesis of normal distribution or of normal correlation may be regarded as established, there are different methods of treating the statistical data; and these may lead to different results. I have therefore given formulæ for comparing the relative accuracy of different methods of calculating the frequency-constants which are required.

The application of the formulæ to actual cases is postponed until certain tables are completed. In the absence of these tables, KRAMP's and ENCKE's tables (printed at the end of DE MORGAN's article on the "Theory of Probabilities" in the 'Encyclopædia Metropolitana') may be used for cases of a single attribute. For cases of correlated attributes, I have given two methods of making a rough calculation of the "theoretical" distribution, for comparison with the "observed" distribution. These methods depend on theorems which can be conveniently expressed in a geometrical form. As the normal curve lends itself to geometrical treatment, and as the fundamental formulæ in the theory of error can be obtained by the use of ordinary algebra, I have attempted to make the memoir complete in itself by starting with a simple definition of the normal curve, and adopting GALTON's definition of normal correlation; and by deducing the necessary theorems without the direct use of the differential or integral calculus.

The normal curve may be defined in various ways, *e.g.* :—

(1.) *Functional Equation*, $z = f(x^2)$, where $f(x^2) \times f(y^2) = f(x^2 + y^2)$.

(2.) *Ordinary Cartesian Equation*, $z \propto e^{-\frac{1}{2}(x^2/a^2)}$.

(3.) *Differential Equation*, $a^2 (dz/dx) + xz = 0$.

(4.) *Geometrical Equation*, abscissa × sub-tangent = constant. This follows at once from (3); for if O is the foot of the central ordinate, and if MP is any other ordinate, and the tangent at P meets OM in T, then sub-tangent MT $= - z\,dx/dz$.

(5.) *Statistical Equation*, $\lambda_{k+2} = (k+1)\lambda_2\lambda_k$, where λ_k denotes the mean kth power of the deviation from the mean in a distribution whose curve of frequency is a normal curve; k being any positive integer. This relation follows from (3). Since, by the definition, $\lambda_1 = 0$, it gives λ_k in terms of λ_2 for all positive integral values of k; and it may therefore be regarded as the equation to the curve, the position of the central ordinate being arbitrary.

Of these different equations the first is in some respects the most important, as it is the direct expression of the relation on which the special property of normal distributions depends; the property, that is to say, that if the measures of a number of independent attributes are normally distributed, any linear function of these measures is also normally distributed. The second equation is, of course, essential for any numerical calculations. The last two, however, have certain conveniences when an elementary investigation is desired. I have therefore adopted the *geometrical definition* of the curve, and have deduced the statistical equation; and then have used either or both of these as occasion might require.

The memoir is divided into four parts. Part I. deals with elementary theorems; most of these are well known, but it is convenient to have them collected, and established by comparatively simple methods.* Part II. contains the investigation of the principal formulæ in the theory of error as applied to numerical statistics. In Part III. these formulæ are applied to cases of normal distribution. Part IV. deals with normal correlation, and is subdivided into two portions. The first consists of a discussion of the more important phenomena which occur when two attributes are normally correlated; while the second contains the applications of the theory of error. Some of the formulæ given in Parts III. and IV. have already been obtained by Professor KARL PEARSON, but by a different method.

PART I.—GENERAL PROPERTIES OF THE NORMAL CURVE AND OF NORMAL DISTRIBUTIONS.

The Normal Curve.

§ 1. *Definition of Normal Curve.*—Let O be a fixed point in a straight line X′OX, and let a point P move so that, if MP is the ordinate to P from X′OX, and PT the tangent at P, intersecting X′OX in T, the rectangle OM.MT is constant and $= a^2$. Then the path of P is a *normal curve*.

Let OZ be drawn at right angles to X′OX, intersecting the curve in H, and let points A′ and A be taken in X′OX, such that A′O = OA = a. Then OZ will be

* It will be seen that some of the proofs are only expressions, in geometrical form, of familiar methods of differentiation or integration.

called the *median* of the curve, X′OX the *base*, OH the *central ordinate*, and A′A the *parameter*.

The curve is obviously symmetrical about the median, and asymptotic to the base in both directions.

The area bounded by the curve and the base will be called a *normal figure*.

§ 2. *Formation of Family of Curves by Projection.*—Let a new curve be formed by orthogonal projection of a normal curve with regard to the base in any ratio. Let MP and NQ be ordinates to the original curve, and MP′ and NQ′ the corresponding ordinates to the new curve (fig. 1). Then MP : MP′ :: NQ : NQ′. Hence PQ and P′Q′ will intersect on the base. Let N move up to and coincide with M. Then PQ and P′Q′ become the tangents at P and P′ to the two curves, and therefore these tangents meet the base in the same point T. Hence for the second curve we have also $OM.MT = OA^2$, and therefore this is also a normal curve of parameter A′A.

Similarly, if the curve is projected with regard to OZ in the ratio $a : b$, the new curve will be a normal curve of parameter $2b$, having the same median.

Fig. 1.

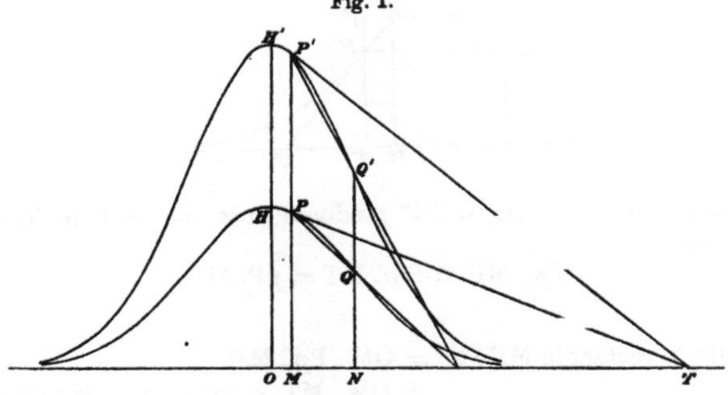

§ 3. *Limitation to Curves so obtained.*—Thus, by projection of a single normal curve with respect to the base and the median, we can get an indefinite number of normal curves of different parameters and different central ordinates. Conversely, if S and S′ are two normal curves placed so as to have the same base and the same median, either can be got from the other by projection. Let the parameters be $2a$ and $2b$ respectively. Project S into a curve S″ of parameter $2b$, and let Σ denote the family of projections of S″ with regard to the base. Then the tangent at each point of S′ coincides with the tangent to the particular curve of Σ which passes through this point. Hence S′ is one of the curves Σ, or else is the envelope of these curves. But the curves have no envelope at a finite distance. Hence S′ is a projection of S″.

§ 4. *Standard Normal Curve.*—It is, therefore, convenient to take a standard normal curve, and to consider all other normal curves as obtained from it by projection.

For the standard form we take the curve whose semi-parameter is unity, and area unity. The central ordinate of this curve will for the present be denoted by C; we shall show later that $C = 1/\sqrt{2\pi}$. It is clear that if A is the area of a curve of parameter $2a$, its central ordinate is CA/a.

The curve may be traced by means of Table I. (p. 153). The second column of that table gives the ordinate of the standard curve in terms of the abscissa; the third gives its ratio to the central ordinate. Table II. (p. 155) is formed by inverting this latter table; it gives the abscissa in terms of the ratio of the ordinate to the central ordinate.

§ 5. *Moment-formulæ.*—Let MP, M′P′, be any two consecutive ordinates to a normal curve whose parameter is $2a$. Draw Pm and P′m' perpendicular to the central ordinate OH, and let p and p' be the intersections of MP, m'P′ and of M′P′,

Fig. 2.

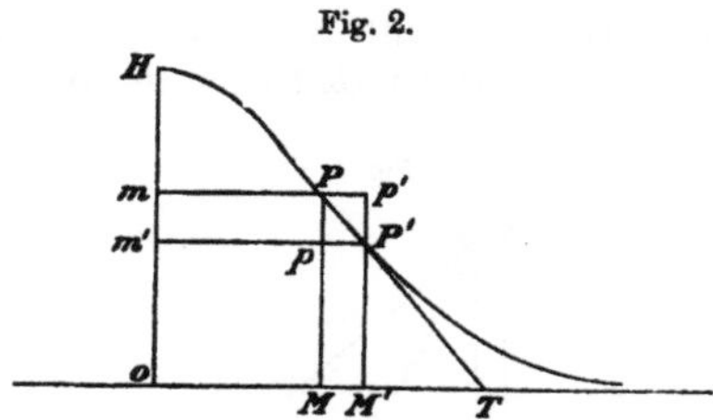

mP respectively (fig. 2). Then, if PP′ produced cuts the base in T, we have, by similar triangles,

$$Pp' . MP = P'p' . MT = pP . MT.$$

Hence

$$\text{(1.)}\quad OM \times \text{rectangle } MPp'M' = OM . Pp' . MP = OM . MT \times pP = OM . MT\,(MP - M'P');$$

$$\text{(2.)}\quad OM^2 \times \text{rectangle } MPp'M' = OM . MT \times m'p . pP = OM . MT \times \text{rectangle } m'pPm;$$

$$\text{(3.)}\quad OM^{k+2} \times \text{rectangle } MPp'M' = OM . MT \times mP^k \times \text{rectangle } m'pPm.$$

The kth moment of the rectangle $m'p$Pm about OH is $\frac{1}{k+1} . mP^k \times m'pPm$. Also when MM′ becomes indefinitely small, $OM . MT = a^2$. Hence, by summation, we see that

(i.) If MP and NQ are any two ordinates, the moment of the area MPQN about OH is $a^2\,(MP - NQ)$;

(ii$a$.) If Pm and Qn are the perpendiculars from P and Q on OH, the second moment of MPQN about OH is $a^2 \times$ area $nQPm$;

(ii*b*.) For the complete normal figure, the mean square of deviation from the mean is a^2;

(iii.) If λ_k denote the mean kth power of the deviation from the mean,

$$\lambda_{k+2} = (k + 1) a^2 \lambda_k = (k + 1) \lambda_2 \lambda_k,$$

which is the statistical equation to the curve.

This equation gives

$$\left.\begin{array}{l}\lambda_{2s-1} = 0 \\ \lambda_{2s} = (2s - 1)(2s - 3) \ldots 1 \, . \, \lambda_2^s = \dfrac{\lfloor 2s}{2^s \lfloor s} \lambda_2^s\end{array}\right\}.$$

The Surface of Revolution of the Normal Curve.

§ 6. *Projective Solids and Surfaces.*—Let Σ be a surface whose equation referred to three rectangular axes OX, OY, OZ, is of the form $z = \phi(x) . \phi(y)$. Then if we take sections of Σ by a system of planes parallel to OZX, and project these sections on OZX, we obtain a system of curves which are the orthogonal projections of one another with regard to their common base OX. Similarly if we take sections by planes parallel to OZY. On this account it is convenient to call such a surface a *projective surface.* If the surface is terminated in all directions by the base-plane OXY, the volume included between this plane and the surface will be called a *projective solid.*

For the geometrical definition of a projective solid it is sufficient that the solid should be bounded by a plane base OXY, and that two lines OX, OY in this plane, at right angles to one another and to a line OZ, should be related to the solid in such a way that the sections of the surface by planes parallel to OZX, when projected on OZX, form a system of curves in orthogonal projection. If this is the case, it follows at once, from the elementary properties of projection, that the same property holds for sections by planes parallel to OZY.

The sections of the solid by the two sets of planes parallel to OZX and to OZY will be called *principal sections.*

The following properties of a projective solid are easily obtained from the geometrical definition.

(i.) Let WR and MP be any two ordinates, and let the other ordinates in which the principal sections through WR and MP intersect be NQ and *nq*. Then WR.MP = NQ.*nq*.

(ii.) In one of the principal sections through an ordinate WR, take any two ordinates NQ and N′Q′; and in the other take any two ordinates *nq* and *n′q′* (fig. 3). Draw the principal sections through these ordinates, and let them enclose (with the base and the upper surface) a volume V. Then WR.V = area NQQ′N′ × area *nqq′n′*.

(iii.) From (ii.) it follows that if we fix a principal section S, and take variable ordinates NQ and N′Q′, the volume of the solid bounded by the other principal sections through NQ and N′Q′ is proportional to the area NQQ′N′.

(iv.) From (ii.) it also follows that if V is the whole volume of the solid, WR any ordinate, and A and A′ the areas of the principal sections through WR, then WR.V = A.A′.

Fig. 3.

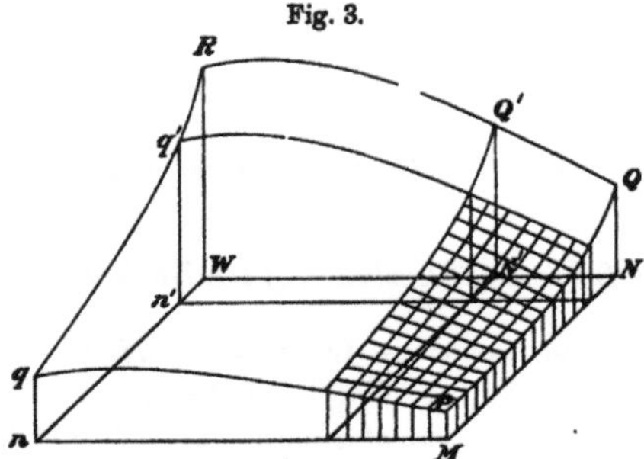

(v.) Let OH be the ordinate passing through the centre of gravity of the solid, and let S and S′ be the principal sections through OH. Then the central ordinates of all sections parallel to S (*i.e.*, the ordinates through their respective centres of gravity) lie in S′, and the central ordinates of all sections parallel to S′ lie in S.

§ 7. *Normal Solid and Normal Surface.*—Let the half of a normal figure of parameter A′A = $2a$, lying on one side of the central ordinate OH, be rotated about this ordinate through four right angles. The solid so formed will be called a *normal solid*, and its surface will be called a *normal surface.* The plane traced out by the base will be called the *base-plane.* A section of the solid by a plane perpendicular to the base-plane will be called a *vertical* section.

§ 8. *Normal Solid is Projective Solid.*—Let S be any vertical section of the solid, and MP any ordinate in this section. Draw ON perpendicular to the plane of the section, and let NQ be the ordinate at N. Let the tangents at P to the section S, and to the central section through MP (*i.e.*, the section through MP and the axis), cut the base-plane in T and T′ respectively (fig. 4).

Since PT and PT′ are tangents to sections through P, the plane PTT′ is the tangent plane to the solid at P. But the solid is a solid of revolution, and therefore this plane is perpendicular to the plane OMP. The base-plane is also perpendicular to the plane OMP, and therefore the intersection TT′ is perpendicular to this latter plane. Hence OT′T is a right angle, and therefore a circle goes round ONT′T, so that NM.MT = OM.MT′.

But the section by the plane OMP is a normal figure of parameter $2a$, and therefore OM.MT′ = a^2. Hence also NM.MT = a^2; *i.e.*, the section S is a normal figure of parameter $2a$, having NQ for its central ordinate.

Fig. 4.

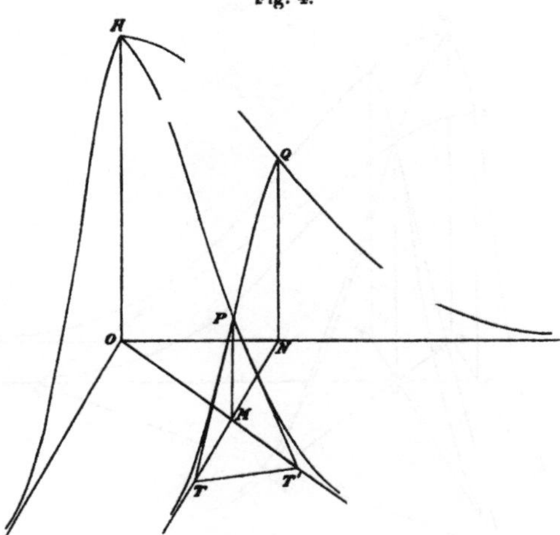

Thus every vertical section of the solid is a normal figure of the same parameter, having its central ordinate in the plane through the axis at right angles to the plane of the section.

It follows from § 3 that the solid is a projective solid, any two vertical sections at right angles to one another being regarded as principal sections.

§ 9. *Converse Propositions.*—There are two converse propositions.

(i.) If two principal sections of a projective solid are normal figures of equal parameter, the solid is one of revolution.

Let this parameter be $2a$. From § 2 it follows that every principal section is a normal figure of parameter $2a$. The solid will obviously have a maximum ordinate OH; and each of the two principal sections through OH will contain the central ordinates of all sections by planes perpendicular to it. Take any other section through OH; and let MP be any ordinate in this section. Draw planes through MP cutting the principal sections through OH in ordinates NQ and nq. Then the sections NQPM and nqPM are normal figures of parameter $2a$, having NQ and nq for their central ordinates. Let the tangents to these sections and to the section OHPM cut the respective bases in T, t, T′ (fig. 5). Then PT, PT′, Pt all lie in the tangent plane to the surface at P, and therefore TT′t is a straight line. Also $\mathrm{NM.MT} = a^2 = n\mathrm{M.M}t$, so that ON : NM :: TM : Mt. Hence the triangles ONM, TMt are similar, and angle MTt = angle NOM; and therefore a circle goes round NOTT′. Hence $\mathrm{OM.MT'} = \mathrm{NM.MT} = a^2$, and therefore the section OHPM is a normal figure of parameter $2a$, having OH for its central ordinate. This is true for every section through OH, and therefore the solid is one of revolution.

Fig. 5.

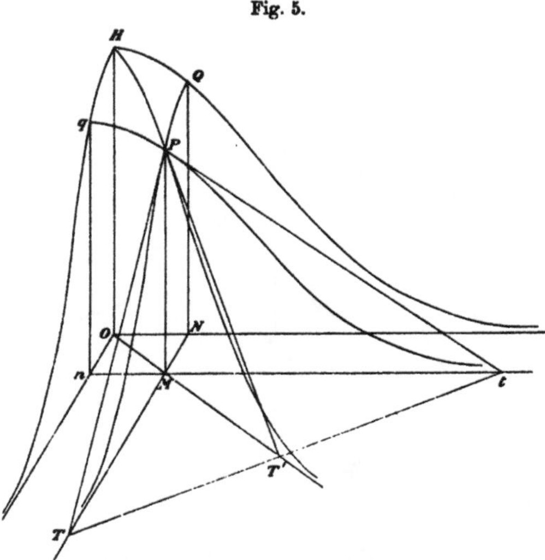

(ii.) If a solid of revolution is also a projective solid, the generating figure is a normal figure.

Let OH be the central ordinate. Then every vertical section is symmetrical about the plane through OH perpendicular to it, and any two vertical sections, if arranged so as to have their central ordinates coincident, will be interconvertible by projection. Let S be any section through OH, and let NQ and N′Q′ be any two ordinates in this section, ON being greater than ON′. Let the tangents to S at Q and Q′ cut ON′ N in T and T′.

Describe a circle in the base-plane on ON as diameter, and draw the chord NM = ON′. Draw the ordinate MP, and let the tangent at Q to the section MPQN cut MN produced in R (fig. 6). Then MP is the central ordinate of the section MPQN; and therefore, since this section and the section OHQ′N′ are interconvertible by projection, it follows that NR = N′T′.

Since QR and QT are tangents to sections through NQ, QRT is the tangent plane at Q. The solid being a solid of revolution about OH, this tangent plane must be perpendicular to the plane OQT. The base-plane is also perpendicular to the plane OQT, and therefore TR, which is the line of intersection of the tangent plane and the base-plane, is perpendicular to the plane OQT. Hence OTR is a right angle, and therefore a circle goes round OMTR, so that ON . NT = MN . NR = ON′ . N′T′. In other words, the rectangle ON . NT is constant for different positions of N, and therefore the central section is a normal figure.

Fig. 6.

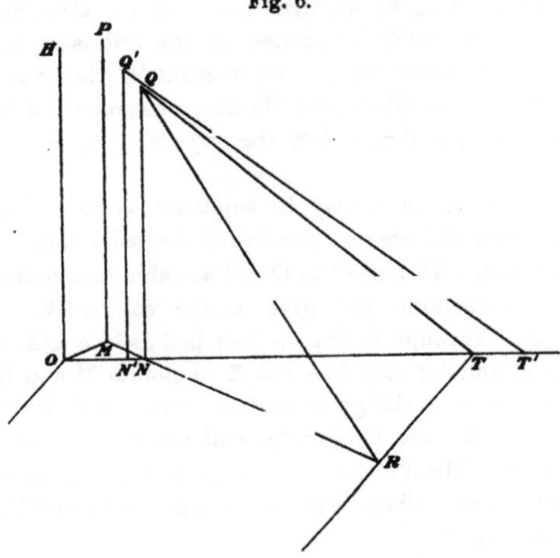

§ 10. *Value of* C.—Let A and A′ be the areas of two sections through OH at right angles to one another; and let V be the whole volume of the solid. Then, since the solid is a projective solid, $\mathrm{OH}.\mathrm{V} = \mathrm{A}.\mathrm{A}' = \mathrm{A}^2$ (§ 6 (iv.)); and, since it is a solid of revolution, $\mathrm{V} = 2\pi a^2.\mathrm{OH}$ (§ 5 (i.), and GULDINUS' theorem). But $\mathrm{OH} = \mathrm{CA}/a$ (§ 4). Hence $\mathrm{C} = 1/\sqrt{2\pi}$.

It is convenient to consider the solid as obtained from a standard form by an orthogonal and an axial* projection. As the standard solid we shall take the solid whose volume is unity and whose vertical sections are normal figures of semi-parameter unity. The central ordinate of this solid is $1/2\pi$.

§ 11. *Representation of Segment of Normal Solid by an Area.*—Let Σ be any closed curve in the base of a normal solid, whose principal ordinate is OH, and whose parameter is $2a$; and let V be the portion of the solid which lies above Σ, *i.e.*, which is bounded by Σ, by the surface of the solid, and by a cylinder K of which Σ is a normal section. We require a method of determining the volume V.

Let Σ′ be the upper boundary of V, *i.e.*, the area cut out of the surface of the normal solid by the cylinder K. Describe a circular cylinder of radius b, and of height OH, having OH as axis; and project Σ′ on this cylinder by lines perpendicular to OH. The projection will be a closed curve σ. Now the volume V can be divided into elements by a series of planes through OH at indefinitely small angular distances from one another. Let Π and Π′ be two consecutive planes of the system,

* By an axial projection of a surface or a solid with regard to a straight line is meant the surface or solid obtained by projecting every point orthogonally with regard to this straight line in a definite ratio.

the angle between them being θ; let them cut σ in the straight lines pq and $p'q'$, and let Π cut V in the area MPQN, bounded by the ordinates MP and NQ. Then $pq =$ NQ $\backsim$ MP; and therefore, by § 5, the moment of the area MPQN about OH is equal to $a^2 . pq$. Hence, by GULDINUS' theorem, the portion of V included between Π and Π' is equal to $a^2 . pq . \theta = a^2/b \times$ area $pq\,q'p'$. By summation, we see that $\mathrm{V} = a^2/b \times$ area σ.

The cylinder, with the curve σ, may be supposed to be unwrapped on a plane. Hence when we are given the central section of the solid, and a plan showing the form of Σ and its position with regard to O, we are able to construct, by geometrical methods, a curve whose area will give us the volume V. Take a standard line OX on the plan. Through O draw a line inclined to OX at an angle whose circular measure is α, and let this line cut Σ in points M and N. Take abscissæ OM and ON along the base of the given central section, and draw the ordinates MP and NQ. On a line O'X' take $\mathrm{O'L'} = b\alpha$, and draw an ordinate $\mathrm{L'}qp$ such that $\mathrm{L'}p = \mathrm{MP}$, $\mathrm{L'}q = \mathrm{NQ}$. The different points p and q corresponding to different values of α will form a curve, whose area can be measured; and this area, multiplied by a^2/b, is the volume required.*

If the curve Σ encloses the base of the principal ordinate OH, the continuity of the boundary of σ will be broken when the cylinder is unwrapped. The locus of the points p is then the top of the rectangle representing the complete cylinder, and the area to be taken is the area between this, the sides of the rectangle, and the curve which is the locus of q. Similarly, if any portion of the boundary of Σ is at infinity, the corresponding part of the boundary of σ will lie along the base of the rectangle representing the complete cylinder.

The area σ is unaltered by projecting it at right angles to O'X' in the ratio $1 : \lambda$, and parallel to O'X' in the ratio $\lambda : 1$. Thus we shall have $\mathrm{L'}p = \lambda . \mathrm{MP}$, $\mathrm{L'}q = \lambda . \mathrm{NQ}$, the point L' being taken so that $\mathrm{O'L'} = b\alpha/\lambda$. When the solid is the standard solid, it is convenient to take $b = a\ (= 1)$, and $\lambda = 2\pi$; the unwrapped cylinder then becomes a square whose base is unity and height unity; and the values of $\mathrm{L'}p$ and $\mathrm{L'}q$ are given by the third column of Table I. (p. 153).

If, for example, we divide the standard solid into twenty equal portions by nineteen parallel vertical planes, and if the cylinder is supposed to be divided along one of the lines in which it is cut by the central plane, and then unwrapped, and projected vertically in the ratio of $1 : 2\pi$ and horizontally in the ratio of $2\pi : 1$, we

* Generally, let V be a portion cut out of a solid of revolution by a closed cylinder K, whose generating lines are parallel to the axis of revolution. Let F denote the section of the solid by a plane through the axis of revolution; and let S be a curve lying in the plane of F and related to it in such a way that any ordinate MP (drawn to S from a base at right angles to the axis of revolution) is proportional to the moment, about the axis, of that portion of F which lies beyond MP. Then, if F is given geometrically, and if the section of the cylinder K and its position with regard to the axis are given, we can construct a figure whose area will be proportional to the volume V.

shall obtain the figure shown in fig. 7. The figure consists of two similar portions, each of which is divided into ten equal parts by nine curves; each curve touching the corresponding half of the base at its extremities, and being symmetrical about

Fig. 7.

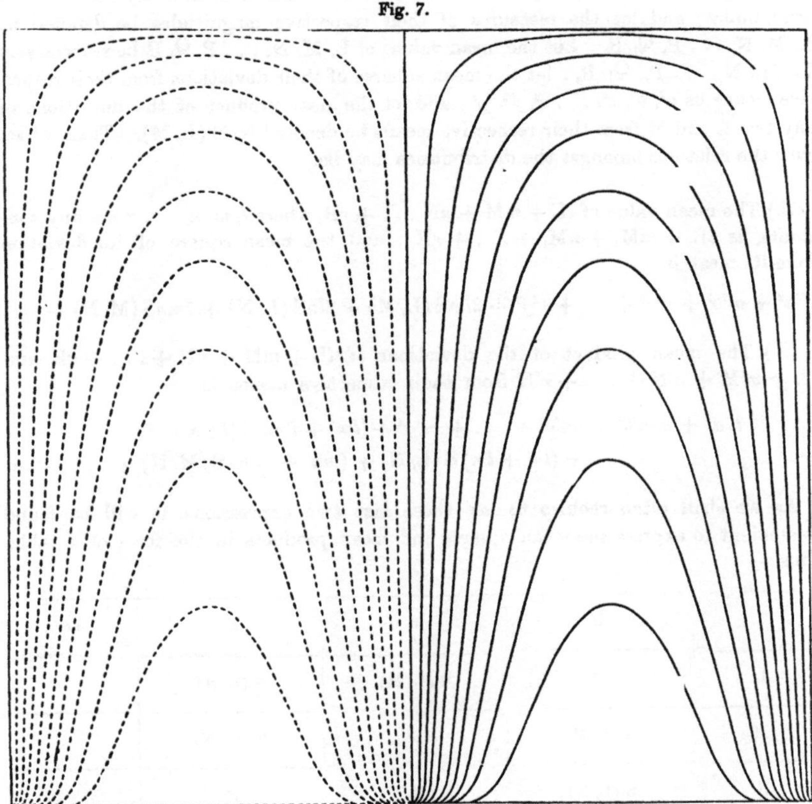

its central ordinate. The curves may be traced by means of Tables III. and IV. (pp. 156–158); Table III. gives the ordinates in terms of the abscissa, measured from the extremity of the base of the figure; and Table IV. is a converse table, giving the abscissæ of the different curves in terms of the ordinate.*

* The values in Table IV. were calculated by means of CALLET's tables, in which the quadrant is divided centesimally.

General Theorems Relating to Normal Distributions.

§ 12. *Mean Squares and Mean Products of Composite Measures.*—Let A, B, C, ... E, F, G be a number of attributes, all of which exist in every member of a community; and let the measures of their respective magnitudes be denoted by L, M, N, ... P, Q, R. Let the mean values of L, M, N, ... P, Q, R be respectively L_1, M_1, N_1, ... P_1, Q_1, R_1; let the mean squares of their deviations from their respective means be a^2, b^2, c^2, ... e^2, f^2, g^2; and let the mean product of the deviations of any two L and M from their respective means be denoted by S (L, M). Then, whatever the relations amongst the distributions may be,

(i.) The mean value of $lL + mM + nN \ldots + rR$, where $l, m, n, \ldots r$ are any constants, is $lL_1 + mM_1 + nM_1 + \ldots + rR_1$; and the mean square of its deviation from its mean is

$$l^2a^2 + m^2b^2 + n^2c^2 + \ldots + r^2g^2 + 2lm\mathrm{S}(L, M) + 2ln\mathrm{S}(L, N) + 2mn\mathrm{S}(M, N) + \ldots$$

(ii.) The mean product of the deviations of $lL + mM + nN + \ldots + rR$ and $l'L + m'M + n'N + \ldots + r'R$ from their respective means is

$$ll'a^2 + mm'b^2 + nn'c^2 + \ldots + rr'g^2 + (lm' + l'm)\mathrm{S}(L, M) + (ln' + l'n)\mathrm{S}(L, N) + (mn' + m'n)\mathrm{S}(M, N) + \ldots$$

As we shall often require to use these last two expressions, it will be found convenient to express the mean squares and mean products in the form of a table, thus:—

	L	M	N	&c.
L	a^2	S (L, M)	S (L, N)	
M	S (L, M)	b^2	S (M, N)	
N	S (L, N)	S (M, N)	c^2	
&c.				

§ 13. *Independent Normal Distributions.*—If the different values of L, in the class distinguished by particular values of M, N, ... P, Q, R, are distributed in the same way, whatever these particular values may be, the distribution of L is said to be independent of the distributions of M, N, ... P, Q, R.

If the distribution of Q is independent of that of R; the distribution of P independent of those of Q and R; and so on, for L, M, N, ... P, Q, R: then the distributions of L, M, N, ... P, Q, R may be said to be mutually independent.

Now suppose that each distribution, taken separately, is normal; we require to find the distribution of $lL + mM + nN + \ldots + pP + qQ + rR$, where $l, m, n, \ldots p, q, r$ are any constants.

Consider first the case of two measures L and M. Let their mean values be L_1 and M_1, and let their mean squares of deviation from the mean be a^2 and b^2. Let $L = L_1 + ax$, $M = M_1 + by$. Then the values of x and of y are distributed normally about mean values zero with mean squares unity, and the distribution of x is independent of the distribution of y. Take two lines OX, OY at right angles to one another, and on OXY as base-plane construct the solid of frequency of values of x and y, these values being measured parallel to OX and OY respectively. Let OZ be drawn at right angles to OXY; and let K_1 and K_2 be two planes whose equations referred to OX, OY, OZ as axes are $la.x + mb.y = \xi_1$ and $la.x + mb.y = \xi_2$ respectively, where ξ_1 and ξ_2 have any values. Then the portion of the solid lying between K_1 and K_2 includes all elements representing individuals for which $la.x + mb.y$ lies between ξ_1 and ξ_2; and therefore the number of these individuals is proportional to the volume of this portion of the solid. Denote this volume by V.

Since the distribution of x is independent of the distribution of y, the sections of the solid of frequency by planes parallel to OZX are figures which when projected on OZX are orthogonal projections of one another with regard to OX; in other words, the solid is a projective solid. Since the values of x are distributed normally with mean value zero and mean square unity, it follows from (iii.) of § 6 that the sections by planes parallel to OZX are normal figures whose semi-parameters are unity, and whose central ordinates lie in OZY; and similarly the sections by planes parallel to OZY are normal figures whose semi-parameters are unity and whose central ordinates lie in OZY. Hence, by § 9 (i.), the solid is a normal solid; and therefore it may be regarded as a projective solid whose principal sections are parallel and perpendicular to the planes K_1 and K_2. Through OZ draw a plane at right angles to K_1 and K_2, cutting them in ordinates W_1R_1 and W_2R_2, and cutting the solid in a normal figure S. Then the volume V is proportional to the area $W_1R_1R_2W_2$ of the figure S. Also $OW_1 = \xi_1/\sqrt{l^2a^2 + m^2b^2}$, $OW_2 = \xi_2/\sqrt{l^2a^2 + m^2b^2}$. Hence the number of individuals for which $la.x + mb.y$ lies between ξ_1 and ξ_2 is proportional to the area, comprised between ordinates at distances $\xi_1/\sqrt{l^2a^2 + m^2b^2}$ and $\xi_2/\sqrt{l^2a^2 + m^2b^2}$ from the median, of a normal figure of semi-parameter unity; and therefore, by § 2, it is proportional to the area, comprised between ordinates at distances ξ_1 and ξ_2 from the median, of a normal figure of semi-parameter $\sqrt{l^2a^2 + m^2b^2}$. In other words, the values of $la.x + mb.y$ are distributed normally with mean square $l^2a^2 + m^2b^2$ about a mean value zero, and therefore the values of $lL + mM$ are distributed normally with this mean square* about a mean value $lL_1 + mM_1$.

Next take the more general case. Since the distributions of Q and of R are independent and normal, the distribution of $qQ + rR$ is normal. Again, since the distribution of P is independent of the distributions of Q and R, it is independent of the distribution of $qQ + rR$; and therefore, since the distribution of P is normal, the distribution of $pP + qQ + rR$ is normal. Proceeding in this way, we see that if the distributions of L, M, N, ... P, Q, R are mutually independent, and if each distribution, taken separately, is normal, the distribution of $lL + mM + nN + \ldots + pP + qQ + rR$ is also normal.

We might have obtained this result from the *statistical equation* of the normal curve (§ 5). Let $L - L_1 = L'$, $M - M_1 = M'$, $N - N_1 = N'$, ... Also let $S(L'^{\alpha}M'^{\beta}N'^{\gamma}\ldots)$ denote the mean value of $L'^{\alpha}M'^{\beta}N'^{\gamma}\ldots$, and let λ_k denote the mean value of $(lL' + mM' + nN' + \ldots)^k$. Then, since the distributions are independent, $S(L'^{\alpha}M'^{\beta}N'^{\gamma}\ldots) = S(L'^{\alpha}).S(M'^{\beta}).S(N'^{\gamma})\ldots$. Also, by § 5, $S(L'^{2s-1}) = 0$, and $S(L'^{2s}) = \frac{\lfloor 2s}{2^s \lfloor s} a^{2s}$; and similarly for M', N', ... Hence we see that—

(i.) Every term in the expansion of $(lL' + mM' + nN' + \ldots)^{2s-1}$ must contain an odd power of one at least of the quantities L', M', N', ...; and therefore, by taking the mean, $\lambda_{2s-1} = 0$;

(ii.) $\lambda_2 = l^2a^2 + m^2b^2 + n^2c^2 + \ldots$

(iii.) λ_{2s} = mean value of $(lL' + mM' + nN' + \ldots)^{2s}$

$$= \Sigma\Sigma\Sigma \ldots \frac{\lfloor 2s}{\lfloor 2\alpha \lfloor 2\beta \lfloor 2\gamma \ldots} S\{(lL')^{2\alpha}\}.S\{(mM')^{2\beta}\}.S\{(nN')^{2\gamma}\}\ldots$$

(the summation being made for all positive integral values of α, β, γ, ... satisfying the condition $\alpha + \beta + \gamma + \ldots = s$)

$$= \Sigma\Sigma\Sigma \ldots \frac{\lfloor 2s}{\lfloor 2\alpha \lfloor 2\beta \lfloor 2\gamma \ldots} l^{2\alpha}m^{2\beta}n^{2\gamma}\ldots \frac{\lfloor 2\alpha}{2^{\alpha}\lfloor \alpha} a^{2\alpha} . \frac{\lfloor 2\beta}{2^{\beta}\lfloor \beta} b^{2\beta} . \frac{\lfloor 2\gamma}{2^{\gamma}\lfloor \gamma} c^{2\gamma}\ldots$$

$$= \frac{\lfloor 2s}{2^s \lfloor s} \Sigma\Sigma\Sigma \ldots \frac{\lfloor s}{\lfloor \alpha \lfloor \beta \lfloor \gamma \ldots} (l^2a^2)^{\alpha} . (m^2b^2)^{\beta} . (n^2c^2)^{\gamma}\ldots$$

$$= \frac{\lfloor 2s}{2^s \lfloor s}(l^2a^2 + m^2b^2 + n^2c^2 + \ldots)^s = \frac{\lfloor 2s}{2^s \lfloor s} \lambda_2^s;$$

and therefore, for all positive integral values of k,

$$\lambda_{k+2} = (k + 1)\lambda_2\lambda_k.$$

* The expression "mean square" may generally be used, without confusion, to denote the mean square of deviation from the mean.

Hence the values of $lL' + mM' + nN' + \ldots$ are normally distributed; and therefore the values of $lL + mM + nN + \ldots$ are normally distributed.

§ 14. *Correlated Normal Distributions.*—If L, M, N, ... R are the measures of coexistent attributes A, B, C, ... G; and if the values of L, in every class distinguished by particular values of M, N, ... R, are distributed normally with constant mean square about a mean value $L_1 + \mu(M - M_1) + \nu(N - N_1) + \ldots + \rho(R - R_1)$, where L_1, M_1, N_1, ... R_1 are the respective mean values of L, M, N, ... R taken separately, and $\mu, \nu, \ldots \rho$ are constants: then the distribution of L is said to be correlated with the distributions of M, N, ... R.

If the distribution of R is normal; the distribution of Q correlated with that of R; the distribution of P correlated with those of Q and R; and so on, for L, M, N, ... P, Q, R: then the distributions of L, M, N, ... P, Q, R may be said to be mutually correlated. We require to find, in this case, the distribution of $lL + mM + nN + \ldots + pP + qQ + rR$, where $l, m, n, \ldots p, q, r$ are any constants.

For convenience, consider only the case of four attributes L, M, N, R. From the definition, we see that $L - L_1$ is equal to $\mu(M - M_1) + \nu(N - N_1) + \rho(R - R_1) + L'$, where L' is independent of $M - M_1$, $N - N_1$, and $R - R_1$, and is distributed normally with mean value zero. Similarly $M - M_1$ is equal to $\nu'(N - N_1) + \rho'(R - R_1) + M'$, where M' is independent of $N - N_1$ and $R - R_1$; and $N - N_1$ is equal to $\rho''(R - R_1) + N'$, where N' is independent of $R - R_1$; the values of M' and of N' being distributed normally with mean values zero. Since M' is independent of $N - N_1$ and $R - R_1$, and $N - N_1$ is equal to $\rho''(R - R_1) + N'$, it follows that M' is independent of N' and $R - R_1$; and similarly L' is independent of M', N', and $R - R_1$. Thus the distributions of L', M', N', and $R - R_1$ are mutually independent. Also each of the measures $L - L_1$, $M - M_1$, $N - N_1$, $R - R_1$, is a linear function of the measures L', M', N', $R - R_1$; and therefore $l(L - L_1) + m(M - M_1) + n(N - N_1) + r(R - R_1)$ is a linear function of these measures It follows, from § 13, that the values of $l(L - L_1) + m(M - M_1) + n(N - N_1) + r(R - R_1)$ are normally distributed; *i.e.*, the values of $lL + mM + nN + rR$ are normally distributed. The argument obviously applies to any number of correlated distributions.

This result might also be obtained by the second of the two methods given in the last section.

II. Theory of Error.

§ 15. *Distribution of linear function of errors of random selection.*—Let the individuals comprised in an indefinitely great community be divided into any number of classes A, B, C, ..., and let the numbers in these classes be proportional to $\alpha, \beta, \gamma, \ldots$, so that $\alpha + \beta + \gamma + \ldots = 1$. Suppose a random selection of n individuals to be made, and let the numbers drawn from the different classes be respectively $n\alpha'$, $n\beta'$, $n\gamma'$, ..., so that $\alpha' + \beta' + \gamma' + \ldots = 1$. Then $\alpha' - \alpha$, $\beta' - \beta$, $\gamma' - \gamma$, ... are the *errors* in $\alpha, \beta, \gamma, \ldots$ We require to investigate the distribution of the different values of $a(\alpha' - \alpha) + b(\beta' - \beta) + c(\gamma' - \gamma) + \ldots$ for different random selections of n individuals, $a, b, c, \ldots$ being any constants.

(1.) If we only require the mean and the mean square, we can most conveniently use the formulæ of § 12. Suppose an indefinitely great number of random selections to be made. Then the proportion of cases in which p come from A and the remaining $n - p$ from the other classes is

$$\frac{\lfloor n}{\lfloor p \lfloor n-p} \alpha^p (1-\alpha)^{n-p}.$$

Hence

(i.) the mean value of α' is

$$\sum_{p=0}^{p=n} \frac{\lfloor n}{\lfloor p \lfloor n-p} \alpha^p (1-\alpha)^{n-p} \cdot \frac{p}{n} = \alpha \sum_{p=1}^{p=n} \frac{\lfloor n-1}{\lfloor p-1 \lfloor n-p} \alpha^{p-1} (1-\alpha)^{n-p} = \alpha;$$

so that the mean value of $\alpha' - \alpha$ is zero; and

(ii.) the mean square of α' is

$$\sum_{p=0}^{p=n} \frac{\lfloor n}{\lfloor p \lfloor n-p} \alpha^p (1-\alpha)^{n-p} \cdot \frac{p^2}{n^2} = n^{-2} \sum_{p=0}^{p=n} \frac{\lfloor n}{\lfloor p \lfloor n-p} \alpha^p (1-\alpha)^{n-p} \{p(p-1) + p\}$$

$$= n^{-2} \{n(n-1)\alpha^2 + n\alpha\} = \alpha^2 + \alpha(1-\alpha)/n;$$

so that the mean square of $\alpha' - \alpha$ is $\alpha(1-\alpha)/n$.

(iii.) Similarly the mean value of $\alpha' \beta'$ is

$$\sum_{p=0}^{p=n} \sum_{q=0}^{q=n} \frac{\lfloor n}{\lfloor p \lfloor q \lfloor n-p-q} \alpha^p \beta^q (1-\alpha-\beta)^{n-p-q} \cdot \frac{p}{n} \cdot \frac{q}{n}$$

$$= \frac{n(n-1)}{n^2} \alpha\beta \sum_{p=1}^{p=n} \sum_{q=1}^{q=n} \frac{\lfloor n-2}{\lfloor p-1 \lfloor q-1 \lfloor n-p-q} \alpha^{p-1} \beta^{q-1} (1-\alpha-\beta)^{n-p-q}$$

$$= \alpha\beta - \alpha\beta/n;$$

and therefore the mean product of $\alpha' - \alpha$ and $\beta' - \beta$ is $-\alpha\beta/n$. From these three results it follows that

(iv.) the mean value of $a(\alpha' - \alpha) + b(\beta' - \beta) + c(\gamma' - \gamma) + \ldots$ is zero;

(v.) the mean square is

$$a^2 \alpha(1-\alpha)/n + b^2 \beta(1-\beta)/n + c^2 \gamma(1-\gamma)/n + \ldots$$
$$- 2ab\alpha\beta/n - 2ac\alpha\gamma/n - 2bc\beta\gamma/n - \ldots$$
$$= \{(a^2\alpha + b^2\beta + c^2\gamma + \ldots) - (a\alpha + b\beta + c\gamma + \ldots)^2\}/n;$$

(vi.) the mean product of $a(\alpha'-\alpha)+b(\beta'-\beta)+c(\gamma'-\gamma)+\ldots$ and $a'(\alpha'-\alpha)+b'(\beta'-\beta)+c'(\gamma'-\gamma)+\ldots$ is

$$\{(aa'\alpha+bb'\beta+cc'\gamma+\ldots)-(a\alpha+b\beta+c\gamma+\ldots)(a'\alpha+b'\beta+c'\gamma+\ldots)\}/n.$$

(2.) Let λ_k denote the mean kth power of $a(\alpha'-\alpha)+b(\beta'-\beta)+c(\gamma'-\gamma)+\ldots$ The proportion of cases in which the numbers drawn from the different classes are $p, q, r\ldots$, where $p+q+r+\ldots=n$, is

$$\frac{\lfloor p+q+r+\ldots}{\lfloor p\lfloor q\lfloor r\ldots}\alpha^p\beta^q\gamma^r\ldots$$

Hence the mean kth power of $a\alpha'+b\beta'+c\gamma'+\ldots$ is

$$n^{-k}\Sigma\Sigma\Sigma\ldots\frac{\lfloor p+q+r+\ldots}{\lfloor p\lfloor q\lfloor r\ldots}\alpha^p\beta^q\gamma^r\ldots(ap+bq+cr+\ldots)^k$$

$$=n^{-k}\lfloor k\times\text{coefficient of }\theta^k\text{ in }\Sigma\Sigma\Sigma\ldots\frac{\lfloor p+q+r+\ldots}{\lfloor p\lfloor q\lfloor r\ldots}\alpha^p\beta^q\gamma^r\ldots e^{(ap+bq+cr+\ldots)\theta}$$

$$=n^{-k}\lfloor k\times\text{co. }\theta^k\text{ in }\Sigma\Sigma\Sigma\ldots\frac{\lfloor p+q+r+\ldots}{\lfloor p\lfloor q\lfloor r\ldots}(\alpha e^{a\theta})^p.(\beta e^{b\theta})^q.(\gamma e^{c\theta})^r\ldots$$

$$=\lfloor k\times\text{co. }\theta^k\text{ in }(\alpha e^{a\theta/n}+\beta e^{b\theta/n}+\gamma e^{c\theta/n}+\ldots)^n.$$

Denote $a\alpha+b\beta+c\gamma+\ldots$ by ω. Then, since $\alpha'+\beta'+\gamma'+\ldots=1$,

$$\begin{aligned}a(\alpha'-\alpha)+b(\beta'-\beta)+c(\gamma'-\gamma)+\ldots\\ &=a\alpha'+b\beta'+c\gamma'+\ldots-\omega(\alpha'+\beta'+\gamma'+\ldots)\\ &=(a-\omega)\alpha'+(b-\omega)\beta'+(c-\omega)\gamma'+\ldots\end{aligned}$$

Hence, writing $a-\omega, b-\omega, c-\omega, \ldots$ for $a, b, c, \ldots$, in the above result, we see that

$$\lambda_k=\lfloor k\times\text{coefficient of }\theta^k\text{ in }\{\alpha e^{(a-\omega)\theta/n}+\beta e^{(b-\omega)\theta/n}+\gamma e^{(c-\omega)\theta/n}+\ldots\}^n.$$

§ 16. *Tendency of Distribution to become Normal.*—We have now to prove that, when n becomes very great, the distribution of values of $a(\alpha'-\alpha)+b(\beta'-\beta)+c(\gamma'-\gamma)+\ldots$ tends to become normal. To do this, we can use either the geometrical or the statistical definition of the normal curve. Of the two methods, the latter is the simpler.

(1.) Since the mean square of $a(\alpha'-\alpha)+b(\beta'-\beta)+c(\gamma'-\gamma)+\ldots$ varies inversely as n, it is more convenient to find the distribution of

$$\sqrt{n}\{a(\alpha'-\alpha)+b(\beta'-\beta)+c(\gamma'-\gamma)+\ldots\}.$$

Let the mean kth power of this last expression be denoted by μ_k, so that

$$\mu_2 = (a^2\alpha + b^2\beta + c^2\gamma + \ldots) - (a\alpha + b\beta + c\gamma + \ldots)^2.$$

By expanding the expression at the end of § 15, and writing $n\theta$ for θ, we see that

$$\mu_k = n^{-\frac{1}{2}k} \lfloor k \times \text{ coefficient of } \theta^k \text{ in } \{1 + \tfrac{1}{2}\mu_2\theta^2 + C_3\theta^3 + C_4\theta^4 + \ldots\}^n,$$

where $C_3, C_4, \ldots$ are functions of $a, b, c, \ldots, \alpha, \beta, \gamma, \ldots$ Denote $\frac{1}{2}\mu_2\theta^2 + C_3\theta^3 + C_4\theta^4 + \ldots$ by Θ, and expand $(1 + \Theta)^n$ by the binomial theorem. Then the highest power of n contained in μ_k comes from the term involving $\Theta^{\frac{1}{2}k}$ when k is even, or from the term involving $\Theta^{\frac{1}{2}(k-1)}$ when k is odd. Hence, when n is made indefinitely great,

$$\left.\begin{aligned} \mu_{2s} &= n^{-s} \lfloor 2s \times \frac{n^s}{\lfloor s} (\tfrac{1}{2}\mu_2)^s = \frac{\lfloor 2s}{2^s \lfloor s} \mu_2^{s} \\ \mu_{2s+1} &= n^{-s-\frac{1}{2}} \lfloor 2s+1 \times \frac{n^s}{\lfloor s} \cdot s\, (\tfrac{1}{2}\mu_2)^{s-1} C_3 = 0 \end{aligned}\right\},$$

and therefore the distribution is ultimately normal.

It follows that the distribution of values of $a(\alpha' - \alpha) + b(\beta' - \beta) + c(\gamma' - \gamma) + \ldots$ is also normal.

It will be noticed that, when n is finite, the number of terms in μ_{2s} or μ_{2s+1} increases with s, and becomes infinite when s is infinite. Thus the approximation of the actual distribution to the ultimate normal distribution is close as regards the low moments, n being supposed to be moderately great, but is not close as regards very high moments. The difference between the two distributions is therefore due mainly to the values of $\sqrt{n}\{a(\alpha' - \alpha) + b(\beta' - \beta) + c(\gamma' - \gamma) + \ldots\}$ which are great in comparison with $\sqrt{\mu_2}$. But these are values which only occur very rarely; and therefore, for practical purposes, we may regard the two distributions as identical.

(2.) To obtain the same result from the geometrical definition of the curve, we must use § 14.

(i.) To find the distribution of values of $\sqrt{n}(\alpha' - \alpha)$, we take a series of points $M_0, M_1, \ldots M_n$, at equal distances $1/\sqrt{n}$ along a straight line X'X; and then draw ordinates $M_0P_0, M_1P_1, \ldots M_nP_n$ equal to the coefficients in the expansion of $\sqrt{n}(\beta x + \alpha y)^n$, where $\alpha + \beta = 1$. Thus

$$M_pP_p = \sqrt{n} . \alpha^p \beta^{n-p} C_p^n,$$

where C_p^n stands for $\dfrac{\lfloor n}{\lfloor p \lfloor n-p}$. Then, if n is increased indefinitely, the locus of the points $P_0, P_1, \ldots P_n$ will be a curve, which will be the curve of frequency of values of $\sqrt{n}(\alpha' - \alpha)$.

To find this curve, take a second series of points $N_0, N_1, \ldots N_{n+1}$, also at equal distances $1/\sqrt{n}$, and in such a position with regard to the former series that

$$M_{p-1}N_p = \alpha/\sqrt{n}, \qquad N_pM_p = \beta/\sqrt{n};$$

and at the points $N_1, N_2, \ldots N_n$ erect ordinates $N_1Q_1, N_2Q_2, \ldots N_nQ_n$ (fig. 8) equal to the coefficients in the expansion of $\sqrt{n}\,(\beta x + \alpha y)^{n-1}$. Thus

$$\left.\begin{aligned} N_pQ_p &= \sqrt{n}.\alpha^{p-1}\beta^{n-p}C^{n-1}_{p-1} \\ N_{p+1}Q_{p+1} &= \sqrt{n}.\alpha^{p}\beta^{n-p-1}C^{n-1}_{p} \end{aligned}\right\}.$$

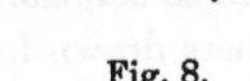

Fig. 8.

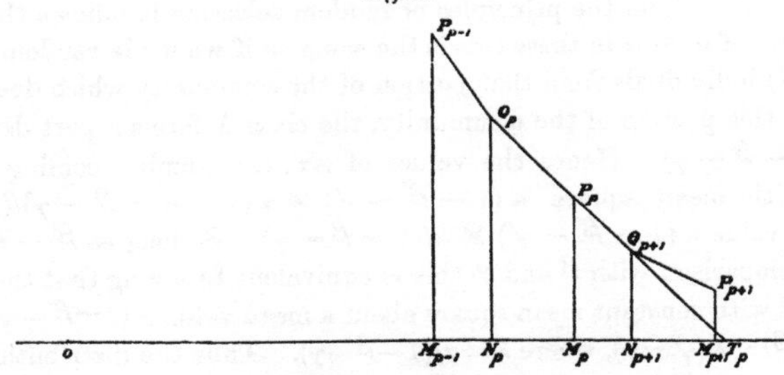

These ordinates lie in the successive intervals between the ordinates M_0P_0, M_1P_1, $\ldots M_nP_n$; and it is easily shown that N_pQ_p (except where it is the maximum ordinate) is intermediate in magnitude between $M_{p-1}P_{p-1}$ and M_pP_p. Also we have

$$\alpha.N_pQ_p + \beta.N_{p+1}Q_{p+1} = \sqrt{n}.\alpha^p\beta^{n-p}\,(C^{n-1}_{p-1} + C^{n-1}_{p}) = \sqrt{n}.\alpha^p\beta^{n-p}C^n_p = M_pP_p.$$

But $N_pM_p : M_pN_{p+1} :: \beta : \alpha$; and therefore P_p lies in Q_pQ_{p+1}. It follows that, in the limit, Q_pQ_{p+1} becomes the tangent at P_p.

Let Q_pQ_{p+1} meet X'X in T_p. Then

$$\frac{M_pP_p}{M_pT_p} = \frac{N_pQ_p - N_{p+1}Q_{p+1}}{N_pN_{p+1}} = n\,.\,\alpha^{p-1}\,\beta^{n-p-1}\,\{\beta C^{n-1}_{p-1} - \alpha C^{n-1}_{p}\}$$

$$= \alpha^{p-1}\beta^{n-p-1}C^n_p\,\{p\beta - (n-p)\,\alpha\}.$$

Hence if we choose the point O so that

$$\sqrt{n}.OM_p = -n\alpha + p = p\beta - (n-p)\,\alpha,$$

we have

$$OM_p.\,M_pT_p = \alpha\beta.*$$

* When n is not infinite, the relation $OM_p.M_pT_p = \alpha\beta$ shows that, if Σ denote any one of the family of normal curves of parameter $2\sqrt{\alpha\beta}$ having their median at O, the sides of the polygon $N_0Q_1Q_2 \ldots$

Now let n become indefinitely great, the point O remaining fixed. Then this relation holds all along the curve which is the limit of the polygon $P_0P_1 \ldots P_n$, and therefore this curve is a normal curve of parameter $2\sqrt{\alpha\beta}$, having its central ordinate at O. The mean value of α' is found by putting $OM = 0$, which gives $\alpha' = p/n = \alpha$. Thus the values of $\sqrt{n}(\alpha' - \alpha)$ are distributed normally with mean square $\alpha\beta = \alpha(1 - \alpha)$ about a mean value zero; and therefore the values of $\alpha' - \alpha$ are distributed normally with mean square $\alpha(1 - \alpha)/n$.

(ii.) Next, consider the distribution of values of $\alpha' - \alpha$ when certain other errors, as $\beta' - \beta$ and $\gamma' - \gamma$, have particular values. This distribution is found by taking an indefinitely great number of random selections, each containing n individuals, and isolating those sets in which the numbers drawn from the classes B and C are respectively $n\beta'$ and $n\gamma'$. From the principles of random selection it follows that the distribution of values of $\alpha' - \alpha$ in these sets is the same as if we made random selections of $n(1 - \beta' - \gamma')$ individuals from that portion of the community which does not involve B and C. Of this portion of the community, the class A forms a part denoted by the fraction $\alpha/(1 - \beta - \gamma)$. Hence the values of $n\alpha'$, the number coming from A, are distributed with mean square $n(1 - \beta' - \gamma') \times \alpha(1 - \alpha - \beta - \gamma)/(1 - \beta - \gamma)^2$ about a mean value $n(1 - \beta' - \gamma') \times \alpha/(1 - \beta - \gamma)$. So long as $\beta' - \beta$ and $\gamma' - \gamma$ are small in comparison with β and γ, this is equivalent to saying that the values of α' are distributed with constant mean square about a mean value $\alpha(1 - \beta' - \gamma')/(1 - \beta - \gamma) = \alpha - \lambda(\beta' - \beta) - \lambda(\gamma' - \gamma)$, where $\lambda = \alpha/(1 - \beta - \gamma)$. Thus the distributions of $\alpha' - \alpha$, $\beta' - \beta$, $\gamma' - \gamma, \ldots$ are normally correlated; and therefore, since the separate distributions are normal, the values of $a(\alpha' - \alpha) + b(\beta' - \beta) + c(\gamma' - \gamma) + \ldots$ are normally distributed.

Since this argument only applies when $\alpha' - \alpha$, $\beta' - \beta$, $\gamma' - \gamma, \ldots$ are small, the result is subject to the limitation pointed out in (1) (above).

§ 17. *Probable Error and Probable Discrepancy.*—Let X be any magnitude which is determined by observation of the ratios $\alpha', \beta', \gamma', \ldots$ Then X can be written in the form $f(\alpha', \beta', \gamma', \ldots)$. Now suppose n to be very great. Then the values of $\alpha' - \alpha$, $\beta' - \beta$, $\gamma' - \gamma, \ldots$ are distributed normally with mean values zero and mean squares $\alpha(1 - \alpha)/n$, $\beta(1 - \beta)/n$, $\gamma(1 - \gamma)/n, \ldots$; and therefore it may be supposed that in any particular case the values of $\alpha' - \alpha$, $\beta' - \beta$, $\gamma' - \gamma, \ldots$ will be very

Q_nN_{n+1} have the same slope at the points $P_1P_2 \ldots P_{n+1}$ as the respective curves Σ which pass through those points. Professor KARL PEARSON has arrived at a different result ('Phil. Trans.,' A, vol. 186 (1895) p. 357) by forming the polygon $P_1P_2 \ldots P_{n+1}$ and finding the "slope" at the middle points of its sides. There is of course no discrepancy between the two results, since they deal with different polygons, and with points having different relative positions on these polygons. The curve found by Professor PEARSON becomes the normal curve when n is made indefinitely great.

To prevent misunderstanding, it should be pointed out that, in either case, the slope of the polygon at the points in question is not the same as the slope of any *one* curve of the family considered. Professor PEARSON's statement (*op. cit.*, p. 356) as to the existence of a close relation between the binomial polygon (for $\alpha = \beta$) and "the" normal curve seems to require some qualification.

small. Thus X is of the form $f(\alpha,\beta,\gamma\ldots)+f_\alpha(\alpha'-\alpha)+f_\beta(\beta'-\beta)+f_\gamma(\gamma'-\gamma)+\ldots$; and therefore, by § 16, its mean value is $f(\alpha,\beta,\gamma,\ldots)$, and the different possible values are distributed normally about this mean value with mean square

$$\{(\alpha f_\alpha^2+\beta f_\beta^2+\gamma f_\gamma^2+\ldots)-(\alpha f_\alpha+\beta f_\beta+\gamma f_\gamma+\ldots)^2\}/n.$$

If we denote the expression in curled brackets by σ^2, the quartile deviation from the mean is $Q\sigma/\sqrt{n}$, where Q is the deviation of the quartile ordinate from the central ordinate in the standard normal curve (= ·67449 approximately*).

The applications are of two kinds. In one class of cases X is a "frequency-constant" whose value is required. Its observed value $f'(\alpha',\beta',\gamma',\ldots)$ differs from its true value $f(\alpha,\beta,\gamma,\ldots)$ by an *error* due to the paucity of observations, and $Q\sigma/\sqrt{n}$ is then the *probable error*. In the other class of cases the theory is applied to the testing of any hypothesis with regard to numerical statistics. The difference between the observed and the calculated values of X is a *discrepancy*, and we test the hypothesis that this discrepancy is due to paucity of observations by comparing it with the *probable discrepancy* $Q\sigma/\sqrt{n}$. If the comparison is made for several different values of X, we ought to find that for about half of them the discrepancy ($=d$) is less than the probable discrepancy ($=q$), and that, amongst the remaining values, d is in no case a very large multiple of q. The following considerations will enable us to determine whether, in any particular case, the values of d/q are or are not greater than we might reasonably expect.

Let the different values of a magnitude δ be distributed normally, with quartile deviation q, about a mean value zero; and let m values be taken at random. Then, if the area of the standard normal figure lying between the ordinates at the points $x=-\rho/q$ and $x=+\rho/q$ is ϕ, the probability of one at least of the values of δ being numerically greater than ρ is $1-\phi^m$. If we choose ϕ so that this probability may be equal to $\frac{1}{2}$, the corresponding value of ρ may, by analogy with the "probable error," be called the *probable limit* of δ. The following table gives the values of ρ/q determined by this condition, for values of m from 1 to 20† :—

m	ρ/q	m	ρ/q	m	ρ/q	m	ρ/q
1	1·000	6	2·375	11	2·777	16	3·009
2	1·559	7	2·481	12	2·832	17	3·046
3	1·874	8	2·570	13	2·882	18	3·080
4	2·088	9	2·648	14	2·928	19	3·112
5	2·248	10	2·716	15	2·970	20	3·142

* The value of Q to 20 places of decimals is ·67448 97501 96081 74320, and its logarithm to 13 places is $\bar{1}$·82897 53543 532. The successive convergents to Q are $\frac{1}{2}, \frac{2}{3}, \frac{27}{40}, \frac{29}{43}, \frac{201}{298}, \frac{230}{341}, \ldots$

† For larger values of m, the value of ρ/q may be taken as equal to that given by CHAUVENET'S criterion for the rejection of one out of $m/\log_e 4+\frac{1}{4}$ observations.

If m values of X were observed, and if the discrepancies were independent, it would be an even chance that in one case at least the ratio of the discrepancy to the probable discrepancy would exceed the value given by the above table. As a matter of fact, the discrepancies are usually correlated; but, if we bear this in mind, the table may be used to decide whether the greatest value of the ratio is such as to negative the hypothesis under consideration.

For calculating $Q\sigma/\sqrt{n}$, in either class of cases, it will not always be necessary to express σ^2 in terms of $\alpha, \beta, \gamma, \ldots$ If the value of X depends solely on the values of certain frequency-constants, and if $\varsigma, \eta, \theta, \ldots$ are the errors in these frequency-constants, then $f(\alpha', \beta', \gamma', \ldots) - f(\alpha, \beta, \gamma, \ldots)$ may be written in the form $k\varsigma + l\eta + m\theta + \ldots$ The errors $\varsigma, \eta, \theta, \ldots$ being of the form $a(\alpha' - \alpha) + b(\beta' - \beta) + c(\gamma' - \gamma) + \ldots$, their mean squares and mean products can be found; and thence the mean square of $k\varsigma + l\eta + m\theta + \ldots$ can be obtained by the general formula given in § 12. The expressions for the mean squares and mean products of the errors in frequency-constants of certain particular forms will be found in §§ 18 and 19.

The true values of $\alpha, \beta, \gamma, \ldots$, or of the frequency-constants on which X depends, are not known; and therefore, in calculating $Q\sigma/\sqrt{n}$, we can only use the observed values $\alpha', \beta', \gamma', \ldots$ But, n being great, the mistake so introduced in $Q\sigma/\sqrt{n}$ is small in comparison with $Q\sigma/\sqrt{n}$ itself. In general, it is sufficient to determine $Q\sigma/\sqrt{n}$ within about 1 per cent. of its true value. It will therefore be found simplest to calculate σ^2/n, and then to take out the corresponding value of $Q\sigma/\sqrt{n}$ from Table V. (p. 159). This table gives $Q\sqrt{N}$, for any given value of N, within from ·8 to ·08 per cent. of its true value.

§ 18. *Error in Mean, Mean Square, &c.*—Let the mean value of a measure L (in an indefinitely great community), and the pth power of the deviation from the mean, be denoted by L_1 and λ_p respectively. Also let the actual values of L be $L_1 + x_1, L_1 + x_2, L_1 + x_3, \ldots$; and let the relative frequencies of these values be $z_1, z_2, z_3, \ldots$ Thus we have $\Sigma z = 1, \Sigma zx = 0, \Sigma zx^p = \lambda_p$. Now let a random selection of n individuals be made, and let the numbers for which L has the values $L_1 + x_1, L_1 + x_2, L_1 + x_3, \ldots$, be respectively $n(z_1 + \epsilon_1), n(z_2 + \epsilon_2), n(z_3 + \epsilon_3), \ldots$ Then (§§ 15, 16) the mean value of $A_1\epsilon_1 + A_2\epsilon_2 + A_3\epsilon_3 + \ldots \equiv \Sigma A\epsilon$ is zero; its mean square is $\{\Sigma A^2 z - (\Sigma Az)^2\}/n$; the mean product of $\Sigma A\epsilon$ and $B_1\epsilon_1 + B_2\epsilon_2 + B_3\epsilon_3 + \ldots \equiv \Sigma B\epsilon$ is $(\Sigma ABz - \Sigma Az . \Sigma Bz)/n$; and, n being supposed to be great, the values of $\Sigma A\epsilon$ or of $\Sigma B\epsilon$ are normally distributed.

Hence we obtain the following results :—

(i.) The calculated value of L_1 is $L_1 + (x_1\epsilon_1 + x_2\epsilon_2 + x_3\epsilon_3 + \ldots)$. Thus the error in L_1 is $x_1\epsilon_1 + x_2\epsilon_2 + x_3\epsilon_3 + \ldots$, and therefore this error is distributed normally with mean square $\{\Sigma zx^2 - (\Sigma zx)^2\}/n = \lambda_2/n$.

(ii.) Denote the error in L_1 by ω. Then the calculated value of λ_p is

$$\Sigma (z + \epsilon)(x - \omega)^p = \Sigma (z + \epsilon)(x^p - px^{p-1}\omega);$$

and therefore the error in λ_p is

$$\Sigma x^p \epsilon - p\Sigma z x^{p-1}\omega = \Sigma x^p \epsilon - p\lambda_{p-1}\omega = \Sigma (x^p - p\lambda_{p-1} x)\epsilon.$$

Hence this error is distributed normally with mean square

$$[\Sigma z(x^p - p\lambda_{p-1}x)^2 - \{\Sigma z(x^p - p\lambda_{p-1}x)\}^2]/n = (\lambda_{2p} - 2p\lambda_{p+1}\lambda_{p-1} + p^2\lambda_{p-1}^2\lambda_2 - \lambda_p^2)/n.$$

In particular, the mean square of the error in λ_2 is $(\lambda_4 - \lambda_2^2)/n$.

(iii.) The mean product of the errors in L_1 and in λ_p is

$$\{\Sigma zx(x^p - p\lambda_{p-1}x) - \Sigma zx . \Sigma z(x^p - p\lambda_{p-1}x)\}/n = (\lambda_{p+1} - p\lambda_{p-1}\lambda_2)/n.$$

In particular, the mean product of the errors in the mean and in the mean square of deviation is λ_3/n.

(iv.) The mean product of the errors in λ_p and in λ_q is

$$\{\Sigma z(x^p - p\lambda_{p-1}x)(x^q - q\lambda_{q-1}x) - \Sigma z(x^p - p\lambda_{p-1}x) . \Sigma z(x^q - q\lambda_{q-1}x)\}/n$$
$$= (\lambda_{p+q} - p\lambda_{p-1}\lambda_{q+1} - q\lambda_{p+1}\lambda_{q-1} + pq\lambda_{p-1}\lambda_{q-1}\lambda_2 - \lambda_p\lambda_q)/n.$$

§ 19. *Error in Class-Index.*—Let the values $L_1 + x_1$, $L_1 + x_2$, $L_1 + x_3 \ldots$, in § 18, be supposed to be in order of magnitude, $L_1 + x_1$ being least; and let X be any possible value of L, not coinciding with any one of these actual values.* Let the two classes for which L is respectively less and greater than X be denoted by C′ and C, and let the numbers in these classes be in the ratio of $1 + \alpha : 1 - \alpha$; then α will be called the *class-index* of X for classification according to values of L. Its value ranges from -1 to $+1$.

If a representative selection of n individuals were made, the numbers coming from the two classes would be $n_1 = \frac{1}{2}n(1 + \alpha)$ and $n_2 = \frac{1}{2}n(1 - \alpha)$; so that $\alpha = (n_1 - n_2)/(n_1 + n_2)$. Suppose however that the selection is a random one, the errors being as in § 18. Then, if we take X as lying between X_r and X_{r+1}, the observed value of α is $(z_1 + \epsilon_1) + (z_2 + \epsilon_2) + \ldots + (z_r + \epsilon_r) - (z_{r+1} + \epsilon_{r+1}) - \ldots$, and therefore the "error" in α is $\epsilon_1 + \epsilon_2 + \ldots + \epsilon_r - \epsilon_{r+1} - \ldots$ Hence :—

(i.) By considering the division of the community into the two classes C′ and C, we see from § 15 (i.) and (ii.) that the error in α is distributed normally with mean square $(1 - \alpha^2)/n$ about a mean value zero.

* This limitation does not introduce any difficulty in the case of continuous variation, since the frequency of any single value is then indefinitely small. (Cases in which the curve of frequency has an infinite ordinate are excluded from consideration.)

(ii.) Let β be another class-index. The lines of division corresponding to these two class-indices divide the community into three classes, whose numbers are proportional to quantities Z_1, Z_2, Z_3, where $Z_1 + Z_2 + Z_3 = 1$. From § 15 (iii.) it will be seen that the mean product of the errors in α and in β is

$$4Z_1Z_3/n = \{(1 - \alpha\beta) - (\alpha \smile \beta)\}/n.$$

(iii.) Let the values of Σzx^p for the classes C′ and C be respectively ν_p' and ν_p, so that $\nu_p + \nu_p' = \lambda_p$. Then it will be found from § 15 (vi.) that the mean product of the errors in α and in L_1 is $-(\nu_1 - \nu_1')/n$; and that the mean product of the errors in α and in λ_p is

$$-\{(\nu_p - \nu_p') - (\nu_1 - \nu_1')\,p\lambda_{p-1} + \alpha\lambda_p\}/n.$$

The following table shows the general results obtained in this and the last section; for convenience, the divisor n is omitted throughout.

	L_1	λ_p	α
L_1	λ_2	$\lambda_{p+1} - p\lambda_{p-1}\lambda_2$	$-(\nu_1 - \nu_1')$
λ_p		$\lambda_{2p} - 2p\lambda_{p+1}\lambda_{p-1} + p^2\lambda_{p-1}^2\lambda_2 - \lambda_p^2$	$-\{(\nu_p - \nu_p') - (\nu_1 - \nu_1')\,p\lambda_{p-1} + \alpha\lambda_p\}$
λ_q		$\lambda_{p+q} - p\lambda_{p-1}\lambda_{q+1} - q\lambda_{p+1}\lambda_{q-1} + pq\lambda_{p-1}\lambda_{q-1}\lambda_2 - \lambda_p\lambda_q$	(Similar expression)
α			$1 - \alpha^2$
β			$(1 - \alpha\beta) - (\alpha \smile \beta)$

§ 20. *Mean Squares and Products of Errors in Case of Two Attributes.*—Let M be the measure of a second attribute, M_1 its mean value, and μ_q the mean qth power of the deviation from the mean; and suppose that each z in § 18 denotes the proportion of individuals for which L and M jointly have certain specified values. Let $S_{p,q}$ denote the mean value of $(L - L_1)^p (M - M_1)^q$, so that $S_{p,0} = \lambda_p$, $S_{0,q} = \mu_q$. Then it will be found that the error in $S_{p,q}$ (*i.e.*, the error produced by taking $S_{p,q}$ as equal to the average of $x^p y^q$, where x and y are the respective deviations of L and M from their averages for the n individuals) is of the form $\Sigma A\epsilon$, and therefore is distributed normally; its mean square being

$$\begin{aligned}&[\Sigma z\,(x^py^q - pS_{p-1,q}x - qS_{p,q-1}y)^2 - \{\Sigma z\,(x^py^q - pS_{p-1,q}x - qS_{p,q-1}y)\}^2]/n \\ &= (S_{2p,2q} - 2pS_{p+1,q}S_{p-1,q} - 2qS_{p,q+1}S_{p,q-1} + p^2S_{p-1,q}^2\lambda_2 \\ &\qquad + 2pqS_{p-1,q}S_{p,q-1}S_{1,1} + q^2S_{p,q-1}^2\mu_2 - S_{p,q}^2)/n.\end{aligned}$$

Let X and Y be the values of L and M corresponding to class-indices α and β; and let $\frac{1}{2}(1-\chi)$ be the proportion of individuals for which L exceeds X and M exceeds Y: thus χ is necessarily greater than either α or β. Let the constituent parts of $S_{p,q}$ corresponding to $\frac{1}{2}(1-\chi)$ and $\frac{1}{2}(1+\chi)$ be $\sigma_{p,q}$ and σ'_{pq} respectively, so that, if a representative selection of N individuals is made, the value of $\Sigma(L-L_1)^p(M-M_1)^q$ for the $\frac{1}{2}N(1-\chi)$ individuals for which L exceeds X and M exceeds Y is $N\sigma_{p,q}$, while for the remaining $\frac{1}{2}N(1+\chi)$ it is $N\sigma'_{p,q}$. Then it can be shown by the methods of §§ 18 and 19 that the following tables give the mean products of the errors in the quantities concerned, the divisor n being omitted:—

	L_1	λ_l	$S_{p,q}$
M_1	$S_{1,1}$	$S_{l,1} - l\lambda_{l-1}S_{1,1}$	$S_{p,q+1} - pS_{p-1,q}S_{1,1} - qS_{p,q-1}\mu_2$
μ_m	$S_{1,m} - m\mu_{m-1}S_{1,1}$	$S_{l,m} - l\lambda_{l-1}S_{1,m} - mS_{l,1}\mu_{m-1}$ $+ lm\lambda_{l-1}\mu_{m-1}S_{1,1} - \lambda_l\mu_m$	$S_{p,q+m} - pS_{p-1,q}S_{1,m} - q\mu_{m+1}S_{p,q-1}$ $- m\mu_{m-1}S_{p,q+1} + mp\mu_{m-1}S_{p-1,q}S_{1,1}$ $+ mq\mu_{m-1}S_{p,q-1}\mu_2 - \mu_m S_{p,q}$
$S_{r,s}$	$S_{r+1,s} - rS_{r-1,s}\lambda_2$ $- sS_{r,s-1}S_{1,1}$	$S_{l+r,s} - l\lambda_{l-1}S_{r+1,s} - r\lambda_{l+1}S_{r-1,s}$ $- sS_{l,1}S_{r,s-1} + lr\lambda_{l-1}S_{r-1,s}\lambda_2$ $+ ls\lambda_{l-1}S_{r,s-1}S_{1,1} - \lambda_l S_{r,s}$	$S_{p+r,q+s} - pS_{p-1,q}S_{r+1,s} - qS_{p,q-1}S_{r,s+1}$ $- rS_{p+1,q}S_{r-1,s} - sS_{p,q+1}S_{r,s-1}$ $+ prS_{p-1,q}S_{r-1,s}\lambda_2 + qrS_{p,q-1}S_{r-1,s}S_{1,1}$ $+ psS_{p-1,q}S_{r,s-1}S_{1,1} + qsS_{p,q-1}S_{r,s-1}\mu_2$ $- S_{p,q}S_{r,s}$
χ	$-(\sigma_{1,0} - \sigma'_{1,0})$	$-\{(\sigma_{l,0} - \sigma'_{l,0}) - l\lambda_{l-1}(\sigma_{1,0} - \sigma'_{1,0}) + \chi\lambda_l\}$	$-\{(\sigma_{p,q} - \sigma'_{p,q}) - pS_{p-1,q}(\sigma_{1,0} - \sigma'_{1,0})$ $- qS_{p,q-1}(\sigma_{0,1} - \sigma'_{0,1}) + \chi S_{p,q}\}$
α, β	(similar expressions)		

	χ	α	β
χ	$1-\chi^2$	$(1-\chi)(1+\alpha)$	$(1-\chi)(1+\beta)$
α		$1-\alpha^2$	$1+\alpha+\beta-\alpha\beta-2\chi$
β			$1-\beta^2$

Suppose, for instance, that we are considering the error in $S_{1,1}/\sqrt{\lambda_2\mu_2} \equiv k$. Let the errors in λ_2, in $S_{1,1}$, and in μ_2 be θ, ϕ, and ψ respectively; then the error in k is $(-\theta/2\lambda_2 + \phi/S_{1,1} - \psi/2\mu_2)\,k$. For the mean squares and mean products of θ, ϕ, ψ, we have the table—

	λ_2	$S_{1,1}$	μ_2
λ_2	$\lambda_4 - \lambda_2^2$	$S_{3,1} - \lambda_2 S_{1,1}$	$S_{2,2} - \lambda_2\mu_2$
$S_{1,1}$		$S_{2,2} - S_{1,1}^2$	$S_{1,3} - \mu_2 S_{1,1}$
μ_2			$\mu_4 - \mu_2^2$

from which it will be found that the mean square of the error in k is

$$\left\{\frac{\lambda_4}{4\lambda_2^2} + \left(\frac{1}{S_{1,1}^2} + \frac{1}{2\lambda_2\mu_2}\right)S_{2,2} + \frac{\mu_4}{4\mu_2^2} - \frac{S_{3,1}}{\lambda_2 S_{1,1}} - \frac{S_{1,3}}{\mu_2 S_{1,1}}\right\}\Big/ n.$$

§ 21.—*Test of Independence of Two Distributions.*—For an illustration of the application of the theory of error to testing statistical hypotheses, let us take the case of two independent distributions. The criterion of independence of the distributions of two measures L and M is that, if α denotes the proportion of individuals, in the complete community, for which L lies between any two values L′ and L″, and if β denotes the proportion for which M lies between any two values M′ and M″, then the proportion for which both these conditions are satisfied is $\alpha\beta$. Hence, in order to test the hypothesis of independence when n individuals have been obtained by random selection, we must arrange them in a table of double entry, thus:—

Values of L.	Values of M.			Total.
	M′ to M″.	M″ to M‴.	&c.	
L′ to L″	n_{11}	n_{12}	&c.	p_1
L″ to L‴	n_{21}	n_{22}		p_2
⋮	⋮	⋮		⋮
Total . . .	q_1	q_2	. . . . .	n

then form a new table by dividing each number in this table by n, so as to show the proportions in the different classes; and then consider whether the discrepancies between these proportions and the corresponding proportions in a table showing independent distribution are such as might be accounted for by random selection.

Let the following table represent the proportions, in the original community, of the individuals specified :—

Values of L.	Values of M.	
	M″ to M‴.	Remainder.
L′ to L″	V	V″
Remainder	V′	V‴

and let ψ, ψ', ψ'', ψ''' be the errors in V, V′, V″, V‴. Thus $n_{12} = n(\mathrm{V} + \psi)$, $p_1 = n(\mathrm{V} + \mathrm{V}'' + \psi + \psi'')$, $q_2 = n(\mathrm{V} + \mathrm{V}' + \psi + \psi')$. If the distributions are independent, $\mathrm{V} = (\mathrm{V}+\mathrm{V}')(\mathrm{V}+\mathrm{V}'')$; *i.e.* (since $\mathrm{V}+\mathrm{V}'+\mathrm{V}''+\mathrm{V}'''=1$), $\mathrm{V}\mathrm{V}'''=\mathrm{V}'\mathrm{V}''$. Hence (since $\psi + \psi' + \psi'' + \psi''' = 0$)

$$n_{12} - p_1q_2/n = n\{\psi - (\mathrm{V} + \mathrm{V}')(\psi + \psi'') - (\mathrm{V} + \mathrm{V}'')(\psi + \psi')\}$$
$$= n\{(\mathrm{V}'''\psi + \mathrm{V}\psi''') - (\mathrm{V}''\psi' + \mathrm{V}'\psi'')\}.$$

By § 15 (v.) it will be found that the mean square of this discrepancy is $n\mathrm{V}\mathrm{V}''' = n\mathrm{V}'\mathrm{V}''$; and therefore the "probable discrepancy" is $\mathrm{Q}\sqrt{n\mathrm{V}\mathrm{V}'''} = \mathrm{Q}\sqrt{n\mathrm{V}'\mathrm{V}''}$. By calculating this expression for each number in the table, and comparing the actual discrepancies, as $n_{12} - p_1q_2/n$, with the values so obtained, we have data for deciding as to the validity of the hypothesis of independence.

The following example of a case in which, on *a priori* grounds, we should expect to find independence, will serve as an illustration. The table is compiled from a list of school-teachers who passed a certain examination.

List.	First letter of name.				Total.
	A–D.	E–J.	K–R.	S–Z.	
Men	166	174	180	164	684
Women, 1st year	427	379	411	366	1583
" 2nd " . . .	549	493	577	492	2111
Total	1142	1046	1168	1022	4378

By multiplying each total of a row by each total of a column, and dividing each product by $n = 4378$, we get the "calculated" table

178·4	163·4	182·5	159·7
413·0	378·2	422·3	369·5
550·6	504·4	563·2	492·8

showing discrepancies in the actual table amounting to

− 12·4	+ 10·6	− 2·5	+ 4·3
+ 14·0	+ 0·8	− 11·3	− 3·5
− 1·6	− 11·4	+ 13·8	− 0·8

If nV represents any number in the calculated table, the corresponding values of nV‴ will be found to be

2730·4	2811·4	2708·5	2831·7
2066·0	2127·2	2049·3	2142·5
1675·6	1725·4	1662·2	1737·8

Multiplying each number in this table by the corresponding number in the "calculated" table, and dividing by 4378, we get the values of nVV‴

111·26	104·93	112·91	103·29
194·90	183·76	197·67	180·83
210·73	198·79	213·83	195·61

Whence, from Table V. (p. 159) the probable discrepancies are

7·1	6·9	7·2	6·9
9·4	9·1	9·5	9·1
9·8	9·5	9·9	9·4

The ratios of the actual discrepancies to these probable discrepancies are

− 1·7	+ 1·5	− 0·3	+ 0·6
+ 1·5	+ 0·1	− 1·2	− 0·4
− 0·2	− 1·2	+ 1·4	− 0·1

Thus six out of the twelve ratios are numerically less than unity, and six numerically greater, while the greatest ratio is well within the probable limit (§ 17). The hypothesis of independence in this case is therefore justified by the data.*

PART III.—APPLICATION TO NORMAL DISTRIBUTIONS.

§ 22. *Probable Errors in Mean and in Semi-parameter by Different Methods.*—If the values of a measure L are known to be distributed normally, the distribution is

* The method of this section is an extension of the ordinary method (used largely by Professor LEXIS and Professor EDGEWORTH) for testing the "stability of statistical ratios."

determined when the mean value L_1 and the semi-parameter a are determined. When the values of L for n individuals obtained by random selection are given, the values of L_1 and of a can be found in either of two different ways.

(1.) We can find the average and the standard deviation (square root of average square of deviation from the average*) of the n individuals. The average will differ from L_1 by an error whose mean square (§ 18 (i.)) is a^2/n, so that the probable error of L_1 as found in this way is $Qa/\sqrt{n}$; and (§ 18 (ii.)) the square of the standard deviation will differ from a^2 by an error whose mean square is $(\lambda_4 - \lambda_2^2)/n = 2a^4/n$ (§ 5); so that the probable error in a will be $Qa/\sqrt{2n}$. These are familiar results.

(2.) The other method is that which has been mainly used by Mr. GALTON.† Let α and β be any two class-indices, and let X and Y be the corresponding values of L in the complete community. Then, if x and y are the abscissæ corresponding to class-indices α and β in the standard normal figure (*i.e.*, if ordinates at distances x and y from the central ordinate divide the figure into areas whose ratios are $1 + \alpha : 1 - \alpha$ and $1 + \beta : 1 - \beta$ respectively), we have

$$\left.\begin{aligned} X &= L_1 + ax \\ Y &= L_1 + ay \end{aligned}\right\} \quad \text{(i.)}.$$

Whence

$$\left.\begin{aligned} L_1 &= (xY - yX)/(x - y) \\ a &= (X - Y)/(x - y) \end{aligned}\right\} \quad \text{(ii.)}.$$

Now let ξ and η be the errors in the observed values of X and of Y; *i.e.*, let α and β be the class-indices of $X + \xi$ and $Y + \eta$ in the collection of n individuals. Then, if we deduce the values of L_1 and of a from (ii.), the resulting errors are $-(y\xi - x\eta)/(x - y)$ and $(\xi - \eta)/(x - y)$ respectively. Now the errors ξ and η are due to errors $-2z\xi/a$ and $-2z'\eta/a$ in the class-indices of X and Y, where z and z' are the ordinates of the standard normal figure corresponding to abscissæ x and y; and therefore (§ 19) the mean squares and mean product of ξ and η are $a^2(1 - \alpha^2)/4nz^2$, $a^2(1 - \beta^2)/4nz'^2$, and $a^2\{(1 - \alpha\beta) - (\alpha \smile \beta)\}/4nzz'$. Hence the probable errors in L_1 and in a, as found from (ii.), are respectively $Q.E/\sqrt{n}$ and $Q.H/\sqrt{n}$, where

$$\left.\begin{aligned} E^2 &= a^2\left\{\left(\frac{y}{x-y}\right)^2\frac{1-\alpha^2}{4z^2} - \frac{2xy}{(x-y)^2}\,\frac{(1-\alpha\beta)-(\alpha\smile\beta)}{4zz'} + \left(\frac{x}{x-y}\right)^2\frac{1-\beta^2}{4z'^2}\right\} \\ H^2 &= \frac{a^2}{(x-y)^2}\left\{\frac{1-\alpha^2}{4z^2} - 2\,\frac{(1-\alpha\beta)-(\alpha\smile\beta)}{4zz'} + \frac{1-\beta^2}{4z'^2}\right\} \end{aligned}\right\} \quad \text{(iii.)}.$$

* It seems convenient to use the term "standard deviation" in this sense, as denoting a quantity which has a definite value for the particular data.

† GALTON, 'Natural Inheritance,' p. 62.

(3.) As an extension of this last result, let X, Y, U, ... be values of L corresponding (in the complete community) to class-indices $\alpha, \beta, \gamma, \ldots$, and let the corresponding abscissæ in the standard figure be $x, y, u, \ldots$ Then $X = L_1 + ax$, $Y = L_1 + ay$, $U = L_1 + au, \ldots$; and therefore

$$\left.\begin{aligned} L_1 &= (lX + mY + pU + \ldots) / (l + m + p + \ldots) \\ a &= (l'X + m'Y + p'U + \ldots) / (l'x + m'y + p'u + \ldots) \end{aligned}\right\} \quad . \quad . \quad . \quad \text{(i.)},$$

where $l, m, p \ldots, l', m', p' \ldots$ are any quantities which satisfy the conditions

$$\left.\begin{aligned} lx + my + pu + \ldots &= 0 \\ l' + m' + p' + \ldots &= 0 \end{aligned}\right\} \quad . \quad . \quad . \quad . \quad . \quad . \quad . \quad . \quad \text{(ii.)}.$$

Suppose that we fix on the values of $\alpha, \beta, \gamma, \ldots$ beforehand, and choose $l, m, p, \ldots$, $l', m', p', \ldots$ to satisfy (ii.), and then observe the values of L whose class-indices in the collection of n individuals are $\alpha, \beta, \gamma, \ldots$ If the errors in these values are $\xi, \eta, \theta, \ldots$, the resulting errors in L_1 and in a will be $(l\xi + m\eta + p\theta + \ldots)/(l + m + p + \ldots)$ and $(l'\xi + m'\eta + p'\theta + \ldots) / (l'x + m'y + p'u + \ldots)$; and therefore the probable errors in L_1 and in a, as deduced from (i.), are $Q.E / \sqrt{n}$ and $Q.H / \sqrt{n}$, where

$$\left.\begin{aligned} E^2 &= \tfrac{1}{4}a^2 \left\{\frac{l^2(1-\alpha^2)}{z^2} + \frac{m^2(1-\beta^2)}{z'^2} + \ldots \right. \\ &\qquad \left. + \frac{2lm\{(1-\alpha\beta) - (\alpha \backsim \beta)\}}{zz'} + \ldots \right\} \Big/ (l + m + \ldots)^2 \\ &= \tfrac{1}{4}a^2 \left\{\left(\Sigma \frac{l}{z}\right)^2 - \left(\Sigma \frac{l\alpha}{z}\right)^2 - 2\Sigma \frac{lm(\alpha \backsim \beta)}{zz'}\right\} \Big/ (\Sigma l)^2 \\ H^2 &= \tfrac{1}{4}a^2 \left\{\left(\Sigma \frac{l'}{z}\right)^2 - \left(\Sigma \frac{l'\alpha}{z}\right)^2 - 2\Sigma \frac{l'm'(\alpha \backsim \beta)}{zz'}\right\} \Big/ (\Sigma l'x)^2 \end{aligned}\right\} \quad . \quad \text{(iii.)}.$$

For any particular values of $\alpha, \beta, \gamma, \ldots$, the values of $l, m, p, \ldots, l', m', p', \ldots$ can be chosen so as to reduce E^2 or H^2 to a minimum.

§ 23. *Relative Accuracy of the Different Methods.*—Now let ω and ρ be the errors in L_1 and in a as obtained by the *average-and-average-square* method; *i.e.*, the errors due to taking them as equal to the average and the standard deviation of the n individuals. Also let the class-index of X, in the n individuals, be $\alpha + \theta$, the true class-index of X being α. Then, with the notation of § 19, the mean values of $\omega\theta$ and of $2a\rho\theta$ are respectively $-(\nu_1 - \nu'_1)/n$ and $-(\nu_2 - \nu'_2 + a^2\alpha)/n$. But, by § 5, $\nu_1 = az$, $\nu'_1 = -az$, $\nu_2 = \frac{1}{2}(1-\alpha)a^2 + a^2xz$, $\nu'_2 = \frac{1}{2}(1+\alpha)a^2 - a^2xz$. Also the error ξ in X is due to the error θ, and is equal to $-a\theta / 2z$. Thus we have the following table of mean squares and mean products of errors, the divisor n, as usual, being omitted :—

	L_1	a	X
L_1	a^2	0	a^2
a	0	$\frac{1}{2}a^2$	$\frac{1}{2}a^2x$
X	a^2	$\frac{1}{2}a^2x$	$a^2(1-\alpha^2)/4z^2$

and thence

	L_1	a	$X-(L_1+ax)$
L_1	a^2	0	0
a	0	$\frac{1}{2}a^2$	0
$X-(L_1+ax)$	0	0	$a^2(1-\alpha^2)/4z^2 - a^2 - \frac{1}{2}a^2x^2$

The true value of $X-(L_1+ax)$, of course, is zero; so that the "error" in $X-(L_1+ax)$ is the difference between X as determined by direct observation of the value whose class-index is α, and L_1+ax, as determined by calculating the average and the standard deviation. This error is $\xi-(\omega+x\rho)$; and therefore, if we write $\xi=\omega+x\rho+\phi$, the last table shows that the mean products of ω, ρ, and ϕ, taken in pairs, are zero. Hence we deduce the following conclusions :—

(1.) The mean square of ξ is greater than the mean square of $\omega+x\rho$.* Hence, if we fix a class-index α, corresponding to abscissa x in the standard normal figure, and if X denote the unknown value of L whose class-index is α, the probable error in X as obtained by direct observation is greater† than the probable error in the value obtained by calculating the average and the standard deviation, and deducing X from the formula $X=L_1+ax$. The following table, for instance, gives the probable errors in certain values which are often chosen for exhibiting the frequency-constants in any particular case :—

* This shows that $a^2(1-\alpha^2)/4z^2 > a^2(1+\frac{1}{2}x^2)$. Hence, if OH is the central ordinate, and MP any other ordinate, of a normal figure of parameter $2a$, and if A_1 and A_2 are the areas into which the figure is divided by MP, the product A_1A_2 is greater than $MP^2(a^2+\frac{1}{2}OM^2)$.

† The result, of course, only holds when we *know* that the distribution is normal. When we know nothing about it, the value corresponding to any particular class-index can only be obtained by direct observation.

Value of L.	Value of α.	Probable error in L by direct observation.	Probable error by average-and-average-square method.	Ratio of probable errors.
Median	·0	$\cdot84535\ a/\sqrt{n}$	$\cdot67449\ a/\sqrt{n}$	1·25
Quartiles	$\pm$ ·5	$\cdot91968\ a/\sqrt{n}$	$\cdot74728\ a/\sqrt{n}$	1·23
Deciles	$\pm$ ·2	$\cdot85528\ a/\sqrt{n}$	$\cdot68523\ a/\sqrt{n}$	1·25
	$\pm$ ·4	$\cdot88897\ a/\sqrt{n}$	$\cdot71937\ a/\sqrt{n}$	1·24
	$\pm$ ·6	$\cdot96869\ a/\sqrt{n}$	$\cdot78489\ a/\sqrt{n}$	1·23
	$\pm$ ·8	$1\cdot15298\ a/\sqrt{n}$	$\cdot91023\ a/\sqrt{n}$	1·27

(2.) If we take L_1 as equal to the average for the n individuals, and find X and Y by observing the values of L whose class-indices are α and β respectively, the mean square of the resulting error in $L_1 - (x\,Y - y\,X)/(x - y)$ is

$$a^2/n - 2\,(x\,a^2/n - y\,a^2/n)/(x - y) + E^2/n = (E^2 - a^2)/n,$$

where E^2 has the value given in § 22 (2.); and similarly, if we take a as equal to the standard deviation of the n individuals, the mean square of the error in $a - (Y - X)/(x - y)$ is $(H^2 - \frac{1}{2}a^2)/n$. Hence E^2 and H^2 are respectively greater than a^2 and $\frac{1}{2}a^2$; in other words, the probable errors in the values of L_1 and of a as determined by the formulæ (ii.) of § 22 (2.), are greater than the probable errors in their values as determined by the average-and-average-square method of § 22 (1.).

If, for instance, $\alpha = -\beta = \pm\frac{1}{2}$, so that the observed values are the two quartiles, the probable error in L_1 as determined by (ii.) of § 22 (2.) is $\cdot75043\ a/\sqrt{n}$, which is 11 per cent. greater* than the probable error $\cdot67449\ a/\sqrt{n}$ due to the average-and-average-square method; and the probable error in a is $\cdot78672\ a/\sqrt{n}$, which is nearly 65 per cent. greater than the probable error $\cdot47694\ a/\sqrt{n}$ due to the average-and-average-square method.

If we are unable to calculate the average and the standard deviation, we should

* When the quartiles are observed, it is also usual to observe the "median," for which $\alpha = 0$. If we take the arithmetic mean of the median and the two quartiles, the probable error due to taking this as the value of L_1 is reduced to $\cdot72736\ a/\sqrt{n}$, which is less than 8 per cent. in excess of the probable error due to taking the average. If X and Y are the quartiles and M the median, it may be shown that the best result from these data is obtained by giving to $\frac{1}{2}(X + Y)$ and M weights in the ratio of $2\,(\exp. - \frac{1}{2}Q^2) - 1 : (\exp. \frac{1}{2}Q^2) - 1$, and the probable error in the mean is then $[\frac{1}{2}Qa\sqrt{\pi}/\{1 - 2\,(\exp. - \frac{1}{2}Q^2) + 2\,(\exp. - Q^2)\}^{\frac{1}{2}}]/\sqrt{n}$. The first two convergents to the above ratio are 2 : 1 and 7 : 3, so that $\{7\,(X + Y) + 6\,M\}/20$ is a slightly better value than $(X + Y + M)/3$.

I have assumed that the quartiles, &c., are found by actual observation. But there is reason to believe that their values are sometimes obtained by faulty methods of interpolation. This does not affect the magnitude of the probable error, but it affects the calculated values of L_1 and of a.

choose α and β so as to make the values of E^2 and of H^2 as small as possible. It is obvious that one of the class-indices must be positive and the other negative. Suppose α to be negative, and equal to $-\gamma$; then it will be found from KRAMP's tables that E^2 is a minimum when β and γ are each taken a little greater than ·459, the probable error in the mean being then ·74951 $a/\sqrt{n}$, which is about the same as the probable error due to using the quartiles; and that H^2 is a minimum when β and γ are each taken a little less than ·862, the probable error in the semi-parameter being then ·59055 $a/\sqrt{n}$, which is about 25 per cent. less than the probable error due to using the quartiles, but nearly 24 per cent. greater than that due to the average-and-average-square method.

(3.) Suppose the values of the mean and of the semi-parameter to be found by the extended class-index method of § 22 (3.). Then, with the notation used above, the errors in the observed values of X, Y, U, . . . are of the form $\omega + x\rho + \phi$, $\omega + y\rho + \psi$, $\omega + u\rho + \chi$, . . . where ϕ, ψ, χ, . . . are errors whose mean products with ω, and also with ρ, are zero. Substituting in (i.) of § 22 (3.), and taking account of (ii.), we see that the resulting errors in L_1 and in a due to this method are respectively

$$\omega + (l\phi + m\psi + p\chi + \ldots)/(l + m + p + \ldots)$$

and

$$\rho + (l'\phi + m'\psi + p'\chi + \ldots)/(l'x + m'y + p'u + \ldots).$$

Hence if Φ^2/n and Φ'^2/n are the mean squares of

$$(l\phi + m\psi' + p\chi + \ldots)/(l + m + p + \ldots)$$

and of

$$(l'\phi + m'\psi + p'\chi + \ldots)/(l'x + m'y + p'u + \ldots),$$

the mean squares of the errors in L_1 and in a, due to the use of the class-index method, are $(a^2 + \Phi^2)/n$ and $(\frac{1}{2} a^2 + \Phi'^2)/n$. Since these are necessarily greater than a^2/n and $\frac{1}{2} a^2/n$ respectively, the probable errors in L_1 and in a due to this method are greater than the probable errors due to the average-and-average-square method. In other words, we cannot, by observation of the values corresponding to particular class-indices, obtain such good results for L_1 and a as by calculating the average and the standard deviation.*

(4.) Generally, let R be any quantity which would be known if the true mean and mean square of the distribution were known; let R_1 be the value obtained by taking the mean and mean square as equal to the average and the average square for the n observations, and let R_2 be the value obtained by any other method involving observation of the class-indices of any finite number of values of L, with or without the

* Professor EDGEWORTH's contrary statement ('Phil. Mag.,' vol. 36, 1893, p. 100) appears to be based on neglect of the correlation of errors.

use of the average and the average square. Let Θ_1^2/n and Θ_2^2/n be the mean squares of the errors in R as determined by the two methods. Then it may be shown that the mean square of the error involved in taking $pR_1 + qR_2$ as the value of $(p+q)R$ is $\{(p^2 + 2pq)\Theta_1^2 + q^2\Theta_2^2\}/n = \{(p+q)^2\Theta_1^2 + q^2(\Theta_2^2 - \Theta_1^2)\}/n$. Since this must be positive, it follows, by taking $p + q = 0$, that Θ_2^2 must be greater than Θ_1^2; and therefore R_1 gives a better value of R than R_2. By taking $p = -q = \pm 1$ we see that the quartile of $R_1 \backsim R_2$ is $Q(\Theta_2^2 - \Theta_1^2)^{\frac{1}{2}}/\sqrt{n}$.

§ 24. *Test of Hypothesis as to Normal Distribution.*—To test whether any particular distribution is normal, we use the result obtained at the beginning of the last section. Having found the average and the standard deviation of the n individuals, we calculate $L_1 + ax$, the value which should correspond to class-index α. The difference between this and the observed value X is a discrepancy whose mean square is $a^2\{(1-\alpha^2)/4z^2 - (1 + \frac{1}{2}x^2)\}/n$, so that the probable discrepancy is $Qa\{(1-\alpha^2)/4z^2 - (1+\frac{1}{2}x^2)\}^{\frac{1}{2}}/\sqrt{n}$; and the actual discrepancy has to be compared, for as many values of x as possible, with this probable discrepancy.

Suppose, for instance, that we take the chest-measurements of Scotch soldiers,* to which QUETELET refers in the work quoted above :—

CHEST-MEASUREMENTS, to the nearest inch, of 5,732 Scotch soldiers.

Inches.	Number.	Inches.	Number.
33	3	41	935
34	19	42	646
35	81	43	313
36	189	44	168
37	409	45	50
38	753	46	18
39	1062	47	3
40	1082	48	1

The values of the average and of the standard deviation cannot, of course, be calculated exactly; as the most probable values we find† $L_1 = 39{\cdot}8489$ inches, $a = 2{\cdot}05301$ inches. Thus we get the following results :—

* 'Edinburgh Medical Journal,' vol. 13, pp. 260–262. QUETELET made some mistakes, which I have corrected, in transcribing the figures.

† The formula for calculating the standard deviation has been given by me in a paper "On the Calculation of the most Probable Values of Frequency-Constants," in vol. 29 of the 'Proceedings of the London Mathematical Society' (p. 353).

The values given in the text are obtained by a first approximation. A second approximation might be made by assuming that the data represent the result of a random selection from the normal distribution given by the first approximation; but this correction would not alter any discrepancy shown in the table by as much as 1 per cent., and it may therefore be omitted.

Value of L.	α.	x*.	$L_1 + a x$.	Discrepancy.	Probable discrepancy.	Ratio of actual to probable discrepancy.
32·5	−1·00000					
33·5	−0·99895					
34·5	−0·99232					
35·5	−0·96406	−2·09762	35·5425	−·0425	·0442	0·96
36·5	−0·89812	−1·63579	36·4906	+·0094	·0263	0·36
37·5	−0·75541	−1·16359	37·4600	+·0400	·0177	2·26
38·5	−0·49267	−0·66301	38·4877	+·0123	·0145	0·85
39·5	−0·12212	−0·15366	39·5334	−·0334	·0138	2·42
40·5	+0·25541	+0·32578	40·5177	−·0177	·0139	1·27
41·5	+0·58165	+0·80928	41·5104	−·0104	·0150	0·69
42·5	+0·80705	+1·30190	42·5217	−·0217	·0195	1·11
43·5	+0·91626	+1·72938	43·3993	+·1007	·0290	3·47
44·5	+0·97488	+2·23952	44·4467	+·0533	·0525	1·02
45·5	+0·99232					
46·5	+0·99860					
47·5	+0·99965					
48·5	+1·00000					

The extremities of the range are not considered, as the values of $\frac{1}{2} n (1 + \alpha)$ or $\frac{1}{2} n (1 - \alpha)$ are small when α is nearly equal to ± 1, so that the law of normal distribution does not hold with regard to the errors in these values; and, moreover, z is changing rapidly, so that ξ is not exactly proportional to θ. For the ten values considered, the actual discrepancy is less than the probable discrepancy in four cases, and greater in six; and for nine of them the ratio of the two is within the probable limit (§ 17). The remaining ratio is rather large (3·47); but otherwise the data appear to justify the hypothesis of normal distribution.†

* The values of x shown in this column correspond to the fractional values of α given by the data (− 2763/2866, − 2574/2866, &c.), not to the nearest decimal values as shown in the second column (− ·96406, − ·89812, &c.).

The quantities shown in the final column are the ratios of the quantities given in the preceding columns. If these were taken to the fifth place of decimals, the last figure in some of the ratios might be altered; but it is not necessary to make such exact calculations (§ 17).

† It should be remembered that when the probable discrepancy is small, the possibility of errors of scale must be considered; thus an inaccuracy of one-hundredth of an inch in a division of the scale near 40 inches would make an appreciable difference in the ratio of the actual to the probable discrepancy. Also it should be noted in the present case that the observed individuals came from different parts of Scotland, so that the "original community" was really heterogeneous; and it is likely that the measurements in different regiments were taken by different observers, with different personal equations, and were not taken with as great care as would be observed at the present day. On the other hand, as the exact measurements are not given, but only the measurements to the nearest inch, the values of L_1 and of a are fitted more closely to the class-indices than they should be; and the probable discrepancy should therefore be slightly less than that given by the theoretical formula.

PART IV.—APPLICATION TO NORMAL CORRELATION.

(1.) *Correlation-Solid of Two Attributes.*

§ 25. *Correlation-Solid in General.*—Let the values of L and of M, the measures of two coexistent attributes A and B, be distributed in any manner whatever. Let L_1 and M_1 be the means, and a^2 and b^2 the mean squares of deviation from the mean. Then we know that the mean value of $(L - L_1)(M - M_1)$ is less than ab. Let this mean value be $ab \cos D$; then the angle D will be called the *divergence* of the two distributions.

Take two lines OX, OY, including an angle $\pi - D$, and on OXY as base-plane construct the solid of frequency of values of $(L - L_1)/a \sin D$ and $(M - M_1)/b \sin D$, these values being measured parallel to OX and OY respectively. Thus if we draw Ox at right angles to OY, and Oy at right angles to OX, and if on Ox and Oy respectively we take $ON' = x'$, $ON'' = x''$, and $On' = y'$, $On'' = y''$, then the portion of the solid included between planes through N′ and N″ at right angles to ON′N″ and planes through n' and n'' at right angles to $On'n''$ includes all the elements representing individuals for which L lies between $L_1 + ax'$ and $L_1 + ax''$, and M between $M_1 + by'$ and $M_1 + by''$. This solid will be called the *correlation-solid* of the two distributions. The ordinates are supposed to be measured on such a scale that the total volume of the solid is unity.

Let $L' = lL + mM$, $M' = l'L + m'M$, and let the means, mean squares of deviation, and mean product of deviation of L′ and M′ be respectively L'_1, M'_1, a'^2, b'^2, and $a'b' \cos D'$. Then

$$L' = lL_1 + mM'_1, \qquad M' = l'L_1 + m'M_1,$$
$$a'^2 = l^2a^2 + 2lmab \cos D + m^2b^2,$$
$$b'^2 = l'^2a^2 + 2l'm'ab \cos D + m'^2b^2,$$
$$a'b' \cos D' = ll'a^2 + (lm' + l'm)\, ab \cos D + mm'b^2.$$

Let WR be any ordinate of the correlation-solid, the co-ordinates of W with regard to OX and OY being x cosec D and y cosec D; and let $a'x' = lax + mby$, $b'y' = l'ax + m'by$. Then WR is proportional to the number of individuals for which $L = L_1 + ax$ and $M = M_1 + by$, and therefore it is proportional to the number for which $L' = L'_1 + a'x'$, $M' = M'_1 + b'y'$. Through O draw the lines OY′, OX′, whose equations referred to OX and OY as axes are $lax + mby = 0$, $l'ax + m'by = 0$; and draw WN parallel to Y′O, meeting OX′ in N (fig. 9). Then ON sin X′OY′ $= (lax + mby)/\{l^2a^2 + 2lmab \cos D + m^2b^2\}^{\frac{1}{2}} = x'$; and similarly NW sin X′OY′ $= y'$. Hence the solid is the solid of frequency of values of $(L' - L'_1)/a'$ sin X′OY′ and $(M' - M'_1)/b'$ sin X′OY′, these values being measured parallel to OX′ and OY′ respectively. Also

$$\cos(\pi - X'OY') = \{ll'a^2 + (lm' + l'm)ab\cos D + mm'b^2\}$$
$$/\,\{(l^2a^2 + 2lmab\cos D + m^2b^2)(l'^2a^2 + 2l'm'ab\cos D + m'^2b^2)\}^{\frac{1}{2}}$$
$$= \cos D',$$

and therefore $X'OY' = \pi - D'$. Hence the solid is the correlation-solid of the distributions of $lL + mM$ and $l'L + m'M$, OX' and OY' being taken as axes.

Thus the correlation-solid of the distributions of L and M is the same as the correlation-solid of the distributions of $lL + mM$ and $l'L + m'M$, where l, m, l', m' are any constants whatever.*

Fig. 9.

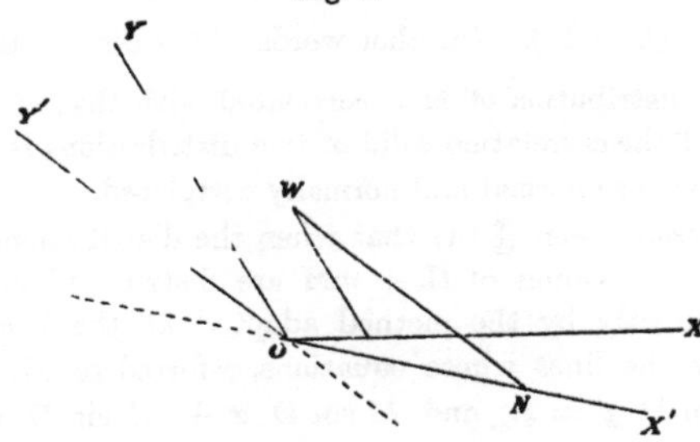

It may be noted that if D_1 and D_2 are the divergences of the distribution of $lL+mM$ from the distributions of L and of M, we have $D = D_1 + D_2$. Or, generally, if the divergence may be supposed to be either positive or negative, and if L, M, N are measures connected by a linear relation $lL + mM + nN = 0$, their divergences D, D', D'' from one another are subject to the relation $D + D' + D'' = 0$.

§ 26. *Correlation-Solid for Normal Distributions.*—(i.) Now suppose that the distribution of L is correlated with that of M, *i.e.*, that the values of M are distributed normally with mean square b^2, and that for any particular value of M the values of L are distributed normally with constant mean square β^2 about a mean value $L_1 + \lambda(M - M_1)$, where λ is a constant. Then (§ 14) we may write $L - L_1 = \lambda(M - M_1) + L'$, where L' is a measure whose values are distributed normally with mean square β^2 about a mean value zero, this distribution being independent of that of M. Hence the mean square of $L - L_1$ is $\lambda^2b^2 + \beta^2$, and the mean product of $L - L_1$ and $M - M_1$ is λb^2; so that, if a^2 is the mean square of $L - L_1$, we have $\lambda = a/b \,.\, \cos D$, $\beta^2 = a^2\sin^2 D$. Thus for any particular value of $(M - M_1)/b\sin D$ the values of $(L - L_1)/a\sin D$ are distributed normally with mean

* We must, of course, allow for the possibility of two solids, which really are identical, appearing to be the "reflexions" of one another.

square unity about a mean value $\{(M - M_1)/b \sin D\} \cos D$. Hence the correlation-solid is a projective solid whose vertical sections by planes parallel to OX are normal figures of semi-parameter unity; and since the values of $(M - M_1)/b$ are distributed normally with mean square unity, the sections by planes at right angles to OX are also normal figures of semi-parameter unity; *i.e.*, the correlation-solid is the standard normal solid.

(ii.) By taking vertical sections parallel to OY, we see that the values of $(L - L_1)/a$ are normally distributed, so that the values of L are normally distributed; and that in any class distinguished by a particular value of L the values of M are distributed normally with mean square $b^2 \sin^2 D$ about a mean value $M_1 + \frac{b}{a} \cos D . (L - L_1)$. In other words, if the distribution of L is correlated with that of M, the distribution of M is correlated with that of L.

(iii.) Conversely, if the correlation-solid of two distributions is the standard normal solid, the distributions are normal and normally correlated.

(iv.) We have already seen (§ 14) that when the distributions of L and of M are normally correlated, the values of $lL + mM$ are distributed normally. We might obtain this result directly by the method adopted at the beginning of § 13. In the base-plane draw the lines whose equations, referred to OX and OY as axes, are $la \sin D . x + mb \sin D . y = \xi_1$, and $la \sin D . x + mb \sin D . y = \xi_2$. Then the vertical planes through these lines will include between them the elements representing individuals for which $l(L - L_1) + m(M - M_1)$ lies between ξ_1 and ξ_2. Draw the central vertical plane at right angles to these planes, cutting the two sections in the ordinates W_1R_1 and W_2R_2. Then the number of these individuals is proportional to the area $W_1R_1R_2W_2$, *i.e.*, it is proportional to the area of the standard normal figure included between ordinates at distances $\xi_1/\{l^2a^2 + 2lmab \cos D + m^2b^2\}^{\frac{1}{2}}$ and $\xi_2/\{l^2a^2 + 2lmab \cos D + m^2b^2\}^{\frac{1}{2}}$ from the median; and therefore the values of $lL_1 + mM_1$ are distributed normally with mean square $l^2a^2 + 2lmab \cos D + m^2b^2$ about the mean value $lL_1 + mM_1$.

(v.) Since (§ 25) the correlation-solid of the distributions of $lL + mM$ and of $l'L + m'M$ is also the standard normal solid, it follows (see (iii.) above) that these two distributions are normally correlated.

§ 27. *Determination of Divergence by Double Median Classification.*—The portion of the solid which lies on the positive side of each of the two planes OZY and OZX (OZ being the axis of the solid) represents all the individuals for which L and M are greater than L_1 and M_1 respectively; and the portion which lies on the negative side of OZY and the positive side of OZX represents those for which L is less than L_1 and M greater than M_1. But, since the solid is a solid of revolution, these volumes are in the ratio of $\pi - D : D$. Hence, if we arrange the whole number of individuals in four classes, thus :—

	Below L_1.	Above L_1.
Below M_1 . .	P	R
Above M_1 . .	R	P

the divergence is equal to $\frac{R}{P+R}\pi$.*

§ 28. *Calculation of Table of Double Classification.*—In the base-plane draw Ox, Oy at right angles to OX, OY, and therefore including an angle D. In Ox take $ON = (X - L_1)/a$, $ON' = (X' - L_1)/a$; and in Oy take $On = (Y - M_1)/b$, $On' = (Y' - M_1)/b$. Through these points draw vertical planes at right angles to Ox and Oy respectively; then (§ 25) the volume of the portion of the standard solid included between these four planes represents the proportion of individuals for which L lies between X and X′ and M between Y and Y′.

The calculation of this volume requires the use of the integral calculus. For a rough calculation we may use either of two methods.

(1.) The planes by which the volume is bounded will meet the base-plane in lines forming a parallelogram, two of the sides of the parallelogram being at right angles to Ox, at distances $(X - L_1)/a$ and $(X' - L_1)/a$ from O, and the other two at right angles to Oy, at distances $(Y - M_1)/b$ and $(Y' - M_1)/b$ from O. Now suppose that the base-plane is divided up into very small areas such that the portions of the solid lying above these areas are all equal. Then the ratio of the number of these areas which lie inside the parallelogram to the total number will be the proportion of individuals for which L lies between X and X′, and M between Y and Y′. For effecting this division of the base-plane into small areas we can use either of the two characteristic properties of the normal solid.

(i.) The solid is a projective solid. Hence if we find the values of x corresponding to $\alpha = \pm 1/m$, $\alpha = \pm 2/m$, ... $\alpha = \pm (m-1)/m$, and if we take the corresponding points on each of two rectangular axes $\xi' O \xi$, $\eta' O \eta$ in the base-plane, and draw lines through these points parallel to $\eta' O \eta$ and to $\xi' O \xi$ respectively, the two sets of lines will divide the base into $4m^2$ areas, corresponding to the division of the solid into $4m^2$ equal portions. Fig. 10 shows the arrangement of these lines for $m = 50$; thus the figure contains 10,000 rectangles (one or two of the sides of some of them being at infinity), and each rectangle represents 1/10,000 of the whole volume of the solid. The centre O of the figure is shown by a small circle. The larger circle is introduced to show the scale; its radius is the semi-parameter of the solid, and is therefore the unit for measuring the distances $(X - L)/a$, &c.

The values of x corresponding to $m = 100$ are given in Table VI. (p. 167); so that

* This formula obviously applies in any case in which the correlation-solid is a solid of revolution.

by means of this table we can divide the base into 40,000 areas, each representing 1/40,000 of the whole volume. To simplify the counting of the areas, every tenth line should be drawn in ink, the others being in pencil; a dot should be placed in each area, and the pencil lines should then be erased. There will thus be 400 larger areas, each containing 100 dots. It will be found convenient to replace the circle shown in fig. 10 by a larger graduated circle; if the radius of this circle is ρ, and if Ox cuts the circumference of the circle in F, the line at right angles to Ox at a distance x from O will cut the circumference in points at an angular distance $\cos^{-1} x/\rho$ from F.

The lines Ox, Oy, &c., may be shown on tracing-paper, instead of on the figure itself; and the paper may then be turned round O into two or three different positions, so as to minimise inaccuracies of counting. Or the figure may be copied on to a glass plate, and the lines Ox, Oy, &c., drawn on ordinary paper.

(ii.) The solid is a solid of revolution, and therefore can be divided into mm' equal portions by a set of m planes through the central ordinate at successive angular distances $2\pi/m$, and a set of concentric cylinders enclosing portions $1/m'$, $2/m'$, ... $(m' - 1)/m'$ of the whole volume. Let the rth cylinder cut a central section in the ordinate MP. Then, if OH is the central ordinate, $r/m' = (\text{OH} - \text{MP})/\text{OH}$ (§§ 5, 11). Hence the radii of the successive cylinders are the abscissæ of the standard curve corresponding to ordinates whose ratios to the central ordinate are respectively $(m' - 1)/m'$, $(m' - 2)/m'$, ... $1/m'$. Thus for $m' = 100$ the values are given by Table II. (p. 155).

This method of division of the base-plane is not so convenient as the method explained in (i.), but it may be used for testing the accuracy of a figure constructed according to that method. If on such a figure we draw circles with the radii given by Table II., each of the rings so formed should contain one-hundredth of the total number of dots in the figure. Or, if we draw circles with radii ·05, ·10, ·15, . . ., the numbers in the successive rings should be proportional to the differences shown in the fourth column of Table I.

(2.) A more accurate method can be adopted when the values of X and X′, and also those of Y and Y′, have been chosen so as to correspond to particular class-indices. Let these be α, α', β, and β' respectively, and let the corresponding abscissæ of the standard normal figure be x, x', y, and y'. Thus $(\text{X} - \text{L}_1)/a = x$, $(\text{X}' - \text{L}_1)/a = x'$, $(\text{Y} - \text{M}_1)/b = y$, $(\text{Y}' - \text{M}_1)/b = y'$. Now if, by the method of § 11, we construct a figure representing the division of the standard solid by parallel vertical planes at distances x and x' from OH, and also a corresponding figure for distances y and y', the bases of the two figures being in the same straight line, and the distance between corresponding extremities being equal to $\text{D}/2\pi$ of either base, the area formed by the two pairs of curves will give the proportion of individuals for which L lies between X and X′, and Y between Y and Y′. The most important case is that in which the class-indices for each distribution separately correspond to the division of the community into p numerically equal classes. The table of double classification of values of L and of M will then contain p^2 compartments; and if we draw the figure corresponding to the division of the standard solid into p equal portions by parallel vertical planes, and shift this figure along its base through a

Fig. 10.

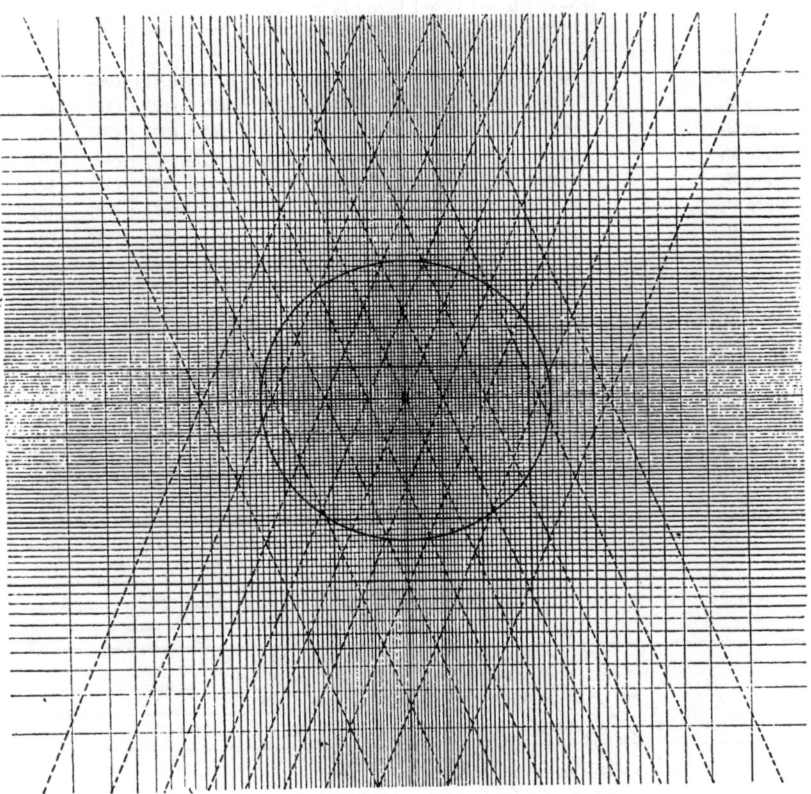

distance equal to $D/2\pi$ of its whole breadth (the part of the figure which projects on one side being superposed on the other side, so as to leave the whole breadth unaltered), we obtain a diagram with p^2 compartments, whose areas are proportional to the numbers in the corresponding compartments of the table of double classification.

Suppose, for instance, that $p = 10$. If $X_1, X_2, \ldots X_9$ and $Y_1, Y_2, \ldots Y_9$ denote the "decile" values of L and of M respectively, the table of double classification will be of this form:—

Values of M.	Values of L.									
	$-\infty$ to L_1.	L_1 to L_2.	L_2 to L_3.	L_3 to L_4.	L_4 to L_5.	L_5 to L_6.	L_6 to L_7.	L_7 to L_8.	L_8 to L_9.	L_9 to $+\infty$.
$-\infty$ to M_1	(00)	(01)	(02)	(03)	(04)	(05)	(06)	(07)	(08)	(09)
M_1 to M_2	(10)	(11)	(12)	(13)	(14)	(15)	(16)	(17)	(18)	(19)
M_2 to M_3	(20)	(21)	(22)	(23)	(24)	(25)	(26)	(27)	(28)	(29)
M_3 to M_4	(30)	(31)	(32)	(33)	(34)	(35)	(36)	(37)	(38)	(39)
M_4 to M_5	(40)	(41)	(42)	(43)	(44)	(45)	(46)	(47)	(48)	(49)
M_5 to M_6	(50)	(51)	(52)	(53)	(54)	(55)	(56)	(57)	(58)	(59)
M_6 to M_7	(60)	(61)	(62)	(63)	(64)	(65)	(66)	(67)	(68)	(69)
M_7 to M_8	(70)	(71)	(72)	(73)	(74)	(75)	(76)	(77)	(78)	(79)
M_8 to M_9	(80)	(81)	(82)	(83)	(84)	(85)	(86)	(87)	(88)	(89)
M_9 to $+\infty$	(90)	(91)	(92)	(93)	(94)	(95)	(96)	(97)	(98)	(99)

The corresponding portions of the standard solid will be bounded by planes whose intersections with the base-plane will form a "plan" such as the following (fig. 11*):—

Fig. 11.

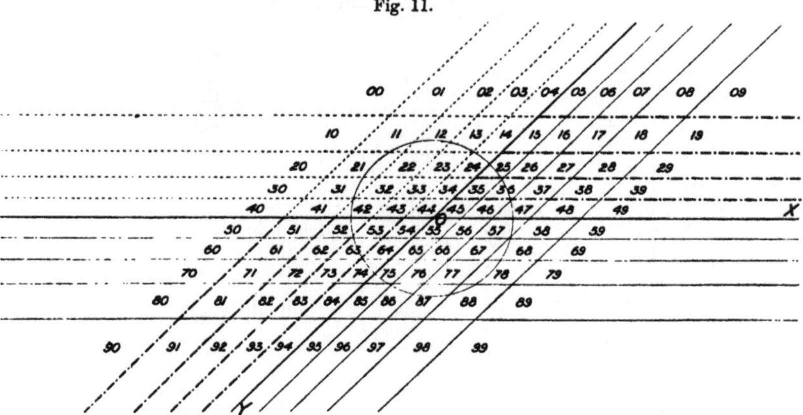

and the volumes of these portions are equal to the one hundred compartments in the diagram formed by shifting fig. 7 (omitting the alternate curves, which correspond to the values ·1, ·3, ·5, ·7, and ·9 of a) through the required distance.

* In this figure, as in fig. 10, the base-plane is supposed to be seen from above. In fig. 9 it is seen from below.

Fig. 12 shows the form of this diagram for the case of $D = \frac{1}{4}\pi$, so that it is produced by shifting fig. 7 from right to left through one-eighth of its whole breadth. The dotted lines in fig. 10 (p. 143) show the position of the corresponding planes dividing the standard solid into 100 portions; the angle between the two sets of lines

Fig. 12.

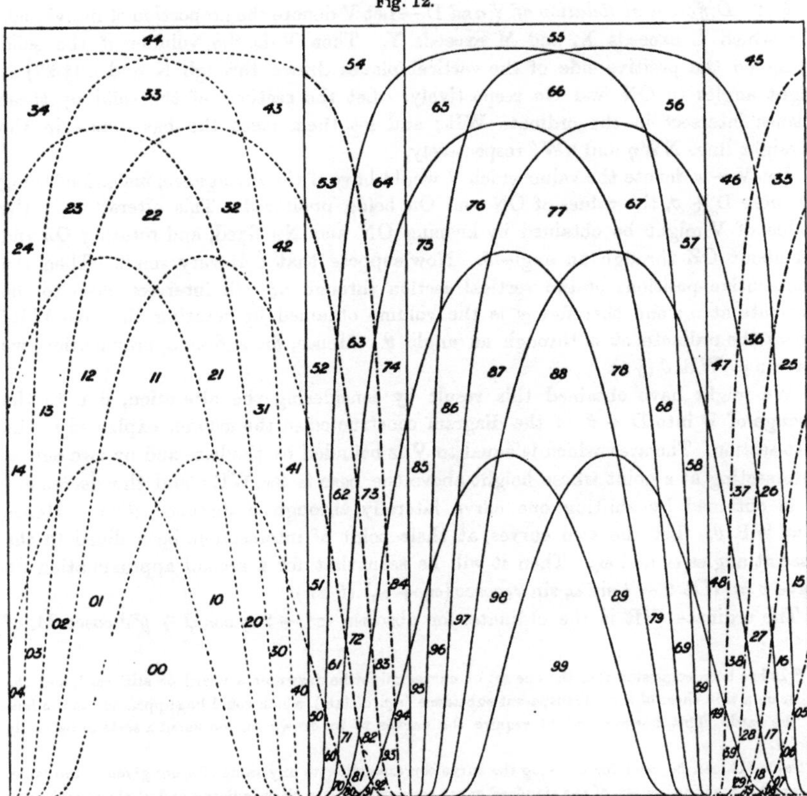

is $\frac{1}{4}\pi$ (or $\frac{3}{4}\pi$), and the distances of the lines in either set from the foot of the central ordinate are respectively ·25335, ·52440, ·84162, and 1·28155, the radius of the circle shown in the figure being the unit. If instead of dividing up the base-plane, as in fig. 10, in the manner explained in (1.) (i.) above, we had divided it up by 100 radial lines and 99 circles as explained in (1.) (ii.), these lines and circles would become vertical and horizontal straight lines dividing the diagram (fig. 12) into 10,000 equal squares.

For practical applications of this method, it is sufficient to have the single figure as shown in fig. 7. The curves representing the displacement of the figure through the distance $D/2\pi$ can then be traced by means of a double-barred parallel rule or an antigraph. But it is better to draw the curves directly from Tables III. and IV.*

§ 29. *Differential Relation of* V *and* D.—Let V denote the proportion of individuals for which L exceeds X, and M exceeds Y. Then V is the volume of the solid lying on the positive side of the vertical planes drawn through N and n (§ 28) at right angles to ON and On respectively. Let the sections of the solid by these planes intersect in the ordinate WR; and let them meet the base-plane in the straight lines NWη and nWξ respectively.

Let $V - v$ denote the value which V would have if the divergence, instead of being D, were $D + \theta$, the values of ON and On being unaltered. This alteration in the value of V might be obtained by keeping ON and Nη fixed, and rotating On and $n\xi$ about OZ through an angle θ. Now suppose that θ is very small. Then the consecutive positions of the vertical section through $n\xi$ will intersect close to the ordinate at n; and therefore v is the volume obtained by rotating the area WRξ about the ordinate at n through an angle θ. Hence, for a first approximation, we have $v = \text{WR}.\theta$ (§ 5).

We might have obtained this result by considering the alteration, due to the change of D into $D + \theta$, of the diagram constructed in the manner explained in the last section. The area which is equal to V is bounded by the base and by two curves intersecting at a point whose height above the base is $2\pi.\text{WR}$; and the decrement v is obtained by shifting one curve laterally through a distance $\theta/2\pi$. Hence $v = \text{WR}.\theta$. Let the two curves, at their point of intersection, be inclined to the base at angles ω_1 and ω_2. Then it will be seen that for a second approximation we have $v = \text{WR}.\theta + \frac{1}{2}\sin\omega_1 \sin\omega_2 \operatorname{cosec}(\omega_1 + \omega_2).(\theta/2\pi)^2$.

The ordinate WR is the ordinate, for abscissa $(x^2 - 2xy\cos D + y^2)^{\frac{1}{2}} \operatorname{cosec} D$, of

* It has been suggested that the one set of curves might be drawn on a board or stiff card, and the other on a thin sheet of some transparent substance (*e.g.*, of talc), which could be slipped across the face of the card. This, however, might require the curves to be drawn on too small a scale to be really useful.

Table III. can be used for drawing the curve corresponding to any value of α not given in the table. If x and x' are the abscissæ of the standard curve corresponding to class-indices α and α', the equations to the corresponding curves of the divergence-diagram are $z = \exp(-\frac{1}{2}x^2\sec^2 2\pi\theta)$ and $z' = \exp(-\frac{1}{2}x'^2\sec^2 2\pi\theta)$. Hence, for any particular value of θ, we have $\log z'/\log z = x'^2/x^2$. The value of z being given by the table, the value of z' may be deduced by means of an ordinary slide-rule and a pair of proportional compasses.

The methods described in the text can be extended to the problems which occur in the theory of the error in the position of a point in a plane (as in BRAVAIS' memoir, referred to by Professor PEARSON). Thus the condition that the point lies within an area limited by the curve $f(x, y) = 0$ is found by taking the curve Σ of § 11 to be the curve whose equation, referred to axes including an angle $\pi - D$, is $f(x/a \sin D, y/b \sin D) = 0$, and then counting the dots or measuring the corresponding cylinder-area.

the normal curve of semi-parameter unity and central ordinate $1/2\pi$ (area $= 1/\sqrt{2\pi}$); where $x = (X - L_1)/a$, $y = (Y - M_1)/b$.

Applications of the Theory of Error.

§ 30. *Probable Error in Value of Divergence, as Obtained by Different Methods.*—Let the distributions of L and of M be normally correlated, the means, mean squares of deviation, and mean product of deviation, being L_1, M_1, a^2, b^2, and $ab \cos D$. If a random selection of n individuals is made, the divergence can be found by any one of several different methods. We require to find the probable error in D, due to the use of each method.

(1.) Suppose that we take the averages, average squares, and average product, as equal to the means, mean squares, and mean product for the complete community. The general expression for the resulting probable error in $\cos D \equiv S_{1,1}/\sqrt{\lambda_2\mu_2}$ has been found in § 20. To find the values of $S_{3,1}$, $S_{2,2}$, and $S_{1,3}$, in the case of normal correlation, we write $L - L_1 = (a/b) \cos D . (M - M_1) + L'$; then $M - M_1$ and L' are independent, and their mean squares are respectively b^2 and $a^2 \sin^2 D$. The mean fourth power of $M - M_1$ is $3b^4$; and thus we find $S_{3,1} = 3a^3b \cos D$, $S_{2,2} = a^2b^2 (1 + 2 \cos^2 D)$, $S_{1,3} = 3ab^3 \cos D$. The table in § 20 becomes

	λ_2	$S_{1,1}$	μ_2
λ_2	$2a^4$	$2a^3b \cos D$	$2a^2b^2 \cos^2 D$
$S_{1,1}$		$a^2b^2 (1 + \cos^2 D)$	$2ab^3 \cos D$
μ_2			$2b^4$

and hence we find that the probable error in D, due to adopting this method, is $Q \sin D/\sqrt{n}$.

(2.) Let D be determined by the method of § 27. Let the medians as given by the data be respectively L'_1 and M'_1, and let the result of the double median classification be

	Below L'_1.	Above L'_1.
Below M'_1	P′	R′
Above M'_1	R′	P′

so that $n = 2(P' + R')$. Let $P = n(\pi - D)/2\pi$, $R = nD/2\pi$; and let the classification of the observed individuals with regard to the true means of the complete community be

	Below L_1.	Above L_1.
Below M_1	(D) $P - \theta - \phi - \psi$	(C) $R + \psi$
Above M_1	(B) $R + \phi$	(A) $P + \theta$

Then the erroneous values L'_1 and M'_1 are obtained by shifting the medians so that this table may present the appearance of the former table. Thus L_1 is shifted so as to transfer $\theta + \psi$ individuals, and M_1 is shifted so as to transfer $\theta + \phi$. In the first case the particular individuals are in the class for which $L = L_1$ (to a first approximation); and the median of M for this class is at M_1, so that half of the $\theta + \psi$ are put from class (C) into class (D), and half from class (A) into class (B). Similarly half of the $\theta + \phi$ are put from (A) into (C), and half from (B) into (D). Hence

$$P' = P - \tfrac{1}{2}\phi - \tfrac{1}{2}\psi, \qquad R' = R + \tfrac{1}{2}\phi + \tfrac{1}{2}\psi;$$

and the error in D is

$$\tfrac{1}{2}\pi(\phi + \psi)/(P + R) = \pi(\phi + \psi)/n.$$

This error is distributed with mean square $D(\pi - D)/n$; and therefore the probable error in D as obtained by the second method is $Q\sqrt{D(\pi - D)}/\sqrt{n}$.

This probable error is of course greater than the probable error due to using the method of (1.), since $\sqrt{D(\pi - D)} > \sin D$.

(3.) Suppose that, instead of taking the medians, we fix on any two class-indices α and β, and divide the total community into four classes (A), (B), (C), and (D) by a double classification with regard to the corresponding values X and Y of L and M respectively, thus:—

	Below X.	Above X.	Total.
Below Y	(D) $\frac{1}{2}(\alpha + \beta) + V = V'''$	(C) $\frac{1}{2}(1 - \alpha) - V = V''$	$\frac{1}{2}(1 + \beta)$
Above Y	(B) $\frac{1}{2}(1 - \beta) - V = V'$	(A) V	$\frac{1}{2}(1 - \beta)$
Total	$\frac{1}{2}(1 + \alpha)$	$\frac{1}{2}(1 - \alpha)$	1

The value of V is different for different values of D. But, if x and y are the abscissæ of the standard normal figure corresponding to the class-indices α and β, it is easily seen that V depends solely on x, y, and D. Hence, if we choose α and β, and observe V, D is (theoretically) determined.

Let the errors in the values of X and of Y be ξ and η. Then the observed value of V is the proportion of individuals for which L exceeds $X+\xi$ and M exceeds $Y+\eta$. Let the actual nubmers coming from the four classes (A), (B), (C), and (D) be $n(V+\psi)$, $n(V'+\psi')$, $n(V''+\psi'')$, and $n(V'''+\psi''')$; thus $\psi+\psi'+\psi''+\psi'''=0$. Let the areas of the sections of the standard solid by the planes $NWR\eta$ and $nWR\xi$ (§ 29) be Γ and Δ, and let these areas be divided by WR in the ratios of $1+\gamma : 1-\gamma$ and $1+\delta : 1-\delta$ respectively. Thus Γ and Δ are equal to the ordinates of the standard figure corresponding to abscissæ x and y (class-indices α and β); while γ is the class-index of Y in the class for which $L = X$, and δ is the class-index of X in the class for which $M = Y$, these being the class-indices corresponding to abscissæ $(y - x\cos D)/\sin D$ and $(x - y\cos D)/\sin D$ in the standard figure.

The erroneous values $X+\xi$ and $Y+\eta$ are obtained by transferring $n(\psi+\psi'')$ individuals from (A) and (C) to (B) and (D), and $n(\psi+\psi')$ from (A) and (B) to (C) and (D). The first transfer takes place (to our order of approximation) in the class for which $L = X$, and the second in the class for which $M = Y$; so that the proportion appearing to fall in (A) is

$$\begin{aligned} V &+ \psi - \tfrac{1}{2}(1-\gamma)(\psi+\psi'') - \tfrac{1}{2}(1-\delta)(\psi+\psi') \\ &= V + \tfrac{1}{2}(1+\gamma).\tfrac{1}{2}(1+\delta).\psi - \tfrac{1}{2}(1+\gamma).\tfrac{1}{2}(1-\delta).\psi' \\ &\quad - \tfrac{1}{2}(1-\gamma).\tfrac{1}{2}(1+\delta).\psi'' + \tfrac{1}{2}(1-\gamma).\tfrac{1}{2}(1-\delta)\psi''' \\ &= V + \Psi. \end{aligned}$$

Let $WR = Z$. Then the error Ψ in V produces (§ 29) an error $-\Psi/Z$ in D, and therefore the probable error in D, as determined by this method, is

$$Q.\Theta/\sqrt{n},$$

where

$$\begin{aligned} \Theta^2 = \tfrac{1}{16}Z^{-2}[\{&V(\overline{1+\gamma}.\overline{1+\delta})^2 + V'(\overline{1+\gamma}.\overline{1-\delta})^2 + V''(\overline{1-\gamma}.\overline{1+\delta})^2 + V'''(\overline{1-\gamma}.\overline{1-\delta})^2\} \\ &- \{V.\overline{1+\gamma}.\overline{1+\delta} - V'.\overline{1+\gamma}.\overline{1-\delta} - V''.\overline{1-\gamma}.\overline{1+\delta} + V'''.\overline{1-\gamma}.\overline{1-\delta}\}^2]. \end{aligned}$$

Since $V+V'+V''+V'''=1$, $V+V'=\tfrac{1}{2}(1-\beta)$, $V+V''=\tfrac{1}{2}(1-\alpha)$, this probable error can be expressed in terms of V, α, β, γ, δ. But the above is the most symmetrical form, and the most convenient for calculation.

(4.) By taking a number of different values of α and β, and observing the corresponding values of V, we get a series of values of D; and then we can take the weighted mean of these, the weights being assigned in such a way as to make the probable error as small as possible.

§ 31. *Relative Accuracy of the Different Methods.*—By means of § 5 it may be shown that, with the notation of § 20 and § 30 (3.),

$$\sigma_{1,0} = a \{\tfrac{1}{2}\Gamma(1-\gamma) + \tfrac{1}{2}\Delta(1-\delta)\cos D\},$$
$$\sigma_{0,1} = b \{\tfrac{1}{2}\Gamma(1-\gamma)\cos D + \tfrac{1}{2}\Delta(1-\delta)\},$$
$$\sigma_{2,0} = a^2\{V + Z\sin D\cos D + \tfrac{1}{2}\Gamma(1-\gamma)x + \tfrac{1}{2}\Delta(1-\delta)y\cos^2 D\},$$
$$\sigma_{1,1} = ab\{V\cos D + Z\sin D + \tfrac{1}{2}\Gamma(1-\gamma)x\cos D + \tfrac{1}{2}\Delta(1-\delta)y\cos D\},$$
$$\sigma_{0,2} = b^2\{V + Z\sin D\cos D + \tfrac{1}{2}\Gamma(1-\gamma)x\cos^2 D + \tfrac{1}{2}\Delta(1-\delta)y\}.$$

Thus from § 20 we have the following table :—

	L_1	M_1	a	b	D	V
L_1	a^2	$ab\cos D$	0	0	0	$a\{\frac{1}{2}\Gamma(1-\gamma) + \frac{1}{2}\Delta(1-\delta)\cos D\}$
M_1		b^2	0	0	0	$b\{\frac{1}{2}\Gamma(1-\gamma)\cos D + \frac{1}{2}\Delta(1-\delta)\}$
a			$\frac{1}{2}a^2$	$\frac{1}{2}ab\cos^2 D$	$-\frac{1}{2}a\sin D\cos D$	$\frac{1}{2}a\{Z\sin D\cos D + \frac{1}{2}\Gamma(1-\gamma)x + \frac{1}{2}\Delta(1-\delta)y\cos^2 D\}$
b				$\frac{1}{2}b^2$	$-\frac{1}{2}b\sin D\cos D$	$\frac{1}{2}b\{Z\sin D\cos D + \frac{1}{2}\Gamma(1-\gamma)x\cos^2 D + \frac{1}{2}\Delta(1-\delta)y\}$
D					$\sin^2 D$	$-[Z\sin^2 D + \frac{1}{2}\{\frac{1}{2}\Gamma(1-\gamma)x + \frac{1}{2}\Delta(1-\delta)y\}\sin D\cos D]$
V						$V(1-V)$

Let the errors in L_1, M_1, a, b, D, be ω, ω', ρ, ρ', θ. The error in V is ψ; if we write this $= \frac{1}{2}\Gamma(1-\gamma)(\omega + x\rho)/a + \frac{1}{2}\Delta(1-\delta)(\omega' + y\rho')/b - Z\theta + \phi$, then it may be shown by the above table that the mean products of ϕ with ω, ω', ρ, ρ', and θ are zero. By writing in the one case $\Delta = 0$, $\gamma = -1$, $Z = 0$, and in the other $\Gamma = 0$, $\delta = -1$, $Z = 0$, we see that $\psi + \psi''$ and $\psi + \psi'$ are of the forms $\Gamma(\omega + x\rho)/a + \chi$ and $\Delta(\omega' + y\rho')/b + \chi'$, where the mean products of χ or χ' with ω, ω', ρ, ρ', and θ, are zero (*cf.* § 23). Hence we obtain the following results :—

(1.) Suppose that we fix on definite values X and Y of L and M, and that we require the proportion of individuals for which L exceeds X, and M exceeds Y. If we determine L_1, M_1, a, b, and D from the averages, average squares, and average product, and then calculate the value of V, the resulting error is $\frac{1}{2}\Gamma(1-\gamma)(\omega + x\rho)/a + \frac{1}{2}\Delta(1-\delta)(\omega' + y\rho')/b - Z\theta$. The mean square of this error is less than the mean square of ψ, the error in V as obtained by direct observation;* and therefore we obtain a better result for V by the calculation than by observation.

(2.) Suppose that we fix on particular class-indices α and β, and that we require the corresponding value of V. The error in V, as determined by calculating the averages, average squares, and average product, is $-Z\theta$; while the error for direct observation is (§ 30 (3.)) $\psi - \frac{1}{2}(1-\gamma)\{\Gamma(\omega + x\rho)/a + \chi\} - \frac{1}{2}(1-\delta)\{\Delta(\omega' + y\rho')/b + \chi'\} = -Z\theta + \phi - \frac{1}{2}(1-\gamma)\chi - \frac{1}{2}(1-\delta)\chi'$. Since the mean products of ϕ, χ, χ' with θ are zero, the mean square of this last error is greater than the mean square of $-Z\theta$.

This result, of course, is identical with (1.); for if the observed class-index of X′ is α', we may consider that we are observing either the class-index of X′ or the value of L corresponding to class-index α'.

(3.) If we determine D by the method of § 30 (3.), the resulting error is $\theta - Z^{-1}\{\phi - \frac{1}{2}(1-\gamma)\chi - \frac{1}{2}(1-\delta)\chi'\}$. The mean product of θ and $\phi - \frac{1}{2}(1-\gamma)\chi - \frac{1}{2}(1-\delta)\chi'$ is zero; hence the probable error due to the method of § 30 (3.) is greater than that due to the method of § 30 (1.).

(4.) Similarly, if we take the weighted mean of a number of different values of D, as in § 30 (4.), we shall still get an error of the form $\theta + \Phi$, where the mean value of $\theta\Phi$ is zero. Hence, if the averages, average squares, and average product can be determined, the value of D so obtained cannot be improved by direct observation of the values of V corresponding to selected pairs of class-indices.†

(5.) Generally, let R be any quantity which would be known if the true means, mean squares, and mean product of L and M were known. Let R_1 be the value obtained by taking these as equal to the averages, average squares, and average product, for the n individuals; and let R_2 be the value obtained by any other method involving observation of the numbers occurring in any set of classes determined by a finite number of class-indices of L and M, with or without the use of the averages, average squares, and average products. Let Θ_1^2/n and Θ_2^2/n be the mean squares of the errors in R as determined by the two methods. Then the propositions stated in § 23 (4.) hold good. The theorem may be extended to the case of any number of mutually correlated attributes.

§ 32. *Test of Hypothesis as to Normal Correlation.*—To test whether the distributions of L and of M, in any particular case, may be regarded as normally correlated. we use the method of § 24, with the necessary modifications.

(1.) With the notation of § 31 (5.), let R denote the proportion of individuals for which L exceeds X and M exceeds Y, the values of X and Y being fixed beforehand. Then, writing $\frac{1}{2}\Gamma(1-\gamma) = A$, $\frac{1}{2}\Delta(1-\delta) = B$, we have

* This shows that $V(1-V)$ is greater than $A^2(1+\frac{1}{2}x^2) + 2AB(1+\frac{1}{2}xy\cos D)\cos D + B^2(1+\frac{1}{2}y^2) + (Ax+By)Z\sin D\cos D + Z^2\sin^2 D$, where $A = \frac{1}{2}\Gamma(1-\gamma)$, $B = \frac{1}{2}\Delta(1-\delta)$.

† *Cf.* KARL PEARSON, in 'Phil. Trans.,' A, vol. 187 (1896), p. 265.

$$\Theta_1^2 = A^2(1+\tfrac{1}{2}x^2) + 2AB(1+\tfrac{1}{2}xy\cos D)\cos D + B^2(1+\tfrac{1}{2}y^2) + (Ax+By)Z\sin D\cos D + Z^2\sin^2 D,$$

and

$$\Theta_2^2 = V(1-V).$$

Thus the "discrepancy" is the difference between V as calculated by finding the means, mean squares, and mean product, and V as found by direct observation; and the probable discrepancy is $Q(\Theta_2^2 - \Theta_1^2)^{\frac{1}{2}}/\sqrt{n}$.

In adopting this method we are testing both the normal distribution of each measure separately and also the normal correlation of the two distributions; and therefore it is not necessary to test first whether the separate distributions are normal.

(2.) Suppose that we are satisfied that the separate distributions are normal, and that we require to test whether, on this assumption, they may be regarded as normally correlated. Then R, in § 31 (5.), will denote the proportion for which L exceeds the value found to correspond to class-index α, and M exceeds the value found to correspond to class-index β. The discrepancy is (§ 30 (3.)) the difference between the errors $-Z\theta$ and $\psi - \frac{1}{2}(1-\gamma)(\psi+\psi'') - \frac{1}{2}(1-\delta)(\psi+\psi')$. (This difference, by § 31 (2.), may be written in the form $\phi - \frac{1}{2}(1-\gamma)\chi - \frac{1}{2}(1-\delta)\chi'$.) The mean square of the discrepancy is $Z^2(\Theta^2 - \sin^2 D)/n$, where Θ^2 has the value given in § 30 (3.); so that the probable discrepancy is $Q.Z(\Theta^2 - \sin^2 D)^{\frac{1}{2}}/\sqrt{n}$. When this method is adopted, the sum of all the discrepancies in any row or in any column of the table of double classification is zero.

(3.) In some cases we are not able either to calculate the averages, average squares, and average product, or to test whether the separate distributions are normal. We must then determine D by some other method, and proceed as in (2.). Suppose, for instance, that D is determined by the double-median-classification method of § 27. Then, as in (2.), the discrepancy is the difference between the value of V, calculated for particular class-indices α and β, and the observed value of V for these class-indices; and the probable discrepancy is $Q.\Phi/\sqrt{n}$, where Φ^2 has different forms according as α and β are positive or negative. If α and β are both positive, it may be shown that

$$\Phi^2 = D(\pi - D)Z^2 - 2(\pi - D)Z\{\tfrac{1}{2}(1-\alpha).\tfrac{1}{2}(1-\gamma) + \tfrac{1}{2}(1-\beta).\tfrac{1}{2}(1-\delta)\} - 2DZV + 2\pi Z\{\tfrac{1}{2}(1-\gamma)W + \tfrac{1}{2}(1-\delta)W'\} + \Theta^2 Z^2;$$

Θ^2 having the value given in § 30 (3.), and W and W′ denoting what V would become if we put $\beta = 0$ and $\alpha = 0$ respectively, without altering the value of D.

TABLES.

Table I.—Ordinate of Standard Normal Curve in terms of Abscissa.

Abscissa $= x$. Ordinate $= z$. $z = \mathrm{C}e^{-\frac{1}{2}x^2}$, where $\mathrm{C} = \frac{1}{\sqrt{2\pi}}$.

x.	z.	z/C.	Differences of z/C.	x.	z.	z/C.	Differences of z/C.
·00	·39894	1·00000	125	1·25	·18265	·45783	2827
·05	·39844	·99875	374	1·30	·17137	·42956	2754
·10	·39695	·99501	620	1·35	·16038	·40202	2671
·15	·39448	·98881	861	1·40	·14973	·37531	2581
·20	·39104	·98020	1097	1·45	·13943	·34950	2485
·25	·38667	·96923	1323	1·50	·12952	·32465	2383
·30	·38139	·95600	1541	1·55	·12001	·30082	2278
·35	·37524	·94059	1747	1·60	·11092	·27804	2170
·40	·36827	·92312	1941	1·65	·10226	·25634	2059
·45	·36053	·90371	2121	1·70	·09405	·23575	1948
·50	·35207	·88250	2287	1·75	·08628	·21627	1837
·55	·34294	·85963	2436	1·80	·07895	·19790	1726
·60	·33322	·83527	2570	1·85	·07206	·18064	1617
·65	·32297	·80957	2687	1·90	·06562	·16447	1509
·70	·31225	·78270	2786	1·95	·05959	·14938	1404
·75	·30114	·75484	2869	2·00	·05399	·13534	1304
·80	·28969	·72615	2935	2·05	·04879	·12230	1205
·85	·27798	·69680	2982	2·10	·04398	·11025	1111
·90	·26609	·66698	3015	2·15	·03955	·09914	1022
·95	·25406	·63683	3030	2·20	·03547	·08892	936
1·00	·24197	·60653	3030	2·25	·03174	·07956	855
1·05	·22988	·57623	3016	2·30	·02833	·07101	780
1·10	·21785	·54607	2986	2·35	·02522	·06321	708
1·15	·20594	·51621	2946	2·40	·02239	·05613	641
1·20	·19419	·48675	2892	2·45	·01984	·04972	578

TABLE I.—Ordinate of Standard Normal Curve in terms of Abscissa (continued).

x.	z.	z/C.	Differences of z/C.	x.	z.	z/C.	Differences of z/C.
2·50	·01753	·04394	521	3·60	·00061	·00153	25
2·55	·01545	·03873	468	3·65	·00051	·00128	22
2·60	·01358	·03405	419	3·70	·00042	·00106	18
2·65	·01191	·02986	374	3·75	·00035	·00088	15
2·70	·01042	·02612	333	3·80	·00029	·00073	13
2·75	·00909	02279	295	3·85	·00024	·00060	10
2·80	·00792	·01984	261	3·90	·00020	·00050	9
2·85	·00687	·01723	231	3·95	·00016	·00041	7
2·90	·00595	·01492	203	4·00	·00013	·00034	12
2·95	·00514	·01289	178	4·10	·00009	·00022	7
3·00	·00443	·01111	156	4·20	·00006	·00015	5
3·05	·00381	·00955	136	4·30	·00004	·00010	4
3·10	·00327	·00819	119	4·40	·00002	·00006	2
3·15	·00279	·00700	102	4·50	·00002	·00004	1
3·20	·00238	·00598	89	4·60	·00001	·00003	1
3·25	·00203	·00509	77	4·70	·00001	·00002	1
3·30	·00172	·00432	66	4·80	·00000	·00001	0
3·35	·00146	·00366	57	4·90	·00000	·00001	1
3·40	·00123	·00309	49	5·00	·00000	·00000	
3·45	·00104	·00260	41				
3·50	·00087	·00219	36				100000
3·55	·00073	·00183	30				

TABLE II.—Abscissa of Standard Normal Curve in terms of Ordinate.

(*Converse of Table I.*)

z/C.	x.	z/C.	x.	z/C.	x.
1·00	·0000000	·66	·9116090	·33	1·4890686
·99	·1417768	·65	·9282057	·32	1·5095922
·98	·2010110	·64	·9447615	·31	1·5304790
·97	·2468166	·63	·9612861	·30	1·5517557
·96	·2857341	·62	·9777891	·29	1·5734512
·95	·3202914	·61	·9942800	·28	1·5955975
·94	·3517823	·60	1·0107677	·27	1·6182295
·93	·3809743	·59	1·0272612	·26	1·6413858
·92	·4083665	·58	1·0437693	·25	1·6651092
·91	·4343056	·57	1·0603008	·24	1·6894475
·90	·4590436	·56	1·0768644	·23	1·7144538
·89	·4827708	·55	1·0934688	·22	1·7401883
·88	·5056350	·54	1·1101226	·21	1·7667189
·87	·5277539	·53	1·1268347	·20	1·7941226
·86	·5492229	·52	1·1436140	·19	1·8224880
·85	·5701209	·51	1·1604693	·18	1·8519171
·84	·5905140	·50	1·1774100	·17	1·8825285
·83	·6104582	·49	1·1944454	·16	1·9144615
·82	·6300015	·48	1·2115851	·15	1·9478809
·81	·6491857	·47	1·2288390	·14	1·9829840
·80	·6680472	·46	1·2462173	·13	2·0200103
·79	·6866183	·45	1·2637307	·12	2·0592540
·78	·7049275	·44	1·2813903	·11	2·1010830
·77	·7230004	·43	1·2992075	·10	2·1459660
·76	·7408601	·42	1·3171944	·09	2·1945139
·75	·7585276	·41	1·3353637	·08	2·2475447
·74	·7760220	·40	1·3537287	·07	2·3061917
·73	·7933609	·39	1·3723036	·06	2·3720922
·72	·8105604	·38	1·3911032	·05	2·4477468
·71	·8276356	·37	1·4101434	·04	2·5372725
·70	·8446004	·36	1·4294413	·03	2·6482288
·69	·8614681	·35	1·4490149	·02	2·7971496
·68	·8782511	·34	1·4688837	·01	3·0348543
·67	·8949610				

TABLE III.—Ordinates of curves of Divergence-diagram in terms of Abscissa.

Abscissa = ·25 ± θ or ·75 ± θ.

Ordinate = z.

$z = e^{-\frac{1}{2}x^2 \sec^2 2\pi\theta}$, the value of x being given by $\alpha = \sqrt{\frac{2}{\pi}} \int_0^x e^{-\frac{1}{2}x^2} dx$.

VALUES of x.

$\alpha = \cdot 0$; $x = 0\cdot 00000\ 00000$
$\alpha = \cdot 1$; $x = 0\cdot 12566\ 13469$
$\alpha = \cdot 2$; $x = 0\cdot 25334\ 71031$
$\alpha = \cdot 3$; $x = 0\cdot 38532\ 04664$
$\alpha = \cdot 4$; $x = 0\cdot 52440\ 05127$
$\alpha = \cdot 5$; $x = 0\cdot 67448\ 97502$
$\alpha = \cdot 6$; $x = 0\cdot 84162\ 12336$
$\alpha = \cdot 7$; $x = 1\cdot 03643\ 33895$
$\alpha = \cdot 8$; $x = 1\cdot 28155\ 15655$
$\alpha = \cdot 9$; $x = 1\cdot 64485\ 36270$

VALUES of z.

	α = ·0	α = ·1	α = ·2	α = ·3	α = ·4	α = ·5	α = ·6	α = ·7	α = ·8	α = ·9
θ	z	z	z	z	z	z	z	z	z	z
·00	1·00000	·99214	·96842	·92845	·87154	·79655	·70176	·58444	·43991	·25852
·01	1·00000	·99210	·96829	·92818	·87106	·79583	·70078	·58320	·43848	·25714
·02	1·00000	·99201	·96792	·92735	·86963	·79366	·69781	·57945	·43418	·25300
·03	1·00000	·99185	·96729	·92595	·86719	·78998	·69277	·57813	·42696	·24610
·04	1·00000	·99162	·96637	·92392	·86367	·78469	·68557	·56411	·41673	·23647
·05	1·00000	·99131	·96514	·92120	·85898	·77765	·67601	·55222	·40338	·22412
·06	1·00000	·99091	·96356	·91771	·85295	·76865	·66386	·53725	·38677	·20912
·07	1·00000	·99040	·96156	·91332	·84540	·75742	·64883	·51891	·36677	·19161
·08	1·00000	·98977	·95907	·90785	·83606	·74363	·63053	·49687	·34322	·17177
·09	1·00000	·98899	·95598	·90110	·82459	·72682	·60848	·47076	·31603	·14993
·10	1·00000	·98801	·95215	·89277	·81052	·70642	·58210	·44016	·28517	·12658
·11	1·00000	·98679	·94738	·88246	·79326	·68172	·55071	·40467	·25078	·10243
·12	1·00000	·98525	·94139	·86962	·77202	·65177	·51351	·36395	·21324	·07842
·13	1·00000	·98329	·93381	·85349	·74571	·61544	·46964	·31785	·17336	·05575
·14	1·00000	·98076	·92405	·83301	·71291	·57130	·41826	·26663	·13251	·03581
·15	1·00000	·97741	·91129	·80665	·67168	·51768	·35876	·21128	·09284	·01993
·16	1·00000	·97288	·89424	·77216	·61946	·45282	·29126	·15402	·05726	·00899
·17	1·00000	·96655	·87086	·72625	·55298	·37527	·21740	·09884	·02906	·00294
·18	1·00000	·95738	·83776	·66399	·46839	·28515	·14176	·05168	·01078	·00057
·19	1·00000	·94340	·78914	·57822	·36254	·18665	·07328	·01900	·00234	·00005
·20	1·00000	·92064	·71457	·45960	·23695	·09236	·02451	·00361	·00018	·00000
·21	1·00000	·88015	·59517	·30110	·10826	·02528	·00326	·00017	·00000	·00000
·22	1·00000	·79862	·40091	·12072	·01992	·00154	·00004	·00000	·00000	·00000
·23	1·00000	·60494	·12964	·00886	·00016	·00000	·00000	·00000	·00000	·00000
·24	1·00000	·13499	·00029	·00000	·00000	·00000	·00000	·00000	·00000	·00000
·25	Indeterminate	·00000	·00000	·00000	·00000	·00000	·00000	·00000	·00000	·00000

TABLE IV.—Abscissæ of curves of Divergence-diagram in terms of Ordinate.

(*Converse of Table III.*)

	$\alpha = \cdot 0$	$\alpha = \cdot 1$	$\alpha = \cdot 2$	$\alpha = \cdot 3$	$\alpha = \cdot 4$	$\alpha = \cdot 5$	$\alpha = \cdot 6$	$\alpha = \cdot 7$	$\alpha = \cdot 8$	$\alpha = \cdot 9$
z	θ	θ	θ	θ	θ	θ	θ	θ	θ	θ
1·00	Indeterminate	..	..	..	..	..	..	..	..	..
·99	·25000	·07662	..	..	..	..	..	..	..	..
·98	·25000	·14252	..	..	..	..	..	..	..	..
·97	·25000	·16498	..	..	..	..	..	..	..	..
·96	·25000	·17753	·07651	..	..	..	..	..	..	..
·95	·25000	·18583	·10478	..	..	..	..	..	..	..
·94	·25000	·19186	·12203	..	..	..	..	..	..	..
·93	·25000	·19650	·13422	..	..	..	..	..	..	..
·92	·25000	·20022	·14349	·05372	..	..	..	..	..	..
·91	·25000	·20328	·15087	·07632	..	..	..	..	..	..
·90	·25000	·20587	·15695	·09145	..	..	..	..	..	..
·89	·25000	·20809	·16207	·10291	..	..	..	..	..	..
·88	·25000	·21003	·16647	·11210	..	..	..	..	..	..
·87	·25000	·21174	·17031	·11973	·01795	..	..	..	..	..
·86	·25000	·21326	·17369	·12624	·04803	..	..	..	..	..
·85	·25000	·21463	·17671	·13189	·06417	..	..	..	..	..
·84	·25000	·21587	·17943	·13686	·07603	..	..	..	..	..
·83	·25000	·21700	·18189	·14128	·08553	..	..	..	..	..
·82	·25000	·21804	·18413	·14526	·09349	..	..	..	..	..
·81	·25000	·21900	·18619	·14886	·10033	..	..	..	..	..
·80	·25000	·21988	·18809	·15215	·10634	..	..	..	..	..
·79	·25000	·22071	·18985	·15517	·11168	·02996	..	..	..	..
·78	·25000	·22148	·19149	·15796	·11649	·04694	..	..	..	..
·77	·25000	·22220	·19302	·16054	·12085	·05863	..	..	..	..
·76	·25000	·22287	·19445	·16295	·12484	·06788	..	..	..	..
·75	·25000	·22351	·19580	·16520	·12851	·07563	..	..	..	..
·74	·25000	·22411	·19707	·16730	·13191	·08233	..	..	..	..
·73	·25000	·22468	·19827	·16929	·13507	·08825	..	..	..	..
·72	·25000	·22523	·19941	·17116	·13801	·09356	..	..	..	..
·71	·25000	·22574	·20049	·17293	·14079	·09838	..	..	..	..
·70	·25000	·22623	·20151	·17460	·14339	·10279	·01337	..	..	..
·69	·25000	·22670	·20249	·17620	·14584	·10686	·03423	..	..	..
·68	·25000	·22715	·20343	·17771	·14816	·11063	·04613	..	..	..
·67	·25000	·22758	·20432	·17916	·15036	·11415	·05523	..	..	..
·66	·25000	·22799	·20518	·18054	·15245	·11744	·06277	..	..	..
·65	·25000	·22839	·20600	·18187	·15445	·12054	·06929	..	..	..
·64	·25000	·22877	·20679	·18314	·15635	·12346	·07506	..	..	..
·63	·25000	·22914	·20755	·18436	·15817	·12622	·08026	..	..	..
·62	·25000	·22949	·20829	·18553	·15991	·12885	·08500	..	..	..
·61	·25000	·22983	·20899	·18666	·16158	·13134	·08936	..	..	..
·60	·25000	·23016	·20968	·18775	·16319	·13372	·09341	..	..	..
·59	·25000	·23048	·21034	·18881	·16473	·13600	·09718	..	..	..
·58	·25000	·23079	·21098	·18982	·16622	·13818	·10073	·01888	..	..
·57	·25000	·23109	·21160	·19081	·16766	·14027	·10406	·03383	..	..
·56	·25000	·23139	·21220	·19176	·16905	·14227	·10722	·04375	..	..
·55	·25000	·23167	·21279	·19269	·17040	·14421	·11021	·05163	..	..
·54	·25000	·23195	·21336	·19358	·17170	·14607	·11306	·05832	..	..
·53	·25000	·23221	·21391	·19446	·17296	·14787	·11577	·06419	..	..
·52	·25000	·23248	·21445	·19531	·17419	·14961	·11837	·06946	..	..

TABLE IV.—Abscissæ of curves of Divergence-diagram in terms of Ordinate (continued).

	$\alpha = \cdot 0$	$\alpha = \cdot 1$	$\alpha = \cdot 2$	$\alpha = \cdot 3$	$\alpha = \cdot 4$	$\alpha = \cdot 5$	$\alpha = \cdot 6$	$\alpha = \cdot 7$	$\alpha = \cdot 8$	$\alpha = \cdot 9$
z	θ	θ	θ	θ	θ	θ	θ	θ	θ	θ
·51	·25000	·23273	·21497	·19613	·17538	·15129	·12086	·07426	..	..
·50	·25000	·23298	·21548	·19694	·17653	·15292	·12326	·07868	..	..
·49	·25000	·23323	·21598	·19772	·17766	·15450	·12556	·08280	..	..
·48	·25000	·23346	·21647	·19849	·17876	·15603	·12778	·08664	..	..
·47	·25000	·23370	·21695	·19924	·17983	·15753	·12992	·09027	..	..
·46	·25000	·23392	·21742	·19997	·18088	·15898	·13200	·09369	..	..
·45	·25000	·23415	·21788	·20069	·18190	·16040	·13401	09695	..	..
·44	·25000	·23437	·21832	·20139	·18290	·16178	·13595	·10005	..	..
·43	·25000	·23458	·21876	·20208	·18387	·16312	·13785	·10301	·02627	..
·42	·25000	·23479	·21920	·20275	·18483	·16444	·13968	·10586	·03711	..
·41	·25000	·23500	·21962	·20341	·18577	·16573	·14147	·10859	·04534	..
·40	·25000	·23520	·22004	·20406	·18669	·16699	·14322	·11122	·05221	..
·39	·25000	·23541	·22045	·20470	·18760	·16822	·14492	·11376	·05821	..
·38	·25000	·23560	·22085	·20533	·18848	·16944	·14659	·11621	·06359	..
·37	·25000	·23580	·22125	·20595	·18936	·17062	·14821	·11859	·06850	..
·36	·25000	·23599	·22164	·20656	·19022	·17179	·14980	·12091	·07304	..
·35	·25000	·23618	·22203	·20716	·19106	·17294	·15137	·12315	·07727	..
·34	·25000	·23637	·22241	·20776	·19190	·17407	·15290	·12534	·08126	..
·33	·25000	·23655	·22279	·20834	·19272	·17518	·15440	·12747	·08503	..
·32	·25000	·23674	·22316	·20892	·19354	·17628	·15588	·12956	·08862	..
·31	·25000	·23692	·22353	·20949	·19434	·17736	·15733	·13160	·09205	..
·30	·25000	·23710	·22390	·21006	·19513	·17843	·15876	·13359	·09534	..
·29	·25000	·23728	·22426	·21062	·19592	·17949	·16018	·13555	·09851	..
·28	·25000	·23745	·22462	·21118	·19670	·18054	·16157	·13748	·10157	..
·27	·25000	·23763	·22498	·21174	·19748	·18157	·16295	·13937	·10453	..
·26	·25000	·23780	·22534	·21229	·19824	·18260	·16431	·14123	·10741	..
·25	·25000	·23798	·22569	·21283	·19901	·18362	·16566	·14307	·11022	·02485
·24	·25000	·23815	·22604	·21338	·19977	·18464	·16700	·14489	·11295	·03665
·23	·25000	·23832	·22640	·21392	·20053	·18565	·16833	·14668	·11563	·04550
·22	·25000	·23850	·22675	·21446	·20128	·18665	·16966	·14846	·11825	·05293
·21	·25000	·23867	·22710	·21501	·20204	·18766	·17098	·15022	·12083	·05946
·20	·25000	·23884	·22745	·21555	·20279	·18866	·17229	·15198	·12337	·06538
·19	·25000	·23902	·22780	·21609	·20355	·18966	·17360	·15372	·12588	·07085
·18	·25000	·23919	·22816	·21664	·20431	·19067	·17492	·15547	·12836	·07598
·17	·25000	·23937	·22852	·21719	·20507	·19168	·17623	·15721	·13082	·08084
·16	·25000	·23955	·22888	·21775	·20584	·19270	·17756	·15895	·13327	·08549
·15	·25000	·23973	·22924	·21831	·20662	·19372	·17889	·16070	·13572	·08997
·14	·25000	·23991	·22961	·21888	·20740	·19476	·18024	·16247	·13816	·09432
·13	·25000	·24009	·22999	·21945	·20820	·19582	·18160	·16425	·14062	·09857
·12	·25000	·24028	·23037	·22004	·20902	·19689	·18299	·16606	·14309	·10274
·11	·25000	·24048	·23076	·22065	·20985	·19799	·18441	·16790	·14560	·10688
·10	·25000	·24068	·23117	·22127	·21071	·19911	·18586	·16978	·14814	·11100
·09	·25000	·24088	·23159	·22191	·21160	·20028	·18736	·17171	·15075	·11514
·08	·25000	·24110	·23202	·22258	·21252	·20149	·18891	·17372	·15343	·11933
·07	·25000	·24132	·23248	·22328	·21349	·20276	·19054	·17582	·15623	·12361
·06	·25000	·24156	·23297	·22403	·21452	·20411	·19227	·17803	·15916	·12805
·05	·25000	·24183	·23350	·22484	·21564	·20557	·19414	·18041	·16230	·13272
·04	·25000	·24211	·23408	·22574	·21687	·20718	·19619	·18303	·16573	·13775
·03	·25000	·24245	·23475	·22676	·21827	·20901	·19853	·18600	·16960	·14334
·02	·25000	·24285	·23557	·22801	·21998	·21124	·20136	·18959	·17425	·14995
·01	·25000	·24341	·23670	·22974	·22236	·21433	·20528	·19453	·18062	·15884
·00	·25000	·25000	·25000	·25000	·25000	·25000	·25000	·25000	·25000	·25000

TABLE V.—Table for Calculation of Probable Error.

This table gives $Q\sqrt{N}$ in terms of N, where $Q = \cdot 67448975\ldots$, and N has any value. The values of N in the first column are the values corresponding to values of $Q\sqrt{N}$ intermediate between those in the second column. Thus $Q\sqrt{N} = 93\cdot 5$ gives $N = 19216$, and $Q\sqrt{N} = 94\cdot 5$ gives $N = 19630$; and therefore for any value of N between 19216 and 19630 the value of $Q\sqrt{N}$ to the nearest integer is 94. The figures in N are arranged in pairs, since the result of dividing $\sqrt{N}$ by 10 is to divide N by 100. Thus for $N = \cdot 01\ 93\ 00$ the value of $Q\sqrt{N}$ to three places of decimals is $\cdot 094$; and similarly, if $N = \cdot 00\ 00\ 01\ 93$, $Q\sqrt{N} = \cdot 00094$, correct to five places of decimals. Thus the table gives $Q\sqrt{N}$ within from ·8 to ·08 per cent. of its value, without the necessity for any interpolation. This is accurate enough for ordinary purposes.

N.	Q√N.	N.	Q√N.	N.	Q√N.
00 97 21	067	01 56 95	085	02 30 94	103
01 00 15	068	60 69	086	35 47	104
03 14	069	64 47	087	40 04	105
06 17	070	68 29	088	44 66	106
09 25	071	72 16	089	49 32	107
12 37	072	76 07	090	54 02	108
15 54	073	80 03	091	58 77	109
18 75	074	84 03	092	63 56	110
22 00	075	88 08	093	68 39	111
25 30	076	92 16	094	73 27	112
28 64	077	96 30	095	78 20	113
32 02	078	02 00 47	096	83 17	114
35 45	079	04 69	097	88 18	115
38 93	080	08 96	098	93 23	116
42 44	081	13 27	099	98 33	117
46 00	082	17 62	100	03 03 48	118
49 61	083	22 01	101	08 66	119
53 26	084	26 45	102	13 90	120

TABLE V.—Table for Calculation of Probable Error (continued).

N.	Q√N.	N.	Q√N.	N.	Q√N.
03 19 17	121	04 84 73	149	06 84 76	177
24 49	122	91 28	150	92 54	178
29 85	123	97 88	151	07 00 37	179
35 26	124	05 04 52	152	08 24	180
40 71	125	11 20	153	16 15	181
46 21	126	17 92	154	24 11	182
51 75	127	24 69	155	32 11	183
57 33	128	31 51	156	40 15	184
62 96	129	38 37	157	48 24	185
68 63	130	45 27	158	56 37	186
74 34	131	52 21	159	64 55	187
80 10	132	59 20	160	72 77	188
85 91	133	66 24	161	81 04	189
91 75	134	73 32	162	89 35	190
97 64	135	80 44	163	97 70	191
04 03 58	136	87 60	164	08 06 10	192
09 56	137	94 81	165	14 54	193
15 58	138	06 02 07	166	23 02	194
21 65	139	09 37	167	31 55	195
27 76	140	16 71	168	40 12	196
33 91	141	24 09	169	48 74	197
40 11	142	31 52	170	57 40	198
46 35	143	39 00	171	66 10	199
52 64	144	46 51	172	74 85	200
58 97	145	54 07	173	83 65	201
65 35	146	61 68	174	92 48	202
71 76	147	69 33	175	09 01 36	203
78 23	148	77 02	176	10 29	204

TABLE V.—Table for Calculation of Probable Error (continued).

N.	Q√N.	N.	Q√N.	N.	Q√N.
09 19 25	205	11 88 22	233	14 91 64	261
28 27	206	98 46	234	15 03 12	262
37 32	207	12 08 75	235	14 63	263
46 42	208	19 08	236	26 20	264
55 57	209	29 45	237	37 80	265
64 76	210	39 87	238	49 45	266
73 99	211	50 33	239	61 15	267
83 26	212	60 84	240	72 88	268
92 58	213	71 39	241	84 67	269
10 01 95	214	81 99	242	96 49	270
11 36	215	92 63	243	16 08 36	271
20 81	216	13 03 31	244	20 28	272
30 30	217	14 04	245	32 23	273
29 84	218	24 81	246	44 24	274
49 43	219	35 62	247	56 28	275
59 05	220	46 48	248	68 37	276
68 73	221	57 38	249	80 50	277
78 44	222	68 33	250	92 68	278
88 20	223	79 32	251	17 04 90	279
98 01	224	90 35	252	17 17	280
11 07 85	225	14 01 43	253	29 48	281
17 74	226	12 55	254	41 83	282
27 68	227	23 72	255	54 23	283
37 66	228	34 93	256	66 67	284
47 68	229	46 19	257	79 16	285
57 75	230	57 48	258	91 68	286
67 86	231	68 83	259	18 04 26	287
78 02	232	80 21	260	16 87	288

TABLE V.—Table for Calculation of Probable Error (continued).

N.	Q√N.	N.	Q√N.	N.	Q√N.
18 29 54	289	22 01 90	317	26 08 72	345
42 24	290	15 83	318	23 89	346
54 99	291	29 81	319	39 10	347
67 78	292	43 84	320	54 35	348
80 62	293	57 90	321	69 65	349
93 50	294	72 02	322	85 00	350
19 06 43	295	86 17	323	27 00 38	351
19 39	296	23 00 37	324	15 81	352
32 41	297	14 61	325	31 29	353
45 46	298	28 90	326	46 81	354
58 56	299	43 23	327	62 37	355
71 71	300	57 61	328	77 98	356
84 90	301	72 03	329	93 63	357
98 13	302	86 49	330	28 09 32	358
20 11 41	303	24 01 00	331	25 06	359
24 73	304	15 55	332	40 84	360
38 09	305	30 15	333	56 67	361
51 50	306	44 79	334	72 54	362
64 95	307	59 47	335	88 45	363
78 45	308	74 20	336	29 04 41	364
91 99	309	88 97	337	20 41	365
21 05 57	310	25 03 78	338	36 46	366
19 20	311	18 64	339	52 55	367
32 87	312	33 55	340	68 68	368
46 59	313	48 49	341	84 86	369
60 35	314	63 48	342	30 01 08	370
74 16	315	78 52	343	17 35	371
88 00	316	93 60	344	33 66	372

TABLE V.—Table for Calculation of Probable Error (continued).

N.	Q√N.	N.	Q√N.	N.	Q√N.
30 50 01	373	35 25 77	401	40 36 00	429
66 41	374	43 40	402	54 86	430
82 85	375	61 07	403	73 76	431
99 34	376	78 79	404	92 71	432
31 15 87	377	96 55	405	41 11 70	433
32 44	378	36 14 36	406	30 74	434
49 06	379	32 21	407	49 82	435
65 72	380	50 10	408	68 94	436
82 43	381	68 03	409	88 11	437
99 18	382	86 02	410	42 07 32	438
32 15 97	383	37 04 04	411	26 57	439
32 81	384	22 11	412	45 87	440
49 69	385	40 22	413	65 22	441
66 62	386	58 38	414	84 60	442
83 59	387	76 58	415	43 04 04	443
33 00 60	388	94 82	416	23 51	444
17 66	389	38 13 11	417	43 03	445
34 76	390	31 44	418	62 59	446
51 90	391	49 82	419	82 20	447
69 09	392	68 24	420	44 01 85	448
86 32	393	86 70	421	21 55	449
34 03 60	394	39 05 21	422	41 29	450
20 92	395	23 76	423	61 07	451
38 29	396	42 36	424	80 90	452
55 70	397	61 00	425	45 00 77	453
73 15	398	79 68	426	20 68	454
90 65	399	98 41	427	40 64	455
35 08 19	400	40 17 18	428	60 64	456

TABLE V.—Table for Calculation of Probable Error (continued).

N.	Q$\sqrt{N}$.	N.	Q$\sqrt{N}$.	N.	Q$\sqrt{N}$.
45 80 69	457	51 59 85	485	57 73 47	513
46 00 78	458	81 17	486	96 02	514
20 91	459	52 02 53	487	58 18 62	515
41 09	460	23 94	488	41 26	516
61 32	461	45 40	489	63 95	517
81 58	462	66 90	490	86 67	518
47 01 89	463	88 44	491	59 09 45	519
22 25	464	53 10 02	492	32 26	520
42 65	465	31 65	493	55 12	521
63 09	466	53 32	494	78 03	522
83 57	467	75 04	495	60 00 98	523
48 04 11	468	96 80	496	23 97	524
24 68	469	54 18 61	497	47 00	525
45 30	470	40 46	498	70 08	526
65 96	471	62 35	499	93 21	527
86 67	472	84 29	500	61 16 38	528
49 07 42	473	55 06 27	501	39 59	529
28 21	474	28 29	502	62 84	530
49 05	475	50 36	503	86 14	531
69 93	476	72 48	504	62 09 49	532
90 86	477	94 63	505	32 88	533
50 11 83	478	56 16 83	506	56 31	534
32 84	479	39 08	507	79 78	535
53 90	480	61 37	508	63 03 30	536
75 00	481	83 70	509	26 87	537
96 15	482	57 06 08	510	50 48	538
51 17 34	483	28 50	511	74 13	539
38 57	484	50 96	512	97 82	540

TABLE V.—Table for Calculation of Probable Error (continued).

N.	Q√N.	N.	Q√N.	N.	Q√N.
64 21 56	541	71 04 12	569	78 21 14	597
45 35	542	29 13	570	47 39	598
69 17	543	54 19	571	73 68	599
93 04	544	79 29	572	79 00 01	600
65 16 96	545	72 04 44	573	26 39	601
40 92	546	29 63	574	52 81	602
64 92	547	54 87	575	79 27	603
88 97	548	80 14	576	80 05 78	604
66 13 06	549	73 05 47	577	32 34	605
37 20	550	30 83	578	58 93	606
61 38	551	56 24	579	85 57	607
85 60	552	81 70	580	81 12 26	608
67 09 87	553	74 07 19	581	38 99	609
34 18	554	32 74	582	65 76	610
58 53	555	58 32	583	92 58	611
82 93	556	83 95	584	82 19 44	612
68 07 37	557	75 09 63	585	46 34	613
31 86	558	35 34	586	73 29	614
56 39	559	61 11	587	83 00 29	615
80 97	560	86 91	588	27 32	616
69 05 59	561	76 12 76	589	54 40	617
30 25	562	38 66	590	81 53	618
54 96	563	64 59	591	84 08 70	619
79 71	564	90 57	592	35 91	620
70 04 50	565	77 16 60	593	63 17	621
29 34	566	42 67	594	90 47	622
54 22	567	68 78	595	85 17 81	623
79 15	568	94 94	596	45 20	624

TABLE V.—Table for Calculation of Probable Error (continued).

N.	$Q\sqrt{N}$.	N.	$Q\sqrt{N}$.	N.	$Q\sqrt{N}$.
85 72 63		90 45 71		95 31 49	
	625		642		659
86 00 11		73 93		60 46	
	626		643		660
27 63		91 02 20		89 48	
	627		644		661
55 19		30 51		96 18 54	
	628		645		662
82 80		58 87		47 64	
	629		646		663
87 10 45		87 27		76 79	
	630		647		664
38 15		92 15 71		97 05 98	
	631		648		665
65 89		44 20		35 21	
	632		649		666
93 67		72 73		64 49	
	633		650		667
88 21 50		93 01 31		93 81	
	634		651		668
49 37		29 92		98 23 18	
	635		652		669
77 29		58 59		52 59	
	636		653		670
89 05 25		87 30		82 05	
	637		654		671
33 25		94 16 05		99 11 54	
	638		655		672
61 30		44 84		41 09	
	639		656		673
89 39		73 68		70 67	
	640		657		674
90 17 53		95 02 56		100 00 30	
	641		658		

TABLE VI.—Abscissa of Standard Normal Curve in terms of Class-Index.

Class-Index = α.

Abscissa = x.

$$\alpha = \sqrt{\frac{2}{\pi}}\int_0^x e^{-\frac{1}{2}x^2}\,dx\,.$$

α.	x.	α.	x.	α.	x.
·00	·00000	·34	·43991	·67	·97411
·01	·01253	·35	·45376	·68	·99446
·02	·02507	·36	·46770	·69	1·01522
·03	·03761	·37	·48173	·70	1·03643
·04	·05015	·38	·49585	·71	1·05812
·05	·06271	·39	·51007	·72	1·08032
·06	·07527	·40	·52440	·73	1·10306
·07	·08784	·41	·53884	·74	1·12639
·08	·10043	·42	·55338	·75	1·15035
·09	·11304	·43	·56805	·76	1·17499
·10	·12566	·44	·58284	·77	1·20036
·11	·13830	·45	·59776	·78	1·22653
·12	·15097	·46	·61281	·79	1·25357
·13	·16366	·47	·62801	·80	1·28155
·14	·17637	·48	·64335	·81	1·31058
·15	·18912	·49	·65884	·82	1·34076
·16	·20189	·50	·67449	·83	1·37220
·17	·21470	·51	·69031	·84	1·40507
·18	·22754	·52	·70630	·85	1·43953
·19	·24043	·53	·72248	·86	1·47579
·20	·25335	·54	·73885	·87	1·51410
·21	·26631	·55	·75542	·88	1·55477
·22	·27932	·56	·77219	·89	1·59819
·23	·29237	·57	·78919	·90	1·64485
·24	·30548	·58	·80642	·91	1·69540
·25	·31864	·59	·82389	·92	1·75069
·26	·33185	·60	·84162	·93	1·81191
·27	·34513	·61	·85962	·94	1·88079
·28	·35846	·62	·87790	·95	1·95996
·29	·37186	·63	·89647	·96	2·05375
·30	·38532	·64	·91537	·97	2·17009
·31	·39886	·65	·93459	·98	2·32635
·32	·41246	·66	·95417	·99	2·57583
·33	·42615				

—V. *On the Recon-*
s.

*ondon.**

INDEX SLIP.

work of this paper from
table on p. 180, and the
epared the diagrams and

30.12.98

IV. *Mathematical Contributions to the Theory of Evolution.—*V. *On the Reconstruction of the Stature of Prehistoric Races.*

By KARL PEARSON, *F.R.S., University College, London.**

Received June 6,—Read June 16, 1898.

[PLATES 3 AND 4]

CONTENTS.

* I have received constant aid and assistance in the laborious arithmetical work of this paper from Mr. LESLIE BRAMLEY-MOORE. To Miss ALICE LEE I owe the fundamental table on p. 180, and the preparation of the test tables on pp. 188 and 189. Mr. G. U. YULE has prepared the diagrams and repeatedly assisted me with suggestion and criticism.

(1.) THE object of this paper is to show, by the use of a special case as illustration, the true limits within which it is possible to reconstruct the parts of an extinct race from a knowledge of the size of a few organs or bones, when complete measurements have been or can be made for an allied and still extant race. The illustration I have taken is one of considerable interest in itself, and has been considered from a variety of standpoints by a long series of investigators. But I wish it to be considered purely as an illustration of a general method. What is here done for stature from long bones is equally applicable to other organs in Man. We might reconstruct in the same manner the dimensions of the hand from a knowledge of any of the finger bones, or the bones of the upper limbs from a knowledge of the bones of the lower limbs. Further, we need not confine our attention to Man, but can predict, with what often amounts to a remarkable degree of accuracy, the dimensions of the organs of one local race of any species from a knowledge of a considerable number of organs in a second local race, and of only one or two organs of the first. The importance of this result for the reconstruction of fossil or prehistoric races will be obvious.

What we need for any such reconstruction are the following data :—

(*a.*) The mean sizes, the variabilities (standard-deviations), and the correlations of as many organs in an extant allied race as it is possible conveniently to measure. When the correlations of the organs under consideration are high (*e.g.*, the long bones in Man), fifty to a hundred individuals may be sufficient; in other cases it is desirable that several hundred at least should be measured.

(*b.*) The like sizes or characters for as many individual organs or bones of the extinct race should then be measured as it is possible to collect. It will be found always possible to reconstruct the *mean* racial type with greater accuracy than to reconstruct a single individual.

(*c.*) An appreciation must be made of the effect of time and climate in producing changes in the dimensions of the organs which have survived from the extinct race.

(2.) Supposing the above data to exist in any particular instance, we have next to ask what is theoretically the best method of dealing with them. There cannot be a doubt about the answer to be given. If we know an organ A, then the most probable value of an organ B is that given by the regression formula for the two organs. Let m_a, m_b be the mean sizes of A and B, σ_a, σ_b their standard deviations, r_{ab} their coefficient of correlation, then the most probable value of B for a given value of A is,

$$B - m_b = \frac{\sigma_b}{\sigma_a} r_{ab} (A - m_a)$$

or

$$B = \left(m_b - \frac{\sigma_b}{\sigma_a} r_{ab} m_a\right) + \frac{\sigma_b}{\sigma_a} r_{ab} A$$

$$= c_1 + c_2 A \quad \ldots \quad \text{(i.)}$$

where c_1 and c_2 are constants for the pair of organs under consideration. The probable error of such a determination is $\cdot 67449\, \sigma_b \times \sqrt{(1 - r_{ab}^2)}$.

Now there are several points to be noticed here.

(i.) If r_{ab} be small, the probable error of reconstruction will be large, if the organ B is to be reconstructed for a single individual. No ingenuity in constructing other formulæ can in the least get over this difficulty; it is simply an expression of the fact that races are variable. Any formula which professes to reconstruct individuals with extreme accuracy may at once be put aside as unscientific. On the other hand, if A be known for p individuals, the corresponding *mean* value of the unknown organ B may be found with a probable error of $\cdot 67449\, \sigma_b \times \sqrt{(1 - r_{ab}^2)}/\sqrt{p}$, and thus with increasing accuracy as p increases.

(ii.) Anthropologists and anatomists have frequently assumed that the ratio of two organs, B/A, is the measure to be ascertained in a reconstruction problem. They were soon compelled to admit, however, that this varies with A, and accordingly have tabulated the ratio B/A for three or four ranges of the organ A. Such a table, for example is given by M. MANOUVRIER* for the ratio of stature to the length of the six long bones. He gives the ratio for three values of each long bone. He also in a second table gives values of the ratios which are to be taken when the long bones exceed or fall short of certain values, *i.e.*, in cases of what he terms *macroskely* and *microskely*. The regression formula shows us that:

$$B/A = c_2 + c_1/A,$$

and since c_1 is never small as compared with A, this ratio can never be treated as constant. Accordingly, while a table can be constructed which will give quite good reconstruction values, by determining the mean value of B/A for each value of A, we see that it is theoretically an erroneous principle to start from; no constancy of the ratio B/A ought to be expected. The theory of regression shows us that the most probable value of B is expressible, so long as the correlation is normal (or at least "linear"), as a *linear* function of A.†

* 'Mémoires de la Société d'Anthropologie de Paris,' vol. 4, pp. 347–402

† Sir GEORGE HUMPHRY gives a table of the ratio B/A for stature in his "Treatise on the Human

(3.) So far we have dealt only with the reconstruction of the most probable value of B from one organ A, but we may propose to find the most probable value of B from n organs $A_1, A_2, A_3 \ldots A_n$. Let r_{0q} represent the correlation coefficient of B and the organ A_q, $r_{qq'}$ the correlation coefficient of A_q and $A_{q'}$; σ_0 the S.D. of B, and σ_q of the organ A_q, m_0 the mean of B, and m_q of A_q; let R be the determinant

$$\begin{vmatrix} 1 & r_{01} & r_{02} & r_{03} & \cdot & \cdot & \cdot & r_{0n} \\ r_{10} & 1 & r_{12} & r_{13} & \cdot & \cdot & \cdot & r_{1n} \\ r_{20} & r_{21} & 1 & r_{23} & \cdot & \cdot & \cdot & r_{2n} \\ r_{30} & r_{31} & r_{32} & 1 & \cdot & \cdot & \cdot & r_{3n} \\ \cdot & \cdot & \cdot & \cdot & \cdot & \cdot & \cdot & \cdot \\ \cdot & \cdot & \cdot & \cdot & \cdot & \cdot & \cdot & \cdot \\ r_{n0} & r_{n1} & r_{n2} & r_{n3} & \cdot & \cdot & \cdot & 1 \end{vmatrix}$$

and R_{pq}, the minor corresponding to r_{pq}. Then the general theory of correlation shows us that

$$B - m_0 = -\frac{R_{01}}{R_{00}}\frac{\sigma_0}{\sigma_1}(A_1 - m_1) - \frac{R_{02}}{R_{00}}\frac{\sigma_0}{\sigma_2}(A_2 - m_2) \ldots - \frac{R_{0n}}{R_{00}}\frac{\sigma_0}{\sigma_n}(A_n - m_n). \quad \text{(ii.)}$$

is the most probable value of B, and that there is a probable error $= \cdot 67449\, \sigma_b \sqrt{(R/R_{00})}$ in this determination.

Thus we reach again a formula of the character

$$B = c_0 + c_1 A_1 + c_2 A_2 + c_3 A_3 + \ldots + c_n A_n,$$

or, B is expressible as a *linear* function of the organs from which its value is to be predicted. This again supposes normal, or at least "linear" correlation. Now there are several points to be noticed here.

(i.) The linear function which will give the best value for B is unique. For example, some anthropologists have attempted to reconstruct stature by adding together the lengths of femur and tibia. The proportions in which femur and tibia are to be combined are given once for all by the regression formula, and they are not those of equality. I have succeeded in proving the following general theorem, which settles this point conclusively. Given any linear function of the n organs $A_1, A_2, A_3 \ldots A_n$, say

$$b_0 + b_1 A_1 + b_2 A_2 + b_3 A_3 + \ldots + b_n A_n,$$

Skeleton," Cambridge, 1858, p. 108. Many others have been given by French writers, in some cases with several values of B/A for three ranges of stature or of long bone (TOPINARD, ROLLET, etc.). Dr. BEDDOE has given a rule which really amounts to making B a linear function of A, but his values for c_1 and c_2 are widely divergent from what I have obtained by applying the theory of correlation. 'Journal of the Anthropological Institute,' vol. 17, 1888, p. 205.

and let ρ be the correlation of this expression with B, then ρ will be greatest or the probable error of the determination of B by means of its correlation with such an expression will be least, *i.e.*, $\cdot 67449\, \sigma_0 \sqrt{(1-\rho^2)}$ will be least, when the b's are proportional to the corresponding c's of the regression formula.

Let Σ be the standard-deviation of the quantity

$$Q = b_0 + b_1 A_1 + b_2 A_2 + \ldots + b_n A_n.$$

Then

$$\Sigma^2 = S_1^n (b_1^2 \sigma_1^2) + 2S (b_1 b_2 \sigma_1 \sigma_2 r_{12})$$

and

$$\rho = S_1^n (b_1 r_{01} \sigma_1)/\Sigma.$$

The best value of B as determined from Q is

$$B = m_0 + \frac{\sigma_0 \rho}{\Sigma} \{b_1 (A_1 - m_1) + b_2 (A_2 - m_2) + \ldots + b_n (A_n - m_n)\} \quad . \quad \text{(iii.)},$$

with a probable error $\cdot 67449\, \sigma_0 \sqrt{(1 - \rho^2)}$.

This may be taken to be any linear function of the A's, since so far $b_1, b_2 \ldots b_n$ are n quite arbitrary constants, and the constant b_0 has to satisfy the condition that B takes its mean value when the A's take their mean values.

Now select such a value of the b's as to give the greatest value to ρ. By differentiating ρ with regard to the b's in succession we find the system of equations

$$\begin{aligned}
r_{01}\Sigma/\rho &= b_1\sigma_1 + b_2\sigma_2 r_{12} + b_3\sigma_3 r_{13} + \ldots + b_n\sigma_n r_{1n} \\
r_{02}\Sigma/\rho &= b_1\sigma_1 r_{12} + b_2\sigma_2 + b_3\sigma_3 r_{23} + \ldots + b_n\sigma_n r_{2n} \\
r_{03}\Sigma/\rho &= b_1\sigma_1 r_{13} + b_2\sigma_2 r_{23} + b_3\sigma_3 + \ldots + b_n\sigma_n r_{3n} \\
&\ldots\ldots\ldots\ldots\ldots\ldots \\
&\ldots\ldots\ldots\ldots\ldots\ldots \\
r_{0n}\Sigma/\rho &= b_1\sigma_1 r_{1n} + b_2\sigma_2 r_{2n} + b_3\sigma_n r_{3n} + \ldots + b_n\sigma_n.
\end{aligned}$$

The solutions of these equations are

$$b_1\sigma_1 = -\frac{R_{01}}{R_{00}}\frac{\Sigma}{\rho}, \qquad b_2\sigma_2 = -\frac{R_{02}}{R_{00}}\frac{\Sigma}{\rho}\ldots \qquad b_n\sigma_n = -\frac{R_{0n}}{R_{00}}\frac{\Sigma}{\rho};$$

or, the equation to the best value of B, (iii.) above, reduces to the regression formula (ii.). In other words, no attempt to reconstruct the organ B from a linear relation to the organs $A_1, A_2 \ldots A_n$ will give such a good result as the ordinary regression formula.* This, of course, excludes all attempts to form type ratios of

* I note that what is here demonstrated is only a special case of Mr. YULE's general theorem. See 'Roy. Soc. Proc.,' vol. 60, p. 477.

A/B or B/A as a method of prediction. We may, in fact, at once dismiss all reconstruction formulæ as insufficient which are not based on the theory of correlation. The theory as here applied, be it noted, depends on the *linearity* of the proposed formula and not on any special form of the distribution of variations.

(ii.) The accuracy of a prediction will not be indefinitely increased by increasing the number of organs upon which the prediction is based. This fundamental fact of the application of the theory of correlation to prediction has already been noticed by Miss ALICE LEE and myself in the case of barometric prediction.* The choice of organs upon which to base the prediction is far more important. Thus, to illustrate this from stature I may remark that the probable error of a prediction of male stature from radius is to a prediction from femur in the ratio of 2·723 to 2·174; that if one takes both femur and tibia for the prediction, the probable error is only reduced to 2·030, and further, if one takes femur, tibia, humerus, and radius, we only reach 1·961. This latter reduction is so small as to be well within the errors of the determination of our means, variations, and correlations, and accordingly scarcely worth making. To pass from the radius to the femur is a real gain; to pass from femur and humerus, say, to femur, humerus, tibia, and radius, is no sensible gain. Hence, one or two organs well selected are worth much more for prediction than a much larger number selected less carefully.

(iii.) It is the custom of French writers, when determining stature, to predict it from several single types of bones, say from femur, tibia, humerus, and radius, and then to take the mean of these results for the true stature. This is not the best *theoretical* procedure. Suppose the regression formulæ for the prediction of B from A_1, A_2, A_3, A_4 separately to be

$$B = c_0' + c_1'A_1,$$
$$B = c_0'' + c_1''A_2,$$
$$B = c_0''' + c_1'''A_3,$$
$$B = c_0'''' + c_1''''A_4.$$

Then the mean of all these results would give

$$B = \tfrac{1}{4}(c_0' + c_0'' + c_0''' + c_0'''') + \tfrac{1}{4}c_1'A_1 + \tfrac{1}{4}c_1''A_2 + \tfrac{1}{4}c_1'''A_3 + \tfrac{1}{4}c_1''''A_4,$$

that is to say, B has been really found from a linear relationship between B and the four organs in question. But the best linear relationship for the four organs is

$$B = c_0 + c_1A_1 + c_2A_2 + c_3A_3 + c_4A_4,$$

where the c's are the true regression coefficients. But the slightest acquaintance with the theory of regression shows that the *partial* regression coefficient c_1 is as a

* "On the Distribution of Frequency (Variation and Correlation) of the Barometric Height at Divers Stations," 'Phil. Trans.,' A, vol. 190, p. 456 *et seq.*

rule not just $\frac{1}{4}$ of the value of the total regression coefficient c_1'. For example, if A_4 were the radius and B stature, $c_1'''' = 3 \cdot 271$, while c_4 is a *negative* quantity $- \cdot 187$. This process of taking means may accordingly screen some most important element, like the negative value of the partial regression coefficient of the radius. *Theoretically*, therefore, as well as from the standpoint of discovery, the regression formula for n organs will give more valuable results than the mean of the results of the n regression formulæ for the n organs. A *practical* modification of this principle will be referred to below (p. 178).

(4.) The theory of regression will thus enable us to determine the best value to be assigned to an unknown organ, when the values of any other n organs are known, *supposing the individual to which these organs belong is a member of a race or group for which the regression coefficients have been ascertained.*

On what principle, however, can we extend the regression formulæ for one race to a second? The regression coefficients depend upon two things, the variability of the organs under consideration and their correlation. Now the change in variability as we pass from one race to a second has never been questioned. It has been suggested that the correlations were racial characters, but the divergences in correlations between local races are far beyond the probable errors of the observations.* Mr. FILON and I have shown that every *random* selection from a race changes both variation and correlation.† I have shown in a memoir not yet published that all natural and all artificial selection also changes these quantities. How then can we hope that a regression formula as applied from one local race to another will give accurate results? Why should the stature formula obtained from measurements on modern Frenchmen apply to palæolithic man?

I think M. MANOUVRIER somewhat lightly skips this difficulty in the following sentences:—"Enfin les *variations ethniques* des proportions du corps seront dans le même cas que les précédentes [les variations individuelles]. Il y a des races macroskèles et des races microskèles, comme il y des individus de ces deux sortes, et les variations individuelles sont bien plus grandes que les variations ethniques les plus accusées. Or les coefficients moyens des os de grande longueur tendant à abaisser la taille et ceux des os de faible longueur tendant à l'élever, il s'ensuit qu'il sera tenu compte dans une certaine mesure de la macroskélie des races comme de celle des individus dont les os seront absolument longs et de la microskélie des races comme de celle des individus ayant des os absolument courts."‡ If we admit for the moment, which I should not be prepared to do generally,|| that the individual variations in a local race are greater than the "ethnic variations" or divergences between the means of local races, M. MANOUVRIER's conclusion by no means follows.

* See 'Phil. Trans," A, vol. 187, pp. 266, 280, and 'Roy. Soc. Proc.,' vol. 61, p. 350.

† "On Random Selection," see 'Phil. Trans,' A, vol. 191, p. 229, and 'Roy. Soc. Proc.,' vol. 62, p. 173.

‡ *Loc. cit.*, on my page 171.

|| See the results as to the radius referred to on p. 176 below.

The formulæ for stature reconstruction, whether obtained with a consciousness of the theory of regression, as in the present paper, or indirectly by taking the means of small groups, as by M. MANOUVRIER, are based upon averages, and involve the standard-deviations, the variabilities of distribution of each organ. Hence, the fact that individual variations may be greater than ethnic variations does not touch the real point at issue, for the formulæ depend on the proportions of macroskely and microskely in each race, and these undoubtedly change. The individual variation being greater than the ethnic, is not a valid argument for applying a formula based on the observation of one local race straight away to a second.

The validity of applying the formula for one local race to a second depends, I think, upon very different considerations. In the first place, the validity is not general. If we endeavoured to reconstruct the radius, for example, of Aino or Naqada races from the femur or tibia by a regression formula obtained from measurements on the French, the results would, we might *à priori* expect, not be so satisfactory as for stature.*

The validity depends on our conceptions as to "local races." While the problem of local races is dealt with at length in my memoir on artificial and natural selection, and I do not want to anticipate the results there stated, it is still needful to cite here a theorem reached in that memoir. When a sub-race is established by the selection out of a primary race of a group having p organs distributed with given variabilities and given correlations about given means, we shall speak of its establishment as due to a *direct* selection of these p organs. But this direct selection is shown to alter also the sizes of all the remaining organs of the organism, the variabilities of all those organs, and the correlations among themselves of the non-directly selected as well as their correlations with the selected organs. We shall speak of this result as

* Allowing, as in my page 193, for cartilage and shrinking, I find the following formulæ from the French measurements for the reconstruction of radius in centimetres:

$$R = 7{\cdot}839 + {\cdot}367F,$$
$$R = 5{\cdot}715 + {\cdot}508T.$$

	Aino race.		Naqada race.	
	Calculated.	Observed.	Calculated.	Observed.
Reconstruction of R from F .	22·799	22·913	24·692	25·697
Reconstruction of R from T .	22·934	22·913	25·494	25·697

In the case of the Ainos, the prediction is within ·5 per cent. of the observed value. In the case of the Naqada race, the prediction from the femur differs by 1 centim., or 4 per cent. from its true value. An error of 6 to 7 centims. in the prediction of stature of a local race which would correspond in magnitude is hardly likely to occur. The explanation is that the radius is a much differentiated bone.

indirect selection. The changes due to indirect selection are shown in the memoir referred to to be in many cases of considerable importance; every mean, every standard deviation, every correlation may be altered; but the following theorems govern the changes in the regression formulæ :—

(i.) The regression formula of a directly selected organ on any number of other organs, whether directly or indirectly selected, will change.

(ii.) The regression formula of an indirectly selected organ on *all* the directly selected organs, and any number of the indirectly selected organs, does *not* change.

(iii.) The regression formula of an indirectly selected organ on some, but *not all* the directly selected organs, will change, unless the selection happens to be one of size only, and not of variability and correlation at the same time, in which case the formula remains unchanged.

(iv.) Most local races show sensible but small differences in both variability and correlation; if we call these differences quantities of the first order of small quantities. then the changes in the regression formulæ between two or more indirectly selected organs will be of this order of small quantities × the squares and products of correlations, quantities which are themselves less than unity, or what we may term a quantity of the third order; further, the changes in the regression formulæ between an indirectly selected organ and some but *not all* the directly selected organs will be of the first order of small quantities × the correlation, or what we may term a quantity of the second order.

To sum up, then, it would appear that the regression formulæ in general will change from local race to local race, but that a particular set (see (ii.) above) exist which would not be changed at all, while many others, supposing *size** to be the chief character selected, would only be changed by quantities of the second or third order. It will be obvious then that a knowledge of a considerable series of regression formulæ of two local races will enable us to ascertain to some extent the nature and amount of differentiation which has gone on from a common ancestral stock. Further, if we have not sufficient data for one local race to find the variabilities and correlations of its organs, but if we can find fairly closely the mean size of its organs, then the degree of consistency of the results obtained when these means are inserted in the regression formulæ for the second local race is an indication of the amount of differentiation which has taken place. The larger the number of organs we include in a regression formula the more likely we are to embrace *all* the directly selected organs, and so to obtain a formula which remains unchanged for the two races.

Thus we see that the extension of the stature regression formulæ from one local race—say, modern French—to other races—say, palæolithic man—must be made with very great caution. The extension assumes (i.) that stature itself has not been

* A selection of the mean sizes of two organs, which would alter their relative proportions, does not of course involve a selection of correlation; in other words, selection of mean relationship does not necessarily connote a selection of differential relationship.

directly selected, however widely changed by indirect selection, (ii.) that the formulæ involve *all* the directly selected organs closely correlated with stature, or that the selection has been principally one of size, and not of variability of, or correlation between, these organs. The real test of the applicability of the formulæ is whether or not they give for another local race of which we know *à priori* the stature, results in agreement with themselves and with the known stature. I take it that the justification required for applying our formulæ to palæolithic man is not the statement that ethnic are less than individual intra-racial variations, but is to be drawn from the fact that our formulæ, based upon measurements on the French, give results very fairly consistent among themselves and with observation for such a divergent race as the Aino. Such results seem to indicate that racial differences in stature are not the result of direct selection of stature, and that the selection of the long bones has been rather a selection of their absolute and relative sizes than a selection, in the first place, of their degrees of variation and correlation, although these have to some extent undoubtedly changed.

Our general theorems will to some degree indicate the manner in which differentiation has taken place. Suppose there has been a selection of femur and tibia, but not of humerus and radius. Then the regression formulæ for stature on femur and tibia, and for stature on femur and tibia together with one or both of the other two, humerus and radius, ought to give identical results; but these results ought to differ from those given by the formulæ for stature on humerus or on radius, or on both together. Practically, however, we have in many cases so few bones to obtain our means from (and these bones themselves parts of different skeletons), that the probable errors of these means quite obscure the deviations in stature as obtained from various formulæ and due to the influence of selection. From this standpoint a partial practical justification can be found for taking the mean of the divergent reconstructions of stature given by a series of regression formulæ, at any rate for the case when the divergences are not very large.

These divergences may be due to errors in the mean lengths of the long bones, or to selection directly of one or more of the long bones, or even to some small direct selection of stature. But as in our ignorance of these sources of errors we can only suppose some positive and some negative, the mean of all the formulæ may to some extent eliminate these quite unknown and unascertainable divergences (see p. 175). Generally, however, I should expect the stature in which two or more formulæ agree, to be more probable than the mean of several divergent formulæ.

(5.) *On the Data available for Stature Regression Formulæ.*—The only data available for the calculation of the correlation between stature and long bones occur in the measurements made by Dr. ROLLET on 100 corpses in the dissecting room at Lyons.* This material has already been made use of by Miss ALICE LEE and myself in our memoir, "On the Relative Correlation of Civilised and Uncivilised Races,"† so that

* 'De la Mensuration des Os Longs des Membres,' par Dr. ETIENNE ROLLET, Lyons, 1889.

† 'Roy. Soc. Proc.,' vol. 61, p. 343

all the coefficients of correlation and all the variations of the long bones have already been calculated.

I owe to Miss ALICE LEE the knowledge of the additional constants required for this further investigation, and embodied in Tables I. and II. below, which embrace all that is needed to fully determine the correlation of stature and long bones.

The treatment of Dr. ROLLET's material was not to be briefly settled. He had measured only 50 bodies of each sex, and this number included a great variety of ages. M. MANOUVRIER in determining his table of statures has at once excluded from his calculations all the males but 24 as senile, and all the females but 25. Now, although the correlations between stature and long bones are high, it would be quite hopeless to attempt to calculate them from 25 cases ; 50 cases are hardly sufficient, 25 impossible. It seemed, therefore, necessary to include all Dr. ROLLET's cases, and the question now arises how far the inclusion of the senile ones will affect our results. Taking 50 as the age at which stature begins to decrease, we notice that of the 25 lowest statures recorded by ROLLET, 18 are of men over 50, and of the 25 highest statures, 17 are of men over 50. In other words, there appear sensibly as many senile statures above as below the median stature. Of women there are 16 over 50 years old with a stature greater than the median, and only 14 women over 50 under the median stature. Turning to means, we notice that 24 males under 60 years had for mean stature 167·17 centims., and 26 males over 59 years had 165·4 centims., 25 females under 60 had for mean stature 154·04 centims., and 25 females over 59 had 154·00 centims. 37 females under 70 had a stature 153·94 centims., and 13 over 70 gave 154·23 centims., an absolutely greater stature. 24 years was the minimum age. From this it would appear that whatever shrinkage may be due to old age, it is not of a very marked character in these data, or largely disappears when a body is measured after death on a flat table ; the senile stoop may then be largely eliminated.

But there is another point to be noted : we shall not directly make use of the mean stature as obtained from ROLLET's data, except to test how far our formulæ will reproduce ROLLET's results. What we shall make use of from ROLLET's data are the standard-deviations and coefficients of correlation, and these will hardly have their values sensibly influenced by such comparatively small senile changes as are to be found indicated in ROLLET's measurements.* Accordingly our constants are calculated by including all ROLLET's measurements, namely, on 50 of each sex.

The following results were found :—

* If the bones shrink with old age, like the stature, the correlation would not be altered. The length of a bone varies with the amount of moisture in it (see below), and such shrinkage is itself a possibility. The bones of the aged will of course be included among those of extinct races, and cannot easily be eliminated.

TABLE I.—Correlation between Stature and Long Bones.

Pairs of organs.	Male.	Female.
Stature and tibia	·7769 ± ·0378	·7963 ± ·0349
Stature and radius	·6956 ± ·0492	·6717 ± ·0523
Stature and humerus *	·8091 ± ·0329	·7706 ± ·0387
Stature and femur	·8105 ± ·0327	·8048 ± ·0336
Stature and humerus + radius	·7973 ± ·0347	·7547 ± ·0411
Stature and femur + tibia	·8384 ± ·0283	·8268 ± ·0302

The means, standard deviations, and correlations of femur, tibia, humerus, and radius, for ROLLET's measurements, are given in the 'Roy. Soc. Proc.,' vol. 61, pp. 347–350. The means and variability of the remaining organs not there recorded were found to be as follows:—

TABLE II.

	Mean.		Standard deviation.	
	Male.	Female.	Male.	Female.
Stature	166·260 ± ·525	154·020 ± ·520	5·502 ± ·371	5·450 ± ·368
Humerus + radius	57·368 ± ·242	51·240 ± ·241	2·536 ± ·171	2·526 ± ·170
Femur + tibia	82·028 ± ·380	75·024 ± ·382	3·979 ± ·268	4·001 ± ·270

Without reproducing the full tables of the memoir referred to, it is of value to form the correlation tables, which serve as the determinants from which the regression formulæ have been calculated. It is only in the case of stature in terms of the four long bones that the numerical work proved lengthy.

The general formula used is (ii.) on p. 172. S, F, H, T, R stand for Stature, Femur, Humerus, Tibia, Radius, all measured in ROLLET's manner, which will be discussed at length below.

* The somewhat low value of the correlation for female stature and humerus was tested by means of the formula

$$r_{zu} = \frac{\sigma_x}{\sigma_z} r_{xu} + \frac{\sigma_y}{\sigma_z} r_{yu},$$

where $z = x + y$, x, y, and u are organs, σ_x, σ_y, σ_u their standard deviations, and r a coefficient of correlation. Hence putting x = humerus, y = radius, and u = stature, I found the correlation between stature and humerus + radius indirectly; it was ·7564. The table shows that the directly-calculated value was ·7547, a difference well within the errors of observation. Thus the correlations as given for female humerus and stature and female radius and stature must be correct, *i.e.*, the somewhat lengthy arithmetic involved is not at fault.

TABLE III.

MALES.—Stature and Long Bones Correlation.

	S.	F.	H.	T.	R.
S.	1	·8105	·8091	·7769	·6956
F.	·8105	1	·8421	·8058	·7439
H.	·8091	·8421	1	·8601	·8451
T.	·7769	·8058	·8601	1	·7804
R.	·6956	·7439	·8451	·7804	1

TABLE IV.

FEMALES.—Stature and Long Bones Correlation.

	S.	F.	H.	T.	R.
S.	1	·8048	·7706	·7963	·6717
F.	·8048	1	·8718	·8904	·7786
H.	·7706	·8718	1	·8180	·8515
T.	·7963	·8904	·8180	1	·8053
R.	·6717	·7786	·8515	·8053	1

The following cases of reconstruction were then dealt with :—

(*a*) Reconstruction of mean stature from a knowledge of the femur of *p* individuals.
(*b*) " " " " " humerus " "
(*c*) " " " " " tibia
(*d*) " " " " " radius
(*e*) " " " " " femur + tibia " "
(*f*) " " " " " femur and tibia " "
(*g*) " " " " " humerus + radius " "
(*h*) " " " " " humerus and radius " "
(*i*) " " " " " femur and humerus " "
(*k*) " " " " " {femur, humerus, tibia, and radius} " "

In the formulæ M denotes a mean, and e the probable error of the estimate.

TABLE V.—Male.

(a) $S - M_S = 1{\cdot}880\,(F - M_F)$, $e = 2{\cdot}174/\sqrt{p}$.

(b) $S - M_S = 2{\cdot}894\,(H - M_H)$, $e = 2{\cdot}181/\sqrt{p}$.

(c) $S - M_S = 2{\cdot}376\,(T - M_T)$, $e = 2{\cdot}337/\sqrt{p}$.

(d) $S - M_S = 3{\cdot}271\,(R - M_R)$, $e = 2{\cdot}666/\sqrt{p}$.

(e) $S - M_S = 1{\cdot}159\,(F + T - M_{F+T})$, $e = 2{\cdot}023/\sqrt{p}$.*

(f) $S - M_S = 1{\cdot}220\,(F - M_F) + 1{\cdot}080\,(T - M_T)$, $e = 2{\cdot}030/\sqrt{p}$.

(g) $S - M_S = 1{\cdot}730\,(H + R - M_{H+R})$, $e = 2{\cdot}240/\sqrt{p}$.

(h) $S - M_S = 2{\cdot}769\,(H - M_H) + {\cdot}195\,(R - M_R)$, $e = 2{\cdot}179/\sqrt{p}$.

(i) $S - M_S = 1{\cdot}030\,(F - M_F) + 1{\cdot}557\,(H - M_H)$, $e = 1{\cdot}962/\sqrt{p}$.

(k) $$\left.\begin{array}{l} S - M_S - {\cdot}913\,(F - M_F) + {\cdot}600\,(T - M_T) \\ \quad + 1{\cdot}225\,(H - M_H) - {\cdot}187\,(R - M_R) \end{array}\right\}, \quad e = 1{\cdot}961/\sqrt{p}.$$

TABLE VI.—Female.

(a) $S - M_S = 1{\cdot}945\,(F - M_F)$, $e = 2{\cdot}182/\sqrt{p}$.

(b) $S - M_S = 2{\cdot}754\,(H - M_H)$, $e = 2{\cdot}343/\sqrt{p}$.

(c) $S - M_S = 2{\cdot}352\,(T - M_T)$, $e = 2{\cdot}245/\sqrt{p}$.

(d) $S - M_S = 3{\cdot}343\,(R - M_R)$, $e = 2{\cdot}723/\sqrt{p}$.

(e) $S - M_S = 1{\cdot}126\,(F + T - M_{F+T})$, $e = 2{\cdot}068/\sqrt{p}$.

(f) $S - M_S = 1{\cdot}117\,(F - M_F) + 1{\cdot}125\,(T - M_T)$, $e = 2{\cdot}085/\sqrt{p}$.

(g) $S - M_S = 1{\cdot}628\,(H + R - M_{H+R})$, $e = 2{\cdot}412/\sqrt{p}$.

(h) $S - M_S = 2{\cdot}582\,(H - M_H) + {\cdot}281\,(R - M_R)$, $e = 2{\cdot}340/\sqrt{p}$.

(i) $S - M_S = 1{\cdot}339\,(F - M_F) + 1{\cdot}027\,(H - M_H)$, $e = 2{\cdot}120/\sqrt{p}$.

(k) $$\left.\begin{array}{l} S - M_S = {\cdot}782\,(F - M_F) + 1{\cdot}120\,(T - M_T) \\ \quad + 1{\cdot}059\,(H - M_R) - {\cdot}711\,(R - M_R) \end{array}\right\}, \quad e = 2{\cdot}024/\sqrt{p}.$$

(6.) Now these tables require a good deal of comment. In the first place they must not be considered as extending beyond the range of data on which they are based, thus R, F and H are the maximum lengths of bones measured with the cartilage attached, and in a humid state, T is the tibia length excluding spine. All the constants were worked out for the *right* members, except in one or two cases in which they were missing. The stature is the stature measured on the corpse. Further the measurements are made on the French race.

We shall now proceed to generalise these formulæ. In the first place, the

* It may appear strange that the probable error of (e) is less than (f), but the difference is really less than the probable error of the observations. If $r_{S,F+T}$ be calculated from the known values of σ_F, σ_T and r_{FT}, &c., we find it equals ·8369 instead of ·8384 the directly calculated value, while σ_{F+T} thus calculated $= 3{\cdot}967$ instead of 3·979, whence $e = 2{\cdot}031/\sqrt{p}$ instead of $2{\cdot}023/\sqrt{p}$, which is in agreement with the general theorem on p. 173.

numerical factors are functions only of the standard deviations and the correlation coefficients, and will accordingly be unchanged if these be unchanged.

Let O_1 and O_2 be any organs and M_1 and M_2 their means, n_1 and n_2 their numbers, and r_{12} their coefficient of correlation. Suppose that any hygrometric changes, different method of measurement, amount of animal matter in the organs at time of measurement, etc., cause us to measure $\beta_1 O_1 + \beta_2 = O'_1$ and $\gamma_1 O_2 + \gamma_2 = O'_2$ instead of O_1 and O_2, and let σ'_1, σ'_2, M'_1, M'_2, and r'_{12} be the resulting characters, then clearly, S standing for summation :—

$$M'_1 = \beta_1 M_1 + \beta_2, \qquad M'_2 = \gamma_1 M_2 + \gamma_2,$$

$$\sigma'^2_1 = S(O'_1 - M'_1)^2 = \beta_1^2 S(O_1 - M_1)^2 = \beta_1^2 \sigma_1^2, \text{ or } \sigma'_1 = \beta_1 \sigma_1,$$

$$\sigma'^2_2 = S(O'_2 - M'_2)^2 = \gamma_1^2 S(O_2 - M_2)^2 = \gamma_1^2 \sigma_2^2, \text{ or } \sigma'_2 = \gamma_1 \sigma_2,$$

$$r'_{12} = \frac{S(O'_1 - M'_1)(O'_2 - M'_2)}{\sigma'_1 \sigma'_2} = \beta_1 \gamma_1 \frac{S(O_1 - M_1)(O_2 - M_2)}{\sigma'_1 \sigma'_2} = r_{12}.$$

Thus a correlation coefficient will be quite unchanged. A regression coefficient will be changed or not according as the ratio of two standard deviations is changed or not, or according as to whether β_1/γ_1 sensibly differs from unity. Now in stature or any of the long bones with which we have to deal quantities corresponding to β_2, γ_2 may amount to 1 per cent. of the value of O_1 or O_2, but the multipliers like β_1 and γ_1 are not only quantities differing in the second order from unity, but probably very nearly equal to each other. Hence it is reasonable to suppose that changes in the condition of the bones, and stature measured on the living or on the corpse, while sensibly affecting M_S, M_F, M_H, M_T, and M_R will produce little or no effect on the numerical constants of the regression formulæ (*a*) to (*k*). We shall find that this *à priori* conclusion is borne out by actual measurements. Hence we conclude that Tables V. and VI. may be applied to stature measured on the living or the corpse, to bones measured humid or dry, with or without the cartilage, provided proper modifications are made in the values of the five means. We might even go so far as to predict that provided M_T be properly altered, the stature from tibia reconstruction formulæ will not be much modified, even if the tibia be measured with instead of without the spine. The change, however, in the regression formulæ when the femur is measured in the oblique position is more likely to be of importance, and the correlation between stature and oblique femur has accordingly been worked out. If F′ denote oblique femur we have :—

Male	$M_{SF'} = 44{\cdot}938,$	$\sigma_{F'} = 2{\cdot}331,$	$r_{SF'} = {\cdot}8025,$
Female	$M_{SF'} = 41{\cdot}240,$	$\sigma_{F'} = 2{\cdot}205,$	$r_{SF'} = {\cdot}8007,$

whence for (*a*) we find :

Male $\quad S - M_S = 1{\cdot}894\,(F' - M_{F'}),$

Female $\quad S - M_S = 1{\cdot}979\,(F' - M_{F'}).$

Thus the regression coefficient is not changed more than ·55 per cent. for males and 1·7 per cent. for females, even in this case where the difference between the maximum and oblique lengths of the femur has been much insisted upon as very significant with regard to stature. Putting in the lengths of the means as found on the corpse, we have:

$$\left.\begin{array}{ll}\text{Male} & S = 81{\cdot}147 + 1{\cdot}894\,F' \\ \text{Female} & S = 72{\cdot}406 + 1{\cdot}979\,F'\end{array}\right\} \quad \ldots\ldots\ldots \quad \text{(i.)}.$$

The corresponding formulæ for the stature in terms of the maximum length of femur are, as we shall see later:

$$\left.\begin{array}{ll}\text{Male} & S = 81{\cdot}231 + 1{\cdot}880\,F \\ \text{Female} & S = 73{\cdot}163 + 1{\cdot}945\,F\end{array}\right\} \quad \ldots\ldots\ldots \quad \text{(ii.)}.$$

The extreme oblique femur lengths are for males 39·6 and 49·8, and for females 37·4 and 48·0. Let us calculate the stature of these individuals directly from (i.) and indirectly from (ii.), by putting $F = F' + {\cdot}32$ for males and $F' + {\cdot}33$ for females. We find

	(i.)	(ii.)
Male min.	156·15	156·21
Female min.	146·42	146·53
Male max.	175·47	175·46
Female max.	167·40	167·17

The differences here in these extreme cases are absolutely unimportant for the determination of stature. In other words, the changes in the regression equation are insignificant, when we even make such a change as from oblique to maximum femur length. Accordingly we have the rule, if the oblique length of femur be given, the equations for the maximum length can always be safely used if we add ·32 for the male and ·33 for female to the oblique length in centimetres before using equations of type (ii.).

So far we have generalised Tables V. and VI., having regard to the nature and condition of the organs when measured. We see that the regression coefficients will remain sensibly constant. Our general considerations on pp. 177 and 178 indicate the limits under which these regression coefficients may be considered constant for different local races. But the constancy of the regression coefficients is not sufficient to preserve the constancy of the linear reconstruction formulæ for stature. It would be of no service if M_S, M_F, M_H, M_T, M_R varied from local race to local race absolutely independently. Now if m_0 be the mean of a not directly selected organ, and m_1, m_2, m_3 . . . the means of any other organs, the constant part in a reconstruction formula will with the notation of p. 172, be:

$$m_0 + \frac{R_{01}}{R_{00}}\frac{\sigma_0}{\sigma_1}m_1 + \frac{R_{02}}{R_{00}}\frac{\sigma_0}{\sigma_2}m_2 + \ldots + \frac{R_{0n}}{R_{00}}\frac{\sigma_0}{\sigma_n}m_n.$$

It is shown in the memoir on selection to which I have previously referred, that this expression remains the same for all local races, and equal to its value in the original stock under precisely the same conditions (stated on p. 177) as the regression coefficients themselves remain constant. Hence we have the same degree of justification in applying our whole stature reconstruction formula from one race to a second, as in applying the regression coefficients.

(7.) Re-examining Tables V. and VI. with a view to drawing one or two general conclusions before we proceed further, we notice :

(i.) The probable error of the reconstruction of the stature of a single individual is never sensibly less than two centimetres, and if we have only the radius to predict from may amount to $2\frac{3}{8}$ centims.

Hence no attempt to reconstruct the stature of an individual from the four chief long bones can possibly exceed this degree of accuracy on the average, at any rate no *linear* formula.* No other linear formulæ will give a better, or indeed as good a result as the above.

The reconstruction of racial stature is naturally more accurate, since if we reconstruct the mean from p bones of one type, the probable error is reduced by the multiplier $1/\sqrt{p}$. At the same time we must bear in mind that possibly a definite, if small amount of direct, selection by stature has actually taken place in the differentiation of human races, and accordingly the values of e given in Tables V. and VI. are not absolutely true measures of the probable error of racial reconstruction, even when one or more of the long bones have not been directly selected. A direct selection of the long bones is usually evidenced by one or more of the formulæ giving discordant results. When, as will be seen later to be usually the case, several of the formulæ give results well in accordance with each other, then we may assume that $2/\sqrt{p}$ centims. is an approximate† measure of the probable error of the reconstructed stature.

(ii.) The four long bones give for males the least probable error, but with sensibly equal accuracy and less arithmetic we may use F & H, F + T or F & T ; then follow fairly close together H & R, F or H alone ; T alone is sensibly worse, and R is worst of all. It is noteworthy that H is better than T, and the H & R is sensibly as good as F alone.

Turning to female stature reconstruction, we notice that the order of probable errors is considerably altered. Tibia and radius now play a more important part in the determination of stature. The four long bones still give the best result ; F & T, and F + T follow closely ; then come F & H, and F alone ; followed at some distance by H & R, and H alone, but both these are now worse than T alone ; last of

* I shall return to the question of the linearity of the formula, when dealing later with the stature of giants and dwarfs, see p. 222.

† It must be remembered that we have, as a rule, a number of long bones which in part do not even belong to the same skeletons. This result accordingly is the probable error of a group to whom one kind of long bones belonged, rather than the probable error of the racial stature as reconstructed.

all comes R, as before. Thus while in the case of men the humerus, in the case of women the tibia is the better bone of the two to predict stature from. A simple examination shows the emphasising of the tibia coefficients in the case of woman. The same holds for the radius coefficients, but in a still more marked degree.

Both male and female show in the regression formula for the four long bones a remarkable feature which they have in common with the anthropomorphous apes, namely, the *negative* character of the partial regression coefficient. *The longer the radius for the same value of femur, humerus, and tibia, the shorter will be the stature.* In this point women are more akin to the anthropomorphous apes than man, for the negative radius coefficient in formula (k) is nearly four times as large. The tibia also has a coefficient almost double that of the male, and pointing in the same direction.

(iii.) A comparison of Table V. with Table VI. shows us that man and woman are in all probability not only differentiated from a common stock directly with regard to stature, but also directly with regard to all other long bones. If we use female to construct male stature, or male to reconstruct female, we get surprisingly bad results. The fact that the formula (k) for female diverges in a direction from that of man, which approximates to that of at least one species of anthropomorphous ape, is only of course a round-about quantitative manner of indicating, what is obvious on other grounds, that a substantial part of the differentiation of male and female took place in that part of the history of man's evolution which preceded his differentiation from the stock common to him and certain of the anthropomorphous apes.

(8.) Before we modify our formulæ in Tables V. and VI. to suit the reconstruction of stature by measurements on prehistoric and other bones, we will put the numerical values for M_S, M_F, M_T, M_H, M_R into these formulæ. This will serve a double purpose (i.), it will enable us to verify our formulæ on ROLLET'S material, and (ii.) it will place at the disposal of the criminal authorities the best formulæ yet available for the reconstruction of the stature of an adult of whom one or more members have been found under suspicious circumstances.

FORMULÆ for the Reconstruction of the Stature as Corpse, the Maximum Lengths of F, H, R, and of T without Spine being measured with the Cartilage on and in a Humid State.*

TABLE VII.—Male.

(a) $S = 81{\cdot}231 + 1{\cdot}880\,F.$
(b) $S = 70{\cdot}714 + 2{\cdot}894\,H.$
(c) $S = 78{\cdot}807 + 2{\cdot}376\,T.$
(d) $S = 86{\cdot}465 + 3{\cdot}271\,R.$
(e) $S = 71{\cdot}164 + 1{\cdot}159\,(F + T).$
(f) $S = 71{\cdot}329 + 1{\cdot}220\,F + 1{\cdot}080\,T.$

* The probable error in these and later tables are not reproduced; they may be considered to be substantially the same as in V. and VI.

TABLE VII.—Male (continued).

(g) $S = 67{\cdot}025 + 1{\cdot}730\,(H + R)$.
(h) $S = 69{\cdot}870 + 2{\cdot}769\,H + {\cdot}195\,R$.
(i) $S = 68{\cdot}287 + 1{\cdot}030\,F + 1{\cdot}557\,H$.
(k) $S = 66{\cdot}918 + {\cdot}913\,F + {\cdot}600\,T + 1{\cdot}225\,H - {\cdot}187\,R$.

TABLE VIII.—Female.

(a) $S = 73{\cdot}163 + 1{\cdot}945\,F$.
(b) $S = 72{\cdot}046 + 2{\cdot}754\,H$.
(c) $S = 75{\cdot}369 + 2{\cdot}352\,T$.
(d) $S = 82{\cdot}189 + 3{\cdot}343\,R$.
(e) $S = 69{\cdot}525 + 1{\cdot}126\,(F + T)$.
(f) $S = 69{\cdot}939 + 1{\cdot}117\,F + 1{\cdot}125\,T$.
(g) $S = 70{\cdot}585 + 1{\cdot}628\,(H + R)$.
(h) $S = 71{\cdot}122 + 2{\cdot}582\,H + {\cdot}281\,R$.
(i) $S = 67{\cdot}763 + 1{\cdot}339\,F + 1{\cdot}027\,H$.
(k) $S = 67{\cdot}810 + {\cdot}782\,F + 1{\cdot}120\,T + 1{\cdot}059\,H - {\cdot}711\,R$.

Should the stature of the living be required from the corpse stature, then 1·26 centim. should be subtracted for the male and 2 centims. for the woman.* If a *left* member has been measured instead of a right, a small allowance might be made for this on the basis of ROLLET's means for the left side, but such refinement is hardly of service when we look at the probable error of an individual reconstruction, *i.e.*, about 2 centims. We shall return to the point later as a second order error in racial reconstruction.

In order to indicate to the reader the degree of confidence he may place in the above formulæ of reconstruction, and also their relative value, I give below a table of observed and reconstructed statures in the case of 20 out of ROLLET's 100 cases. The individuals, in order to avoid any bias, were taken at random as the 5th, 10th, 15th, &c. entries through ROLLET's Tables. The observed statures are recorded and the differences as obtained by the formulæ (*a*)–(*k*) Under the heading M, I give the differences which would be yielded by M. MANOUVRIER's Table. It is formulæ (*f*), (*h*), (*i*), and (*k*) on which I should lay most weight, and which should be used whenever the material is available.

* For the reasons for these numbers, see p. 191 below.

TABLE IX.—Table of Differences of Actual and Reconstructed Male Stature.

No.	5		10		15		20		25		30		35		40		45		50		Mean error.	
	P.	M.	P.	M.	P.	M.	P.	M.	P.	M.	P.	M.	P.	M.	P.	M.	P.	M.	P.	M.	P.	M.
(a)	+5	+6	−1	0	+7	+7	−1	0	+1	+1	−2	−1	−5	−5	−8	−7	−1	−1	−2	0	3·3	2·8
(b)	+6	+6	−2	−3	0	+1	0	+1	+3	+4	−3	−2	0	+2	−7	−6	−1	+1	−1	+5	2·3	3·1
(c)	+2	+4	0	+2	0	+1	−1	0	+3	+3	0	0	−1	−1	−8	−7	+1	+3	0	+4	1·6	2·5
(d)	+4	+5	−2	−3	0	+1	0	+1	+3	+4	−3	−2	−2	−2	−6	−5	−5	−5	0	+7	2·5	3·5
(e)	+3	..	−1	..	+4	..	−1	..	+2	..	−1	..	−3	..	−9	..	0	..	0	..	2·4	..
(f)	+4	..	−1	..	+5	..	−1	..	+2	..	−1	..	−3	..	−9	..	0	..	0	..	2·6	..
(g)	+5	..	−3	..	0	..	0	..	+4	..	−3	..	−1	..	−7	..	−2	..	−1	..	2·6	..
(h)	+5	..	−3	..	0	..	0	..	+3	..	−3	..	0	..	−7	..	−1	..	−1	..	2·3	..
(i)	+5	..	−2	..	+4	..	0	..	+2	..	−2	..	−2	..	−8	..	−1	..	0	..	2·6	..
(k)	+4	+5	−2	−1	+3	+3	−1	+1	+2	+3	−2	−1	−2	−2	−8	−6	0	0	0	+4	2·4	2·6
M.D.	−4·3	..	−1·7	..	+2·1	..	− ·5	..	+2·5	..	−2·0	..	−1·9	..	−7·7	..	−1	..	− ·5	..	2·46	2·975
Actual stature	159		161		163		165		166		167		170		171		173		177			

TABLE X.—Table of Differences of Actual and Reconstructed Female Stature.

No.	5		10		15		20		25		30		35		40		45		50		Mean error.	
	P.	M.	P.	M.	P.	M.	P.	M.	P.	M.	P.	M.	P.	M.	P.	M.	P.	M.	P.	M.	P.	M.
(a)	+2	+2	−1	+2	+2	+3	+3	+4	+3	+4	−2	−2	+5	+7	0	+1	−1	0	− 3	+1	2·2	2·6
(b)	0	−2	−1	−2	+1	+3	+3	+5	+1	+3	+1	+3	+3	+5	0	+1	+1	+3	− 7	−4	1·8	3·1
(c)	0	+1	+3	+4	+1	+3	+3	+5	+2	+3	−2	0	+5	+7	+4	+6	−2	−1	− 5	−1	2·7	3·1
(d)	+2	+1	+3	+3	+1	+3	+2	+5	+2	+3	+2	+4	+3	+5	−3	+1	0	+3	−10	−7	2·8	3·5
(e)	+1	..	0	..	+2	..	+3	..	+3	..	−2	..	+6	..	+2	..	−1	..	− 3	..	2·3	..
(f)	+1	..	0	..	+2	..	+3	..	+3	..	−2	..	+6	..	+2	..	−1	..	− 4	..	2·4	..
(g)	+1	..	+1	..	+1	..	+3	..	+1	..	+2	..	+3	..	−1	..	+1	..	− 7	..	2·1	..
(h)	0	..	0	..	+1	..	+3	..	+1	..	+1	..	+3	..	0	..	+1	..	− 7	..	1·7	..
(i)	+1	..	−1	..	+2	..	+3	..	+3	..	−1	..	+5	..	0	..	0	..	− 4	..	2·0	..
(k)	0	0	0	+1	+1	+3	+3	+5	+2	+3	−2	+1	+5	+6	+3	+2	−1	+1	− 3	−3	2·0	2·4
M.D.	+ ·8	..	− ·4	..	+1·4	..	+2·9	..	+2·1	..	− ·5	..	+4·4	..	+ ·7	..	− ·3	..	−5·3	..	2·20	3·075
Actual stature	148		149		152		152		153		155		156		158		160		171			

The first point with regard to these tables is to note how, even with only ten cases, the mean errors accord closely with their theoretical values. For example, the mean error of k is 2·31 centims. for male and 2·35 centims. for female when deduced from the probable errors in Tables V. and VI.; the observed mean errors in the two cases are 2·4 centims. for male and 2·0 centims. for female. The mean of the mean errors is for male 2·57 centims., and for females 2·66 centims.; the observed values are 2·46 centims. and 2·2 centims. for the two sets of ten cases respectively. We conclude at once that our formulæ, and therefore certainly any other linear formulæ, will not give results with a probable error of less than 2 centims. for the individual stature. In our case the worst error is one of 8 centims. (about 3 inches) in the stature of a man of 47 years of age, who must have had a remarkably long trunk in proportion to his leg and arm-lengths. It would be impossible to have predicted his stature any closer without taking into account the correlation between stature and trunk. The preservation of the vertebral column is comparatively rare, and at present there are absolutely no statistics on the relationship between the dimensions of any part of it and living stature. We must therefore content ourselves with a probable error of 2 centims., and expect, but rarely, to make an error of as much as 8 centims. in the reconstructing of the stature of an individual.

We have placed in the above tables M. MANOUVRIER's results as calculated from his 'Table-barême.' They give somewhat larger mean errors than our formulæ, which would have been probably reduced somewhat if we had excluded, as he has done, the aged. We have seen, however (p. 179), that there seems no reason to exclude the aged women, and in the case of the seven men over 60, he actually in three cases under-estimates their stature. In other words, while in four cases his table might have given better results for adult stature, in three it would have given worse results. If we allow a mean old-age shrinkage of 3 centims.*—an amount hardly justified by averaging the adult and old-age portions of ROLLET's returns—we should find that MANOUVRIER's method would have made a total error of 17 centims. in estimating the stature of these seven old men in youth, whereas it gives a total error of 16 centims. in estimating their old-age stature. Thus there might, perhaps, be a small, but it would not be a very sensible, reduction of the mean errors of the results given by MANOUVRIER's 'Table-barême' had we excluded the old age cases.

What deserves special notice is that our formula (k) gives a better result than the mean of all the formulæ (a)–(k), and a better result than the mean of the values obtained by MANOUVRIER's method for the four long bones.

(9.) The next stage in our work is to so modify Tables IX. and X. that they will serve for the reconstruction of the *living* stature from bones *out of which all the animal matter has disappeared, and which are dry and free of all cartilage.* This

* This value is that given by M. MANOUVRIER himself, 'Mémoires de la Société d'Anthropologie de Paris,' vol. 4, p. 356, 1892.

is either the condition in which we find the bones of a prehistoric or early race, or it is one to which they are soon reduced on being preserved in museum or laboratory.

The first question which arises is the difference between the *mean* stature of the living and the mean stature of the corpse for both sexes. It is impossible to measure this difference satisfactorily on a sufficiently large number of individuals, and then take the mean difference. If we suppose ROLLET'S individuals to be an average sample of the French race, then we must place in Tables V. and VI. for M_s on the left the mean heights of French men and French women.

Now there is a considerable amount of evidence to show that the mean height of Frenchmen is 165 centims. almost exactly. The anthropometric service of M. BERTILLON gives 164·8 centims., and this is the stature furnished by the measurements for military recruiting.* M. MANOUVRIER takes 165 centims. as the mean height, and as by selecting only twenty of ROLLET'S cases he gets a mean height of about 167 centims. for the corpse, he concludes that 2 centims. must be deducted from the corpse length to get the living stature. In our case all we have to do is then to put $M_s = 165$ centims. At the same time, BERTILLON'S numbers probably include many men over 50, and the recruiting service many men not yet fully grown; hence it seems to me doubtful whether 2 centims. really represents the difference between living and dead stature. 165 centims. is probably a good mean height for the whole adult population,† and should accordingly be compared with ROLLET'S whole adult population, which has a mean of 166·26 centims. I accordingly conclude that 1·26 centims. is on the average a more reasonable deduction to make in order to pass from the dead to the living stature of the general population. In the course of my investigations, however, no use is made of this difference, but M_s given its observed living value.

The value for women is far less easy to obtain, as a good series of French statistics entirely fails. The mean given in the footnote below is clearly only that of a special class. MANOUVRIER has found from 130 women, between 20 and 40 years of age, inscribed in BERTILLON'S registers the mean height 154·5 centims., and RAHON holds that this is the best result yet obtained.‡ But the mean height of ROLLET'S material is 154·02 centims. (see my p. 180), and, as we have seen, this is not sensibly increased by taking only the women in the prime of life (see p. 179, above). If 154·5 centims. were the mean living stature of ROLLET'S women, we should have to suppose a shrinkage of stature in women when the corpse is measured, whereas in the case of men the corpse length is greater than the living stature. RAHON, disregarding his own statement as to 154·5 centims. being the best value, follows MANOUVRIER in deducting 2 centims. from the stature as corpse to get the living stature. MANOUVRIER'S

* 'Mémoires de la Société d'Anthropologie de Paris,' vol. 4, p. 413, 1893.

† For special classes the stature is considerably greater. See the values 166·8 centims. for male and 156·1 centims. for female given in the 'Mém. Soc. d'Anthrop.,' vol. 3, 1888.

‡ *Loc. cit.*, p. 413.

rule for deducting 2 centims. seems based partly on a comparison of BERTILLON'S measurements for men, with his own selection from ROLLET'S material, which give mean heights 165 centims. and 167 centims. respectively, and partly on the measurement standing and reclining of six men and four women.* Now the reader should notice that in our method of reaching the reconstruction equations, we are not concerned with the amount to be subtracted from an individual stature, but with the mean living stature of the population which ROLLET has sampled. Now there is a quantity which has very remarkable constancy, namely, the sexual ratio for stature. The mean male is to the mean female stature in a great variety of races and classes as 13 to 12.† If, therefore, ROLLET'S women are the same class as his men, we should expect their living stature to have had a mean $= \frac{12}{13}$, that of the men $= \frac{12}{13}(165) = 152{\cdot}3$ centims. We have seen that from the registers of BERTILLON the mean stature of women between 20 and 40 was 154·5 centims.; these probably include a considerable number of stout tramps or vagabonds, not a fair sample of those who would find their way into the Lyons Hospital. TENON measured in 1783 60 women of the village of Mussey, and obtained a mean stature of 150·6 centims.‡ If we take the mean of these groups we find 152·55 centims. as the mean stature for French women of the lower classes; this differs by less than 3 millims. from the result already suggested by using the sex ratio. I am, accordingly, inclined to hold that the best that can be done at present is to take 152·3 centims. as the mean stature of Frenchwomen of the class sampled by ROLLET.

The next stage in our work is to consider the difference in length of the long bones, as measured in the dissecting room by ROLLET and his assistants, and as they would be measured in the case of a primitive race whose bones had been exhumed, and then been preserved and dried before measuring. ROLLET merely observes that he kept several of his bones for some months, and, the cartilage being then dry, they measured on the average 2 millims. less.§ On the strength of this, MANOUVRIER,‖ and he is followed by RAHON, add 2 millims. to the length of each prehistoric bone when reconstructing the stature. Now I am doubtful whether this gives a really close enough result. ROLLET measured the bones in the dissecting room, the cartilages were still on, and the animal matter in the bones, but in the case of prehistoric and ancient bones this does not at all represent the state of affairs. Nor are they merely such bones with the cartilage dry; the cartilage, together with the animal matter, has entirely gone. There are accordingly two allowances to be made (*a*) for the cartilage, and (*b*) for the disappearance of the animal matter and drying of the bone.

* 'Mémoires de la Société d'Anthropologie de Paris,' vol. 4, p. 384, 1892.

† ROLLET'S corpse statures give a sexual ratio = 1·079.

‡ "Notes manuscrites relatives à la stature de l'homme, recueillies par VILLERMÉ," 'Annales d'Hygiène,' 1833.

§ ROLLET, *loc. cit.*, p. 24.

‖ MANOUVRIER, *loc. cit.*, p. 386.

(a.) *Allowance for the Cartilage.**

The thicknesses of the cartilages here cited are taken from HEINRICH WERNER'S Inaugural Dissertation, 'Die Dicke der menschlichen Gelenkknorpel,' Berlin, 1897. They are only discussed for the cases required for the long bones as measured by ROLLET and used in my reconstruction formulæ.†

Femur.—(i.) Maximum length ("straight") from top of head to bottom of internal condyle (F).

(ii.) "Oblique" length from top of head to plane in contact with both condyles (F′).

For both we have for articular cartilage at upper end 2 millims., at lower end 2·5 millims., or the total together of 4·5 millims. This is more than double MANOUVRIER'S allowance.

Humerus.—Length from top of head to lowest point of internal margin of trochlea (H). At upper end we must allow 1·5 millims., and at lower 1·3 millims., altogether 2·8 millims. for articular cartilage.

Tibia.—The spine is excluded by ROLLET. The length is from plane of upper surfaces (margins) to tip of internal malleolus (T). In this case the articular cartilage has only to be allowed for at the upper end, and is here 3 millims.

Radius.—The length is measured from top of head to tip of styloid process (R). The allowance must be for articular cartilage at upper end only, and is 1·5 millims.

(b.) *Allowance for Animal Matter in Bones.*

Here unfortunately I had not the same amount of data to guide me. The best hypothesis to go upon seemed to be that a thoroughly dry bone, free from all animal matter, would, if it were thoroughly soaked, approximate to the condition of the bones measured by ROLLET. BROCA, who has written a very elaborate memoir on the effect of humidity in altering the capacity and dimensions of skulls, has referred incidentally to the extension of the femur by humidity.‡ He took three femurs, one macerated in 1873, one of the 15th century, and one of the polished stone age. After soaking for seven days, he found an increase of 1·5 millims. in the first, 1·5 millims. in the second, and 1 millim. in the third. These results, he says, compare very well with WELCKER'S,§ who gives 1·2 millims. for increase of length of femur with humidity.

It was somewhat difficult to make fresh experiments on a considerable number of

* The details of this section I owe entirely to my colleague, Professor GEORGE THANE, who in this matter, as in many others, has given me most ready and generous assistance.

† On another occasion I may take into consideration the ulna and fibula, but they have nothing like the importance for stature of the bones here dealt with.

‡ 'Mémoires d'Anthropologie de PAUL BROCA,' vol. 4, pp. 163 *et. seq.*; p. 195.

§ 'Ueber Wachstum und Bau des menschlichen Schädels,' p. 30, 1862. WELCKER only dealt with one male femur, and soaked it for three days.

long bones of each kind, but it seemed worth while to measure dry and thoroughly humid a bone of each type. A bone of each type was placed at my disposal by Professor THANE, and they were measured independently on each occasion by Mr. BRAMLEY-MOORE and myself. In the one or two instances in which we did not agree within ·02 millim., the bone was again independently measured. Our results were as follows :—

TABLE XI.—Lengths of Long Bones, Dry and Wet, in Centimetres.

	Dry as received.	24 hours in water.	120 hours in water.	72 hours drying.
F.	42·58	42·79	42·84	42·50
T.	37·41	37·52	37·58	37·37
H.	34·52	34·62	34·65	34·48
R.	23·11	23·20	23·19	23·00

The bones themselves were between 200 and 300 years old.* They were only allowed to stand two hours for the water to run off before they were measured after soaking. In the case of the final 72 hours' drying, it concluded with six hours in the neighbourhood of a stove. The first column may be considered to represent the average humidity of bones preserved in a museum ; the last column complete dryness. It seems to me that the difference between the first and third column is what we in general have to deal with. In this case we have a difference of

F.	T.	H.	R.
2·6 millims.	1·7 millims.	1·3 millims.	·7 millim.

between dry and humid bones.

The difference between this result for the femur and BROCA's is very considerable. I think it is due to the fact that he allowed his bones to dry for 24 hours in a room before measuring them. I was much impressed by the rapidity with which the bones dried, and their conditions, of course, are very unlike what they would be if containing or surrounded by animal matter. It is clear that the extensions due to humidity are not by any means proportional to the length of the bone, and it would be quite futile to attempt any percentage allowance for the extension due to this cause, the effect of which clearly differs with the different structure of different parts of the same bone. I have accordingly thought it best to subtract the above quantities from ROLLET's means, M_F, M_T, M_H, and M_R, and to consider the results so derived as giving the means of ROLLET's material on the supposition that the bones were dry and free from animal matter. Even so I do not think we shall err in over-estimating the difference between the lengths of living and dead bone. Making allowances (*a*) and (*b*) we have finally to subtract from ROLLET's results for

* See additional note, p. 244.

M_F.	M_H.	M_T.	M_R.
7·1 millims.	4·1 millims.	4·7 millims.	2·2 millims., respectively.

Making these subtractions (which are sensibly different from MANOUVRIER's allowance of 2 millims. for each bone), we are in a position to find the reconstruction formulæ connecting living stature with dry bone entirely free of animal matter. We have for the French population, if $M_{S''}$ denotes living mean stature, and $M_{F''}$, $M_{H''}$, $M_{T''}$, $M_{R''}$, the mean lengths of the corresponding dry bones in centimetres:

TABLE XII.

	$M_{S''}$.	$M_{F''}$.	$M_{H''}$.	$M_{T''}$.	$M_{R''}$.
Male	165·0	44·52	32·60	36·34	24·17
Female	152·3	40·86	29·36	32·97	21·27

If we want the mean oblique length of the femur $M_{F'}$, we must follow the rule given on p. 184, and we find $M_{F'} = 44·20$ for male and $= 40·53$ for female. M. RAHON has measured the lengths of a large collection of long bones in the Faculty of Medicine of Paris,* and he finds :—

Femur, oblique length,	62 males,	mean	44·1 (44·2).
,, ,, ,,	38 females,	,,	39·6 (40·5).
Humerus, maximum length,	44 males,	,,	32·3 (32·6).
,, ,, ,,	39 females,	,,	29·2 (29·4).

My results are placed in brackets, and it is clear that for these bones the allowances for cartilage and animal matter have been very satisfactory; there has certainly been no over-correction, although in the case of the femur our allowance is more than thrice, and in that of the humerus more than twice M. MANOUVRIER's.

M. RAHON does not give the measurement of the radius, but he does of the tibia, and in this case there is undoubtedly some source of error in his result, or in the collection. He gives :—T for 53 males, mean = 37·7; for 26 female = 35·7. Now ROLLET's material for 50 of either sex gives, male mean = 36·8, and female = 33·4, without allowance for the cartilage or presence of animal matter. Allowing for these, RAHON's measurements are, male, 1·4 centims., and female, 2·7 centims. *too large.* These are errors much beyond those of the determinations, which have probable errors of about ·17 to ·18 centim. RAHON, since he is using MANOUVRIER's method must be supposed to be measuring the tibia in the same manner as ROLLET, *i.e.*, with the malleolus and without the spine. But even supposing he had included the spine,

* *Loc. cit.*, p. 413.

it could not make this great difference.* That there is some substantial error is evidenced by the fact that tibias of these dimensions would give a reconstructed stature for French males of 168·2 centims. instead of 165 centims., and for French females of 158·9 centims. instead of 152·3 centims. RAHON himself, on the basis of MANOUVRIER's method, forms the estimates of 166·8 centims. and 159·5 centims. respectively,—the latter, at any rate, a quite impossible height for the French female population.

(10.) We are now in a position to write down the reconstruction formulæ for living stature from dry long bones; they are the following:—

TABLE XIV.—Male. Living Stature from Dead† Long Bones.

(*a*) $S = 81{\cdot}306 + 1{\cdot}880\ F.$
(*b*) $S = 70{\cdot}641 + 2{\cdot}894\ H.$
(*c*) $S = 78{\cdot}664 + 2{\cdot}376\ T.$
(*d*) $S = 85{\cdot}925 + 3{\cdot}271\ R.$
(*e*) $S = 71{\cdot}272 + 1{\cdot}159\ (F + T).$
(*f*) $S = 71{\cdot}443 + 1{\cdot}220\ F + 1{\cdot}080\ T.$
(*g*) $S = 66{\cdot}855 + 1{\cdot}730\ (H + R).$
(*h*) $S = 69{\cdot}788 + 2{\cdot}769\ H + {\cdot}195\ R.$
(*i*) $S = 68{\cdot}397 + 1{\cdot}030\ F + 1{\cdot}557\ H.$
(*k*) $S = 67{\cdot}049 + {\cdot}913\ F + {\cdot}600\ T + 1{\cdot}225\ H - {\cdot}187\ R.$

TABLE XV.—Female. Living Stature from Dead† Long Bones.

(*a*) $S = 72{\cdot}844 + 1{\cdot}945\ F.$
(*b*) $S = 71{\cdot}475 + 2{\cdot}754\ H.$
(*c*) $S = 74{\cdot}774 + 2{\cdot}352\ T.$
(*d*) $S = 81{\cdot}224 + 3{\cdot}343\ R.$
(*e*) $S = 69{\cdot}154 + 1{\cdot}126\ (F + T).$
(*f*) $S = 69{\cdot}561 + 1{\cdot}117\ F + 1{\cdot}125\ T.$
(*g*) $S = 69{\cdot}911 + 1{\cdot}628\ (H + R).$
(*h*) $S = 70{\cdot}542 + 2{\cdot}582\ H + {\cdot}281\ R.$
(*i*) $S = 67{\cdot}435 + 1{\cdot}339\ F + 1{\cdot}027\ H.$
(*k*) $S = 67{\cdot}469 + {\cdot}782\ F + 1{\cdot}120\ T + 1{\cdot}059\ H - {\cdot}711\ R.$

Remarks.—(i.) If the femur has been measured in the oblique position and not

* Dr. WARREN found for the New Race from Egypt the mean length of spine for 85 males = ·96 centim., and for 115 females = ·87 centim. These numbers should be introduced as an addition to M_T in Tables V. and VI., when the tibia has been measured including spine.

† The word "dead" is here used to denote a bone from which all the animal matter has disappeared, and which is in a dry state.

straight, add ·32 centim. for male and ·33 centim. for female to the length before using the above formulæ.

(ii.) If the tibia has been measured with, and not without, the spine, subtract ·96 centim. for male and ·87 centim. for female from the length before using the above formulæ.

(iii.) The above formulæ have been determined from the *right* members; a small error, of the second order as a rule, arises when the left is used. The following numbers are determined from ROLLET'S measurements; they give the amount to be added to a left bone when it is used in the formulæ:—

	Femur.	Humerus.	Tibia.	Radius.
Male.	−·04	+·42	+·18	+·28
Female	+·03	+·51	+·09	+·19

The femur change is insignificant. In most statements of lengths the rightness or leftness of the bone is not given, and hence, no correction can generally be made for an individual. The error will, however, be hardly sensible except in the case of the humerus and radius. If a considerable number of bones have been averaged, probably half may be looked upon as right and half left, and in this case half the above corrections may be added to the average. In any case, it is probably only the estimate based on the humerus and radius which need to be corrected in this manner.

Even here it is a problem how far there is a racial character in this right and left-sideness. Results due to CALLENDER, ROBERTS, GARSON, HARTING, and RAYMONDAUD are cited by ROLLET (*loc. cit.*, pp. 53–60), but being based either on very few cases, on measurements on the living, or on unsexed material, they are not of much service for our present purpose. Results of much greater value for racial comparison have been given by Dr. WARREN for the Naqada race ('Phil. Trans.,' B, vol. 189, p. 135 *et seq.*). He finds:—

	Femur.	Humerus.	Tibia.	Radius.
Male.	−·11	+·34	−·08	+·20
Female	−·16	+·57	−·105	+·305

Dr. WARREN'S results are for the oblique femur, and from centre to centre of the articulate surfaces in the case of tibia and radius. Thus they are not directly comparable with the results for the French. On the whole, if the bone is stated to be left, we may add ·45 for the humerus and ·25 for the radius, leaving the femur and tibia unaltered. These additions are approximately the same for both sexes.

(11.) Before we proceed to apply the formulæ in Tables XIV. and XV. to the general reconstruction of stature, it is desirable to obtain some measure of confidence in the application of the formulæ. We require to test them by finding what sort of results they give for a second race.* That race ought to be as widely divergent from the French as possible, but one in which the stature as well as the measurement of the long bones is known. There are, I believe, no other measurements than those of ROLLET, in which both the stature and long bones have been measured on the same individuals. A fairly complete series of measurements of the long bones of the Aino have, however, been made by KOGANEI, and he has also determined the mean living

* There is very little detail for verification of our results even in the same race. M. MANOUVRIER gives the dimensions of seven men, six of whom were assassins (see p. 387 of *loc. cit.* in footnote, p. 171 above). I have reconstructed the statures of these seven individuals from our ten formulæ with the following results :—

	Assassins.						A.B. Name unknown.
	MATHELIN.	SELLIER.	KAPS.	RIVIÈRE.	GAMAHUT.	ALORTO.	
Long bones:—							
F.	50·12	45·22	44·52	44·72	42·72	44·82	39·52
H.	35·4	32·6	31·9	32·8	30·5	33·3	29·8
T.	43·3	36·4	37·7	35·3	37·6	36·3	33·4
R.	27·6	24·1	24·4	23·8	24·7	24·5	22·1
Stature :—							
(*a*)	175·5	166·3	165·0	165·4	161·6	165·6	155·6
(*b*)	173·1	165·0	163·6	165·6	158·9	167·0	156·9
(*c*)	181·6	165·2	168·2	162·5	168·0	164·9	158·0
(*d*)	176·2	164·8	165·7	163·8	166·7	166·1	158·2
(*e*)	179·6	165·9	166·6	164·0	164·4	165·3	155·8
(*f*)	179·3	165·9	166·5	164·1	164·2	165·3	155·7
(*g*)	175·9	165·0	164·3	164·8	162·4	166·8	156·6
(*h*)	173·2	164·8	162·9	165·3	159·1	166·8	156·6
(*i*)	175·1	165·7	163·9	165·5	159·9	166·4	154·9
(*k*)	177·0	165·6	164·8	164·8	161·4	166·2	155·5
Mean . . .	176·7	165·4	165·2	164·6	162·7	166·0	156·4
Actual . . .	180·0	173·4	171·7	168·3	165·2	160·9	156·6
Difference . .	−3·3	−8·0	−6·5	−3·7	−2·5	+5·1	−0·2

While the means of the whole series of formulæ agree very closely with the results of (*k*), they differ very markedly from the actual statures. I do not know under what conditions the long bones or the statures were measured. A suggestive but somewhat hasty conclusion (failing more data) would be that the average assassin is tall (170 centims. against the general French population of 165 centims.), but his limbs are relatively short, *i.e.*, he is long of trunk. Anyhow, the divergence is noteworthy.

stature from a fairly large series of living individuals.* Now the Aino are a race widely divergent from the French, and therefore, although the stature and long bones are not measured on the same group, we are likely to get a very good test of the safety with which we can apply our stature results from one local race to a second. The stature, as measured by KOGANEI on 95 living males, was 156·70 centims., and on 71 living females, 147·10 centims. The long bone measurements were made on 20 to 25 female and 40 to 45 male skeletons, not quite from the same districts as the living groups. The maximum length of the long bones is given in the paper by Miss LEE and myself, 'Roy. Soc. Proc.,' vol. 61, pp. 347–8, and accordingly allowance must be made for the spine in the case of the tibia. We then have the following values for insertion in Tables XIV. and XV. :—

	Femur.	Humerus.	Tibia.	Radius.
Male.	40·77	29·50	32·93	22·91
Female	38·20	27·72	30·99	21·08

TABLE XVI.—Reconstruction of Aino Stature.

Formula.	Male.		Female.	
	Calculated value.	Difference.	Calculated value.	Difference.
(a) Male	157·95	+1·25	153·12	+6·02
(b) „	156·01	−0·69	150·86	+3·76
(c) „	156·90	+0·20	152·30	+5·20
(d) „	160·90	+4·20	154·88	+7·78
(e) „	156·69	−0·01	151·46	+4·36
(f) „	156·75	+0·05	151·34	+4·24
(g) „	157·52	+0·82	151·28	+4·18
(h) „	155·94	−0·76	150·65	+3·55
(i) „	156·32	−0·38	150·90	+3·80
(k) „	155·90	−0·80	150·53	+3·43
Observed	156·70	0	147·10	0
(a) Female . . .	152·14	−4·56	147·14	+0·04
(b) „ . . .	152·72	−3·98	147·82	+0·72
(c) „ . . .	152·33	−4·37	147·66	+0·56
(d) „ . . .	157·82	+1·12	151·69	+4·59
(e) „ . . .	152·14	−4·56	147·06	−0·04
(f) „ . . .	152·14	−4·56	147·18	+0·08
(g) „ . . .	155·23	−1·47	149·36	+2·26
(h) „ . . .	153·15	−3·55	148·04	+0·94
(i) „ . . .	152·32	−4·38	147·05	−0·05
(k) „ . . .	151·14	−5·56	146·48	−0·62

* "Mittheilungen aus der Medicinischen Facultät der k. Japanischen Universität," vol. 2, I. and II., Tokio, 1893 and 1894.

Several results may be noted with regard to this table: (i.) In the first place let us compare our results with those which would be given by M. MANOUVRIER's Tableau II.* Corresponding to our cases (*a*), (*b*), (*c*), (*d*) he would obtain:—

	Male.		Female.	
	Calculated value.	Difference.	Calculated value.	Difference.
(*a*)	156·80	+0·10	145·36	−1·74
(*b*)	152·47	−4·23	146·86	−0·24
(*c*)	155·59	−1·11	147·32	+0·22
(*d*)	161·13	+4·43	153·08	+5·98
(*f*)	156·19	−0·51	146·34	−0·76
(*h*)	156·80	+0·10	149·92	+2·82
(*i*)	154·63	−2·07	146·11	−0·99
(*k*)	156·50	−0·20	148·15	+1·05
Observed	156·70	0	147·10	0

Here (*f*), (*h*), (*i*), and (*k*) are obtained by taking means of the results for the single bones. Comparing the first four formulæ with my first four, M. MANOUVRIER has for male a mean error of 2·47 centims. against my 1·58 centims., and for the last four a mean error of ·72 centim. as against my ·50 centim. His error in stature, as deduced from the male humerus, is greater than my error from the radius even. In the male measurements M. MANOUVRIER has a mean error of 2·04 centims. against my 1·48 centims. in the first four results, and one of 1·40 centims. against my ·42 centim. in the last four results.

But these results by no means represent the full advantage of the present theory. An examination of the results shows us the formulæ give good, *i.e.*, consistent results except in the case of the radius. Here it is that the greatest differentiation has taken place, very possibly owing to the direct selection of other long bones. Our general principles (p. 177) accordingly suggest that we should omit the results for this bone from our consideration. The best formulæ then to use will be (*e*), (*f*), and (*i*); we shall then have a mean error of ·15 centim. for male and ·06 centim. for female—a better approximation to the true stature could not possibly be reached. M. MANOUVRIER, by the process of means, would have deduced from the same three bones a male stature with an error of 1·75 centims. and a female stature with one of ·59 centim.

Dr. BEDDOE's rule† would give for male Aino 155·3 centims., and female Aino 146·6 centims., or errors of 1·4 centims. and ·5 centim.; in this case not as great as those of M. MANOUVRIER, but still sensibly greater than our (*e*), (*f*), or (*i*).

The accordance obtained between the formulæ for reconstruction which I have given,

* *Loc. cit.*, tables at end of Memoir.

† 'Journal of the Anthropological Institute,' vol. 17, 1887, p. 205.

and the actually observed stature in the case of such a diverse race as the Aino ought, I think, to give considerable confidence in their use.

(ii.) I have also included in the table the results for the male Aino, calculated from female formulæ, and for the female Aino, calculated from the male formulæ. The reader will perceive at once that sexual differences are immensely greater than racial differences—that it would be perfectly idle to attempt to reconstruct female stature from male formulæ, or *vice versâ*. Exactly the same order of divergences are obtained if we endeavour to reconstruct French female from male formulæ, or *vice versâ*, and we concluded that French men and French women are more differentiated from each other than French of either sex and Aino of the same sex, at any rate, in the relations between stature and the long bones. It is noteworthy that the only instance in which the formula for one sex gives even approximately the stature of the other, is in the case of the female formula applied to find the male stature by means of the length of the radius. In this case we get a better result than from the male formula itself. Now this is peculiarly significant, for it is in the radius that the most marked differentiation between French and Aino has taken place; and in this respect the Aino male approaches nearer to the French female than to the French male. We must therefore conclude that while the sexes are widely differentiated from a common stock, still in respect of radius the females of a highly civilised race like the French, and the males of a primitive race like the Aino, are even closer together than the males or the females of these two races for this special bone. The agreement between the same sex in two different races, however, is generally far closer than between different sexes in highly civilised and primitive races.

(12.) Having taken an extreme case of divergence in man and tested the confidence that may be put in our reconstruction formulæ, it will not be without interest to see the amount of divergence in the formulæ when we apply them to allied species. Stature is, of course, a very difficult character to deal with when we are considering the anthropomorphous apes, and it would be idle to think of going beyond a round number of centimetres. But even here the agreements and disagreements are so remarkable that they appear to furnish material on which certain quantitative statements with regard to the general lines of evolution can be based, and further they suggest that the regression formulæ for the long bones among themselves* open up quite a new method of attacking the problem of the descent of man. Like the rest of the material in this paper, the considerations of the present paragraph must be looked upon as suggestions for new methods of research. I have taken what material was at hand and not endeavoured to form comprehensive statistics. The methods are illustrated on stature, but they are equally applicable to the regression formulæ connecting any characters or organs whatever.

* I hope later to deal at length with the regression formulæ for the long bones of man and apply them to the anthropomorphous apes, placing stature entirely on one side as a quantity very difficult to measure.

The following table, here given in centimetres, is taken from HUMPHRY's work.*

TABLE XVIII.—Stature and Long Bones of Anthropomorphous Apes.

	No.	Stature.	Femur.	Humerus.	Tibia.	Radius.
Chimpanze . .	4	127	31·52	31·01	25·40	27·90
Orang	2	112	26·92	35·60	23·41	35·60
Gorilla	3	147	35·33	42·12	28·70	32·79

The sexes are not stated, and the results are all mean results for the numbers given. The stature is probably exaggerated rather than understated, and must have been difficult to estimate. It might seem at first sight idle to apply the stature reconstruction formulæ for man, to such data, but as we shall soon see it is a question of coming within 10 or 20 centims. of the true values in all but a few cases. I have calculated the following table from the reconstruction formulæ for both sexes in man :—

TABLE XIX.—Reconstructed Stature of Anthropomorphous Apes.

Formula.	Chimpanze.		Orang.		Gorilla.	
(*a*) Male .	141	+14	132	+20	148	+ 1
(*b*) " .	160	+33	174	+62	193	+46
(*c*) " .	139	+12	134	+22	147	0
(*d*) " .	177	+50	203	+91	193	+46
(*e*) " .	137	+10	130	+18	145	− 2
(*f*) " .	137	+10	130	+18	146	− 1
(*g*) " .	169	+42	190	+78	196	+49
(*h*) " .	161	+34	175	+63	193	+46
(*i*) " .	149	+22	152	+40	170	+23
(*k*) " .	144	+17	143	+31	162	+15
Observed .	127	0	112	0	147	0
(*a*) Female	134	+ 7	125	+13	142	− 5
(*b*) "	157	+30	169	+57	188	+41
(*c*) "	135	+ 8	130	+18	142	− 5
(*d*) "	174	+47	200	+88	191	+44
(*e*) "	133	+ 6	126	+14	141	− 6
(*f*) "	133	+ 6	126	+14	141	− 6
(*g*) "	166	+39	186	+74	192	+45
(*h*) "	158	+31	172	+60	189	+42
(*i*) "	141	+14	140	+28	158	+11
(*k*) "	134	+ 7	127	+15	148	+ 1

* 'A Treatise on the Human Skeleton,' Cambridge, 1851, p. 106. It is, perhaps, needless to remark that the gibbon gives stature results quite incomparable with those for man.

Now we see that, if the gorilla be put on one side, there is no approach to accordance between the calculated and observed statures* in the case either of the chimpanze or orang for any of the ten formulæ. We conclude therefore, that if man and the chimpanze and orang have been derived from a common stock, they must have been directly selected with regard to stature and with regard to the lengths of the four chief long bones. In the case of the gorilla we notice, however, a remarkable accordance between the observed stature, and that calculated from the male reconstruction formulæ in the case of man, when we use only formulæ involving the femur and tibia. It would thus appear that if man and the gorilla have been differentiated from a common stock, they have been directly selected in the same manner so far as femur and tibia are concerned, but in different directions when we consider humerus and radius—we are here referring only to the lengths of these bones. Re-examining the results for the male formulæ from the standpoint of correspondence in the femur and tibia between the gorilla and man, we see that the chimpanze comes nearer to man than the orang; the lengths of the femur and tibia have been modified in the former, but not to such a marked degree as in the case of the latter. Turning to the female reconstruction formulæ we notice in (*a*) to (*k*) for the chimpanze and orang an accordance between the observed and calculated statures which is some 3 centims. to 6 centims. better, although still very poor. The reason for this is obvious, the stature of the woman for the same length of long bone is 3 centims. to 6 centims. shorter than that of man, and accordingly the female formulæ must give slightly better results than the male formulæ when applied to the anthropomorphous apes, which have for the same length of bone a markedly shorter stature than man. In the gorilla we have over-corrected the stature so far as femur and tibia are concerned by using the female formulæ. One point, however, is of very great interest: while the female formulæ for humerus, radius, or for humerus and radius give very bad results, even worse for the gorilla than they do for the chimpanze, yet the female formula for femur and humerus gives a sensibly better, and that for all the long bones a markedly better result for the stature than the corresponding male formulæ. The difference here is not the 3 centims. to 6 centims. due to sex. The improvement in the result when we apply the female formulæ for all four long bones to the estimate of the stature of the gorilla is noticeable also, if to a lesser degree, in the cases of the chimpanze and orang. We may sum up our results as follows:—

(*a*.) Man is apparently differentiated from the chimpanze and orang by direct selection of stature, but this direct selection appears to be small in the case of the gorilla.

* If the chimpanze and orang be treated as "dwarf men," and their statures estimated in the manner indicated on p. 224 below, the femur and tibia give statures, F, 115·5, 105·0; T, 118·0, 112·5 respectively, nearer the actual values, in fact too small, but the radius and humerus still give values far too great. The stature of the gorilla as estimated from femur and tibia in this manner now becomes far too small.

(*b.*) Man and the gorilla appear to have followed common lines of differentiation from a common stock in the case of the femur and tibia, but the differentiation on which they have not followed common lines has not been that of radius and humerus alone, or (k) would have given good results.

(*c.*) Other organs closely correlated with stature beside the four long bones must have been differentially modified in the case of the chimpanze and orang, or (k) would still have given good results.

(*d.*) The accordance between the result given by female (k) and the observed stature of the gorilla, and the want of accordance in all other formulæ, seems to show that woman has been principally differentiated by these four long bones from the common stock, while man has been differentiated in other organs highly correlated with stature. For example, the differentiation in pelvis may be much greater.

So far as I am able to draw a conclusion from the few data at my command, the correlation of radius and humerus with stature appears to be negative for the chimpanze and orang, while it is positive for the gorilla and man. The negative character of the partial correlation coefficient for the radius in (k) seems to be a relic of this stage of evolution, and it is much more marked in woman than in man.

The above statements must not be taken as dogmatic conclusions; they are only *suggestions* of the manner in which the regression formulæ can possibly be applied to the problems of evolution. They are no more weighty than the very slender material* on which they are based. But they may suffice to indicate how a method of quantitative inquiry might be applied to ascertain more about the relationship of man to the anthropomorphous apes, so soon as a sufficient amount of data concerning the dimensions of the organs of adult apes has been collected, and reduced to numerical expression.

* In order to verify Sir G. HUMPHRY'S measurements, I have gone through the catalogues, so far as published, of the German anthropological collections, and extracted the measurements of all *adult* anthropomorphous apes. Unfortunately I could only find one adult chimpanze; the sex was as often as not not given. I find:

	No.	Stature.	Femur.	Humerus.	Tibia.	Radius.
Gorilla	7	144·2	35·51	41·83	28·19	33·91
Orang	9	119·9	26·52	34·34	22·57	34·10

A better agreement with the results cited, p. 202, could not have been expected, or wanted. Thus our data give racial and not random characters.

(13.) *Palæolithic Man.*

I am indebted to the memoir of M. RAHON* for the details of all the individuals that are classed under this heading. I presume that in measuring the tibia he has not included the spine, as his formulæ are, like mine, based on its exclusion. I have further allowed for the fact that he used the oblique length of femur, while I require the maximum length. Unfortunately we have only five cases to base our estimate upon.

Neanderthal Man.

$$F = 44{\cdot}52, \quad H = 31{\cdot}2, \quad R = 24{\cdot}0.$$

We find for stature from:

(*a.*)	(*b.*)	(*d.*)	(*h.*)	(*i.*)	Mean
165·01	160·94	163·46	162·83	161·59	162·96

RAHON gives 161·3 centims. (but I think he ought to have given 165·2 centims., as his femur estimate is incorrect) and SCHAAFFHAUSEN† 160·1 centims., so that our estimate diverges by 2 centims. to 3 centims.

Man from Spy.

$$F = 43{\cdot}32, \quad T = 33{\cdot}0.$$

We find for stature from:

(*a.*)	(*c.*)	(*f.*)	Mean
162·75	157·07	160·26	160·33

RAHON gives 159·0 centims.

Man from Clay at Lahr.

The length of the femur here is doubtful, but it is said to have been between 45·0 centims. and 46·0 centims. If we take the mean value, the probable stature was 166·85 centims., and the maximum value would only be 167·79 centims. RAHON gives 170 centims., using ulna as well as femur. I have not worked out the stature-ulna correlation, but, if this bone is at all akin to the radius, it will give very exaggerated results for primitive man.

Man of Chancelade.

$$F = 40{\cdot}8, \quad H = 30{\cdot}0, \quad R = 23{\cdot}6.$$

(*a.*)	(*b.*)	(*d.*)	(*h.*)	(*i.*)
158·095	157·46	163·125	157·46	157·13.

Here again the radius gives clearly an exaggerated result. The mean is

* 'Mémoires de la Société d'Anthropologie de Paris,' 1893, p. 414 *et seq.*

† "Der Neanderthaler Fund," 'Deutsche Anthropologische Gesellschaft,' 1888.

158·7 centims., but, neglecting (*d.*), I am inclined to take the best value as 157·5 centims. RAHON gives 159·2 centims. MANOUVRIER (*loc. cit.*, p. 391) is inclined from the general character of the bones to consider the stature as determined from the ulna and radius to be the better estimate, and even thinks this troglodyte may have been 165 centims. Judging, however, from other primitive races, I should expect the arm bone estimate to exaggerate the stature, and prefer my estimate of 157·5 centims.

Man of Laugerie.—All we know here is the length of the femur = 45·1 centims. The probable stature is accordingly 166·1 centims. TOPINARD gives it as 168·5 centims., and RAHON at 164·9 centims.

Taking the mean of the best values for the above five cases we have:—

Probable stature of palæolithic man = 162·7 centims. All the above cases are supposed to be males. Considering that it is more probably the massive bones which have survived, we must hold that palæolithic man was shorter than the modern French population, but was taller than the men of Southern Italy (156 centims. to 158 centims.), and about the mean height of the modern Italian male population, *i.e.*, 162·4 centims.

(14.) *Neolithic Man.*

(*a.*) *Great Britain.*

We have not very much data to build upon here. Dr. BEDDOE* gives the length of twenty-five male and five female femora. Converted into centimetres, we have

Male F (25), 45·72 centims. ⎫ hence probable stature ⎧ male, 167·3 centims.
Female F (5), 41·53 „ ⎭ from (*a*) ⎩ female, 153·6 „

Dr. BEDDOE'S estimates, male 170·2 centims., and female 156·3 centims., are, I think, much too high. The sex-ratio is 1·089.

(*b.*) *France and Belgium.*

The following data have been drawn from RAHON (*loc. cit.*, pp. 418 *et seq.*), the numbers in brackets in the left-hand corners denoting the numbers upon which the average lengths of the bones are based.

	F.	H.	T.	R.
Male	(127) 43·99	(127) 31·085	(133) 35·87	(49) 23·54
Female	(53) 40·105	(79) 28·58	(45) 33·11	(18) 21·76

* 'Journal of the Anthropological Institute,' vol. 17, 1887, p. 209.

We find:

STATURE of Neolithic Man.

Formula.	Male.	Female.
(*a*)	164·01	150·85
(*b*)	160·60	150·18
(*c*)	163·89	152·65
(*d*)	162·92	153·97
(*e*)	163·83	151·59
(*f*)	163·85	151·61
(*g*)	161·36	151·86
(*h*)	160·45	150·45
(*i*)	162·11	150·49
(*k*)	162·41	150·71
Mean . .	162·54	151·44

Sexual ratio ♂/♀ = 1·073.

So far, then, as we have material to judge by, there appears to be no sensible difference between Continental palæolithic and neolithic man; they corresponded very closely to the modern Italian in stature.

On the other hand, if we compare British with Continental neolithic man, we find, judging even from femora only, a very sensible difference in stature. Neolithic man in Britain was taller probably than the modern Frenchman, and markedly taller than neolithic man in France.

(*c*.) This leads us to consider one or two special classes of neolithic bones, for it must be remembered that probably as many neolithic races existed in Europe as we find races existing in historic times. In the first place, we have the big bones of the Cro-Magnon man,* F = 48·32 centims., T = 39·5 centims. These give for the stature:

(*a*.)	(*c*.)	(*e*.)	(*f*.)	Mean.
172·15	172·52	173·06	173·05	172·70

which is a centimetre greater than RAHON'S estimate, seven less than ROLLET'S, and seventeen less than TOPINARD'S. This man was undoubtedly tall, but cannot be taken as a type of his race. The second Cro-Magnon skeleton gives us H = 32·1 centims., T = 37·5 centims. from which we find from:

(*b*.)	(*c*.)	Mean.
163·54	167·76	165·68.

This is also taller than the average neolithic man, but much below the other skeleton.

* As carefully determined by RAHON (*loc. cit.*, p. 421).

Two homogeneous series of neolithic bones are given by M. MANOUVRIER in a paper entitled: "Étude des Crânes et Ossements humains recueillis dans la Sépulture Néolithique dite la Cave aux Fées, à Brueil,"* and deserve separate consideration. We find:

	F.	H.	T.	R.
Brueil male	(10) 41·77	(19) 30·86	(4) 35·20	(5) 24·19
" female	(7) 38·63	(8) 28·51	..	(5) 22·08
Mureaux male	(16) 44·51	(10) 31·46	(10) 35·08	(6) 24·63
" female . . .	(2) 40·38	(5) 29·26	(7) 33·84	(3) 21·57

I have deduced the following results:

Formula.	Male.		Female.	
	Brueil.	Mureaux.	Brueil.	Mureaux.
(a)	159·83	164·98	147·98	151·38
(b)	159·95	161·69	149·99	152·06
(c)	162·30	162·01	..	154·37
(d)	165·05	166·59	155·04	153·33
(e)	160·48	163·52	..	152·73
(f)	160·42	163·63	..	152·74
(g)	162·09	163·89	152·27	152·66
(h)	159·96	161·70	150·36	152·15
(i)	159·47	163·23	148·44	151 55
(k)	159·58	162·67	..	152·60
Mean . .	160·91	163·39	150·68	152·56

The corresponding mean values given by M. MANOUVRIER are: 161·2, 163·8, 150·2 and 154·3, of which only the last diverges sensibly from mine.† I should be inclined to omit the results obtained from (*d*) as excessive, only the larger radii surviving. To do so would not much alter my means, based on ten results, although it would more sensibly modify M. MANOUVRIER'S.

The sexual ratios for the two groups are:—

* "Memoires de la Société des Sciences naturelles . . . de la Creuse," 2e Série, vol. 3, 1894 (2e Bulletin).

† The agreement is surprising, considering that M. MANOUVRIER worked only from half my data, and allowed very differently for the drying of the bones.

Brueil ♂/♀ = 1·068. Mureaux ♂/♀ = 1·071,

both less than the result we have obtained for the general averages of neolithic man. Probably we have here to do with local races, but M. MANOUVRIER considers it just possible that the very different environmont at Brueil and Mureaux may account for the differences.

Neither of these groups has a stature equal to that of the modern French commonalty, although the Mureaux group approaches it somewhat closely. The modern British far exceed in stature their neolithic landsmen, and we have thus no evidence at all in favour of a giant stature for prehistoric man. He seems to have been markedly shorter than the taller races (English-Scandinavian) of to-day. Slightly taller than the Aino, he can be compared with the Italians, who appear, as we go southward, to closely represent him in stature.

(15.) *Other Early Races.*

In this group I propose to include a number of prehistoric or protohistoric races of whom we know very little. Their stature is considerably greater than that which we have determined for Continental neolithic man, though sensibly below that of British neolithic man. The data are extracted from RAHON'S memoir, and modified to suit the formulæ of this investigation (see his pp. 431, 438 *et. seq.*).

Race.	F.	H.	T.	R.
Dolmen-builders, India, male . . .	(3) 45·81	(1) 32·5	(3) 35·3	(1) 24·5
" " " female . .	(1) 42·93	..	(1) 33·3	..
" " Algeria, male . .	(16) 45·32	(16) 31 9	(12) 38·0	(15) 23·8
" " " female . .	(8) 40·43	(5) 28.8	(9) 33·8	..
" " Caucasus, male . .	(7) 44·92	(6) 32·4	(3) 34·6	(4) 24·6
" " " female . .	(1) 41·3	(1) 29·1	..	..
Guanches, Group I., male	(87) 45·52	(60) 32·8	(79) 37·7	(30) 24·7
" " female	(90) 41·33	(92) 30·1	(58) 34·7	(32) 22·1
" Group II., male	(76) 45·22	(81) 32·5	(75) 37·6	(56) 24·6
" " female . . .	(33) 41·03	(34) 29·6	(20) 34·4	(10) 22·1

While the dolmens of India and Algeria appear to belong to the Stone Age, those of the Caucasus belong to the first Iron Age.

The series from these dolmens is very small. On the other hand the Guanch series are both very complete. The first are drawn from the Musée Broca, and the second from the Muséum d'Histoire Naturelle (see RAHON, *loc. cit.*, p. 446), both at Paris. Although the first series comes from a single locality, and the second from several localities, the results are in good agreement. The following statures have been found from our formulæ:—

Formula used.	Dolmens, India.		Dolmens, Algeria.		Dolmens, Caucasus.		Guanches, Group I.		Guanches, Group II.	
	Male.	Female.	Male.	Female.	Male.	Female.	Male.	Female.	Male.	Female.
(*a*)	167·43	156·34	166·51	152·26	165·76	153·17	166·88	153·23	166·32	152·65
(*b*)	164·70	..	162·96	150·79	164·41	151·62	165·56	154·37	164·70	152·99
(*c*)	162·54	153·10	168·95	154·27	160·87	..	168·24	156·39	168·00	155·68
(*d*)	166·06	..	163·77	..	166·39	..	166·72	155·10	166·39	155·10
(*e*)	165·28	154·99	167·84	153·19	163·44	..	167·72	154·76	167·26	154·09
(*f*)	164·86	155·01	168·37	153·19	163·61	..	167·69	154·76	167·22	154·09
(*g*)	165·46	..	163·22	..	165·46	..	166·33	154·92	165·64	154·08
(*h*)	164·56	..	162·76	..	164·30	..	165·43	154·47	164·58	153·18
(*i*)	166·18	..	164·75	151·68	165·11	152·62	166·35	153·69	165·58	152·77
(*k*)	165·29	..	165·85	..	163·91	..	166·79	154·82	166·11	153·72
Mean . .	165·24	154·86	165·50	152·56	164·33	152·47	166·77	154·65	166·18	153·83
Sexual ratio .	1·067		1·085		1·078		1·078		1·081	

The first point to be noticed about this table is the confidence it inspires in formula (*k*). Whenever the series is in the least extended, formula (*k*) gives a result sensibly identical with the mean of all ten formulæ.

M. RAHON's means for the eight groups are not very divergent from mine, he gives:—

166·0, 154·8; 165·7, 153·2; 165·3, 154·4; 166·0, 155·4; and 165·9, 154·3.

He thus does not make quite such a sensible distinction between the Guanches and the Dolmen-builders as my numbers seem to indicate. It is curious that these three groups of Dolmen-builders should stand so close together, and also comparatively close to the Guanches. The Dolmen-builders must have been as tall as the modern French, while the Guanches were probably slightly taller. Both were of greater stature than neolithic man in France, approaching more nearly the neolithic man of Britain.

The sexual ratio in the first and third cases cannot be considered of any weight, as the female data contain only single individuals.

(16.) *Stature of the Naqada Race from Upper Egypt.*

This race dates from about 4000 B.C. Its orgin and locus have been discussed by Professor FLINDERS PETRIE in "Naqada and Ballus," 1895, and an elaborate series of measurements made on the long bones by Dr. WARREN; see 'Phil. Trans.,' B, Vol. 189, pp. 135–227, 1897.

The measurements suited to our reconstruction Tables XIV. and XV. are:*

	F.	H.	T.	R.
Male	(80) 45·93	(82) 32 62	(86) 37·97	(47) 25·70
Female . . .	(113) 42·63	(97) 29·87	(115) 34·96	(66) 23·33

Whence we deduce for the stature:

Male.	Bones used.	Female.
165·04	H	153·74
165·13	H & R	154·23
166·61	H & F	155·19
166·93	H, F, R & T	155·02
167·66	F	155·76
167·79	H + R	156·53
168·49	F & T	156·51
168·5	F + T	156·93
168·88	T	156·99
169·99	R	159·21
167·5	Mean	156·0

Had we used M. MANOUVRIER'S "Tableau-barême," we should have found:†

Male.	Bone.	Female.
166·4	F	155·4
167·0	T	156·0
164·7	H	154·5
171·5	R	161·7
167·4	Mean	156·9

* The numbers in brackets to the left indicate the number of bones used to form the average.

† Here, as in other cases, the reader must remember before entering the "Tableau-barême," to correct from the maximum to oblique femur length.

While M. MANOUVRIER's male mean does not differ widely from ours, his female mean is ·9 centim. greater. His range for male stature covers 6·8 centims., and for female stature 7·2 centims., as compared with our 4·9 and 5·5 centims. respectively. But the amount of this range in both cases is very significant considering the large number of bones averaged. While our formulæ applied to the Aino gave very self-accordant results except in the case of the radius, we notice here considerable divergences. In particular, the order of the bones arranged in order of increasing stature, which is nearly the same in both sexes, is very different for the corresponding order for the Aino. The Naqada people for their stature have a remarkably small humerus, and although the Aino could hardly be separated more from the French by civilisation and locality, yet they could be derived from a common stock with the French by far less direct selection of the long bones, than would be possible in the case of the French and the Naqada races. This Egyptian race was a tall race—not as tall as the English commonalty—but taller than the better French classes and 2·5 centims. taller than the mean of the French army. The sexual ratio, 1·074, was less than that of the modern European (about 1·080), and this is in keeping with the greater equality in size observable in primitive and early races. On the whole it may be questioned whether any two *modern* races would give such divergence in character as the Naqada and French. We see not only the radius, as in the case of the Aino, but the humerus as a source of divergence, and so far as the lengths of those long bones are concerned, it would be easier to look upon the Ainos and French than upon the Naqada people and French as local races deduced from a common stock. If they have sprung ultimately from such a stock, there has been a very significant amount of direct selection. There is, however, an interesting point which the Naqada people share with the Ainos —the judgment of stature from the radius is excessive. This peculiarity of early and primitive races is one which the table on p. 202 shows that they share, of course in a much less marked manner, with the anthropomorphous apes. It will later be seen to be a feature of other primitive and early peoples.

(17.) *Protohistoric Races.*

My next group covers to some extent the ground which precedes 1000 A.D.—roughly, the beginning of the Middle Ages.

(*a*.) Dr. BEDDOE gives femur measurements for the Round Barrow population of Britain,* as follows:

Male F = 47·75 centims., mean for 27,
Female F = 44·91 „ „ 2.

We find at once from (*a*):

Stature Male = 171·1 centims., Female = 160·2 centims.
Sexual ratio ♂/♀ = 1·068.

* 'Journal of the Anthropological Institute,' vol. 17, 1887, p. 209.

These values are immense reductions on Dr. BEDDOE'S 176·2 for males and 166·5 for females. Even with this reduction, the Round Barrow population must still be considered a tall one, as tall as the modern English. It will be remembered that it was also brachycephalic,* a curious and infrequent combination in Europe.

(*b.*) We may next consider the Romano-British, for whom we obtain from Dr. BEDDOE the data:

Male F = 45·42 centims., mean for 10,
Female F = 40·82 ,, ,, 4.

Formula (*a*) gives:

Stature Male = 166·7 centims., Female = 152·2 centims.
Sexual ratio ♂/♀ = 1·090.

Here again we have very sensible reductions on Dr. BEDDOE'S estimates of 169·3 and 154·2.

(*c.*) We may compare these results for the Romano-British with those for the Romano-Gauls, based on data provided by RAHON.† These give:

	F.	H.	T.	R.
Male	(40) 45·52	(18) 32·0	(22) 35·9	(9) 24·1
Female	(5) 40·43	(5) 29·7	(1) 30·7	..

Whence we deduce:

STATURE of Romano-Gauls.

	Male.	Female.
(*a*)	166·88	151·48
(*b*)	163·25	153·27
(*c*)	163·96	146·98
(*d*)	164·76	..
(*e*)	165·64	149·25
(*f*)	165·75	149·26
(*g*)	163·91	..
(*h*)	163·10	..
(*i*)	165·11	152·07
(*k*)	165·84	..
Mean	164·82	150·37 (152·27)

* PEARSON, 'The Chances of Death,' vol. 1, "Variation in Man and Woman," p. 363.

† *Loc. cit.*, p. 441.

The second mean estimate for females is determined by neglecting the *single* tibia measurement, and is probably the best obtainable; it agrees closely with (*i.*):

Sexual Ratio ♂/♀ = 1·082.

A series of 12 femora dug up in Boulogne Harbour* have also been attributed to the Romano-Gauls. They give male F = 45·22 centims., or for the stature 166·32.

My estimate here is about a centimetre larger than RAHON'S. We sensibly agree for the males in the larger series above, while for the females I should take the most probable stature to be a centimetre less than that (153·5) given by RAHON.

We cannot compare the Romano-British with the Romano-Gauls on the basis of all bones, for we have only the results for the femur in the former case. But if we compare the femur estimates for the two cases we see that they are sensibly the same (male 166·9 against 166·7, and female 152·3 against 152·2). It is, therefore, probable that the estimate of the Romano-British male is sensibly too high, and that it would have been nearer 165 centims. had we had other bones than femora to base our estimates upon. The sexual ratio is clearly abnormally high.

(*d.*) *Row-Grave Population of South Germany.*

Dr. R. LEHMANN NITSCHE has published a most interesting series of measurements on the long bones found in the Row-graves of Bavaria.† These interments date from the beginning of the 5th to the end of the 7th century. He divides his material into two groups, "Bajuvars," from the Row-graves of Allach in Upper Bavaria,‡ and Suabians and Alemanns from those of Dillingen, Gundelfingen, Schwetzheim, Memmingen and Fischen.§ The mean lengths of the long bones for these two groups are, however, in such complete accordance, that we are quite justified in following Dr. NITSCHE and combining the two groups.‖ We have then the following results after the proper change in the femur:—

	F.	T.	H.	R.
Male	(41) 46·99	(25) 38·05	(17) 33·71	(11) 25·41
Female	(16) 41·07	(7) 33·71	(9) 30·28	(4) 23·10

The following table gives the reconstructed stature on the basis of the ten formulæ of Tables XIV. and XV.:—

* *Loc cit.*, p. 439.

† "Neue Beiträge zur physischen Anthropologie der Bayern," vol. 11, pp. 205–296, München, 1895.

‡ *Ibid.*, p. 207.

§ *Ibid.*, p. 239.

‖ pp. 260, *et seq.*

Male.	Bones used.	Female.
168·1	H & R	155·2
168·2	H	154·9
169·0	R	158·4
169·1	T	154·1
169·2	H + R	156·8
169·3	F & H	153·5
169·4	F, H, T & R	153·0
169·6	F	152·6
169·8	F + T	153·4
169·9	F & T	153·4
169·2	Mean	154·5

MANOUVRIER's "Tableau-barême" gives us—

Male.	Bone.	Female.
168·1	F	152·6
167·6	T	154·6
167·5	H	155·6
170·1	R	160·5
168·3	Mean	155·6

Clearly MANOUVRIER's method gives results in this case differing almost 1 centim. from mine for both sexes. They have ranges 2·6 centims. and 7·9 centims. for male and female as compared with my 1·8 centim. and 4·2 centims. respectively. Our method of taking the means of the results is not, however, very good. There are very few radii, and the results for that bone have little weight. To properly weight, however, the formulæ involving two or more bones is troublesome, and the increased exactness is so small as to be hardly worth the labour. If we treat F and T, F and H, and F, T, H, and R as likely, *à priori*, to give the best result, we have male stature, 169·5 and female stature 153·3. I doubt whether this is as good as the previous result; it would connote a very high sexual ratio, 1·106, which is contrary to what we generally find with primitive peoples. The sexual ratio of the above results is very high, 1·095, and it seems to me probably that in the difficult matter of sexing rather too large a proportion of large bones have been given to the male and too few to the female group. Further, the smaller radii may probably have disappeared, which accounts for something of the irregularity here—as in other cases—of the estimates from the radius. Allowing, however, for these irregularities we find the Row-grave population by no means so widely differentiated from the French as the Naqada race. They were, however, a tall race, taller than the present French commonalty, almost, but not quite, as tall as the present English commonalty in their men, but sensibly below it as regards their women. The men were *at least* 1 to 3 centims. taller than the present Munich population, which gives 168 centims. as mean of *accepted* recruits, and 166 centims. as a mean based on corpse measurement. (See RANKE, "Zur Statistik der Körpergrösse . . .," in 'Anthropologie der Bayern,' vol. 1, and PEARSON, 'The Chances of Death,' vol. 1, p. 295.)

(18.) *Anglo-Saxons.*

Here my data are extracted from Dr. BEDDOE's paper.*

	Number.	F.	T.
Anglo-Saxons in general, male	65	47·17	(12) 39·05
" " female	26	42·77	..
Wittenham, peasantry, male	23	46·69	..
" " female	17	42·24	..
" with tibia, male	7	48·34	39·43
Ely, bishops, male	5	46·74	38·51

Allowance has been made (see p. 197) for the length of the spine.

STATURE of Anglo-Saxons.

	(a.)	(c.)	(e.)	(f).	Mean.
Anglo-Saxons in general, male	170·0	171·4	171·2	171·2	170·9
" " female	156·0	..	..	..	..
Wittenham, peasantry, male	169·1	..	..	..	..
" " female	155·0	..	..	..	..
" with tibia, male	172·2	172·3	173·0	173·0	172·6
Ely, bishops, male	169·2	170·1	170·1	170·1	169·9

Dr. BEDDOE's results diverge again immensely from mine.† For the Anglo-Saxons in general he finds, for example: male, 174·7 centims., and female, 160·2 centims.; while his estimate, using the tibia for the Wittenham second male group, is 70·86 inches, or 180 centims.!

If his conclusions were correct, the modern English would have degenerated very much from the Anglo-Saxons in stature.

* *Loc. cit.* p. 209.

† I make Earl BRITHNOTH (F = 52·07, T = 41·58) about 180 centims., while Dr. BEDDOE's estimate is 192.

For modern English we have the following results :—

	Galton, Commonalty. I.	Pearson, Middle classes.	
		II.	III.
Male . . .	(811) 172·55	(1000) 172·8	(1077) 175·15
Female . .	(770) 160·85	(1000) 159·9	(135) 162·17

Mr. Galton's results were measured at his South Kensington Laboratory during the Exhibition of 1884. My first group are from my family data cards, and without boots ; my second group are from the measurement cards of the Cambridge Anthropometrical Committee. Subtracting 2·54 centims. for boots from I. and II., we find :—

Male	170·0	172·8	172·6
Female . . .	158·3	159·9	159·6

Thus there is a sensible agreement between the results II. and III., while I. shows just the class distinction we might expect to find. Comparing these results with the Anglo-Saxon statures, we notice an increase of about 2 centims. in the female stature, while the present English commonalty is about 1 centim. less than the mean male stature, and the English male middle classes about 2 centims. more. If the Wittenham skeletons with tibia belong to a class apart, then they were quite equal in stature to the modern English classes, while the Anglo-Saxon bishops were distinctly inferior. Probably the bishops were men unsuited for fighting, and showing a lower degree of physical development. The Anglo-Saxon women are not very many in number, and we have only the femora to base an estimate upon, which in all these cases gives a less stature than the tibia. We may therefore conclude that the average Englishman of to-day is certainly not behind his Anglo-Saxon ancestors ; he may be very slightly taller. The average Englishwoman is probably somewhat taller, but the paucity of data for Anglo-Saxon women hardly allows an estimate of how much. The sexual ratio, 1·096, is so high that I am compelled to consider the Anglo-Saxon women under-estimated, or possibly mixed with a Romano-British element. The modern value is about 1·080.

(19.) *Franks.*

I have put into one group the Frankish remains belonging to both the Merovingian and Carolingian periods, to be found in Rahon's memoir,* the separate smaller groups giving results in close accordance. We have then :—

* *Loc. cit.*, p. 440, *et seq.*

FRANKS 500–800 A.D.

	F.	H.	T.	R.
Male	(47) 45·18	(23) 33·23	(31) 36·81	(7) 25·31
Female	(16) 40·87	(8) 29·39	(7) 32·77	(3) 22·80

This gives us :—

FRANKISH STATURE.

Formula.	Male.	Female.
(*a*)	166·24	152·34
(*b*)	166·81	152·41
(*c*)	166·12	151·85
(*d*)	168·71	157·44
(*e*)	166·30	152·07
(*f*)	166·32	152·08
(*g*)	168·13	154·88
(*h*)	166·74	152·83
(*i*)	166·67	152·34
(*k*)	166·36	151·03
Mean	166·84 (166·42)	152·93 (152·12)

Sexual ratio ♂/♀ = 1·091.

The means in brackets are obtained by omitting the results of formulæ (*d*) and (*g*), which are clearly exaggerated, owing to only the larger radii having survived.

It is clear, accordingly, that the Frankish conquerors of Romano-Gaul were not a tall race—nothing like as tall as the Anglo-Saxons who conquered Romano-Britain.* Further, while the English commonalty have, if anything, slightly progressed on the stature of their Teutonic invaders, the French commonalty have, if anything, regressed.

* Of course, occasionally we find tall Franks, as those buried at Harmignies (Hainaut), RAHON *loc. cit.*, p. 440. These give :—

	F.	H.	T.	R.	(*a.*)	(*b.*)	(*c.*)	(*d.*)	(*e.*)	(*f.*)	(*g.*)	(*h.*)	(*i.*)	(*k.*)	Mean.
Male . .	50·52	34·9	41·0	27·3	176·28	171·64	176·08	175·22	177·34	177·36	174·46	171·75	174·77	170·03	174·5
Female .	45·83	30·5	34·1	2[illegible]·5	161·98	155·47	154·98	159·78	159·16	159·12	157·82	155·90	160·12	157·09	158·1

These are tall as compared with the average French of to-day, but not specially tall from the English standpoint, and certainly not comparable with Earl BRITHNOTH.

(20.) *French of the Middle Ages.*

Two groups are classed under this head by RAHON. The first comes from the cemetery of Saint-Marcel, and is said to belong to the 4th to the 7th century. The second comes from the cemetery of Saint-Germain-des-Prés, and probably belongs to the 10th to the 11th century. If these dates be correct, the former group belongs to the protohistoric rather than the mediæval period, and is directly comparable with the above results for the Merovingian and Carolingian periods. The latter group belongs to the early middle ages. We have:

	F.	H.	T.	R.
Saint-Marcel, male	(71) 45·32	(81) 34·2	(96) 37·8	(21) 24·4
„ female	(19) 41·63	(26) 30·3	(40) 34·0	(9) 22·5
Saint-Germain-des-Prés, male	(44) 45·32	(37) 33·1	(37) 37·3	(6) 23·7
„ female	(10) 41·32	(18) 30·9	(18) 34·0	..

These give us for stature of mediæval French:

Formula.	Saint-Marcel. 4th to 7th century.		Saint-Germain-des-Prés. 10th to 11th century.	
	Male.	Female.	Male.	Female.
(*a*)	166·51	153·81	166·51	153·21
(*b*)	169·62	154·92	166·43	156·57
(*c*)	168·48	154·74	167·29	154·74
(*d*)	165·74	156·44	163·45	..
(*e*)	167·61	154·31	167·03	153·96
(*f*)	167·56	154·31	167·02	153·96
(*g*)	168·23	155·87	165·12	..
(*h*)	169·25	155·10	166·06	..
(*i*)	168·33	154·80	166·61	154·50
(*k*)	168·44	154·12	166·92	..
Mean	167·98	154·79	166·24	154·49

RAHON obtains the values:

165·7 155·5 165·6 155·5.

The first of these differs very considerably from my estimate, but RAHON has made a slip in using MANOUVRIER's table, and thus much underestimated the Saint-Marcel male stature.

I think it impossible to accept RAHON's view that the modern French are sensibly of the same stature as the mediæval French, because the slight apparent difference may be accounted for by a process of selection preserving for us only the larger bones. It is not, as RAHON supposed, a difference of ·7 centim. which has to be accounted for, but one of nearly 3 centims. We have the following series for France, male and female :—

Neolithic man	162·5	151·4
Romano-Gauls	164·8	152·3
Franks	166·4	152·9
French, 4th to 7th century	168·0	154·8
,, 10th to 11th century	166·2	154·5
,, modern	165·0	152·3

These results would seem to indicate that the Gauls were taller than the races they superseded in France, that their Frankish conquerors were taller again than they; but that the stature has been sinking during the last 800 years, and that the French commonalty of to-day is very close in stature to the Romano-Gauls.

This may denote a selection of stature, or it may mean that the Celtic element of the population has superseded the Teutonic element—an explanation in accordance with the recognised greater fertility of the Breton element in France. We should then have an interesting illustration of the manner in which reproductive selection may reverse the results of natural selection. While it might be rash to attribute the decrease in stature which has taken place in France to any one definite cause, it is interesting to note that we do not trace the like decrease in stature in England, yet we should certainly expect to do so, if the result were due simply to a selective process by which the larger bones were preserved. There does appear to be a like decrease in the stature of the Bavarian population, where we have compared (p. 215) the Row-grave population with that of Munich town recruits, which appears to be considerably above the average of recruits from other near districts,* and considerably above the corpse length (166 centims.)—itself greater than the stature of the living—which I have found from BISCHOFF's data.

(21.) *On Giants and Dwarfs.*

If we pass from the consideration of races with mean statures varying from about 157 centims. to 170 centims. to the consideration of individual giants and dwarfs, we very soon discover that our formulæ give statures hopelessly too small in the case of

* The average of the conscripts for the *1st Infanterie Brigade*, which includes Munich, was only 166 centims. The average of the Baden conscripts was 163 centims.

giants, and too large in the case of dwarfs. This defect of the theory is the more serious in that while no prehistoric bones at present discovered give us indications of a race with giant proportions, there are such bones which indicate the existence of dwarf races in neolithic Europe. The reconstruction of individual giants from the skeletons preserved is also of some interest, although, from the standpoint of evolution, it, so far, has nothing like the importance of the reconstruction of the dwarf races.

If our formulæ do not apply to giants and dwarfs, we are forced to one or other of the following conclusions :—

(*a.*) Dwarf and giant races must have been differentiated from normal races by a selection which has partially or totally changed the regression formulæ.

(*b.*) The regression formulæ are not really linear ; they are only apparently linear, because, in dealing with the normal range of stature, we have only to consider a small portion of the regression curve which is sensibly straight.

Both these conclusions may of course be partially true.

In order to consider the validity of one or both of these hypotheses, it might seem that all we have to do is to investigate the relation between the long bones and stature in the case of a sufficient number of giants and dwarfs. But alas! the total material is small, and the quality of it is exceptionally bad. The majority of giants and dwarfs probably prefer a quiet life and a normal burial, so that their bones do not reach the anatomical museum.* Of the dwarfs and giants whose skeletons are to be found in museums, the majority earned their livelihood by exhibition, and accordingly their living stature was a character likely to be under- or over-estimated for the purposes of advertisement. If we put aside all records of the living stature, we are thrown back on the measurement of the length as corpse, or on estimates formed by anatomists of the stature from the articulated skeleton. Unfortunately, authorities differ very widely as to (*a*) the difference between the skeleton (after mounting) and the corpse length—ORFILA makes a difference of 7·5 centims., BRIANT and CHAUDÉ of 8 centims., and TOPINARD of 3·5 centims.—and (*b*) on the difference between the living stature and the corpse length (see p. 191). Even if TOPINARD's estimate, based upon 23 *normal* subjects measured as corpse and skeleton, be correct, it could hardly be safely extended to the cases of giants and dwarfs. Professor CUNNINGHAM, in attempting to reconstruct the stature of the Irish Giant, MAGRATH, goes so far as to discard all records of living stature, and all attempts to reconstruct stature from the articulated skeleton, and would estimate only from the length of the femur.† But this method seems to me fatal, at any rate for our present purpose, the very object of which is to find the relation between stature and femur (or any other long bone) in the case of giants. It cannot too

* In the investigation for conscripts in Bavaria, in 1875, 43 dwarfs were found, and among the 35 measured we have a range of 115 centims. to 139 centims. There were also four giants, or men with statures of 190 centims. and over.

† 'Royal Irish Academy Transactions,' vol. 29, 1891, pp. 553–612.

often be repeated that the idea that there is in any sense a constant proportion between stature and any long bone is misleading. MANOUVRIER makes this ratio decrease from dwarf to giant, and this is correct so long as we suppose the regression formula linear, for example, $S/F = a + b/F$. But this ratio really begins to decrease again as we go from short people to actual dwarfs, and to increase again as we go from tall people to actual giants.

For example, we have the following results for the ratios of long bones and stature :—

Data.	S/F.	S/T.	S/H.	S/R.
50 normal Frenchmen	3·71	4·54	5·06	6·83
MANOUVRIER, { "Coefficients moyens ultimes," stature > 181 }	3·53	4·32	4·93	6·70
TOPINARD, 22 cases, stature > 175	3·61	4·46	5·05	6·94
PEARSON, 12 cases, stature > 200	3·73	4·41	5·01	7·07

It will be at once obvious that MANOUVRIER'S "Coefficients moyens ultimes" are by no means ultimate, but that in the case of giants the coefficients actually tend to return to their values for the mean population. This will be sufficient to show that it is quite impossible to consider any method of determining stature from a presumed constant ratio to femur as satisfactory.

But this table shows an important principle, namely, that as the ratio of stature to long bone first decreases as the bone increases and then begins to increase, it is impossible to consider the regression curve as a straight line when we extend it so far as the region of dwarfs and giants.

Now this is, *à priori*, what might have been expected, for all distributions of zoometric frequency that I have come across seem to possess sensible skewness, and in skew correlation the regression curve is not a straight line. Its actual form is of a somewhat complicated nature,* and it would be purely idle to attempt to determine the constants of it from the data for dwarfs and giants which are at present available. Accordingly it seemed to me desirable to select some empirical curve which would, so far as possible, represent the available material and give results in harmony with certain general principles. The considerations which led me to the choice of this curve were of the following character :—

(*a.*) It must sensibly coincide with the line of regression already found between statures of 155 centims. to 175 centims. It must accordingly have a point of inflexion at the mean stature, at which the tangent should be the already determined line of regression. Referred to this tangent and its perpendicular, the form of the curve in the neighbourhood of the origin must be $y = cx^3$. Away from the origin, c may become a sensible function of x and y, one or both.

* I hope to return to this point in a paper on skew correlation.

(*b.*) So far as the data at my command went, the dwarfs and giants appeared to deviate from the regression line in a remarkably symmetrical manner on opposite sides of it. In other words, the branches of the curve on opposite sides of the axis of y appeared to be centrally symmetrical or congruent. Thus the form of the curve was reduced to $y = x^3\phi(x^2, y^2)$.

(*c.*) It follows from this that the asymptotes of the curve, besides $x = 0$, will be given by $\phi(x^2, y^2) = 0$. The problem then turns on what are the probable asymptotes. Now if we examine the regression formula for an organ A on an organ B, it is of the form:

$$A = \left(A_m - \frac{r_{ab}\sigma_a}{\sigma_b} B_m\right) + \left(\frac{r_{ab}\sigma_a}{\sigma_b}\right) B,$$

where A_m and B_m are the mean organs, σ_a and σ_b the standard deviations, and r_{ab} the coefficient of correlation. Now no amount of selection of either A or B, or any other organs, as to *size* only, would influence in the case of normal correlation $r_{ab}\sigma_a/\sigma_b$, but it would change the constant term $A_m - \frac{r_{ab}\sigma_a}{\sigma_b} B_m$. Hence, if we were to take the line of regression for an extreme population of dwarfs alone, or of giants alone, it would seem quite possible that $r_{ab}\sigma_a/\sigma_b$ might have remained constant, while the term $A_m - \frac{r_{ab}\sigma_a}{\sigma_b} B_m$ changed. But these lines of regression would be the asymptotes of the required curve. It was thus suggested to me that the asymptotes might be parallel to the line of regression of the normal population. On examining the points corresponding to giant and dwarf statures plotted to long bones, this hypothesis seemed to be highly probable. Accordingly the form of the curve finally selected to represent the extended curve of regression was

$$y = cx^3 (b^2 - y^2),$$

where the axis of x is the linear line of regression for normal stature, and the axis of y is the perpendicular to it through the mean normal stature of the French.*

(*d.*) A diagram was now formed by plotting to half life-size ($\frac{1}{2}$ centim. for 1 centim.) the points representing giants and dwarfs, and the lines of regression for the normal population were drawn. The y and x for the point for each giant for each bone were then read off, and these formed the data from which the constants of the four curves of the above type were then determined. For this determination only giants over 200 centims. were selected. The class of what may be termed sub-giants, with statures from 180–200 centims., were put on oneside. Such individuals, termed giants, appear in both the Bonn and Munich anthropological catalogues, but the "Körperlänge" there given can hardly represent the living stature; it is very probably only a skeleton

* Some shifting of the origin would probably have improved my results, but the data were not sufficient to justify such extra labour.

length, and considerably under the real stature. A height, for example, of 185 centims., 6 feet 2 inches, say, would hardly entitle a man, in England at any rate, to rank as a giant.

In the next place, no notice whatever was taken of the dwarfs. I felt that, if the curves were determined from the giant data only, the test that they gave good results for dwarfs would be the most satisfactory one conceivable. As it is, I have been able, on the basis of the long-bone stature relations for giants, to predict the stature of dwarfs to within 2·5 centims. average error. MANOUVRIER's "coefficients moyens ultimes" give a mean error for these dwarfs of 7·25 centims., or 2·9 times as great.

The actual fitting of the curves was conducted in the following manner. Remembering that the curve gives the value of the mean stature for the whole series of long-bones of one size, *i.e.*, the mean of the array of statures for a long bone of given type or size, I recognised that the curve, and accordingly its asymptote, must pass fairly centrally through the group of plotted points. An approximate value of the asymptote constant b was accordingly selected, and the value of c calculated from the mean of the observational values of y and x. If this form of the curve gave, as it generally did, not very satisfactory results, b was modified, and the new c calculated. In this manner, for example, three approximations were made in the case of the radius. The method of least squares was not readily applicable to the data (which were at best not very trustworthy), for it involves the calculation of such expressions as $S(x^6y^2)$ and $S(x^6y^4)$, which, owing to the large values of x involved, give far too great importance to the largest giants.

The curves ultimately determined were the following:—*

For the femur:

$$y = \frac{1}{37136} x^3 (49 - y^2).$$

For the tibia:

$$y = \frac{1}{17170} x^3 (22·5625 - y^2).$$

For the humerus:

$$y = \frac{1}{22196} x^3 (20·25 - y^2).$$

For the radius:

$$y = \frac{1}{16981} x^3 (20·25 - y^2).$$

Here the unit for both y and x is equal to two centims. of stature, or of long-bone. Thus the distances 7, 4·75, 4·5 and 4·5 centims. of the asymptotes from the lines of regression of the normal population are really distances of 14, 9·5, 9 and 9 centims. in actual stature or long-bone length.

* The mathematical reader will bear in mind that it is only the "snake" and not the other two branches of the quintic curve which we require.

I do not suggest for a moment that these curves give a final solution of the problem of determining the stature of any individual in the range of 90 to 250 centims. from the lengths of his long bones, but they seem to me to give the best results obtainable with the data at present available.

Reduced to a formula a curve of this type would be of little service, for both x and y are linear functions of the probable stature and the observed length of the long bone. Hence we should have a quintic equation to find the probable stature from the long bone. But if these curves be plotted once for all, we have a graphical means of at once determining, by simply running the eye along a line, the probable stature corresponding to any given length of long bone. With care we can find the probable stature to ·5 centim., but as a rule to the nearest centimetre is sufficient. As the lines of regression for the normal population are given as part of our curves, it is clear that the diagrams attached to this memoir (Plates 3, 4) will also serve for the determination to a like degree of exactitude of the probable stature of individuals or races falling within the ordinary range of statures. In view of the fact that the diagrams serve all practical purposes, I have not considered it needful to deduce from the above quintics numerical approximations for the value of the stature in terms of the lengths of the various long bones.

(22.) If the reader will examine the diagrams, he will see the twelve giants A, B, C, . . . K, L marked by small dots; from these the curves were determined, and he will notice that they strike fairly well through the groups. The triplet O, M, N contains three pseudo-giants, or sub-giants; these as well as the dwarfs, S, U, V, T, were not used in the determination of the curves. One remarkable feature of the curves must be noted, namely, that in the region of what may be termed sub-giants and super-dwarfs, namely, from about 180 to 200 centims. and 150 to 130 centims., a very small change in the long bone makes a remarkable change in stature. This is specially noteworthy in the case of the radius. Thus between normal individuals on the one hand and giants or dwarfs on the other, there appears to be what may be termed a region of instability, in which an insignificant change in long bone may throw the individual across a considerable range of stature. The points of inflexion of our curves—other than those at the origin—may accordingly have a biological as well as a purely mathematical interest.

The following are all the data which I have been able to collect for giants and dwarfs having any degree of probable truth.

TABLE of Giants.

Letter.	Name.	Locus.	Stature.	F.	T.	H.	R.
A	JOACHIM	Musée Broca	210·0	56·72	47·0	40·4	30·5
B	Berlin Giant I	Berlin Museum	223·0	64·0	53·0	45·5	30·5
C	Berlin Giant II	" "	216·0	55·0	48·0	38·5	29·8
D	O'BYRNE	Royal College of Surgeons	231·0	62·5	54·1	45·0	33·4
E	American Giant	" "	213·0	58·5	47·8	41·3	30·0
F	MAGRATH	R. C. S., Dublin	226·0	62·4	50·6	43·3	33·8
G	"Krainer"	Josephinum Vienna	203·3	53·4	43·5	39·5	27·5
H	"Grenadier"	" "	208·7	55·5	45·6	40·5	29·0
I	Innsbruck Giant	Innsbruck	222·6	61·5	52·0	44·6	34·3
J	St. Petersburg Giant	St. Petersburg	219·5	56·5	50·0	46·0	33·5
K	"Wichsmacher"	Vienna	202·3	52·4	44·9	39·4	27·8
L	Paris Giant	Musée Orfila	236·2	60·96	55·9		
		Sub-Giants.					
M	Bonn Giant	Bonn	188·7	51·0	41·8	35·8	26·0
N	"Gendarme"	Vienna	186·9	51·4	44·0	38·6	26·4
O	Munich Giant	Munich	185·0	50·2	40·8	35·0	25·3
		Dwarfs.					
S	MIKOLAJIK	Anat. Instit., Vienna	112·5	31·0	22·8	20·5	15·1
T	SCHAAFHAUSEN'S Dwarf	Bonn	94·0	22·0	16·0		
U	HIS'S Dwarf		120·0	31·0	25·0	21·5	16·5
V	BÉBÉ	Jardin des Plantes	100·0	24·52	17·61	20·38	12·17

Remarks.—A. The measurements of this giant are given by MANOUVRIER, 'Mémoires de la Sociéte d'Anthropologie de Paris,' vol. 4, p. 387. The femur has been given its maximum instead of oblique length. See also TOPINARD, 'Anthropologie Générale,' p. 1101.

B and C. Details extracted from 'Die Anthropologischen Sammlungen Deutschlands,' V. Berlin.

D and E. Data from the Royal College of Surgeons' Catalogue.

F. I have taken the length of the long bones from Professor CUNNINGHAM'S paper, "Royal Irish Academy Transactions," vol. 29, 1891, pp. 553–612. CUNNINGHAM uses the femur and TOPINARD'S ratio to get the stature. TOPINARD himself gives MAGRATH'S stature as 223 centims. I do not see why Dr. BIANCHI'S measurement of 226 centims. should be rejected. There is no reason to suppose the doctor would have any cause to exaggerate MAGRATH'S stature, and he measured him alive. I have accordingly adopted BIANCHI'S value as the best available. It is in very good accordance with the stature of the Innsbruck giant, and both were probably shorter than O'BYRNE.

G, H, I, J, K and N are all taken from the very valuable memoir by K. LANGER: "Wachstum des menschlichen Skeletes mit Bezug auf den Riesen," 'Denkschriften der k. Akademie der Wissenschaften, Math. Naturwiss. Classe,' vol. 31, Wien, 1872, pp. 1–105. F, H and R are here distinctly stated to be the maximum lengths, and T appears to be measured without spine.* The heights are apparently those of the articulated skeletons.

L. This is the only giant I have ventured to retain out of Sir GEORGE HUMPHRY'S list in "The Human Skeleton," Cambridge, 1858, p. 107, for he indicates that he measured it himself (p. 105). I have not been able to identify his "Russian Giant" at Bonn. His Berlin giants differ considerably from those in the Berlin Catalogue, while his estimates of O'BYRNE and of the Irish giant seem hopelessly too large. As he gives the Musée Orfila giant 17 centims. less stature than TOPINARD (*loc. cit.*, p. 436), I think his estimate on this occasion more probable. M and O are taken from the Anthropological Catalogues of the Museums at Bonn and Munich. I am not clear as to what is meant by *Körperlänge* in these cases. The statures are curiously small as compared with the long bones, if *Körperlänge* is to be thus interpreted. Possibly it is the length of the mounted skeleton without disks.

S, T and U. The details of these dwarfs I have taken from PALTAUF'S work: 'Ueber den Zwergwuchs in anatomischer und gerichtsärztlicher Beziehung,' Wien, 1891.† This book compares unfavourably with the careful memoir of LANGER. The measurements of the long bones of MIKOLAJIK are given several times over, on each occasion with different values; the exact nature of the measurements made is not stated, and results such as those on the author's p. 92, depending on the most elementary arithmetic, are erroneously given. I have taken the values which seem to give the most self-consistent results, but it is impossible to feel sure of their absolute accuracy. SCHAAFFHAUSEN'S account of his dwarf appears in the 'Berichte der Niederrhein. Gesellschaft für Naturkunde in Bonn,' vols. 25 and 39, and HIS'S account of his dwarf in 'Virchow's Archiv,' vol. 22, p. 104.

All the giants and dwarfs in the above list were adults; the ages of the four dwarfs at death were S, 49 years; T, 61 years; U, 58 years; and V, 23 years.

The following table gives the reconstructed statures of these giants and dwarfs as obtained from my diagram and from MANOUVRIER'S "Coefficients moyens ultimes." I have not thought it necessary to publish in the latter case the estimate from each individual bone, but have simply printed the mean of the four results and the differences from the supposed actual stature. It will be noticed that MANOUVRIER'S estimate is in every case too small. Of my differences, 2 are zero, 6 are positive, and 11 negative, but the negative differences are sensibly larger than the positive, so that my curves have rather under than over corrected for giant and dwarf stature.

* "Aus der Mitte der lateralen Condylusfläche in die Incisura fibularis."

† I have verified the dimensions given for HIS'S dwarf from 'Virchow's Archiv für Pathologie u. Anatomie,' vol. 22, 1861, p. 104, *et seq.*

My mean error is only 3·7 centims., however, as against MANOUVRIER's 9·3. Allowing for the doubtful character of some of these measurements, I consider this result fairly satisfactory, and believe my estimate may in several cases be better than the supposed stature.

STATURE of Giants and Dwarfs.

	Estimated stature.						Actual stature.	MANOUVRIER.	
	F.	T.	H.	R.	Mean.	Δ.		Mean.	Δ.
A	213	212	210	213	212	+ 2	210	200	−10
B	229	228	228	215	225	+ 2	223	219	− 4
C	207	215	203	211	209	− 7	216	195	−21
D	226	231	227	224	227	− 4	231	223	− 8
E	218	213	213	211	214	+ 1	213	203	−10
F	226	222	221	226	224	− 2	226	218	− 8
G	200	200	205	182	197	− 6	203	187	−16
H	209	207	210	206	208	− 1	209	195	−14
I	224	225	225	227	225	+ 2	223	221	− 2
J	212	220	229	224	221	+ 2	219	214	− 5
K	193	205	205	197	200	− 2	202	188	−14
L	223	235	..	..	229	− 7	236	226	−10
M	180	184	176	171	178	−11	189	178	−11
N	182	202	200	173	189	+ 2	187	183	− 4
O	178	178	173	170	175	−10	185	174	−11
S	114	111	105	107	109	− 3	112·5	109	− 3
T	95	93	..	..	94	0	94	79	−15
U	114	117	108	112	113	− 7	120	116	− 4
V	100	97	104	97	100	0	100	93	− 7

(23.) *Dwarf Races.*

(*a*) Concerning the curves I have given, much diversity of opinion must naturally exist. For we have made use of giants from a great variety of races in order to produce across a considerable range of stature the regression curves based upon the data for one local race, the French. The justification for this can only be *post-facto*, namely, the capacity of the curves to predict the stature of giants and dwarfs satisfactorily. But it will be seen that in doing this we have proceeded rather on mathematical than anatomical grounds. We have supposed a continuity between the normal population and between giants on the one hand and dwarfs on the other. We have treated these beings as rare variations in a normal population, and not as pathological abnormalities. It is true our curves show a region of marked instability, within which any slight change of long bone is accompanied by a great change in probable stature; but nevertheless we have supposed a mathematical continuity, which in itself is hardly consistent with the theory of "pathological abnormality."

The truth of this theory can only be discussed by anatomists, and many anatomists like Professor CUNNINGHAM and Dr. PALTAUF hold that giants and dwarfs are pathological creations—they are the results of abnormal conditions to which they would give the name of a disease. Such a view would exclude any conception—especially in the case of dwarfs among the normal population—of an atavistic influence. The existence even to-day of dwarf races in both Africa and Asia ought, however, to give ground for pause. When we add to this that Professor SERGI actually considers that he has good evidence of a dwarf racial type still extant in Italy, and that Professor KOLLMANN, after examining SERGI's cranial and other evidence, has been converted from strong disbelief to belief,* when we note the forty-three dwarfs (stature $<$ 140 centims.) actually brought to light by *one* annual conscription in Bavaria alone, and finally when we consider the neolithic dwarf skeletons discovered by NUESCH,† we must undoubtedly hesitate to attribute to pathological causes *all* cases of dwarfs which come under notice. The African, Indian, and Italian dwarfs appear as a distinct racial type as little pathological variations of normal man, as a monkey of the anthropomorphous apes. It is thus possible that the pathological characters found in so many dwarfs may be the result of a conflict between atavistic and normal tendencies, rather than themselves the source of dwarfdom. At any rate, while admitting that our curves are largely based on admittedly pathological instances of both giants and dwarfs, it seems well worth while to consider to what results they lead us when we endeavour to reconstruct the stature of dwarf races.

In making this application we have to bear two points in mind (i.) we must expect a wide range in our prediction of statures lying between 130 and 150 centims., for this is the range for which our curves give very unstable results. We can only hope for a fair degree of approximation in the means. (ii.) Our curves are constructed solely from male data, because female data are practically non-extant. We must accordingly endeavour to find some means of passing from male to female stature. To this we must first devote our attention.

(*b*) I take the following data for sexual ratios for the French and Aino from the material of ROLLET and KOGANEI; for the Naqada race from Dr. WARREN's memoir, and for the Andamanese from Sir W. H. FLOWER's memoir, which is discussed below.

SEXUAL ratio, ♂/♀.

Race.	Stature.	Femur.	Tibia.	Humerus.	Radius.
French	1·083	1·090	1·102	1·110	1·137
Naqada	1·074	1·080	1·088	1·088	1·100
Aino	1·065	1·067	1·064	1·064	1·087
Andamanese . . .	?	1·034	1·034	1·049	1·071

* KOLLMANN in NUESCH, *loc. cit. infra*, p. 238.

† *Ibid.*

Now, there appears from this table to be a very clear rule, namely, that the sexual ratio for stature is certainly not sensibly larger than the least sexual ratio for the long bones. It would seem accordingly improbable that the sexual ratio for the Andamanese can exceed 1·034. If we compare this result with MAN's measurements on 48 male and 41 female Andamanese of which the statures were: male, mean 149·2 centims.; female, mean 140·3 centims., we find ♂/♀ = 1·063, a value much nearer that of the Aino. Sir W. H. FLOWER's own estimated statures* give a sexual ratio of 1·034; the fundamental formulæ for a normal population (p. 196 of this paper) give 1·048; MANOUVRIER's "Coefficients moyens ultimes" give 1·030, and by applying the ratios of stature to long bones as obtained from the average French population we find 1·023. The mean of all these results is 1·038. For the Laps MANTEGAZZA found male = 152·3 and female = 145·0, or the sexual ratio = 1·050. For the Negritos del Monte, or the Aigtas of Luzon in the Philippines, MARCHE and MONTANO give male = 144·1 centims. and female = 138·4 centims., from which we find the sexual ratio of 1·041. TOPINARD gives for races under 150 centims. a mean difference of 4 per cent. between male and female which corresponds to a sexual ratio of 1·042. FRITSCH found a mean difference between male and female Bushmen of 4 centims. which gives (male = 144·4 centims.) a sexual ratio of 1·028; while PARRY's observations on the Esquimaux appear to give a sexual ratio of 1·025, SUTHERLAND's 1·036. From all this it is clear that the dwarfs have a very small sexual ratio for stature as compared with the normal population. At first sight it might seem best to assume this sexual ratio for dwarf races to be TOPINARD's average of 1·042, but as we are going to apply our chart in connection with the sexual ratios found for the long bones of the Andamanese in the table above, I doubt whether it ought to be taken greater than 1·035, say 1·034 in agreement with the value obtained from FLOWER's estimates. Accordingly I formulate the following rule for ascertaining from the chart the probable stature of a female of dwarf race :—

Reduce the female long bones to male long bones by multiplying their lengths by 1·034 in the case of femur and tibia, by 1·049 in the case of the humerus and 1·071 in the case of the radius. Find the corresponding male statures from the chart and multiply it by ·9662 (*i.e.*, the reciprocal of 1·035); these are the probable values of the female stature as estimated from the several long bones, and their mean may be taken as the best result available.

(*c*) It seems very desirable to compare the results thus obtained for male and female of dwarf races with their statures otherwise estimated. If we form a table similar to that on p. 222, but for the case of dwarfs, we have—

* Using the values given, 'Journal of Anthropological Institute,' vol. 14, p. 117.

Data—Male.	S/F.	S/T.	S/H.	S/R.
50 normal French	3·71	4·54	5·06	6·83
MANOUVRIER { "Coefficients moyens ultimes," stature < 153. . . . . }	3·92	4·80	5·25	7·11
Aino stature = 156·7	3·84	4·76	5·31	6·84
TOPINARD, 21 men from 143 to 160	3·68	4·59	5·00	6·70
PEARSON, 4 dwarfs under 120	3·93	5·24	5·33	7·59

Now the tendency here is clearly for the ratios to increase with decrease of stature, if we consider only French, Aino and the group of four dwarfs. TOPINARD's measurements show, however, rather a tendency in the ratios to return to their values for the mean of the normal French population, and as this was closely akin to what we found in the case of giants, we cannot afford to disregard it in the case of dwarfs. Sir W. H. FLOWER has reconstructed the Andamese from their femora on this supposition, and it does not give by any means improbable values of the stature. We have only to look, however, at the line of regression for the normal population to see that for statures between 155 and 175 this hypothesis will give bad results, but it is conceivable that for statures above and below these limits the ratios of stature to the long bones obtained for the means of a normal population give results which are closer to the truth than those found from the lines of regression. Accordingly, on Plates 1, 2, dotted lines give these ratios of stature to long bones, and the statures of giants and dwarfs can be at once read off on this hypothesis. It will be seen that these lines do not give such good results for the four dwarfs under 120 centims. as our curves, but possibly they may give better results for normal dwarf races from 140 to 150 centims. At any rate they do not on the surface exhibit the difficulty as to "instability" to which I have previously referred. Sir W. H. FLOWER writes of the Akka skeletons that:

"They conform in the relative proportions of the head, trunk, and limb, not to dwarfs, but to full-sized people of other races."*

The chief and great difficulty, however, of adopting these lines of normal stature ratios to determine the stature of dwarf races is to fix a limit to their application. At what point are we to fall back on the normal line of regression? There must be such a point, for that line gives excellent results for statures from 155 to 175 centims. Wherever we do fall back upon it there will arise the very sort of instability which we find in our curves, only it will be a far more arbitrary and sudden change. For this reason I cannot consider it satisfactory to obtain the stature of races of less than 155 centims. by a process which is not in any sense continuous with that used

* 'Journal of the Anthropological Institute,' vol. 18, p. 90. By "dwarf" in the sentence cited I think we are to understand "pathological" dwarf.

for races of more than 155 centims. stature. The position and character of the instability is undefined and appears to be quite arbitrary. At the same time, I give the stature of the dwarf races with which I have dealt below on this hypothesis. In order to apply it, I add the additional data for the female stature and long bone ratios required for this and MANOUVRIER's method, putting in the Aino for comparison :—

Data—Female.	S/F.	S/T.	S/H.	S/R.
50 normal French	3·73	4·62	5·19	7·16
MANOUVRIER { "Coefficients moyens ultimes," stature < 140 }	3·87	4·85	5·41	7·44
Aino,* stature = 147·1	3·85	4·75	5·31	6·98

The reader must remember that MANOUVRIER's coefficients are for corpse stature and length of bones when the latter contain animal matter. Hence he first adds 2 millims. to the length of the dead bone to get the bone with animal matter, and then 2 centims. are subtracted by him from the corpse length to get the living stature. In the case of the femur, however, he works with the bone in oblique position, or with a length about 3·2 millims. less in the normal individual than the maximum length. This probably does not amount to more than 2 millims. in the case of dwarf races. Hence, when the femur of the dwarf is given by its maximum length, we need not add or subtract anything before multiplying by the stature-femur coefficient. We have accordingly the following methods of estimating the stature of dwarf races from their long bones :—

(i.) The lines of regression for a normal population, *i.e.*, the formulæ of p. 196 of this paper, or the heavy straight lines of our charts. As we have already seen, this over-estimates the stature of dwarfs as it underestimates that of giants.

(ii.) The curves of regression given by the empirical formulæ of p. 224, or by the heavy curves of our charts. In the case of female dwarfs the lengths of their long bones must first be reduced to male equivalents by the rule on p. 230, and the statures found again reconverted to their female equivalents.

(iii.) The "Coefficients moyens ultimes" of MANOUVRIER may be used. These are given on pp. 231 and 232. Special attention must be paid to the reductions (discussed above) of bones and corpse length.

(iv.) The stature and long bone ratios for the normal population may be used. The values of these ratios are given on pp. 231 and 232, but for most practical purposes it suffices to use the dotted lines of the chart.

I shall refer to these methods as P_{I}, P_{II}, M, and Fl. In the latter case, not

* It will be noticed how close these are to the male coefficients on p. 231, except in the case of the radius, a bone very irregular in primitive and dwarf races.

because Sir W. H. FLOWER was the first* to use a ratio of stature and long bone for the mean population for the reconstruction of stature, but because he has emphasised the fact that, for dwarf races, it does appear to give fairly good results.

(24.) *Bushmen.*

My material is very sparse. Sir GEORGE HUMPHRY, in his work on "The Human Skeleton," gives (p. 106) the mean long-bone lengths for three presumably male Bushmen.

$$F = 38{\cdot}10, \quad H = 27{\cdot}43, \quad T = 32{\cdot}77, \quad R = 21{\cdot}08.$$

I find :—

ESTIMATED Stature of Bushmen.

Bone.	P_I.	P_{II}.	M.	Fl.
F	152·9	150 0	147·4	141·4
H	150·0	141·0	143·1	138·8
T	156·5	156·5	156·2	148·8
R	154·9	153·0	149·3	144·0
F + T	152·4	..	..	..
F & T	152·3	..	..	..
H + R	150·8	..	..	..
H & R	149·8	..	..	..
F & H	150·3	..	..	..
F, T, H & R	150·6	..	..	..
Means . . .	152·05	149·9	149·0	143·25

Now it is clear that neither the chart (P_{II}), nor MANOUVRIER'S "Coefficients moyens ultimes" (M), make in this case much alteration on the estimate given by my normal regression formula (k) for all four long bones. But the value given by Fl is 6 centims. less. Sir GEORGE HUMPHRY gives the average stature of these three Bushmen as 137·1 centims. He does not, however, state where his data are taken from. Curiously enough, his value for stature coincides exactly with the value TOPINARD says BARROW has assigned to the Bushmen. I cannot think that this was the stature in life of the individuals whose bones are averaged by HUMPHRY. FRITSCH gives the average stature of six Bushmen he measured as 144 centims.,† and I should hesitate to place the mean stature of the above three below 145 centims. to 150 centims. At

* It has been used by ORFILA, Sir GEORGE HUMPHRY, and others, and, as we have seen, gives quite incorrect results for races from 155 to 175 centims. in stature.

† See TOPINARD, 'Anthropologie générale,' p. 461.

the same time it must be remembered that the stature falls within the range within which our chart shows that a very slight change in the long bones makes a great difference in stature. In case the reader should be inclined to put too great faith in Fl, I would draw attention to the fact that it underestimates by slightly over 5 centims. the known stature of the fairly short Aino race, while P_I or P_{II} give it almost accurately and M fairly closely.

The only other Bushmen I have been able to find are a male and two females in the Royal College of Surgeon's Catalogue. Selecting the right members as those for which our formulæ and curves are deduced, we have :

Male,	F = 35·6 centims.,	H = 25·5 centims.,	T = 29·9 centims.,	R = 20·8 centims.
Female 1,	F = 38·0 ,,	H = 27·0 ,,	T = 33·2 ,,	R = 21·0 ,,
,, 2,	F = 37·6 ,,	H = 25·7 ,,	T = 28·8 ,,	R = 18·6 ,,

The following table gives the estimated statures :—

Bone.	Male.				Female 1.				Female 2.			
Key letter.	P_I.	P_{II}.	M.	Fl.	P_I.	P_{II}.	M.	Fl.	P_I.	P_{II}.	M.	Fl.
(*a*)	148·2	130	137·6	131·9	146·7	148 8	145·1	141·6	146·0	146·9	143·5	140·1
(*b*)	144·4	124	132·9	129·1	145·8	140·6	145·1	140·1	142·2	131·4	138·1	133·3
(*c*)	149·7	136	142·5	135·7	152·9	154·6	160·0	153·3	142·5	129·4	143·5	133·0
(*d*)	154·0	152	147·3	142 0	151·4	154·3	155·7	150·4	143·4	125·1	137·9	133·2
(*e*)	147·2	..	..	..	149·3	..	..	..	143·9	..	..	..
(*f*)	147·2	..	..	..	149·4	..	..	..	144·0	..	..	..
(*g*)	146·9	..	..	..	148·1	..	..	..	142·0	..	..	..
(*h*)	144·4	..	..	..	146·2	..	..	..	142·1	..	..	..
(*i*)	144·8	..	..	..	146·0	..	..	..	144·2	..	..	..
(*k*)	144·8	..	..	..	148·0	..	..	..	143·1	..	..	..
Mean .	147·2	135·5	140·1	134·9	148·4	149·6	151·5	146·3	143·3	133·2	140·75	134·9

The estimates based on the skeleton height of these three Bushmen are: Male = 133·3, female 1 = 140·0, and female 2 = 139·0 centims. The mean error made by P_{II} is 5·9, by M 6·7, and by Fl 4·1 centims. But it must be noticed that the last gives in one instance *less* than the height estimated from the skeleton—a result which is in itself very improbable. A consideration of the values here given seems to show that with the mean length of bones given by HUMPHRY the mean stature could not possibly have been the 137·1 centims. he states. For whatever estimate we take of the Female 1, she must have been with bones no longer, at least 10 centims. taller than HUMPHRY's mean male. Taking our four males and two females we get from P_{II} estimated statures for male and female Bushmen of about 146 and 142 centims., which I expect are not very far from the truth.

(25.) *Akka Stature.*

In a paper by Sir W. H. FLOWER in the 'Journal of the Anthropological Institute,' vol. 18, 1889, entitled: "Description of two Skeletons of Akkas, a Pygmy Race from Central Africa," the following data are given (p. 14):

	F.	H.	T.	R.
Male . . .	32·6 centims.	23·8 centims.	27·0 centims.	18·2 centims.
Female . .	33·4 ,,	24·4 ,,	27·0 ,,	19·4 ,,

In the following table the reconstructed statures are given on the same four hypotheses as we have considered in the case of Bushmen.

Bone.	Male.				Female.			
Key letter.	P_I.	P_{II}.	M.	Fl.	P_I.	P_{II}.	M.	Fl.
(*a*)	142·6	118·5	125·8	121·0	137·8	120·3	127·3	124·6
(*b*)	139·5	117·5	124·0	120·4	138·7	119·8	131·1	126·6
(*c*)	140·7	122·5	128·6	122·6	136·4	121·2	129·9	124·7
(*d*)	145·5	119·5	128·8	124·3	146·1	135·3	143·8	138·9
(*e*)	139·3	..	..	..	136·3	..	..	..
(*f*)	139·4	..	..	..	136·3	..	..	..
(*g*)	139·5	..	..	..	141·2	..	..	..
(*h*)	139·2	..	..	..	139·0	..	..	..
(*i*)	139·0	..	..	..	137·2	..	..	..
(*k*)	138·2	..	..	..	135·0	..	..	..
Mean .	140·3	119·6	126·8	122·1	138·4	124·1	133·0	128·7

Sir W. H. FLOWER estimates the height of both individuals at about 4 feet, or 122 centims. He gives 121·8 as the estimate of stature from the female skeleton. We could hardly want better results than are given by P_{II}. Fl gives also good results, while M appears to err in excess.*

* EMIN PASHA refers to an Akka woman of 136 centims. stature, who must therefore have been considerably taller than the above woman.

(26.) *Andamanese Stature.*

The stature of the Andamanese is a peculiarly difficult one to estimate. They are taller than Bushmen and Akkas, and fall more markedly into the unstable range of our chart curves. The measurements of a very considerable number of long bones have been given by Sir W. H. FLOWER in two papers in the 'Journal of the Anthropological Institute,' vol. 9, 1879, and vol. 14, 1885. I take the following mean values from the latter paper (p. 116) :—

	No.	F.	H.	T.	R.
Male . . .	25	39·34	27·65	33·21	22·52
Female . .	26	38·04	26·35	32·10	21·01

Constructing as in the previous cases a table of stature as estimated by all four methods we find :—

Bone.	Male.				Female.			
Letter.	P_I.	P_{II}.	M.	Fl.	P_I.	P_{II}.	M.	Fl.
(*a*)	155·3	154	152·2	145·8	146·8	148·8	145·2	141·8
(*b*)	150·7	144	144·2	139·9	144·0	138·1	141·6	136·7
(*c*)	157·6	157	158·4	150·8	150·3	151·7	153·7	148·3
(*d*)	159·6	160	159·5	153·7	151·5	154·3	155·8	150·4
(*e*)	155·4	..	..	..	148·1	..	..	..
(*f*)	155·3	..	..	..	148·2	..	..	..
(*g*)	153·6	..	..	..	147·0	..	..	..
(*h*)	150·7	..	..	..	144·5	..	..	..
(*i*)	152·0	..	..	..	145·4	..	..	..
(*k*)	152·6	..	..	..	146·1	..	..	..
Mean .	154·3	153·7	153·6	147·6	147·2	148·2	149·1	144·3

Now it will be observed that P_I, P_{II}, and M give sensibly the same result: 154 centims. for the male; that for the female, P_{II}, owing to our having first to increase the female bones to reduce them to male lengths, gives a higher result than P_I, for we have got into the unstable range of the curves, and the stature-reducing factor afterwards applied does not undo the excess. There is not much, therefore, to choose between P_I, P_{II}, and M for the Andamanese. They give results 4 centims. greater than Fl in the case of males, and 3 centims. greater in the case of females. From them we should conclude that the stature of Andamanese was given by male = 154 centims., female = 148 centims. MAN,* who measured 48 male and 41 female living Andamanese, gives the stature as, male = 149·2 centims., and female = 140·3 centims.

Sir W. H. FLOWER estimates the stature from his skeletons at male = 143·1 centims., and female = 138·3 centims. This is very much less even than MAN's determination of the living stature. MANTEGAZZA, who possesses a skeleton of an Andamanese, gives its skeleton height at 148·5 centims., and KOLLMANN considers its living stature to have been 150 centims.† The femur in this case is 42·4 centims. long, which would correspond in a normal Frenchman to a stature of 161 centims. I must state that I feel inclined to put entirely on one side estimates of stature based on the height of the articulated or unarticulated skeleton, they appear invariably to underrate the living stature, and often by very large amounts. Even if we suppose the Andamanese to have the relative proportions of full-sized people (*e.g.*, use Fl), we obtain statures considerably above Sir W. H. FLOWER's estimates. On the other hand MAN's measurements, which give results much in excess of the latter, fall considerably short of the results we obtain from P_I, P_{II}, or M. They even fall short of Fl, and in the case of females markedly short of it. If we consider that FLOWER's skeletons and MAN's individuals belong to the same group, then it must be confessed that our estimates are unsatisfactory. The hypothesis Fl gives the least divergent result, but it cannot be considered a particularly good one. It will be seen at once that it is the inferior members in each limb which give the exaggerated stature estimates. If we confined our attention to femur and humerus, then P_{II} (*a*) and (*b*) would give 149·0 for males and 143·4 for females, results better in accordance with MAN's measurements than Fl for all four bones, or than Fl for male femur and humerus only.

When we consider the immense importance of these dwarf races for the problem of evolution, the main result of our investigation is obvious; there ought to be an elaborate investigation—such as KOGANEI has made for the Aino—on the long bones of skeletons and the stature of living individuals, of some extant dwarf race. These races are rapidly becoming extinct, and the possibility of making such an investigation is yearly diminishing. Yet it is only by a careful comparison of the regression formulæ for dwarf and normal races that it seems to me possible that we shall be able quantitatively, and therefore definitively, to fix the relationship of dwarf and normal races in the course of evolution.‡

* See Sir W. H. FLOWER on "Pygmy Races," 'Journ. of Anthropological Institute,' vol. 18, 1889, p. 73.

† NUESCH, *loc. cit.*, *infra*, p. 129.

‡ The reader must bear in mind that nearly all the vagueness involved in our attempts to recon-

(27.) *European Neolithic Dwarfs.*

In the recently published work by NUESCH, 'Die prähistorische Niederlassung beim Schweizersbild,' 1896, is a memoir by KOLLMANN, entitled, "Die menschlichen Skelete, besonders über die fossilen menschlichen Zwerge." This publication for the first time showed us that there existed in neolithic Europe, alongside a normal race, with a stature of about 163 centims., a dwarf race, very similar to the pygmy races, of which we still find traces extant in Africa and Asia. At any rate the discovery in the same group of graves of four skeletons, or rather fragments of skeletons, which must have belonged to individuals who were pygmies, and not "pathological" dwarfs, points very strongly in this direction.

KOLLMANN, who gives a most interesting discussion of these neolithic pygmies, provides the following measurements:—

	F.	H.	T.	R.
1. Female . . .	36·9 centims.	..	..	
2. „ or male.	31·3 „	..	..	
3. „ . . .	35·52 „	25·15 centims.	29·90 centims.	..
4. „ or male.	39·40 „	28·20 „	32·70 „	22·60 centims.

Of these: 1, female, is an adult; 2, female or male, is that of a young person 16 to 18 years old, and, according to KOLLMANN, probably, but not certainly, female; 3, female, and 4, female or male, are adults, but as we see the sex of the latter appears doubtful. Proceeding, as in the earlier cases, we find:—

struct stature, arises from the fact that the regression coefficients for long bones and stature are known for *one* local race only, and that we have nothing else to go upon. Had we endeavoured to reconstruct one long bone from a second, we should have had far more exact material to determine the differential evolution of local races.

Bone.	1, female.				2, as female.				2, as male.				3, female.				4, as female.				4, as male.			
Key letter.	P_I.	P_{II}.	M.	Fl.	P_I.	P_{II}.	M.	Fl.	P_I.	P_{II}.	M.	Fl.	P_I.	P_{II}.	M.	Fl.	P_I.	P_{II}.	M.	Fl.	P_I.	P_{II}.	M.	Fl.
(*a*)	144·6	145·4	140·8	137·5	133·7	114·0	119·1	116·7	140·1	115·0	120·7	116·0	141·9	135·7	135·5	132·4	149·5	152·1	150·5	146·8	155·4	154·5	152·5	146·0
(*b*)	..	..	..	..	..	..	..	..	..	..	..	..	140·7	124·1	135·1	130·4	149·1	150·2	151·6	146·3	152·2	149·0	147·1	142·7
(*c*)	..	..	..	..	..	..	..	..	..	..	..	..	145·1	141·5	144·0	138·1	151·7	153·6	157·6	151·0	156·4	155·0	155·9	148·5
(*d*)	..	..	..	..	..	..	..	..	..	..	..	..	..	..	..	..	156·8	159·4	167·8	161·8	159·9	159·5	160·1	154·3
(*e*)	..	..	..	..	..	..	..	..	..	..	..	..	142·8	..	..	..	150·3	..	..	..	154·8	..	..	..
(*f*)	..	..	..	..	..	..	..	..	..	..	..	..	142·9	..	..	..	150·4	..	..	..	154·8	..	..	..
(*g*)	..	..	..	..	..	..	..	..	..	..	..	..	..	..	..	..	152·6	..	..	..	154·7	..	..	..
(*h*)	..	..	..	..	..	..	..	..	..	..	..	..	..	..	..	..	149·7	..	..	..	152·3	..	..	..
(*i*)	..	..	..	..	..	..	..	..	..	..	..	..	140·8	..	..	..	149·1	..	..	..	152·9	..	..	..
(*k*)	..	..	..	..	..	..	..	..	..	..	..	..	..	..	..	..	148·7	..	..	..	153·0	..	..	..
Mean	144·6	145·4	140·8	137·5	133·7	114·0	119·1	116·7	140·1	115·0	120·7	116·0	142·4	133·8	138·2	133·6	150·8	153·8	156·8	151·5	154·6	154·5	153·9	147·9

If we include the non-adult and suppose the whole series female, we have:

P_I.	P_{II}.	M.	Fl.
142·9 centims.	136·7 centims.	138·7 centims.	134·8 centims.

Without the non-adult, we have:

P_I.	P_{II}.	M.	Fl.
145·9 centims.	144·3 centims.	145·3 centims.	140·9 centims.

The two possible males give:

P_I.	P_{II}.	M.	Fl.
144·2 centims.	134·2 centims.	136·5 centims.	132·3 centims.

The adult male gives:

P_I.	P_{II}.	M.	Fl.
154·6 centims.	154·5 centims.	153·9 centims.	147·9 centims.

The single male here is about identical with the means obtained by the different methods on p. 236 for the male Andamanese, and the adult females give a result somewhat less than that of the female Andamanese as reconstructed from their long bones, but in close accordance with MAN's measurements of living Andamanese stature. The dimensions are somewhat larger than those of Bushmen, or Akkas, or Negritos. We seem, therefore, justified in assuming a neolithic pygmy race in Europe having a stature about the same as that of the Andamanese. Whether the actual stature of this race was for the female nearer to 144 centims. (P_{II}) or 141 centims. (Fl) it seems to me impossible to ascertain definitely until we have more trustworthy and extensive measurements than yet exist of the living stature of extant pygmy races.

(28.) *Conclusion.*

The formulæ and curves for the reconstruction of stature which are given in this memoir, must by no means be taken as final. No scientific investigation can be final; it merely represents the most probable conclusions which can be drawn from the data at the disposal of the writer. A wider range of facts, or more refined analysis, experiment, and observation will always lead to new formulæ and new theories. This is the essence of scientific progress. All, therefore, which is claimed for this paper is (i.) that it exhibits a better theory of the reconstruction of stature than any which has so far existed—it might not be too much to say that nothing which can be called a theory has hitherto existed; (ii.) that it determines the constants of the formulæ given by that theory as well as the existing data allow of; (iii.) that it gives values for the probable statures of prehistoric races, which have far less divergence among themselves, whatever be the bone or combination of bones used, than those suggested by previous investigators; and lastly (iv.) that it indicates what additional data ought to be sought for, and to some extent what is the inner meaning of divergent results, for the great problem of racial differentiation by natural selection.*

Of the general conclusions reached by the author, perhaps two deserve restating and emphasising here. In the first place, although there were individual tall men among the neolithic populations, whose bones have so far been unearthed, yet neolithic man as a whole was short. Of course, it is possible that a tall neolithic type, *i.e.*, one with a stature greater than 168 centims. say, may yet be discovered—witness the discovery within the last two years of a neolithic dwarf. But failing its appearance, the question arises, where and how did the tall Anglo-Saxon and Scandinavian develop? To what extent is this tallness racial, to what extent due to environment? The apparently greater stature of British over Continental neolithic man deserves special consideration from anthropologists.

Secondly, granting that the modern populations in the same district are taller than the neolithic populations, there still appears in both France and Southern Germany some regression of the modern stature on that of the ancient Franks, Bajuvars, and Allemans. I differ from both RAHON and LEHMANN-NITSCHE in considering that the difference is too great to be accounted for as a process of natural selection applied to the long bones. RAHON has made a slip in his arithmetic, and LEHMANN-NITSCHE compares the Row Grave population with the *most* favourable element of Munich town recruits. If the divergence could be accounted for by selection applied to the bones, why is not a similar divergence to be found in the case of Anglo-Saxons and modern English? I think an explanation must be sought elsewhere. One suggestion is, that as the physical struggle for existence has been lessened, reproductive selection has had more play, and the greater fertility of an older pre-Germanic element in the populations of both Southern Germany and France has led to a return of stature to its more ancient value. In the case of Anglo-Saxons and Scandinavians in England there was very probably a more complete destruction of the earlier populations. Whatever may be the real reason for this apparent degeneration, it seems most desirable that there should be a systematic measurement of all long bones dug up anywhere in our own country, and this whether they belong to prehistoric or historic times. Stature is quite as marked a racial character as cephalic index, or any other skull measurement, and its high correlation with the long bones admits even in the present state of our data of its reconstruction with very considerable accuracy, if only a sufficient representation, say twenty to forty long bones, of an ancient population has been measured. It is only by the gradual accumulation of such data that we can

* The influence of *directed* as distinguished from random selection on size, variation, correlation, and regression has been theoretically developed in a memoir not yet published. Having been fully discussed in my college lectures of this Session, much of the recent work of my department, like the present memoir, touches on it.

hope for light on the manner in which our own population has developed and is developing.*

(29.) The following table restates some of the numerical results reached, and further includes, for the purposes of comparison, the stature of certain modern races as given by various authorities. No stress whatever is laid on the latter values, which have often been determined by doubtful observers from very small series.† They are merely given here in order to show the *general* position of the reconstructed races in the order of racial statures.

TABLE of Stature and Sexual Ratio for Divers Races.

Race.	Authority.	Male.	Female.	Ratio ♂/♀.
12 giants > 200	Memoir, p. 226	217·6	..	..
4 sub-giants, Bavarian recruits	RANKE	190·5	..	..
3 sub-giants in Museums	Memoir, p. 226	186·9	..	..
Samoans	TOPINARD	188·3	..	..
Patagonians	MOVENO and LISTER	185·0	..	..
Caribeans	HUMBOLDT	184·0	..	..
Red Indians	TOPINARD	175–180	..	..
Polynesians	,,	170–180	..	..
Flamboro' Head English	PITT RIVERS	175·2	162·5	1·078
Livonians	TOPINARD	173·6	..	..
Americans (born)	GOULD	173·5	..	..
Fellahs (Egypt)	WOLNEY	173·0	..	..
English (Middle classes)	PEARSON	172·8	159·9	1·080
Todas of Nilgherry	MARSHALL	172·7	..	..
Norwegians	HUNT	172·0	..	..
American Scottish	GOULD	171·6	..	..
Bantu	FRITSCH	171·8	..	..
Finns	BONSDORFF	171·4	..	..
American Norse	BAXTER	171·3	..	..
Round Barrow British	Memoir, p. 213	171·1	160·2	1·090
Anglo-Saxons	,, p. 216	170·9	156·0	1·096
American Irish	GOULD	170·5	..	..
Lithuanians	TOPINARD	170·4	..	..
American English	GOULD	170·1	..	..
English Commonalty	GALTON	170·0	158·3	1·074
Sikhs	TOPINARD	170·0	..	..
Bajuvars from Row Graves	Memoir, p. 214	169·2	154·5	1·095
American Germans	BAXTER	169·5	..	..
American Danes	,,	169·2	..	..
American Swedes	GOULD	169·2	..	..
Nubians	TOPINARD	169·0	..	..
Bechuanas	FRITSCH	168·4	..	..
American negroes (pure)	GOULD	168·0	..	..

* For example, no one can say at present what was the stature of Englishmen from A.D. 1000 to 1700, and yet large collections of bones exist, which would suffice to answer this problem.

† TOPINARD, for example, considers the sex ratio for 73 series in "Étude sur la taille considérée suivant . . le sexe . . . et les races," 'Revue d'Anthropolgie,' 1876, p. 34, but he merely gives means for grouped results and does not tell us the details for the individual series.

TABLE of Stature and Sexual Ratio for Divers Races—(continued).

Race.	Authority.	Male.	Female.	Ratio ♂/♀.
4th to 7th cent. mediæval French	Memoir, p. 219	167·98	154·79	1·085
Naqada Race	„ p. 211	167·5	156·0	1·074
Neolithic man in Britain	„ p. 206	167·3	153·6	1·089
Kabyles	PRENGRÜBER	167·3	..	..
Guanches I.	Memoir, p. 210	166·77	154·65	1·078
Romano-British	„ p. 213	166·7	152·2	1·090
Franks	„ p. 218	166·42	152·12	1·091
French (as corpse)	„ p. 180	166·26	154·02	1·079
10th to 11th cent. mediæval French	„ p. 219	166·24	154·49	1·077
Guanches II.	„ p. 210	166·18	153·83	1·081
Mordevins	TOPINARD	166·0	..	..
Munich District conscripts	RANKE	166·0	..	..
Bavarians (as corpse)	BISCHOFF	165·93	153·85	1·078
Russian soldiers (Great Russia)	TOPINARD	165·5	..	..
Dolmens (Algeria)	Memoir, p. 210	165·5	152·56	1·085
French conscripts	MANOUVRIER	165·0	..	..
Italians (Tuscany)	TOPINARD	165·0	..	..
Dolmens (India)	Memoir, p. 210	165·24	154·86	1·067
Romano-Gauls	„ p. 213	164·82	152·27	1·082
Chinese	BRIGHAM	164·5	..	..
Esthonians	TOPINARD	164·2	..	..
Ruthenians	„	164·0	..	..
Dolmens (Caucasus)	Memoir, p. 210	164·33	152·47	1·078
Neolithic man (Mureaux)	„ p. 208	163·39	152·56	1·071
Baden conscripts	ECKER	163·0	..	..
Palæolithic man	Memoir, p. 205	162·7	..	..
Neolithic man, France and Belgium	„ p. 207	162·54	151·44	1·073
Poles	TOPINARD	162·0	..	..
Italians (Piedmont)	„	162·0	..	..
Sicilians	„	161·0	..	..
Neolithic man (Brueil)	Memoir, p. 208	160·91	150·68	1·068
Hottentots	FRITSCH	160·4	..	..
Samoyedes	TOPINARD	159·0	..	..
Annamites	„	158·9	..	..
Esquimaux	SUTHERLAND	158·5	152·8	1·036
Sardinians	TOPINARD	158·0	..	..
Aino	Memoir, p. 199	156·7	147·1	1·065
Juags of Oriva	SHORT	156·0	..	..
Veddahs	BAILEY	153·0	143·3 (?)	1·068
Ostiaks	TOPINARD	153·0	..	..
Siamese	„	152·5	..	..
Laps	MANTEGAZZA	152·3	145·0	1·050
Andamanese I.	MAN	149·2	140·3	1·063
Andamanese II.	Memoir, p. 236	147·6	144·3	1·023
Bushmen I.	FRITSCH	144·4	140·4	1·028
Bushmen II.	Memoir, p. 233	146·0	142·0	..
Aigtas of Luzon	MARCHE and MONTANO	144·1	138·4	1·041
Neolithic dwarfs	Memoir, p. 240	148·0 (?)	141·0 (?)	..
35 Bavarian super-dwarfs	RANKE	133·9	..	..
Akkas	Memoir, p. 235	120·0	124·0	..
4 dwarfs < 125 centims.	„ p. 226	106·6	..	..
Gorilla	Memoir, p. 202	147·0	..	..
Chimpanze	„ „	127·0	..	..
Orang	„ „	112·0	..	..

[Note added November 29, 1898.—Dr. WARREN has made an experiment on two Naqada femora and kindly sent me the following results :—

Femur I. *Oblique Length.*		
Wednesday,	1 P.M. . . .	40·82
Put into water at 1 P.M.		
Wednesday,	7 P.M. . . .	40·97
Thursday,	10 A.M. . .	41·00
„	7 P.M. . . .	41·00
Friday,	10 A.M. . . .	41·01
„	6 P.M. . . .	41·02
Saturday,	10 A.M. . . .	41·03
Monday,	10 A.M. . . .	41·04
Removed from water at 10 A.M.		
Monday,	7 P.M. . . .	41·04
Tuesday,	10 A.M. . . .	41·02
„	7.30 P.M. . . .	41·02
Wednesday,	10 A.M. . . .	40·96
Thursday,	10 A.M. . . .	40·89
Friday,	10 A.M. . . .	40·87
Saturday,	10 A.M. . . .	40·82
Monday,	10 A.M. . . .	40·81
Tuesday,	10 A.M. . . .	40·80
Wednesday,	10 A.M. . . .	40·80
Friday,	10 A.M. . . .	40·80

Femur II. *Oblique Length.*		
Tuesday,	10 A.M. . . .	44·31
Put into water at 10 A.M.		
Tuesday,	12 A.M. . . .	44·38
„	7.30 P.M. . . .	44·42
Wednesday,	1 P.M. . . .	44·47
„	7 P.M. . . .	44·48
Thursday,	10 A.M. . . .	44·50
Saturday,	10 A.M. . . .	44·53
Monday,	10 A.M. . . .	44·53
Removed from water at 11 A.M.		
Monday,	7 P.M. . . .	44·53
Tuesday,	10 A.M. . . .	44·43
Wednesday,	10 A.M. . . .	44·34
Thursday,	10 A.M. . . .	44·32
Friday,	7 P.M. . . .	44·32

Thus there was a difference in the dry and wet states of 2·4 and 2·2 millims. respectively. Considering that the bones were some 3500 years older than those I experimented on, the agreement in result must be considered good. The maximum rate of expansion is reached in the first hour or two, and then gradually diminishes; the maximum rate of contraction is not reached before about the second or third day, without artificial drying as in my case.]

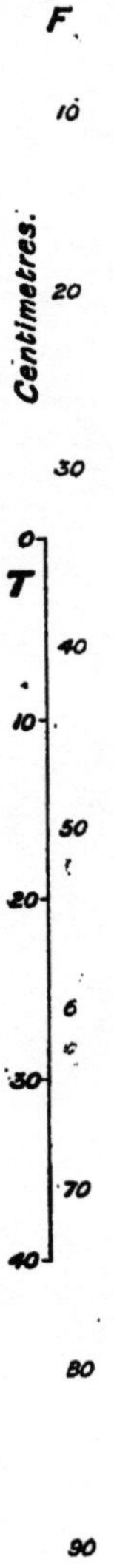

F.

10

Centimetres.

20

30

0

T

40

10

50

20

6

30

70

40

80

90

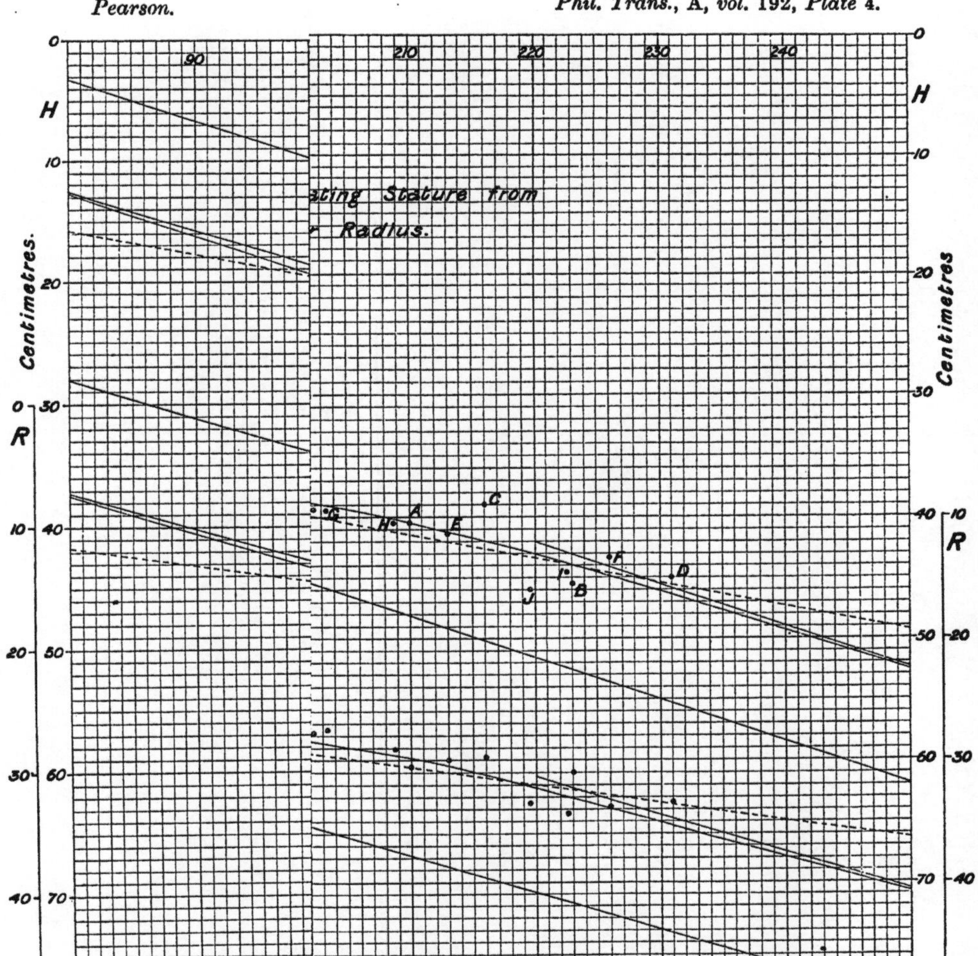

...ating Stature from
...r Radius.
H
R
Centimetres.
Centimetres
90
210
220
230
240
A
B
C
D
E
F
G
H
I
J

INDEX SLIP.

ne Quartz Crystal on another.

'. L. GRAY, *B.Sc.*

r 17, 1898.

differ along the different axes,
llows us to imagine that the
axes. Dr. A. S. MACKENZIE
experiment in which he failed
the Cavendish apparatus, he
lead and on other calc-spar
stalline axes within the limits
total attraction. He further
ghbourhood of a crystal, the
., and 7·421 centims. agreeing

ed out (' The Mean Density of
een two crystal spheres were
axes were parallel or crossed,
n on one sphere in the field of
rowth of a crystal from solu-
lel arrangement—a fact which
hypothesis as showing that,
n.

n of one indicated in the work
and we may say at once that
ce of a directive action of the

pose that the law of the attrac-
l, as in fig. 1 (a), is GMM'/r^2,
the centres, and G a constant
the law of attraction when the
s a constant for this arrange-

). The work done in removing

V. *An Experiment in Search of a Directive Action of one Quartz Crystal on another.*

By J. H. POYNTING, *Sc.D., F.R.S., and* P. L. GRAY, *B.Sc.*

Received September 27,—Read November 17, 1898.

SINCE so many of the physical properties of crystals differ along the different axes, our ignorance of the nature and origin of gravitation allows us to imagine that the gravitative field of crystals may also differ along those axes. Dr. A. S. MACKENZIE ('Phys. Rev.,' vol. 2, 1895, p. 321) has described an experiment in which he failed to find any such difference. Using BOYS'S form of the Cavendish apparatus, he showed that the attraction of calc-spar crystals on lead and on other calc-spar crystals was independent of the orientation of the crystalline axes within the limits of experimental error—about one-half per cent. of the total attraction. He further showed that the inverse-square law holds in the neighbourhood of a crystal, the attractions at distances 3·714 centims., 5·565 centims., and 7·421 centims. agreeing with law to one-fifth per cent.

One of the authors of this paper had already pointed out ('The Mean Density of the Earth,' 1894, p. 7) that if the attraction between two crystal spheres were different for a given distance, according as their like axes were parallel or crossed, such difference should show itself by a directive action on one sphere in the field of the other. This directive action is suggested by the growth of a crystal from solution, where the successive parts are laid down in parallel arrangement—a fact which which we might perhaps interpret on the molecular hypothesis as showing that, within molecular range at least, there is directive action.

The experiment now to be described is a modification of one indicated in the work above referred to, carried out for two quartz spheres, and we may say at once that we have certainly not succeeded in proving the existence of a directive action of the kind sought for.

To bring out the principle of the method, let us suppose that the law of the attraction between two spheres with their like axes parallel, as in fig. 1 (*a*), is GMM'/r^2, where M, M' are the masses, r the distance between the centres, and G a constant for this arrangement. Let us further suppose that the law of attraction when the axes are crossed, as in fig. 1*b*, is $G'MM'/r^2$, where G' is a constant for this arrangement, and different from G.

Let us start with the spheres r apart, as in fig. 1 (*a*). The work done in removing

M′ to an infinite distance, in a line perpendicular to the parallel axes, is GMM′/r. Now turn M′ through 90° to cross the axes, and bring it back to the original position, but with the axes crossed.

Fig. 1.

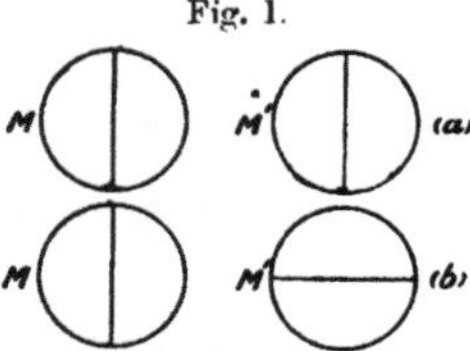

The force will do work G′MM′/r Then turn M′ through 90° into its original orientation. Assuming that the forces are conservative, the total work vanishes, so that there must be a couple acting during the last rotation, which does work equal to the difference between the works done on withdrawal and approach.

If we take the average value of the couple as L, then

$$\frac{\pi}{2}\,L = (G - G')\,\frac{MM'}{r}.$$

Our suppositions as to the law of force are doubtless arbitrary, but they serve to show the probability of the existence of a directive couple accompanying any axial difference in the gravitative field.

In the absence of any distinction between the ends of an axis we may assume that the couple is "quadrantal," that is, that it goes through its range of values with the rotation of the sphere through 180° and vanishing in every quadrant, and we shall suppose that it is zero when the crystals are in the positions shown in fig. 1 (a), and fig. 1 (b).

Taking the couple as a sine function of amplitude F, we have

$$\frac{\pi}{2}\,L = \int_0^{\frac{\pi}{2}} F \sin 2\theta \, d\theta = F,$$

whence

$$F = (G - G')\,\frac{MM'}{r}.$$

But it is conceivable that the two ends of an axis are different, having polarity of the magnetic type. The couple would then be "semicircular," going through its range of values once and vanishing twice in the revolution. We shall suppose that the couple is zero when the axes are parallel. We should now have G and G′ constants for the axes parallel, the one when like ends are in the same direction, the other when they are in opposite directions, and we have

$$\pi L = (G - G')\,\frac{MM'}{r}.$$

But if F is the amplitude of the couple

$$\pi L = \int_0^\pi F \sin\theta \, d\theta = 2F,$$

and

$$2F = (G - G') \frac{MM'}{r}$$

To seek for the directive action we have made use of the principle of forced oscillations, thereby obtaining to some extent a cumulative effect, and at the same time largely eliminating the errors due to accidental disturbances.

Briefly the method was as follows :—A small quartz sphere, about 0·9 centim. in diameter, was carried in a frame to which a light mirror was attached, and suspended by a quartz fibre inside a brass case, the position being determined by the reflection of a scale in the usual way. The complete time of torsional vibration was about 120 seconds.

Outside the case was a larger quartz sphere, about 6·6 centims. in diameter, its centre being level with that of the suspended sphere, and 5·9 centims. from it. The larger sphere could be rotated about a vertical axis through its centre at any desired rate. The crystalline axes of both were horizontal, that of the smaller sphere being perpendicular to the line joining the centres.

To test for the quadrantal couple, the larger sphere was rotated once in 230 seconds—a period nearly double that of the smaller sphere. To test for the semicircular couple, the larger sphere was rotated once in 115 seconds, or nearly the period of the smaller sphere.

Assuming that a couple exists, a continuous rotation of the larger sphere would set up a forced oscillation in the smaller sphere of the same period as the couple, and since the damping was very considerable, this forced oscillation would soon rise to approximately its full value. Meanwhile, any natural vibrations of the suspended system would be rapidly damped out. Though continually renewed by disturbances due to convection-currents and tremors, they would be irregularly distributed, and there was no reason to suspect that their maximum amplitude would recur at any particular phase of the period of the applied couple. To secure the distribution of successive maxima of natural vibrations of the smaller sphere over all phases of the forced period, the latter was made sensibly different from the natural period in the ratio 23 : 24 ; and though the cumulative effect of the forced oscillations was reduced by the largeness of this difference, we did not think it advisable to make the periods more nearly coincident, lest the distribution of the disturbances, which were sometimes large, should not be sufficient. This conclusion was arrived at from the results of preliminary experiments with more nearly equal periods.

During each complete period of the supposed applied couple, the position of the smaller sphere was read ten times at equi-distant intervals of time, and the scale-readings were entered in ten parallel columns, one horizontal line for each period. The observations were continued usually for 70 or 80 periods. Adding up the columns and dividing by the number of periods, any forced oscillation would be indicated by a periodicity in the quotients. The periodicities found were too irregular to be taken as evidence of the existence of a couple.

Description of the Apparatus.

The quartz spheres were placed in a cellar at Mason College, Birmingham, below the room in which the observing telescope and rotating apparatus were fixed.

The smaller sphere, 0·9 centim. diameter and weighing 1·004 grams, was held in an aluminium wire cage, and was suspended by a long, fine quartz fibre in a brass case from a torsion-head at the top of the case.

A light plane mirror was fixed to the cage, and opposite this mirror was a glass window in the case; in front of the window was a plane mirror at 45°, by means of which the light from the scale was reflected into the case and back again to the telescope, as shown in fig. 2.

The case was surrounded by a double-sided wooden box, lined within and without with tin-foil, and with cotton-wool between its inner and outer walls. The box was supported on indiarubber blocks to lessen tremors.

The larger sphere, 6·6 centims. diameter and weighing 399·9 grams, was held at the lower end of a vertical brass tube which terminated in a very carefully turned shallow brass bell, in which the sphere was held by tapes. The tube passed upwards through the top of the wooden casing without contact, a kind of air stuffing-box indicated in the figure serving to prevent currents through the hole. The tube came into the room above, and was there connected with a train of wheels, driven by an electromotor, the rotation of the motor being geared down from 1000 to 1. The observing telescope was fixed to a heavy stone slab resting on indiarubber blocks, standing on a brick-pillar, which was built on the brick arches forming the cellar-roof. A diagonal scale (of half-millimetre graduations, divided into tenths by the diagonal ruling) was clamped to the telescope-tube and illuminated by an incandescent lamp, aided by a concave mirror. A tenth of a division could be read with certainty, and as the distance from scale to mirror was 358 centims., the position of the suspended sphere could be determined within a little more than one second of arc.

The steady rotation of the larger sphere was maintained by a regulator, for which we are indebted to Mr. R. H. HOUSMAN. It consisted of two parts:—(1) the governor proper, which automatically maintained approximate steadiness, and (2) a fine hand-adjustment, by which the motion could be accelerated or retarded when it got "out of time."

One lead to the motor went through two mercury-cups, and the circuit was completed by a fork of platinum-wire dipping into the cups. This wire was fastened to one end of a wooden lever, the other end of which was attached to a sliding collar on the axle of the motor. To this collar were fastened the upper ends of the loaded springs of the governor, as shown in the figure. If the speed increased, the loads

Fig. 2

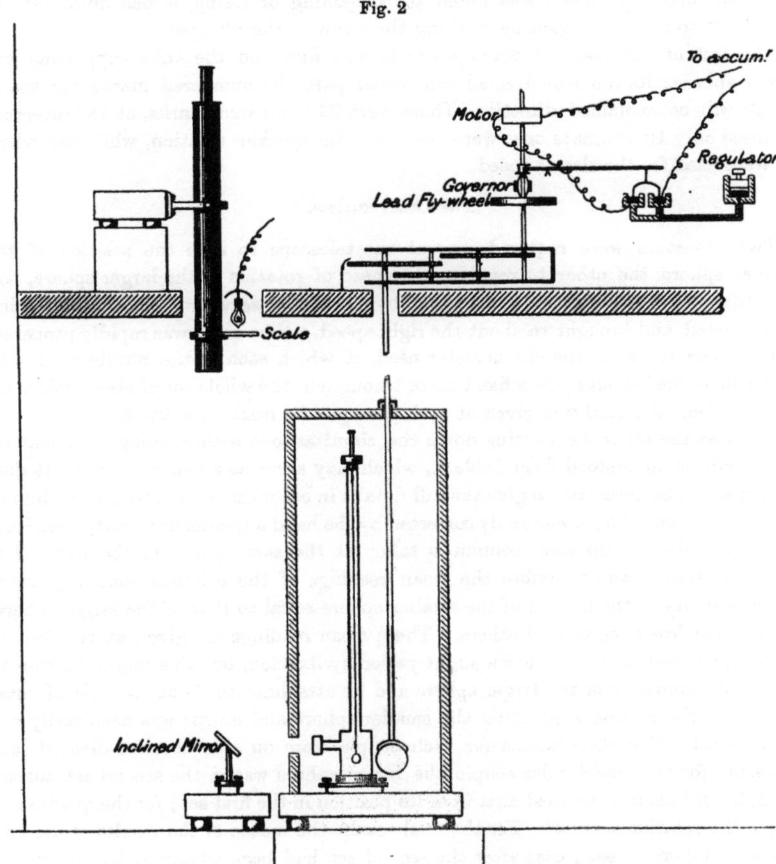

Diagrammatic sketch of the apparatus.

flying out pulled the collar down and so raised the wire out of the mercury-cups, and broke the circuit. As the speed diminished, the wire again dipped into the mercury and re-established the current. To diminish sparking the mercury was covered with alcohol, and the two cups were permanently connected by a high resistance shunt.

The fine hand-adjustment consisted of a small wooden plunger working in a tube connected with one of the mercury-cups; by means of a screw the plunger could be raised or lowered, and the level of the mercury in the cup varied accordingly.

If the revolving sphere was found to be gaining or losing, it was quite easy to bring it "up to time" again by working the screw of the plunger.

The last of the train of driving-wheels was fixed on the tube supporting the larger sphere; its rim was divided into equal parts by numbered marks, the use of which will be explained directly. There were 20 numbered marks, at 18° interval; of these only 10 alternate ones were used for the quicker rotation, while the whole 20 were used for the slower speed.

The Observations.

Two observers were required, one at the telescope to note the position of the smaller sphere, the other to regulate the speed of rotation of the larger sphere, and to notify when readings were to be taken by the first observer. The motion having been started, and brought to about the right speed, a time-table was rapidly prepared, showing the times, on the chronometer used, at which each of the numbered marks above mentioned should pass a fixed mark throughout the whole set of observations for one occasion. A signal was given at each passage of a mark past the fixed point, the observer at the telescope putting down the simultaneous scale-reading in a manner which will be understood from Table I., which may serve as a typical record. It does not appear to be necessary to give the full details in other cases. If the motion did not keep to the time-table, it was easily corrected by the hand adjustment already described.

Every reading in the same column is taken at the same phase in the rotation of the larger sphere, and therefore the mean readings of the columns should preserve any periodicity in the motion of the smaller sphere equal to that of the larger sphere, and more or less eliminate all others. These mean readings are given at the foot of Table I., and appear to indicate a slight periodic vibration, but this might be due to a want of symmetry in the larger sphere and its attachments about its axis of rotation, since the system supporting the smaller sphere and mirror was necessarily not symmetrical. The observations for each couple were on this account divided into two sets: for the semicircular couple the larger sphere was in the second set turned through 180° about a vertical axis from its position in the first set; for the quadrantal couple the rotation was 90°. For the final results the means of the results of the two sets were taken, in each case after the second set had been advanced by an amount corresponding to the change of position of the sphere.

Table II. contains all the mean results obtained in the same way as the figures at the foot of Table I., the greatest range being given in the last column as an indication of the magnitude of the disturbances.

In Table III. are given the means for each azimuth of the larger sphere in its support, the B and D series being advanced as mentioned above.

In combining the results it appeared useless to attempt to weight them according to the number of periods taken, since no accurate conclusion could be expected. It will be seen that in each case there is an outstanding periodicity, but the amplitude is less when the disturbances (as indicated by the greatest range during a period) are less, and it diminishes when the results are combined so as to lessen the effect of want of symmetry.

In the "quadrantal" observations (Series C, D), where the effect of want of symmetry of the apparatus should almost be eliminated, since it is approximately semicircular, the mean range is much smaller than in Series A and B.

For these reasons we do not think that our observations can be taken as indicating the existence of a couple of the kind sought, but only as giving a superior limit to its value, should it exist.

We now proceed to the Calculation of Superior Limit of Couple.

Equation of Motion of the Smaller Sphere.

Let I be the moment of inertia of sphere and cage.

" μ " torsion couple per radian.

" λ " damping couple per unit angular velocity.

" F cos pt be the supposed couple due to the larger sphere, having period $2\pi/p$.

Then

$$I\ddot{\theta} + \lambda\dot{\theta} + \mu\theta = F\cos pt.$$

Putting

$$\kappa = \lambda/I;\ n^2 = \mu/I;\ E = F/I$$

we have

$$\ddot{\theta} + \kappa\dot{\theta} + n^2\theta = E\cos pt \quad . \quad . \quad . \quad . \quad . \quad . \quad . \quad . \quad (1).$$

The solution of this is

$$\theta = \frac{E\sin\epsilon}{p\kappa}\cos(pt - \epsilon) + Ae^{-\frac{1}{2}\kappa t}\cos\{\sqrt{(n^2 - \tfrac{1}{4}\kappa^2)}\,t - \alpha\}. \quad . \quad . \quad . \quad (2)$$

where $\tan\epsilon = \dfrac{p\kappa}{n^2 - p^2}$ and A, α are constants.

The first term in the value of θ in (2) gives the forced, and the second term the natural vibrations, the period of the latter being

$$\frac{2\pi}{\sqrt{(n^2 - \frac{1}{4}\kappa^2)}} = T, \text{ say.}$$

The value of T was always very near to 120 secs., and the mean of various determinations during the observations gave

$$T = \frac{2\pi}{\sqrt{(n^2 - \frac{1}{4}\kappa^2)}} = 120{\cdot}8 \text{ secs.} \quad . \quad . \quad . \quad . \quad . \quad . \quad . \quad (3).$$

Value of κ.—When there are only natural vibrations

$$\frac{\text{any complete swing}}{\text{next complete swing}} = e^{\frac{1}{2}\kappa . \frac{1}{2}T}.$$

The value of this ratio was usually near 1·4. The mean of a number of determinations taken at various times was 1·3953. Putting

$$e^{30 \cdot 2 \kappa} = 1 \cdot 3953,$$

we get

$$\kappa = 0 \cdot 011033.$$

Value of n.—Substituting for κ in the value of T in (3) we get

$$n^2 = 0 \cdot 0027859,$$

and

$$n = 0 \cdot 052306.$$

Value of ϵ.—The forced period $2\pi/p$ was always 115 secs., whence

$$\tan \epsilon = \frac{p\kappa}{n^2 - p^2} = 2 \cdot 420,$$

and

$$\epsilon = 67^\circ \, 33',$$

$$\sin \epsilon = 0 \cdot 9242.$$

From equation (1) it will be seen that the steady deflection due to F is $\frac{E^2}{n}$ while from (2) the amplitude of the forced oscillations is $\frac{E \sin \epsilon}{p\kappa}$ or $\frac{n^2 \sin \epsilon}{p\kappa} \cdot \frac{E}{n^2}$.

Using the values found for $n\kappa$ and ϵ we have

$$\frac{n^2 \sin \epsilon}{p\kappa} = 4 \cdot 196,$$

or the forced oscillations give a cumulative effect, about four times the steady deflection due to the couple at its maximum value.

Value of Moment of Inertia, I.—This was found by vibrating the cage hung by a short quartz fibre, (1) when empty, (2) when containing the sphere, the times of vibration being respectively 8·38 secs. and 11·22 secs. The sphere weighs 1·004 grams, and its radius is 0·45 centim., so that its moment of inertia $\frac{2}{5}$ $Mr^2 = \cdot 08132$.

From this, and the times of vibration, we get

$$I = 0 \cdot 1821.$$

Value of F.—The vibrations were observed in scale divisions, each 0·05 centim., the distance between mirror and scale being 358 centims. If N is the number of scale divisions in the amplitude of vibration, *i.e.*, in half the range, we have from (2)

$$\frac{E \sin \epsilon}{p\kappa} = \frac{5N}{2 \times 35800},$$

whence

$$F = EI = 0{\cdot}8293N \times 10^{-8},$$

using the values already found for ϵ, κ, I.

Taking the limiting values of the amplitudes as half the mean ranges given in Table III., the vibration due to the quadrantal couple has amplitude not greater than 0·033 div., and that due to the semicircular couple, amplitude not greater than 0·095 div. Whence

F (quadrantal) is not greater than $2{\cdot}737 \times 10^{-10}$,

and

F (semicircular) is not greater than $7{\cdot}878 \times 10^{-10}$.

Perhaps some idea of these values may be obtained by noticing that the times of vibration of the small sphere under couple F per radian would be respectively 32 hours and 25 hours. But it is probably best to interpret the value in terms of the assumptions we made as to the force in the introduction. We found for the quadrantal couple

$$F = (G - G') MM'/r,$$

$$= \frac{G - G'}{G} \cdot \frac{GMM'}{r},$$

where MM′ are the masses of the spheres, r the distance between their centres, GG′ the parallel and crossed gravitation constants.

Now M, the mass of the larger sphere, is 399·9, say 400 grams,

M „ „ smaller „ 1·004 grams,

r is 5·9 centims.,

G and G′ are exceedingly near $6{\cdot}66 \times 10^{-8}$,

whence

$$\frac{G - G'}{G} = \frac{Fr}{G \, . \, MM'} = \frac{1}{16500}.$$

On the assumed law of force this implies that the attractions between the two spheres, with distance 5·9 centims. between their centres, do not differ in the parallel and crossed positions by as much as $\frac{1}{16500}$ of the whole attraction.

We may compare this result with RUDBERG'S values of the refractive indices of quartz for the mean D line

$$\frac{\mu_e - \mu_0}{\mu_0} = \frac{1{\cdot}55328 - 1{\cdot}544\ 8}{1{\cdot}54418} = \frac{1}{170} \text{ about.}$$

For the semicircular couple

$$2F = \frac{G - G'}{G} \cdot \frac{GMM'}{r},$$

whence

$$\frac{G - G'}{G} = \frac{1}{2850}.$$

On the assumed law of force, this implies that the attractions between the two spheres, with distance 5·9 centims. between their centres, with their axes parallel and respectively in like and unlike directions, do not differ by as much as $\frac{1}{2850}$ of the whole attraction.

This limit is large, undoubtedly owing to the want of axial symmetry in the apparatus which produced a semicircular couple as already pointed out. This couple was large, and though we attempted to eliminate it by the two sets of observations with the different azimuths of the larger sphere, in all probability we failed.

TABLE I.—Showing Scale-Readings in Tenths of a Division at Phases at Heads of Columns. Time of Revolution of Larger Sphere 115 secs.

0.	1.	2.	3.	4.	5.	6.	7.	8.	9.
..	..	..	..	55	61	61	64	61	42
25	26	31	40	50	55	60	53	52	45
44	40	45	50	54	51	49	49	48	49
52	57	52	54	57	52	40	33	28	25
30	44	57	66	70	64	52	40	38	36
39	46	55	60	63	61	52	44	44	45
44	43	49	50	52	48	42	30	32	37
45	50	62	71	69	62	52	42	39	38
44	55	58	65	65	66	61	51	45	45
41	38	40	49	56	61	62	60	56	50
48	42	39	37	40	42	58	69	69	68
58	48	41	38	38	42	48	54	60	57
55	50	43	41	41	42	47	49	55	57
63	60	58	49	46	47	46	44	51	52
50	54	48	45	44	40	36	40	50	60
67	67	62	54	44	33	35	35	38	50
57	62	68	62	52	45	36	36	39	44
51	56	59	53	47	48	53	51	50	49
50	49	52	50	50	51	53	52	54	55
48	47	44	41	44	52	55	58	60	56
49	41	41	42	43	47	50	55	60	60
60	56	58	47	43	47	49	50	50	50
54	54	54	48	50	51	49	52	52	45
42	43	48	49	55	56	52	52	52	57
56	51	46	42	42	43	49	51	55	55
55	52	49	57	50	50	50	44	43	50
50	50	43	43	46	50	58	54	55	50
49	49	49	48	50	51	54	53	56	56
57	58	56	56	51	43	40	38	41	51
60	60	60	58	52	48	48	48	52	57
58	60	57	47	41	41	51	62	63	59
53	46	40	40	40	43	49	51	61	60
60	56	51	48	42	42	43	51	59	63
62	61	55	52	51	50	51	51	52	56
58	58	53	45	40	41	49	60	70	70
60	52	48	48	50	50	54	55	53	51
50	50	47	50	50	52	53	53	50	48
48	46	48	50	51	51	50	50	52	52
49	46	44	44	49	50	55	59	57	58
51	49	46	43	44	51	59	68	64	56
50	42	40	49	57	68	71	70	59	50

TABLE I. (continued).—Showing Scale-Readings in Tenths of a Division at Phases at Heads of Columns. Time of Revolution of Larger Sphere 115 secs.

	0.	1.	2.	3.	4.	5.	6.	7.	8.	9.
	47	41	43	56	65	66	56	45	39	35
	33	40	48	59	68	70	64	51	43	42
	48	51	60	64	70	67	56	42	39	38
	40	47	52	65	60	61	59	51	51	50
	48	47	50	54	53	60	52	50	41	39
	40	44	51	58	66	70	71	63	50	38
	35	38	41	43	50	56	70	75	70	59
	50	46	45	51	61	70	71	70	62	52
	41	40	40	40	47	54	60	71	71	68
	60	50	48	45	39	39	42	49	51	60
	64	61	50	46	47	49	52	60	72	75
	70	62	57	32	23	21	30	42	57	77
	84	79	63	51	42	33	34	38	49	60
	66	64	57	51	49	44	47	49	52	52
	55	55	52	58	59	56	57	51	42	40
	43	47	55	61	66	64	60	52	50	45
	41	45	49	59	67	67	56	50	49	43
	38	45	48	53	55	56	57	55	54	56
	53	49	42	42	51	61	69	70	65	54
	45	41	40	47	51	56	61	59	55	49
	48	52	60	60	60	58	50	46	44	43
	45	51	53	60	63	67	62	60	55	48
	44	46	49	50	52	54	53	50	50	52
	60	62	63	61	51	41	39	38	42	50
	55	59	54	51	48	47	42	47	48	55
	58	61	62	60	59	54	52	52	50	50
	58	54	55	55	58	56	56	50	51	51
	56	58	51	52	48	48	54	55	50	51
	52	51	51	50	45	44	42	46	51	55
	56	53	56	59	59	60	58	59	59	54
	49	46	46	49	50	52	58	56	57	53
	51	50	50	46	49	51	58	66	67	69
	65	62	51	46	39	39	39	45	51	56
	62	61	53	48	40	38	47	62	67	63
	55	52	57	56	56	53	49	42	38	41
	51	60	65	71	73	72	60	52	50	40
	42	49	52	62	71	73	73	65	59	44
	39	38	40	43	51	51	59	61	61	50
	49	41	49	51	52	58	58	52	52	50
	53	56	57	51	50	49	49	49	51	52
	49	52	53	52	..	..	..	..	..	..
Mean of 80 in divisions. .	5·175	5·163	5·143 min.	5·186	5·246	5·294	5·355 max.	5·284	5·300	5·216

Mean range 5·355 — 5·143 = 0·212 division.

Greatest range in one period 7·5 — 3·5 = 4·0 divisions.

TABLE II.

Series.	Azimuth of large sphere.	Period of revolution.	Number of periods observed.	Mean readings at phases (whole numbers omitted).										Mean range in Scale-divisions.	Greatest range in a period in Scale-divisions.
				0.	1.	2.	3.	4.	5.	6.	7.	8.	9.		
	°	secs.													
A 1	0	115	80	·175	·163	·143	·186	·246	·294	·355	·284	·300	·216	·212	4·0
A 2	0	115	80	·653	·558	·566	·653	·813	·950	1·030	1·008	·929	·769	·472	3·1
B 1	180	115	80	·485	·590	·624	·648	·556	·464	·379	·328	·284	·364	·364	3·0
B 2	180	115	70	·423	·503	·650	·836	·941	·961	·843	·714	·540	·464	·538	7·5
B 3	180	115	54	·650	·632	·619	·648	·656	·680	·717	·739	·785	·739	·166	2·7
	°	secs.													
C 1	0	230	72	·708	·731	·708	·739	·717	·711	·676	·678	·642	·688	·097	1·3
C 2	0	230	80	·370	·400	·358	·326	·271	·214	·173	·158	·253	·310	·242	3·4
C 3	0	230	80	·616	·654	·686	·673	·663	·627	·571	·560	·566	·584	·126	2·0
D 1	90	230	50	1·024	1·042	1·034	1·004	·988	·920	·926	·954	·994	1·010	·122	2·2
D 2	90	230	70	·031	·090	·150	·210	·220	·230	·223	·176	·126	·096	·199	3·1

TABLE III.

Series.	Mean readings at phases.										Mean range.
	0.	1.	2.	3.	4.	5.	6.	7.	8.	9.	
A	·414	·361	·355	·420	·530	·622	·693	·646	·615	·493	·338
B (advanced 180°)	·702	·646	·594	·536	·522	·519	·575	·631	·711	·718	·199
Means of A and B	·558	·503	·474	·478	·526	·570	·634	·638	·663	·605	·189
C	·565	·595	·584	·579	·550	·517	·473	·465	·487	·527	·130
D (advanced 90°)	·575	·575	·565	·560	·553	·528	·566	·592	·607	·604	·079
Means of C and D	·570	·585	·575	·570	·552	·523	·520	·529	·547	·566	·065

VI. *Mathematical Contributions to the Theory of Evolution.—VI. Genetic (Reproductive*) Selection: Inheritance of Fertility in Man, and of Fecundity in Thoroughbred Racehorses.*

By KARL PEARSON, *F.R.S.*, ALICE LEE, *B.A.*, *B.Sc.*, and LESLIE BRAMLEY-MOORE.

Received November 14,—Read December 8, 1898.

CONTENTS.

* The name Reproductive Selection is retained here, although objection has been taken to it, because it has been used in other memoirs of this series. I owe Genetic Selection to Mr. F. GALTON.

Introductory.

I UNDERSTAND by a *factor of evolution* any source of progressive change in the constants—mean values, variabilities, correlations—which suffice to define an organ or character, or the interrelations of a group of organs or characters, at any stage in any form of life. To demonstrate the existence of such a factor we require to show more than the plausibility of its effectiveness, we need that a numerical measure of the changes in the organic constants shall be obtained from actual statistical data. These data must be of sufficient extent to render the numerical determinations large as compared with their probable errors.

In a "Note on Reproductive Selection," published in the 'Roy. Soc. Proc.,' vol. 59, p. 301, I have pointed out that if fertility be inherited or if it be correlated with any inherited character—those who are thoroughly conversant with the theory of correlation will recognise that these two things are not the same—then we have a source of progressive change, a *vera causa* of evolution. I then termed this factor of evolution *Reproductive Selection.* As the term has been objected to, I have adopted *Genetic Selection* as an alternative. I mean by this term the influence of different grades of reproductivity in producing change in the predominant type.

If there be two organs A and B both correlated with fertility, but not necessarily correlated with each other,* then genetic or reproductive selection may ultimately cause the predominance in the population of two groups, in which the organs A and B are widely different from their primitive types—'widely different,' because reproductive selection is a source of *progressive* change. Thus this form of selection can be a source, not only of change, but of differential change. As this differentiation is progressive, it may amount in time to that degree of divergence at which crossing between the two groups begins to be difficult or distasteful. We then reach in genetic or reproductive selection a source of the origin of species.

When I assert that genetic (reproductive) selection is a factor of evolution, I do not intend at present to dogmatise as to the amount it is playing or has played in evolution. I intend to isolate it so far as possible from all other factors, and then measure its intensity numerically. If this be sensible, then the demonstration that it is a factor is complete. How far it may be held in check by other factors—*e.g.*, natural or sexual selection—is a matter for further inquiry. If three forces, F_1, F_2, F_3 hold a system sensibly in equilibrium, then F_1 cannot be asserted to be non-effective because no progressive change is visible; its absence would soon bring to light its effectiveness.

The manner in which genetic (reproductive) selection is to some extent held in check will be clearer when my memoir on the influence of directed selection on

* If r_{ab} be the correlation of two organic characters A and B, and C be a third character, there is a considerable range of values of r_{ac} and r_{bc} for which r_{ab} may be zero (*see* YULE, 'Roy. Soc. Proc.' vol. 60, p. 486).

variation and correlation is published. Meanwhile Mr. FILON and I have shown that even a random selection of one organ alters the whole system of correlated organs.* Hence genetic (reproductive) selection indirectly modifies not only organs A and B, but all correlated organs. These modifications must be consistent with the maintenance of stamina, physique and fitness to the environment, if the change is not to be counteracted by natural selection.

So far as man is concerned, I have shown† that in the case of civilised man, the selective death-rate—*i.e.*, natural selection—does not appear to counteract reproductive selection. A small element of the population produces the larger part of the following generation. I thus concluded that *if* fertility were inherited, reproductive selection was not only a factor of evolution, but in civilised man a very sensible factor, *i.e.*, an apparently incompletely balanced factor.

In the three years which have intervened since writing the essay just referred to, members of the Department of Applied Mathematics in University College, as well as other friends, have occupied their spare time in the collection of data as to fertility and fecundity in the cases of man and of the thoroughbred racehorse. About 16,000 extracts were made in the case of man, and more than 7000 in the case of thoroughbred racehorses. In the course of the work, which proved far more laborious than we had anticipated, many difficulties and pitfalls appeared. But as a general conclusion it seems certain that: *Both fertility and fecundity are inherited, and probably in the manner prescribed by the Law of Ancestral Heredity.*‡

The object of this memoir is to set forth the theory and data by aid of which this conclusion was reached. It will be seen that it completes the establishment of genetic or reproductive selection as a factor of evolution by determining the much disputed point as to whether fertility is or is not inherited.

I. *Theory of Genetic or Reproductive Selection.* By KARL PEARSON, F.R.S.

(1.) While the physical result of fertility in an individual is measurable, the quality of fertility or fecundity in an individual differs from other physical characters in that it does not allow of direct measurements except when the potentiality is exerted and the effects recorded. At present we are not able to measure any series of organs or characters in individuals and so ascertain their fertility or fecundity. At the same time there is little doubt that these characters are functions of the physical and measurable organs and characters of the body. Such organs and characters we have good ground for supposing to be inherited according to the Law

* "Contributions to the Theory of Evolution.—IV. On the Influence of Random Selection on Variation and Correlation," 'Phil. Trans.,' A, vol. 191, p. 234 *et seq.*

† "The Chances of Death and other Studies in Evolution. Reproductive Selection," vol. 1, p. 63.

‡ *See* 'Roy. Soc. Proc.,' vol. 62, p. 386.

of Ancestral Heredity. It seems therefore worth while to prove the following proposition:

Proposition I.—Any character not itself directly measurable, but a function of physically measurable characters and organs inherited according to the Law of Ancestral Heredity, will itself be inherited according to that law.

Thus if we assume intellectual and emotional characters to be ultimately a result of physical conformation, we may be fairly certain that although we know neither the organs of which they are a function, nor the nature of that function, still they will be inherited according to the same law as that which holds for physically measurable organs.

Let y be the character in a parent, and let it be an unknown function f of the unknown physical organs $x_1, x_2, x_3, \ldots x_m$, or let:

$$y = f(x_1, x_2, x_3, \ldots x_m) \quad \ldots \ldots \ldots \quad \text{(i.)}.$$

Let Δy denote the deviation from the mean value of the character y in some special individual, and Δx the deviation from the mean of any x organ in the same individual. Then if these deviations be small compared with the mean values of the organs considered, we have from (i.) above:

$$\Delta y = a_1 \Delta x_1 + a_2 \Delta x_2 + a_3 \Delta x_3 + \ldots \ldots \ldots \quad \text{(ii.)},$$

where $a_1, a_2 \ldots$ are constants independent of the individual variations.

Let σ denote a standard deviation, ρ a coefficient of interorganic correlation, S a summation with regard to all individuals with character y dealt with, and let them be n in number. Then:

$$n\sigma_y^2 = \text{S}(\Delta y)^2 = \text{S}(a_1 \Delta x_1 + a_2 \Delta x_2 + a_3 \Delta x_3 + \ldots)^2$$
$$= n(a_1^2\sigma_{x_1}^2 + a_2^2\sigma_{x_2}^2 + a_3\sigma_{x_3}^2 + \ldots + 2a_1a_2\sigma_{x_1}\sigma_{x_2}\rho_{x_1x_2} + 2a_1a_3\sigma_{x_1}\sigma_{x_3}\rho_{x_1x_3} + \ldots);$$

or

$$\sigma_y^2 = \Sigma(a_1^2\sigma_{x_1}^2) + 2\Sigma(a_1a_2\sigma_{x_1}\sigma_{x_2}\rho_{x_1x_2}) \quad \ldots \ldots \quad \text{(iii.)},$$

where Σ denotes a summation through the group of m organs.

Let y' denote the character in an individual who is the offspring of the individual of character y, and $x'_1, x'_2, x'_3 \ldots$ the corresponding organs. Then, if we do not suppose the nature of the function f to have changed in a single generation, we have:

$$y' = f(x'_1, x'_2, x'_3 \ldots x'_m),$$

and

$$\Delta y' = a_1 \Delta x'_1 + a_2 \Delta x'_2 + a_3 \Delta x'_3 + \ldots \quad \ldots \ldots \quad \text{(iv.)},$$

$$\sigma_{y'}^2 = \Sigma(a_1^2\sigma_{x'_1}^2) + 2\Sigma(a_1a_2\sigma_{x'_1}\sigma_{x'_2}\rho_{x'_1x'_2}) \quad \ldots \ldots \quad \text{(v.)}.$$

Let r be a coefficient of direct heredity expressing the correlation between parent and offspring, and according to the Law of Ancestral Heredity the same for all organs. Then multiplying (ii.) and (iv.) together and summing we have:

$$n\sigma_y\sigma_{y'}\mathrm{R} = \mathrm{S}(\Delta y\,\Delta y') = \Sigma\,(a_1^2\mathrm{S}\,(\Delta x_1\,\Delta x'_1)) + \Sigma\,(a_1 a_2 \mathrm{S}\,(\Delta x_1\,\Delta x'_2 + \Delta x_2\,\Delta x'_1)),$$

where R is the coefficient of correlation between the characters y and y' in parent and offspring. Now:

$$\mathrm{S}\,(\Delta x_1\,\Delta x'_1) = n\sigma_{x_1}\sigma_{x'_1} r$$

$$\mathrm{S}\,(\Delta x_1\,\Delta x'_2 + \Delta x_2\,\Delta x'_1) = n\sigma_{x_1}\sigma_{x'_2} r_{x_1x'_2} + n\sigma_{x'_1}\sigma_{x_2} r_{x'_1x_2},$$

where $r_{x_1x'_2}$ and $r_{x'_1x_2}$ are what I have elsewhere termed coefficients of cross-heredity. Now if the race be stable or sensibly stable for two generations we shall have for all organs $\sigma_{x'} = \sigma_x$. Hence:

$$\mathrm{S}\,(\Delta x_1\,\Delta x'_1) = n\sigma_{x_1}^2 \times r$$

$$\mathrm{S}\,(\Delta x_1\,\Delta x'_2 + \Delta x_2\,\Delta x'_1) = n\sigma_{x_1}\sigma_{x_2}(r_{x_1x'_2} + r_{x'_1x_2}) = n\sigma_{x_1}\sigma_{x_2} \times 2r\rho_{x_1x_2},$$

for it is shown in my memoir on the Law of Ancestral Heredity* that on a probable hypothesis:

$$\tfrac{1}{2}\,(r_{x_1x'_2} + r_{x'_1x_2}) = r \times \rho_{x_1x_2}.$$

Thus we find on substitution:

$$\sigma_y\sigma_{y'}\,\mathrm{R} = r\,(\Sigma\,(a_1^2\sigma_{x_1}^2) + 2\Sigma\,(a_1 a_2\sigma_{x_1}\sigma_{x_2}\rho_{x_1x_2})).$$

But (iii.) and (iv.) show us that $\sigma_y = \sigma_{y'}$, if there be no sensible changes in a generation. Hence:

$$\sigma_y\sigma_{y'} = (\Sigma\,(a_1^2\sigma_{x_1}^2) + 2\Sigma\,(a_1 a_2\sigma_{x_1}\sigma_{x_2}\rho_{x_1x_2})),$$

and

$$\mathrm{R} = r.$$

Thus the character which is a function of physical organs is inherited at the same rate as those organs themselves.

As we may not unreasonably consider fertility and fecundity to be functions of physically measurable organs, even if we cannot specify which organs, we may, *à priori*, expect fertility and fecundity to be inherited characters.

(2.) *Proposition II.—To determine the numerical values of the changes in mean variation and correlation if fertility be inherited.*

Let us first define two terms which will be frequently used in the sequel.

(*a.*) The *fertility* of an individual shall be defined as the total number of actual offspring.

* 'Roy. Soc. Proc.,' vol. 62, p. 411. The hypothesis yet awaits an experimental verification. The need to use it prevents Proposition I. being self-evident.

(*b.*) The *fecundity* of an individual shall be defined as the ratio of the total number of actual offspring to the total number of offspring which might have come into existence under the circumstances.

These definitions are not intended to give precise statistical measures at this stage of our investigations. They are merely meant to convey a general sense of the words, which will be more precisely limited when they are applied to any given species. Fertility and fecundity, as we have thus defined them, leave out of account individual conditions and definite conditions of period, age and environment, which must be fully stated before numerical measures can be made in any special case. When the words are used in this theoretical section the reader must suppose the phrase, "under definite individual and environmental conditions," to be always inserted.

Let M_1 = the mean fertility of parents of one sex; M'_1 = the mean fertility of parents of one sex weighted with their fertility;* N_1 the number of parents considered in the first case, N'_1 the apparent number dealt with in the second case; let σ_1 and σ'_1 be the standard deviations in the two cases, and let x represent the fertility of an individual parent and z its frequency among N_1 parents. Let S denote summation for N_1 parents. Then, without any assumption as to the type of frequency, $N'_1 = S(\lambda xz) = \lambda M_1 N_1$, where λ is a constant such that λx is the weight of a parent of fertility x. This follows at once, since:

$$N_1 = S(z), \qquad M_1 = S(xz)/S(z).$$

Further,

$$M'_1 = S(\lambda x \times xz)/N'_1 = \frac{S(x^2 z)}{M_1 N_1}$$

$$= \frac{S\{(x - M_1)^2 z + 2M_1(xz) - M_1^2 z\}}{M_1 N_1},$$

$$= \frac{N_1\sigma_1^2 + 2M_1^2 N_1 - M_1^2 N_1}{M_1 N_1},$$

by the definition of standard-deviation. Hence, finally:

$$M'_1 = \frac{\sigma_1^2}{M_1} + M_1 \quad . \quad . \quad . \quad . \quad . \quad . \quad . \quad . \quad . \quad . \quad (\text{i.}).$$

Further:

$$\sigma_1'^2 = \frac{S\{\lambda x(x - M'_1)^2 z\}}{N'_1} = \frac{S\{(x - M_1 + M_1)(x - M_1 + M_1 - M'_1)^2 z\}}{M_1 N_1}.$$

Hence, multiplying out, we find after some reductions:

$$\sigma_1'^2 = \sigma_1^2\left(1 - \frac{\sigma_1^2}{M_1^2}\right) + \frac{S\{(x - M_1)^3 z\}}{M_1 N_1} \quad . \quad . \quad . \quad . \quad . \quad . \quad . \quad (\text{ii.}).$$

At first sight it might seem a comparatively easy matter to avoid weighting parents with their fertility, but practically it is almost impossible. For example, if records

* *i.e.*, if f be the fertility of a parent, each parent is repeated λf times, where λ is a constant.

are sought of the fertility of mothers in mankind, the women will appear under their husbands' names, and the labour of ascertaining whether two sisters have been included is enormous, when large numbers are dealt with. But if two or more sisters have been included, their mother has been weighted with her fertility, and when we seek the correlation between mother and daughter, it will be between mothers and daughters when weighted with fertility. But a still more serious difficulty arises from the fact that all records are themselves weighted records; the same number are not married from each family, hence we are more likely to find a member of a large family included than a member of a small. The large families, when we seek a record of two generations, are more likely to appear than small families. Precisely the same difficulty occurs when we are dealing with thoroughbred horses; a mare with large fertility is less likely to have all her offspring colts, or all her progeny sold abroad, some one or more will probably ultimately come to the stud, and thus mares of large fertility are, *à priori*, more likely to contribute to our fecundity correlation cards. We do not get over this difficulty by taking the mother and only one of her offspring. The record is still weighted with fertility. The practical verification of this lies in the experience that the fertility of mothers will always be found to be greater than that of daughters, although the fertility of the community may really be increasing; the weighting, of course, excludes sterility in the generation of mothers, but the mere exclusion of the sterile is far from accounting for the whole difference.

What we actually find from our records are M'_1 and σ'_1, but what we want for the problem of heredity are M_1 and σ_1. Equations (i.) and (ii.) do not suffice to determine these, because we cannot evaluate the third moment $S\{(x - M_1)^3 z\}$. We can hardly, even for a first approximation, assume it zero, for the standard-deviation, and therefore the individual variation is large as compared with the mean in the case of fertility, *i.e.*, the distribution is markedly skew.

Turning to offspring of the same sex as the parents, say: let M_2 be the mean fertility of offspring taking *one* only to one parent for the number N_1 of parents, supposing the parents not weighted with their fertility; let M'_2 be the mean in the same case when the parents are weighted with their fertility; and let M''_2 be the mean of *all* recorded offspring of the second generation. Let σ_2, σ'_2, σ''_2 be the standard deviations in the fertility of the offspring for the same three cases, and r, r', r'' be the corresponding coefficients of correlation between fertility in parent and in offspring. It seems to me that r is the coefficient which actually measures the real inheritance of fertility, but that in any correlation table that we can form we shall get r' or r''.

Let y be the fertility of any individual among the offspring, and x the fertility of the corresponding parent; let λx as before be the weighting of the parent, and $\lambda' x$ the number of offspring included in the record, λ' being supposed a constant.*

* I have been unable so far to find any sensible correlation between size of family and number married in man, but the point is worth a more elaborate investigation.

We have at once the following results for the total numbers dealt with in each case:

$$N_1 = S(z), \qquad N'_1 = S(\lambda x z) = \lambda M_1 N_1,$$

$$N''_1 = S(\lambda x \lambda' x z) = \lambda\lambda' S(x^2 z) = \lambda\lambda'(\sigma_1^2 + M_1^2) N_1 \quad . \quad . \quad . \quad . \quad \text{(iii.)}.$$

Turning to the means:

$$M_2 = S(yz)/N_1 \quad . \quad . \quad . \quad . \quad . \quad . \quad . \quad . \quad . \quad . \quad . \quad \text{(iv.)}.$$

$$M'_2 = S(\lambda x y z)/N'_1 = [S\{(x - M_1)(y - M_2) z\} + M_1 M_2 S(z)]/M_1 N_1 = M_2 + r \frac{\sigma_1\sigma_2}{M_1} \quad \text{(v.)}.$$

$$M''_2 = S(\lambda x \lambda' x y z)/N''_1 = M_2 + \frac{2M_1\sigma_1\sigma_2 r}{\sigma_1^2 + M_1^2} + \frac{S\{(x - M_1)^2 (y - M_2) z\}}{N_1(\sigma_1^2 + M_1^2)}$$

$$= M_2 + r\frac{\sigma_1\sigma_2}{M_1} + r\frac{\sigma_1\sigma_2}{M_1}\frac{M_1^2 - \sigma_1^2}{\sigma_1^2 + M_1^2} + \frac{S\{(x - M_1)^2 (y - M_2) z\}}{N_1(\sigma_1^2 + M_1^2)}$$

after some reductions. Now make use of (ii.) and we have:

$$M''_2 = M_2 + r\frac{\sigma_1\sigma_2}{M_1} + r\frac{\sigma_1\sigma_2}{M_1}\frac{\sigma'^2_1/\sigma_1^2}{1 + \sigma_1^2/M_1^2} + \frac{S\left\{(x - M_1)^2 \left((y - M_2) - r\frac{\sigma_2}{\sigma_1}(x - M_1)\right) z\right\}}{(1 + \sigma_1^2/M_1^2) M_1^2 N_1} \quad \text{(vi.)}.$$

But for normal correlation the equation to the straight line of regression is:

$$y - M_2 = r\frac{\sigma_2}{\sigma_1}(x - M_1).$$

Hence for such correlation the mean value of $y - M_2$ for parents $x - M_1$ is equal to $r\frac{\sigma_2}{\sigma_1}(x - M_1)$ and the summation term would vanish. For skew correlation, Mr. YULE has shown that the line just given is the line of closest fit to the curve of regression. Hence even in the case of fertility, where the correlation is certainly skew, the summation term must be extremely small, or even zero. It follows, therefore, that we may write:

$$M''_2 = M_2 + r\frac{\sigma_1\sigma_2}{M_1}\left(1 + \frac{\sigma'^2_1/\sigma_1^2}{1 + \sigma_1^2/M_1^2}\right) \quad . \quad . \quad . \quad . \quad . \quad . \quad \text{(vii.)}.$$

There is still another mean which ought to be found, namely, that of parents, M''_1, when all their recorded offspring have been entered on the correlation table. We have:

$$M''_1 = S(\lambda x \lambda' x x z)/N''_1 = S(x^3 z)/\{N_1(\sigma_1^2 + M_1^2)\},$$

or, after some reductions:

$$M''_1 = M_1 + \frac{\sigma_1^2}{M_1}\left(1 + \frac{\sigma'^2_1/\sigma_1^2}{1 + \sigma_1^2/M_1^2}\right) \quad . \quad . \quad . \quad . \quad . \quad . \quad \text{(viii.)}.$$

I now proceed to the standard deviations for the three cases, and the additional case (σ''_1) for parents.

$$\sigma_2^2 = S\{(y - M_2)^2 z\}/N_1 \quad \ldots \ldots \ldots \ldots \quad \text{(ix.)}.$$

$$\begin{aligned}\sigma'^2_2 &= S\{\lambda x (y - M'_2)^2 z\}/N'_1 \\ &= S\{x(y - M_2 + M_2 - M'_2)^2 z\}/M_1 N_1 \\ &= \frac{S\{x(y - M_2)^2 z\} + 2S\{x(y - M_2) z\}(M_2 - M'_2) + M_1(M_2 - M'_2)^2 N_1}{M_1 N_1}\end{aligned}$$

Whence, after some reductions, we find :

$$\sigma'^2_2 = \sigma_2^2\left\{1 + r^2\left(\frac{\sigma'^2_1}{\sigma_1^2} - 1\right)\right\} + \frac{S\left\{(x - M_1)\left((y - M_2)^2 - r^2\frac{\sigma_2^2}{\sigma_1^2}(x - M_1)^2\right)z\right\}}{M_1 N_1} \quad . \quad \text{(x.)}.$$

Now for a nearly straight line of regression :

$$y - M_2 = r\frac{\sigma_2}{\sigma_1}(x - M_1) + \eta$$

where η is uncorrelated with $x - M_1$. It follows accordingly that $S\{(x - M_1)^2 \eta z\}$ and $S\{(x - M_1)\eta^2 z\}$ will both vanish, since $S(\eta)$ for an array and $S(x - M_1)$ for the whole correlation surface will be zero. Hence the summation term in (x.) is either absolutely zero or extremely small. We have accordingly :

$$\sigma'^2_2 = \sigma_2^2\left\{1 + r^2\left(\frac{\sigma'^2_1}{\sigma_1^2} - 1\right)\right\} \quad \ldots \ldots \quad \text{(xi.)}.$$

Before we proceed to determine σ''_2 and σ''_1 it seems simplest to find the coefficients of correlation r_1 r' and r''. We have :

$$r = S\{(x - M_1)(y - M_2) z\}/(N_1 \sigma_1 \sigma_2) \quad \ldots \ldots \quad \text{(xii.)}.$$

To find r' we have :

$$r' = S\{\lambda x z (x - M'_1)(y - M'_2)\}/(N'_1 \sigma'_1 \sigma'_2).$$

Now

$$y - M_2 = \frac{r\sigma_2}{\sigma_1}(x - M_1) + \eta,$$

where η is sensibly un-correlated with $x - M_1$. Hence :

$$N'_1 \sigma'_1 \sigma'_2 r' = S\left\{\lambda x z (x - M'_1)\left(r\frac{\sigma_2}{\sigma_1}(x - M_1) + M_2 - M'_2 + \eta\right)\right\}.$$

Expanding, the summations with η vanish, and

$$M_2 - M'_2 - r\frac{\sigma_2}{\sigma_1}M_1 = -\frac{r\sigma_2}{\sigma_1}\left(\frac{\sigma_1^2}{M_1} + M_1\right) \text{ by (v.)}$$

$$= -r\frac{\sigma_2}{\sigma_1}M'_1 \text{ by (i.)}$$

But

$$\sigma'^2_1 = S\{\lambda xz(x - M'_1)^2\}/N'_1,$$

thus :

$$N'_1\sigma'_1\sigma'_2r' = S\{\lambda xz(x - M'_1)\frac{r\sigma_2}{\sigma_1}(x - M'_1)\}$$

$$= \frac{r\sigma_2}{\sigma_1}S\{\lambda xz(x - M'_1)^2\}$$

$$= \frac{r\sigma_2}{\sigma_1}N'_1\sigma'^2_1.$$

Thus we deduce :

$$r' = r\frac{\sigma_2}{\sigma_1}\frac{\sigma'_1}{\sigma'_2}$$

or :

$$r'\sigma'_2/\sigma'_1 = r\sigma_2/\sigma_1 \qquad \text{(xiii.)}.$$

This result has the simple interpretation *that while the coefficient of correlation is changed, the coefficient of regression is unchanged by weighting fertility, or by reproductive selection.*

This important conclusion is only an illustration of a very interesting theorem, which has been referred to in another memoir* and will be proved generally in a memoir on directed selection, written but not yet published, *i.e.*, that in a wide range of cases selection, whether random or directed (natural and artificial) changes correlation but not regression.

Before proceeding further a general remark will enable us to considerably simplify the otherwise lengthy algebra. Namely, the relation of M''_1, M''_2, σ''_1, σ''_2, r'' to M'_1, M'_2, σ'_1, σ'_2, r' is precisely the same as that of M'_1, M'_2, σ'_1, σ'_2, r' themselves to M_1, M_2, σ_1, σ_2, r. Consequently an interchange of symbols in results already found will lead us to the remaining formulæ needful.

As an illustration of this, let us verify the result we have found for M''_2. By an interchange in (v.) :

$$M''_1 = M'_2 + r'\frac{\sigma'_1\sigma'_2}{M'_1},$$

hence using (v.), (i.) and (xiii.), we find :

$$M''_2 = M_2 + \frac{r\sigma_1\sigma_2}{M_1} + r\sigma_2\frac{\sigma'^2_1}{\sigma_1}\frac{1}{M_1 + \frac{\sigma_1^2}{M_1}}$$

$$= M_2 + \frac{r\sigma_1\sigma_2}{M_1}\left(1 + \frac{\sigma'^2_1/\sigma_1^2}{1 + \sigma_1^2/M_1^2}\right),$$

* "Contributions to the Theory of Evolution.—V. On the Reconstruction of Stature," 'Phil. Trans.,' A, vol. 192, p. 177.

exactly the result reached by a longer process in (vii.). Similarly (viii.) may be deduced from (i.). Applying this to find r'' we have from (xiii.):

$$r''\sigma''_2/\sigma''_1 = r'\sigma'_2/\sigma'_1, \text{ and therefore } = r\sigma_2/\sigma_1 \quad . \quad . \quad . \quad . \quad \text{(xiv.)},$$

a result which again extends the constancy of the regression coefficient under the action of reproductive selection.

Next from (xi.):

$$\begin{aligned}\sigma''^2_2 &= \sigma'^2_2 \left\{1 + r'^2\left(\frac{\sigma''^2_1}{\sigma'^2_1} - 1\right)\right\} \\ &= \sigma^2_2 \left\{1 + r^2\left(\frac{\sigma'^2_1}{\sigma^2_1} - 1\right)\right\} + r'^2\sigma'^2_2\left(\frac{\sigma''^2_1}{\sigma'^2_1} - 1\right),\end{aligned}$$

or using (xiii.) and rearranging:

$$\sigma''^2_2 = \sigma^2_2\left\{1 + r^2\left(\frac{\sigma''^2_1}{\sigma^2_1} - 1\right)\right\} \quad . \quad . \quad . \quad . \quad . \quad . \quad . \quad \text{(xv.)}.$$

Again by interchanges in (ii.):

$$\sigma''^2_1 = \sigma'^2_1\left(1 - \frac{\sigma'^2_1}{M'^2_1}\right) + \frac{S\{(x - M'_1)^3 z'\}}{M'_1 N'_1} \quad . \quad . \quad . \quad . \quad . \quad \text{(xvi.)}.$$

Here z' stands for λxz, and we should obtain a fourth moment of the original system of unweighted parents by substitution. But it is practically impossible to obtain a correlation table for such a system. Thus it is better to allow the summation term to stand as it is, where it represents the third moment of a system of parents, weighted for fertility owing to the nature of the record, but not weighted with all their recorded offspring. (xvi.) is then a relation between the standard-deviations of parents weighted solely by forming a record and weighted both by this and by their offspring.

Equations (i.) to (xvi.) contain the chief theoretical relations of our subject,* and I shall consider some points with regard to them in the following section.

(3.) (a.) If we wish to ascertain whether fertility is inherited, we have to discover whether r is or is not zero. Now by (xiv.) r vanishes with both r' and r'', and accordingly either of these will suffice to answer the problem. Still better, we may ascertain the coefficient of regression, and then whether our statistics weight for progeny or not we shall obtain the same value. If there be no secular change taking place in the population, due to something else than reproductive selection, we should expect, provided the Law of Ancestral Heredity holds for fertility, that the regression will be near ·3 for parent and offspring.†

* Two of these formulæ, (v.) and (xi.), were given, but in a less precisely defined manner, in my "Note on Reproductive Selection" of 1896, 'Roy. Soc. Proc.,' vol. 59, p. 303.

† See "Law of Ancestral Heredity," 'Roy. Soc. Proc.,' vol. 62, p. 397.

(*b.*) If no reproductive selection exists, *i.e.*, if fertility be not inherited, then $r = 0$, and

$$\sigma''_2 = \sigma'_2 = \sigma_2, \quad M_2 = M'_2 = M''_2,$$

or, however we form a record of offspring, the mean value and variability of their fertility ought not to be changed. We shall see later that this is very far from the truth, and that these values are in whole or part sensibly affected by the manner in which the record is formed.

(*c.*) Although there be no reproductive selection, M_1, M'_1, and M''_1 will not all be equal, it is impossible that they should be. Further, σ_1, σ'_1 and σ''_1 need not be equal; their degree of sensible divergence will depend on the nature of the primitive frequency distribution for parents.

(*d.*) If fertility be inherited, or reproductive selection be an actual factor of evolution, then we see, by comparing (v.) with (i.) and (vii.) with (viii.), that the mean fertility of mothers will always be *apparently* greater than the mean fertility of daughters. This follows, since r is always less than unity, and if the race be not subjected to secular evolution, other than that due to reproductive selection, σ_2 cannot differ very widely from σ_1.*

(*e.*) An argument from means, as to whether fertility is inherited or not, is very likely to be misleading. We may choose two groups from the record for comparison, neglecting the fact that their frequency in the record is not necessarily that of their frequency in the general population. Thus, if one person, say, in four were married, a marriage record of the community might exhibit the proper frequency of families of four, but it would not do so of families of one. The sort of fallacious arguments we have to be prepared for are, for example:

(i.) That the fertility of the community is diminishing, because M'_2 is less than M'_1.

(ii.) That the fertility of the community is increasing, because M'_2 might be $> M_1$ or M''_2 be $> M'_1$.

(iii.) That fertility is not inherited, because, owing to natural selection, or other factor of evolution, one or other of these means for offspring is sensibly equal to one or other of these means for parents.

Owing to the extreme difficulty of insuring that the method of extracting the record really gives us definitely M'_2, say, and not M''_2 (or M''_2 in part), I have discarded all use of the mean values in attempting to ascertain whether fertility is inherited. The following result, however, is tempting, and might possibly be made

* A difference between σ_1 and σ_2 would mark natural selection, sexual selection, or some other factor of secular evolution at work; of *secular*, not periodic, evolution, as parents and offspring must have reached the same adult stage to have had their fertility measured.

use of in direct experiments on breeding insects, where a record could be kept *ad hoc.* It follows at once from (i.), (v.), (vii.) and (viii.):

$$\frac{M'_2 - M_2}{M'_1 - M_1} = \frac{M''_2 - M'_2}{M''_1 - M'_1} = r\frac{\sigma_2}{\sigma_1} = \text{coefficient of regression} \quad . \quad . \quad \text{(xvii.)}.$$

It is the second ratio which, I think, might with profit be experimentally evaluated.

(*f.*) Since the mean fertility of daughters loaded with the fertility of their mothers is the fertility of the next generation, and we see that this is always greater than M_2, if r be not zero, it follows that the inheritance of fertility marks a progressive change. The only means of counteracting its influence would be the reduction of M_2 to or below M_1 by the action of other equally potent factors of evolution. For the existence of such factors in man I shall later give evidence.

(4.) *Proposition III.—To extend the results obtained for fertility to the problem of fecundity.*

While the fecundity of an individual can often, at any rate approximately, be measured, the fertility is not ascertainable. Thus we can ascertain the number of occasions on which a brood mare has gone to the stallion and the number of foals she has produced, but her fertility, the produce she might have had, if she had throughout her whole career had every facility for breeding, is unknown to us. But if we proceed to form tables for the inheritance of fecundity, we are met by precisely the same difficulties as in the case of fertility. The more fertile individuals are *à priori* more likely to appear in the record, and will be likely to be weighted again with their fertility when we come to deal with their offspring.*

Now it is certain that fertility must be correlated with fecundity; or, if x now represents the fecundity and f the fertility, we shall have for the *mean* fertility for a given fecundity x an expression of the form $\lambda_0 + \lambda_1 x$, always supposing the regression to be sensibly linear. But the fertility must vanish with the fecundity, hence $\lambda_0 = 0$, and λ_1 is really the ratio of mean fertility to mean fecundity. Thus we may write for the fertility f

$$f = \lambda_1 x + \zeta,$$

where ζ may vary widely, but it is not correlated with x.

If now all the symbols we have used with regard to fertility in Section (2) be interpreted as referring to fecundity, we must weight with a factor λf instead of a factor λx, or with a factor $\lambda\lambda_1 x + \lambda\zeta$. So long as this factor is linear, absolutely no change can be made in the results, for, ζ being uncorrelated with x, all summations including

* In the case of sires especially, if we are dealing with thoroughbred horses, their comparative fewness at each period renders it quite impossible to deal with one offspring of each parent only.

$S(\zeta)$ vanish. Thus all the values given for M'_1, M'_2, σ'_1, and σ'_2 remain the same, if their results be interpreted in the sense of fecundity and not fertility. If ρ be the correlation between fecundity and fertility, and σ_1, σ_3 the standard deviations of these quantities, then $\lambda_1 = \rho\sigma_3/\sigma_1$; but we have seen that it is also the ratio of mean fertility to mean fecundity. It follows accordingly that ρ is the ratio of the coefficient of variation in fecundity to the coefficient of variation in fertility. If we may judge by the cases of man and horse, so far as I know the only cases in which fertility and fecundity have yet been examined, a coefficient of variation in fecundity amounts to about 30 per cent., while one in fertility is something like 50 per cent. Thus the correlation of fertility with fecundity would be about ·6· We should expect it to have a high value, perhaps even a higher value than this. In the case of thorough-bred horses, ρ will be the correlation between fecundity and *apparent* fertility. By direct investigation in the case of 1000 brood mares I find its value to be ·5152·

Passing now to the correlations r, r', r'', I observe that the proof given for fertility is valid with but few modifications, if these be fecundity correlations (see p. 266), for the proof involves no expansion of the factor $(\lambda_1 x + \zeta)^2$. Hence we conclude that the regression coefficient for the inheritance of fecundity will not be modified by the nature of the record or the weighting of individuals with their fertility.

When we come to the last series of constants, M''_1, M''_2, σ''_1, σ''_2, we find that these will be modified, owing to the presence of the square factor $(\lambda_1 x + \zeta)^2$, although ζ is not correlated with x. The term ζ^2 now comes in, and $S(\zeta^2)$ will give the standard-deviation of an array of fertilities corresponding to a given fecundity, *i.e.*, $S(\zeta^2) = \sigma_3^2(1 - \rho^2) \times$ number in the array.*

I find after some reductions that M''_2 and M''_1 are given by

$$M''_2 = M_2 + r\frac{\sigma_1\sigma_2}{M_1}\left(1 + \frac{\sigma'^2_1/\sigma^2_1}{1 + \sigma^2_1/(\rho^2 M^2_1)}\right) \quad . \quad . \quad . \quad . \qquad \text{(xviii.)},$$

$$M''_1 = M_1 + \frac{\sigma^2_1}{M_1}\left(1 + \frac{\sigma'^2_1/\sigma^2_1}{1 + \sigma^2_1/(\rho^2 M^2_1)}\right) \quad . \quad . \quad . \quad . \quad . \quad . \qquad \text{(xix.)},$$

the correlation of fertility and fecundity being now introduced into the results.

Clearly the result (xvii.)

$$\frac{M''_2 - M'_2}{M''_1 - M'_1} = \text{coefficient of regression} \quad . \quad . \quad . \quad . \quad . \quad . \qquad \text{(xx.)}$$

still remains true.

For the remaining two constants σ''_2 and σ''_1, I find, after some rather long analysis in the second case, which it seems unnecessary to reproduce,†

* Should the regression not be linear, $\sigma_3\sqrt{(1 - \rho^2)}$ is the mean of the standard-deviations of the arrays.

† In the course of the work the squared standard-deviation of a fertility array is assumed to be the same for all arrays $= \sigma_3^2(1 - \rho^2)$, and λ_1 is given its value $\rho\sigma_3/\sigma_1$. See, however, the previous footnote.

$$\sigma''^2_2 = \sigma^2_2 \left\{1 + r^2 \left(\frac{\sigma''^2_1}{\sigma^2_1} - 1\right)\right\} \quad . \quad . \quad . \quad . \quad . \quad . \quad . \quad (\text{xxi.}),$$

$$\sigma''^2_1 = \gamma\sigma'^2_1 \left(1 - \frac{\gamma\sigma'^2_1}{M'^2_1}\right) + \gamma \frac{1 - \rho^2}{\rho^2} \frac{\sigma^4_1}{M^2_1} + \frac{S\{(x - M'_1)^2 Z'\}}{N'_1 M'_1} \quad . \quad . \quad . \quad (\text{xxii.}),$$

and γ is the factor $\frac{M^2_1 + \sigma^2_1}{M^2_1 + \sigma^2_1/\rho^2}$, or as we can write it

$$\gamma = M'_1 \Big/ \left(M'_1 + \frac{\sigma^2_1}{M_1} \frac{1 - \rho^2}{\rho^2}\right).$$

If ρ be unity or near unity, *i.e.*, fecundity very closely correlated with fertility, $\gamma = 1$, the second term vanishes and (xxii.) becomes identical with the corresponding fertility formula (xvi.), just as (xxi.) is already identical with (xv.).

Thus we see that the whole series of fecundity relations are strikingly like those for fertility, except that in certain of them—those for M''_1, M''_2, σ''_1 and σ''_2—the correlation ρ of fertility and fecundity is introduced. If ρ be considerable, all the remarks we have made on the fertility formulæ may, *mutatis mutandis*, be applied to the measurement of fecundity.

(5.) *Proposition IV.—To deduce formulæ for finding the correlation between any grades of kindred from the means of arrays into which the kindred may be grouped.*

This problem is of very great practical importance. In the case of Man, families are so small that there is comparatively small difficulty in forming all the possible pairs of brethren, say, for any family; but when we come to animals or insects where the fertility may be extremely large, it is practically impossible to form a correlation table involving 50,000 to 100,000 entries.* One thoroughbred sire may have 50 to 80 daughters, and thus give us roughly 1200 to 3200 pairs of sisters to be entered in a correlation table. Still higher results occur in the case of aunts and nieces. It may be asked why we do not content ourselves with one or two pairs from each parent; the answer is simple: we have not (*e.g.*, in the case of thoroughbred animals, pedigree moths, &c.) a great number of sires, and the sire with 50 offspring cannot, for accuracy of result, be put on the same footing as the sire with only 2 to 4. Our process is really an indirect weighting of our results.

(A.) *To find the coefficient of correlation between brethren from the means of the arrays.*

Let x be the measure of any character or organ in one brother (sister), and x' that of a second brother (sister): let m be the mean of one set of brothers, and m' of the

* Even with the reduction in labour, introduced by this proposition and by the use of mechanical calculators, Mr. LESLIE BRAMLEY-MOORE and I took practically a week, of eight-hour days, to deduce two coefficients of correlation, *after* the means of the arrays had already been found.

second set. Let n be the number of brothers in an array, and therefore $\frac{1}{2}n(n-1)$ the number of pairs of brothers in the array. Let σ and σ' be the standard deviations of the two sets of brothers, and r the coefficient of correlation between brothers for the organ in question. Let S denote a summation with regard to all pairs of brothers in the community, and Σ with regard to all brothers in an array. Let N be the total number of brothers in the community. Then if we selected our pairs of brothers for tabulation at random (*e.g.*, not by seniority or other character), we should find $m' = m$ and $\sigma' = \sigma$. Further, by definition of correlation

$$\mathrm{N}r\sigma\sigma' = \mathrm{S}(x-m)(x'-m') = \mathrm{S}\Sigma(x-\mathrm{M}+\mathrm{M}-m)(x'-\mathrm{M}'+\mathrm{M}'-m'),$$

where M and M' are the means of the two sets of brothers in any array and are clearly equal.

Further, $\Sigma(x-\mathrm{M}) = \Sigma(x'-\mathrm{M}') = 0$, when summed for an array, and $\Sigma(x-\mathrm{M})(x'-\mathrm{M}') = 0$, for there is no correlation within the array when the deviations are measured from the mean of the array. Hence:

$$\mathrm{N}r\sigma\sigma' = \mathrm{S}\{\tfrac{1}{2}n(n-1)(\mathrm{M}-m)(\mathrm{M}'-m')\},$$

or

$$\mathrm{N}r\sigma^2 = \mathrm{S}\{\tfrac{1}{2}n(n-1)\mathrm{M}^2\} - 2m\mathrm{S}\{\tfrac{1}{2}n(n-1)\mathrm{M}\} + m^2\mathrm{N};$$

but

$$\mathrm{S}\{\tfrac{1}{2}n(n-1)\mathrm{M}\} = \mathrm{N}m.$$

Thus, finally,

$$r = \frac{\mathrm{S}\{\frac{1}{2}n(n-1)\mathrm{M}^2\}/\mathrm{N} - m^2}{\sigma^2} \quad \text{(xxiii.)}.$$

This can be written

$$r = \sigma_a^2/\sigma^2 \quad \text{(xxiv.)}$$

where σ_a is the standard deviation of the arrays concentrated into their means and loaded with their sizes; σ is the standard deviation of all brethren loaded with the number of times they are counted as brethren; m is the mean of all the offspring loaded with the number of times they are counted as brethren.

Let σ_0 be the standard deviation of offspring, and ρ the correlation between parent and offspring, then the standard deviation of an array of offspring, if correlation be sensibly linear,* will be $\sigma_0\sqrt{(1-\rho^2)}$. We have, further,

$$m = \mathrm{S}(x) = \mathrm{S}\Sigma(x-\mathrm{M}+\mathrm{M}) = \mathrm{S}\{\tfrac{1}{2}n(n-1)\mathrm{M}\},$$

$$\mathrm{N}\sigma^2 = \mathrm{S}(x-m)^2 = \mathrm{S}\Sigma(x-\mathrm{M}+\mathrm{M}-m)^2 = \mathrm{S}\{\Sigma(x-\mathrm{M})^2 + \tfrac{1}{2}n(n-1)(\mathrm{M}-m)^2\}.$$

But

$$\Sigma(x-\mathrm{M})^2 = \tfrac{1}{2}n(n-1)\sigma_0^2(1-\rho^2).$$

* See, however, the first footnote p. 270.

Thus:

$$N\sigma^2 = N\sigma_0^2(1-\rho^2) + N\sigma_a^2,$$

or:

$$\sigma^2 = \sigma_0^2(1-\rho^2) + \sigma_a^2 \quad . \quad . \quad . \quad . \quad . \quad . \quad . \quad . \quad (\text{xxv.}),$$

and r may be written:

$$r = \frac{\sigma_a^2}{\sigma_0^2(1-\rho^2) + \sigma_a^2} \quad . \quad . \quad . \quad . \quad . \quad . \quad . \quad . \quad . \quad (\text{xxvi.}).$$

Here σ_a can be found from the arrays, and σ_0 and ρ will in many cases have been previously ascertained.

(B.) *To find the correlation between "uncles" and "nephews" ("aunts" and "nieces") from the means of the corresponding arrays.*

Let n_1 be the number of uncles in an array, n_2 be the number of nephews in the associated array, so that n_1n_2 is the number of pairs of uncles and nephews provided by the associated arrays. Let $N = S(n_1n_2)$ be the total number of pairs of uncles and nephews in the community under consideration. Let x be the measure of the organ or character in the uncle, x' in the nephew. Let M and M′ be the means of two associated arrays of uncles and nephews respectively. Let m and m' be the means of all uncles weighted with their nephews and all nephews weighted with their uncles respectively, and let $\bar{\sigma}$, $\bar{\sigma}'$ be the corresponding standard deviations under the same circumstances; r' the correlation of uncle and nephew. Then:

$$Nr'\sigma\sigma' = S(x-m)(x'-m') = S\Sigma(x - M + M - m)(x' - M' + M' - m').$$

Now $\Sigma(x-M) = \Sigma(x'-M') = 0$, and within the arrays there is no association of individual uncles with individual nephews, *i.e.*, $\Sigma(x-M)(x'-M') = 0$. Thus:

$$Nr'\sigma\sigma' = S\{n_1n_2(M-m)(M'-m')\} = S(n_1n_2MM') - Nmm',$$

since

$$m = S(n_1n_2M)/N, \qquad m' = S(n_1n_2M')/N.$$

Thus:

$$r' = \frac{S(n_1n_2MM')/N - mm'}{\sigma\sigma'} \quad . \quad . \quad . \quad . \quad . \quad . \quad . \quad (\text{xxvii.}).$$

If $\bar{\sigma}_a$ and $\bar{\sigma}'_a$ be the standard deviations of the means of the arrays of uncles and nephews and R the correlation of these means, the numerator is clearly $R\bar{\sigma}_a\bar{\sigma}'_a$. Thus:

$$r' = R\frac{\bar{\sigma}_a\bar{\sigma}'_a}{\sigma\sigma'} \quad . \quad . \quad . \quad . \quad . \quad . \quad . \quad . \quad (\text{xxviii.}).$$

Here the numerator as a whole or in parts is easily found from the means of the arrays. If $\bar{\sigma}_0$ and $\bar{\sigma}'_0$ be the means of unloaded uncles and nephews, we note that they are arrays owing to common parentage, and hence their array standard deviations* will be $\bar{\sigma}_0\sqrt{1-\rho^2}$ and $\bar{\sigma}'_0\sqrt{1-\rho^2}$, ρ being the standard deviation of parent and offspring. As before we find:

$$\left.\begin{aligned}\bar{\sigma}^2 &= \bar{\sigma}_a^2 + \bar{\sigma}_0^2(1-\rho^2) \quad . \quad . \quad . \quad . \quad . \quad . \\ \bar{\sigma}'^2 &= \bar{\sigma}'^2_a + \bar{\sigma}'^2_0(1-\rho^2) \quad . \quad . \quad . \quad . \quad . \quad .\end{aligned}\right\} \text{(xxviii.)}.$$

If, as will probably be the case, there be no secular change between uncles and nephews, then $\bar{\sigma} = \bar{\sigma}'$, $\bar{\sigma}_a = \bar{\sigma}'_a$, $\bar{\sigma}_0 = \bar{\sigma}'_0$, and accordingly $r' = R\bar{\sigma}_a^2/\bar{\sigma}^2$; whence, using (xxiv.), we have:

$$r' = r \times R \times \frac{\bar{\sigma}_a^2}{\sigma_a^2}\frac{\sigma^2}{\bar{\sigma}^2} \quad . \quad . \quad . \quad . \quad . \quad . \quad . \quad . \quad . \qquad \text{(xxix.)}.$$

If we could assume $\sigma_a = \bar{\sigma}_a$ and $\sigma = \bar{\sigma}$, this result would reduce to the very simple form:

$$r' = r \times R.$$

Now the assumption $\bar{\sigma}_0 = \sigma_0$ is, I think, legitimate, for the distribution for an unloaded array of nephews or uncles should be sensibly that of an array of brethren. But the equality of σ_a and $\bar{\sigma}_a$, which would now involve that of σ and $\bar{\sigma}$, is a much more doubtful point. σ_a and $\bar{\sigma}_a$ mark indeed quite different systems of loading. Both, it is true, are of the form

$$S(nn'M^2)/N - \{S(nn'M)/N\}^2,$$

but in the case of brethren $n' = \frac{1}{2}(n-1)$ or n' has perfect correlation with n, while in the case of uncles and nephews n' is only imperfectly correlated with n. The intensity of this correlation depends upon the correlation between the sizes of arrays of uncles and nephews, a quantity which may be very small, or not, according to the nature of the record. Hence it appears necessary in applying the method to make some attempt to appreciate the value of $\bar{\sigma}_a$ as well as σ_a. If this be done R can be found from (xxix.), if not directly. This value of R is not without importance for the inheritance of characters latent in one or other sex.

We have thus reduced the correlations of individuals to a calculation of the correlation of arrays.

(6.) *Proposition V.—To find a measure of the effect of mingling uncorrelated material with correlated material.*

The importance of this investigation lies in the fact that death, restraint, or other

* Or, again, the means of the standard-deviations of the arrays.

circumstances, completely screen, in a certain number of cases, both the potential fertility and the real fecundity of man. Precisely similar circumstances, which will be considered more at length later, hinder our obtaining in horses a true measure of fecundity for all cases. We are thus really dealing with a mixture of correlated and apparently uncorrelated material. In what manner does the influence of this mixture effect our results?

Let a group N consist of $n_1 + n_2 + n_3 + n_4$ pairs of individuals. Of these, in the case of n_1 pairs, both individuals have the true value of the character under investigation recorded; in the case of n_2 pairs, neither have the true value recorded; in the case of n_3 pairs, it is the first individual of the pair which has a true recorded value, and the second an apparent or fictitious value; lastly, in n_4 cases, let the fictitious value be in the first and the real value in the second individual of the pair. Then there will be no correlation between individuals in the groups n_2, n_3, n_4. Let r be the correlation in the group n_1 and R that observed in the whole group of $N = n_1 + n_2 + n_3 + n_4$. Let x be the measure of a character in the first, x' in the second individual. Let M and M′ be the means of the total groups of the two individuals and Σ, Σ' their standard deviations. In group n_1 let the corresponding quantities be m_1, m'_1, σ_1, σ'_1, and a similar notation hold for the other sub-groups. Then $m_1 = m_3$ and $\sigma_1 = \sigma_3$; $m_2 = m_4$ and $\sigma_2 = \sigma_4$; while $m'_1 = m'_4$ and $\sigma'_1 = \sigma'_4$; $m'_2 = m'_3$, $\sigma'_2 = \sigma'_3$.

We have at once:

$$M = \frac{n_1 m_1 + n_2 m_2 + n_3 m_3 + n_4 m_4}{m_1 + m_2 + n_3 + n_4} = \frac{(n_1 + n_3)\, m_1 + (n_2 + n_4)\, m_2}{n_1 + n_2 + n_3 + n_4},$$

while

$$M' = \frac{(n_1 + n_4)\, m'_1 + (n_2 + n_3)\, m'_2}{n_1 + n_2 + n_3 + n_4}.$$

Further:

$$(n_1 + n_2 + n_3 + n_4)\, \Sigma\Sigma' R = S\,(x - M)\,(x' - M'),$$

by the usual properties of product moments

$$\begin{aligned}
&= n_1\sigma_1\sigma'_1 r + n_1 (m_1 - M)(m'_1 - M') + n_2 (m_2 - M)(m'_2 - M') \\
&\qquad + n_3 (m_3 - M)(m'_3 - M') + n_4 (m_4 - M)(m'_4 - M') \\
&= n_1\sigma_1\sigma'_1 r + n_1 m_1 m'_1 + n_2 m_2 m'_2 + n_3 m_3 m'_3 + n_4 m_4 m'_4 \\
&\qquad - M\,(n_1 m'_1 + n_2 m'_2 + n_3 m'_3 + n_4 m'_4) - M'\,(n_1 m_1 + n_2 m_2 + n_3 m_3 + n_4 m_4) \\
&\qquad + MM'\,(n_1 + n_2 + n_3 + n_4) \\
&= n_1\sigma_1\sigma'_1 r + n_1 m_1 m'_1 + n_2 m_2 m'_2 + n_3 m_3 m'_3 + n_4 m_4 m'_4 - MM'\,(n_1 + n_2 + n_3 + n_4).
\end{aligned}$$

Substituting the values of M and M′ and using the relations between the m's, we find after some reductions:

$$\mathrm{N}\Sigma\Sigma'\mathrm{R} = n_1\sigma_1\sigma'_1 r + \frac{n_3 n_4 - n_1 n_2}{n_1 + n_2 + n_3 + n_4}(m_1 - m_2)(m'_1 - m'_2) \quad . \quad . \quad (\text{xxx.})$$

Let $\frac{1}{p}$ of the N first individuals and $\frac{1}{q}$ of the N second individuals have fictitious values, then $\frac{p-1}{p}$ N and $\frac{q-1}{q}$ N will have their true values. If, now, there is no correlation between the fictitious values in the two cases, we have at once:

$$n_1 = \frac{(p-1)(q-1)}{pq}\mathrm{N}, \qquad n_2 = \frac{1}{pq}\mathrm{N}, \qquad n_3 = \frac{p-1}{pq}\mathrm{N}, \qquad n_4 = \frac{q-1}{pq}\mathrm{N}.$$

From this it follows at once that

$$n_3 n_4 = n_1 n_2,$$

or the second term in (xxx.) vanishes. Thus:

$$\mathrm{R} = \frac{n_1}{\mathrm{N}} \frac{\sigma_1\sigma'_1 r}{\Sigma\Sigma'} \quad . \quad . \quad . \quad . \quad . \quad . \quad . \quad . \quad . \quad (\text{xxx.})\ bis.$$

Thus R vanishes with r, and no spurious correlation could arise from the existence of fictitious values distributed at random through the correlation table. This result might, indeed, (as it often is tacitly) be assumed by some, but it seems very desirable to have a definite proof.

It remains to consider Σ and Σ'. We have:

$$\begin{aligned}\mathrm{N}\Sigma^2 &= n_1\sigma_1^2 + n_2\sigma_2'^2 + n_3\sigma_3^2 + n_4\sigma_4^2 \\ &\quad + n_1(m_1 - \mathrm{M})^2 + n_2(m_2 - \mathrm{M})^2 + n_3(m_3 - \mathrm{M})^2 + n_4(m_4 - \mathrm{M})^2 \\ &= (n_1 + n_3)\sigma_1^2 + (n_2 + n_4)\sigma_2^2 \\ &\quad + (n_1 + n_3)m_1^2 + (n_2 + n_4)m_2^2 - (n_1 + n_2 + n_3 + n_4)\mathrm{M}^2,\end{aligned}$$

or

$$\begin{aligned}\Sigma^2 &= \frac{n_1 + n_3}{\mathrm{N}}\sigma_1^2 + \frac{n_2 + n_4}{\mathrm{N}}\sigma_2^2 + \frac{n_1 + n_3}{\mathrm{N}}\frac{n_2 + n_4}{\mathrm{N}}(m_1 - m_2)^2 \\ &= \left(1 - \frac{1}{p}\right)\sigma_1^2 + \frac{1}{p}\sigma_2^2 + \left(1 - \frac{1}{p}\right)\frac{1}{p}(m_1 - m_4)^2 \\ &= \sigma_1^2 + \frac{1}{p}(\sigma_2^2 - \sigma_1^2) + \left(1 - \frac{1}{p}\right)\frac{1}{p}(m_1 - m_4)^2 \quad . \quad . \quad . \quad . \quad (\text{xxxi.}).\end{aligned}$$

Similarly:

$$\Sigma'^2 = \sigma_1'^2 + \frac{1}{q}(\sigma_2'^2 - \sigma_1'^2) + \left(1 - \frac{1}{q}\right)\frac{1}{q}(m'_1 - m'_2)^2 \quad . \quad . \quad (\text{xxxii.}).$$

Now if the introduction of the fictitious values consisted of anything of the nature of a wrong pairing of certain individuals, we should simply have $\sigma_1 = \sigma_2$, $\sigma'_1 = \sigma'_2$, $m_1 = m_2$, $m'_1 = m'_2$ and, accordingly, $\Sigma = \sigma_1$, and $\Sigma' = \sigma'_1$.

In any case, if the percentage of fictitious values be not large, the second and third terms are of the second order of small quantities, since $\frac{1}{p}$ and $\frac{1}{q}$ are small. The maximum value of the third term cannot be greater than $\frac{1}{4}(m_1 - m_2)^2$, and this will be relatively small in the cases to which we shall apply it.

For example, no great changes are made in σ, when we vary the amount of fictitious cases introduced into our fertility tables. m_1 and m_2 do, however, change. Thus $\sigma_1 = \sigma_2 = 3$ approximately, and the range $m_1 - m_2 = 1{\cdot}2$. Hence:

$$\Sigma^2 = 9 + \tfrac{1}{4}(1{\cdot}2)^2, \text{ at a maximum}, = 9{\cdot}36,$$

or,

$$\Sigma = 3{\cdot}06.$$

Thus in this *extreme* case there is only 2 per cent. change in the value of Σ. In such cases accordingly we may take for rough approximations $\Sigma = \sigma$ and $\Sigma = \sigma'$. This leads us to:

$$R = \frac{n_1}{N} r \quad . \quad . \quad . \quad . \quad . \quad . \quad . \quad . \quad . \quad . \quad \text{(xxxii.)}.$$

Or, *the reduction of correlation, due to the introduction of fictitious values, is obtained by using as a factor the ratio of actual correlated pairs of individuals to the total number of pairs tabulated.*

This result will be of considerable service when we come to deal with the fecundity of thoroughbred racehorses.

(7.) *Proposition VI.—To obtain a measure of the spurious correlation apparently existing between two organs, when a mixture is made of heterogeneous materials.*

Let x and x' be measures of the two organs, and let there be N pairs of organs formed by i heterogeneous groups containing $n_1, n_2, n_3 \ldots$ pairs with means $m_1, m'_1, m_2, m'_2, m_3, m'_3 \ldots$, &c., standard deviations $\sigma_1, \sigma'_1, \sigma_2, \sigma'_2, \sigma_3, \sigma'_3 \ldots$, &c., and correlations $r_1, r_2, r_3 \ldots$, &c. Let M, M' be the means of the whole heterogeneous community, Σ, Σ' the standard deviations, and R the correlation. Then:

$$R\Sigma\Sigma' N = S(n\sigma\sigma' r) + S\{n(m - M)(m' - M')\},$$

where S denotes a summation with regard to all i groups. Now if there were no correlation at all between the organs in any one of the i groups, R for the heterogeneous mixture would still not be zero so long as the second summation did not vanish. This, then, is a measure of the spurious correlation produced by making a mixture of uncorrelated materials.

Now $S\{n(m - M)(m' - M')\}$, remembering the values of M and M′ may be written:

$$S\left\{\frac{n_p n_q}{N}(m_p - m_q)(m'_p - m'_q)\right\} \quad . \quad . \quad . \quad . \quad . \quad . \quad (\text{xxxiv.})$$

where the summation S now refers to every possible pair p and q of the r groups.

Now it is very unlikely, unless i be very large and the numbers $n_1, n_2, n_3 \ldots$ be taken at random, that this expression will vanish. Suppose even that the means of our heterogeneous groups were uncorrelated, *i.e.*, $S(m - M)(m' - M') = 0$, it is unlikely that $S\{n(m - M)(m' - M')\}$ will also be zero, when n is taken at random. With a comparatively few groups, with numbers taken at random, it is extremely improbable that the principal axes of the i points loaded with $n_1, n_2, n_3 \ldots$ will exactly coincide with the directions of the axes of x and x'.

We are thus forced to the conclusion that a mixture of heterogeneous groups, each of which exhibits in itself no organic correlation, will exhibit a greater or less amount of correlation. This correlation may properly be called spurious, yet as it is almost impossible to guarantee the absolute homogeneity of any community, our results for correlation are always liable to an error, the amount of which cannot be foretold. To those who persist in looking upon all correlation as cause and effect, the fact that correlation can be produced between two quite uncorrelated characters A and B by taking an artificial mixture of two closely allied races, must come rather as a shock.*

The better to illustrate this, I take some data recently deduced by Miss C. D. FAWCETT. She finds for 806 male skulls, from the Paris Catacombs, the correlation for length and breadth ·0869 ± ·0236, and for 340 female skulls, from the same locality, − ·0424 ± ·0365. The existence of the negative sign and the comparative smallness of the correlation, as compared with the probable errors, might lead us to assert the correlation between the length and breadth of French skulls to be sensibly zero.

If now the two sexes be mixed, the heterogeneous group has for correlation ·1968 ± ·0192, a value which cannot possibly be considered zero. Thus the mixture exhibits a large spurious correlation.

Whether any given mixture increases or reduces the correlation will depend entirely on the signs of the differences of the means of the sub-groups. But the danger of heterogeneity for the problem of correlation will have been made manifest. If the value of R for any mixture, whose components are known, is to be calculated, then we have only to note that:

$$\Sigma^2 = \frac{S(n\sigma^2)}{N} + \frac{S(n_p n_q (m_p - m_q)^2)}{N^2}, \quad \Sigma'^2 = \frac{S(n\sigma'^2)}{N} + \frac{S(n_p n_q (m'_p - m'_q)^2)}{N^2} \quad (\text{xxxv.}).$$

* Thus the mere fact of breeding from *two or three* individuals selected at random can easily produce a correlation between organs in the offspring, which has no existence in the species at large.

II. *On the Inheritance of Fertility in Mankind.* By Karl Pearson, F.R.S., and Alice Lee, B.A., B.Sc.

(8.) In commencing an investigation of this kind where the results to be expected were quite unknown to us, but where we had reason to believe that the *apparent* strength of inheritance must be very small, we considered that the first thing to be done was to investigate the largest possible amount of material. Thus the probable errors of our results would be very small and any, however small, correlation between fertility in parent and offspring would be brought to light. Attempts might then be made to strengthen any correlation discovered by removing so far as possible one after another the various factors tending to screen the full effect of the inheritance of fertility.

Such factors are for example:

(*a.*) The age of both husband and wife at the time of marriage. The real fertility may be screened by late marriages of one or both parents. The relation of fertility to age at marriage has been dealt with by several writers, notably by Duncan and Ansell.*

(*b.*) The duration of marriage. The data may be taken from a marriage not yet complete, both parents being still alive. Or from a marriage which is complete one or both parents being dead. In the former or the latter case the marriage may be complete so far as fertility is concerned, *i.e.*, details of offspring may be available till the wife has reached the age of 50 years, which for statistical purposes may be taken as an upper limit to fecundity.

(*c.*) Restriction of fertility during marriage. It has been shown in a paper on Reproductive Selection† that there is evidence of the sensible influence of this factor in man. It tends to give fictitious values to the fertility of the younger, rather than the elder generation, and so obscures the correlation.

We have accordingly two problems before us:

(i.) Supposing these and other factors tending to screen the effects of reproductive selection to exist, can we show that it still produces sensible effects in the case of man, and thus demonstrate that fertility is really inherited?

(ii.) Can we by eliminating these factors so far as possible obtain a lower limit to the coefficient of heredity in the case of fertility, and ascertain whether it approximates in value to what we might expect from the Law of Ancestral Heredity?

The first impression of the reader may be that it is only needful to select the

* J. Mathews Duncan, 'Fecundity, Fertility, Sterility and Allied Topics,' second edition, Edinburgh, 1871. Charles Ansell, Junr., 'Statistics of Families in the Upper and Middle Classes,' London, 1874.

† 'The Chances of Death and other Studies in Evolution,' vol. I, pp. 77, 89.

fertility of marriages, which were formed with husband and wife between 20 and 28* say, and which have lasted till the wife is over 50. But these conditions must be true in *two* successive generations, and, had we adopted them, we may safely say that without immense labour it would have been impossible to collect even a thousand cases. From the whole of the peerage, the baronetage, the landed gentry, a variety of family histories, of private pedigrees, and a collection of data formed of families at first hand, it was not possible to extract more than about 4000 cases for the inheritance of fertility in the female line, when the limitations were far less stringent, being applied only to *one* generation, and consisting in our taking marriages entered into at any time of life for either husband or wife, and lasting till the death of one member or for at least fifteen years. Even in this case the pedigree of the wife had to be sought for from one record to another and often in vain. It is the male pedigree with which the recorder in nearly all cases occupies himself.

Only those who have attempted the labour of extracting, as has been done in this case, some 16,000 separate returns, will fully grasp the difficulty of making the limitations of selection more and more complex; the quantity to be obtained becomes dangerously small and the labour immensely increases. Even could with time and patience a sufficient selection of ideal cases have been made, it does not follow that the result would be satisfactory; for, we should have made a narrow *selection*, and this very fact might indicate that possibly we have been selecting one grade or class of fertility. It is possible that the less fertile are the weaker, and so more liable to die early; or again it may be the more fertile women who are subjected to the more frequent risk of childbed, and thus are less likely to appear in the selection of long marriages. Even greater or less risk at birth may be an inherited character in women, and may not unfairly be looked upon in itself as a factor limiting fertility naturally.

Taking these points into consideration, it seemed that if we were to have enough material to draw conclusions from we must entirely drop all attempt to classify by age of parents at marriage. We might make some limitations but they must not be very stringent; they must leave room for an increase of stringency in different directions, so that we could roughly appreciate the influence of the screening factors. Accordingly our plan has been to show that correlation actually does exist between parent and offspring with regard to fertility, and that when we make the conditions more stringent the correlation increases towards the value indicated by the law of ancestral heredity.

(9.) *On the Inheritance of Fertility in Woman.*—(i.) Table I. gives the result for 4418 cases of the fertility of a mother and of her daughter. These were extracted from FOSTER's 'Peerage and Baronetage,' BURKE's 'Landed Gentry,' some family

* As DUNCAN points out, an early marriage on the average means an earlier cessation of fecundity; a somewhat later one does not necessarily connote less fertility.

histories and a collection of family data drawn from private pedigrees and other sources. In the case of the daughter, no marriage was taken which had not lasted at least 15 years, or until the death of husband or wife. In the case of the mother no limitation whatever was made, the number of brothers and sisters of the daughter, including herself, being counted. Weight was given to the fertility of the mother, for every possible case that could be got from the records under the above conditions was extracted. It is quite possible that a certain proportion of offspring dying in early infancy have not been entered in the records.

If M_m, M_d be the mean fertilities of mother and daughter, σ_m, σ_d their standard deviations, and r_{md} their correlation, we found :

$$M_d = 3{\cdot}494, \qquad M_m = 6{\cdot}225,$$
$$\sigma_d = 2{\cdot}975, \qquad \sigma_m = 3{\cdot}052,$$
$$r_{md} = {\cdot}0418.$$

Clearly owing to the near equality of σ_d and σ_m the regression of daughter's on mother's fertility is sensibly equal to the correlation.

The probable error of r_{md} is determined by the formula given by PEARSON and FILON* to be ·0101, or r_{md} is four times its probable error.

We thus conclude :

(i.) That fertility is inherited in the female line.

(ii.) That its effects are very largely screened by the factors to which we have previously referred.

Had we started with no limitation as to the daughter's family, it is highly probable that r_{md} would scarcely have been sensible relatively to its probable error, and, therefore, small series without due regard to screening causes may easily lead the recorder to suppose that fertility is not inherited.

Supposing we exclude from the daughters the 775 barren marriages, we find the mean for 3643 cases of fertile marriages to be 5·237. Comparing this fertility with the observed fertility 6·225 of mothers, a superficial inquirer might at once consider that a diminution of fertility has taken place. The fact is that neither of the results, M_m or M_d gives the actual fertility of the mothers or daughters. These are the means M''_1 and M''_2 of formulæ (viii.) and (vii.) of the theoretical investigation.

Let us apply the theory developed to our statistics. In the first place we note that r is small ; hence r^2 is still smaller, and thus by (xv.) σ''_2 will not differ much from σ_2. Since σ''_1 will be generally less than σ_1 by (xvi.), it follows that σ''_2 will probably be less than σ_2. Approximately, we can take $\sigma_2 = 3$. Turning to (vii.) we see that M''_2 cannot, since r is small, differ widely from M_2. If there be no secular

* "Contributions to Theory of Evolution.—IV." 'Phil. Trans.,' A, vol. 191, p. 242.

evolution in the real fertility sensible in the one generation, then M_1 would equal M_2. Hence to a first approximation we should have:

$$M_1 = M_2 = M''_2 = 3\cdot494.$$

To obtain a second approximation we may substitute this in the small terms of (vii.). Here σ'^2_1 must be found from (ii.); neglecting the cubic term we have:

$$\sigma'^2_1/\sigma^2_1 = 1 - \sigma^2_1/M^2_1 = \cdot2628.$$

Hence:

$$\begin{aligned} M_1 = M_2 &= 3\cdot494 - r\frac{\sigma^2_1}{M_1}\left(1 + \frac{\cdot2628}{1\cdot7372}\right) \\ &= 3\cdot494 - \cdot0418 \times 2\cdot5759 \times 1\cdot1513 \\ &= 3\cdot494 - \cdot124 = 3\cdot370. \end{aligned}$$

We can now substitute this value of M_1 in (viii.), and we find:

$$M''_1 = 3\cdot370 + 2\cdot980 = 6\cdot350.$$

This differs comparatively little from the actually observed value, 6·225, and is satisfactory evidence of the validity of our theory. The fact that the elder generation was in no way limited like the younger, and that we have neglected the third moment—although fertility distributions are never normal—as well as made other approximations, is quite sufficient to account for the difference observed.

We may take it that 3·4 is practically the fertility of the elder generation, and that this is raised to about 3·5 by reproductive selection in the younger generation. The result 6·2 for the elder generation is thus purely a result of weighting due to the nature of the record.

(ii.) Table II. gives the result of 1000 cases taken from the Peerage. Here the conditions of extraction were as follows:—

One member only was taken out of each family, or no weight was given to the fertility of mothers.

The daughters' marriages had all been completed by the death of one parent or had lasted at least 15 years.

There was no limitation with regard to the parents' marriages. We found:

$$M_d = 3\cdot923, \qquad M_m = 5\cdot856,$$
$$\sigma_d = 2\cdot758, \qquad \sigma_m = 2\cdot751,$$
$$r_{dm} = \cdot2096.$$

The coefficient of regression is sensibly equal to that of correlation. The probable error of $r_{dm} = \cdot0204$, or not a tenth of the value of r_{dm} itself. Again we conclude that fertility is certainly inherited in the female line. By selecting fairly homogeneous material with a more definite and complete record than exists for the heterogeneous material of the previous case, we have carried up the correlation to five times its previous value, and within a reasonable distance of the value ·3 which would be required by the law of ancestral heredity. The homogeneity of our material is evidenced by the reduction in both standard deviations; the greater completeness of the record by the rise in the fertility of daughters; and the non-weighting of the fertility of mothers by the fall in their mean fertility.

If the reader will turn back to the theory of the influence of heterogeneity on correlation in section (7), he will notice that the expression in (xxxiv.) will be negative, and therefore the apparent correlation less than the real, if we form a mixture of two groups in which $m_p > m_q$ and $m'_p < m'_q$. Now the entries of women in the Landed Gentry and other records are very often entries of "heiresses," while the entries of women in the Peerage are entries because of class. An "heiress" naturally has fewer brothers and sisters than another woman on an average, or we may expect $m'_q > m'_p$. On the other hand an "heiress" need not have fewer children than other women, unless her heritage is the result of her coming from an infertile stock, and is not a result of the incompleteness of her parents' marriage. If she belongs to a somewhat lower social grade, she may possibly be more fertile than the average of a higher social grade. In this case m_p will be $> m_q$, and when we come to mix records of the Peerage with those of the Landed Gentry and Family Histories, we need not be surprised to find the correlation of fertility much weakened, as it undoubtedly is (as shown by (i.) and (ii.) above) by the mixture.

Let us next apply our theory to the above results. We are now dealing with M'_1, M'_2, σ'_1, σ'_2. Assuming that there is no secular change $\sigma_1 = \sigma_2$, and accordingly since $\sigma'_1 = \sigma'_2$ sensibly, formula (xi.) shows us that both $= \sigma_1$.

Further, if $M_1 = M_2$, formula (v.) is a quadratic equation to find M_1; substituting for M'_1, σ_1^2, and r, we have, on solving and taking the only admissible root, $M_1 = 3·4625$. Then, applying formula (i.) to find M'_1, we have:

$$M'_1 = 5·660.$$

This is not quite as high as the observed value 5·856, but it suffices to show that our theory expresses the main facts. In all probability we have not entirely freed our results from weighting with fertility; because, although every endeavour was made to take only one from each family, it is possible that pairs of sisters have occasionally crept into the record.

(iii.) Table III. gives the result of 1000 cases taken from the Landed Gentry. As we have already noted, the women recorded are largely "heiresses," and we believed this might be one of the chief sources of the heterogeneity of the material in Table I. The conditions of selection were made somewhat more stringent, and were as follows:— Only one daughter was taken from each family, and her marriage must have lasted at least 15 years. No limitation was placed on the duration of the parents' marriage. We found:

$$M_d = 4{\cdot}232, \qquad M_m = 5{\cdot}403,$$
$$\sigma_d = 3{\cdot}292, \qquad \sigma_m = 3{\cdot}241,$$
$$r_{md} = {\cdot}1045.$$

The probable error of $r_{md} = {\cdot}0211$, and again we see that fertility is certainly inherited. The correlation has, however, sunk; probably, as the great increase of variation indicates, because we are dealing with much more heterogeneous material than in the case of the Peerage. While the selection of "heiresses" has largely reduced the number of brothers and sisters, *i.e.*, the fertility of mothers, the limitation to marriages of at least 15 years has increased the apparent fertility of daughters; nor is this increase at all balanced by the fact that heiresses come from small families, and may, therefore, be supposed to be the children of rather sterile mothers. The average number of children of heiresses is sensibly as large as the average number of children of women who are not in the bulk heiresses, and who have, as in the following case, been selected with the same condition as to duration of marriage. The fact is that heiresses are not on the whole the children of sterile mothers; their high fertility and *their small correlation with their mothers* shows us that heiresses in the bulk are rather the daughters of mothers whose apparent fertility is fictitious. They have, owing to the sterility or early death of their husband, to their own marriage late in life, or to some physical disability, or other restraint, never reached their true fertility. If this conclusion be correct, and a comparison of the values of M_d and r_{md} in this and the following cases thrusts it almost irresistibly upon us, then we see that the argument against the inheritance of fertility based upon the fertility of heiresses and non-heiresses is of no validity.* It could not be valid as against the values of the correlation we have found, but the present investigation shows by the value of r_{md} exactly wherein the error lies: the heiress is not infertile, but is the daughter of a fictitiously infertile mother.

Applying our theory to this case, we find from formula (xi.), putting $\sigma_1 = \sigma_2$:

$$\sigma_1^2 = (\sigma'^2_2 - r^2\sigma'^2_1)/(1 - r^2),$$

whence we find $\sigma_1 = 3{\cdot}293$, a result sensibly identical with σ_d. Solving the quadratic (v.) with $M_1 = M_2$ to find M_1 we find:

$$M_1 = 3{\cdot}952.$$

Hence by (i.) we have:

$$M'_1 = 6{\cdot}838,$$

the actually observed value being 5·403. Thus the theory completely fails to give the fertility of the heiresses' mothers; for such a fertility as we find in the daughters,

* See, for example, a recent letter of Mr. HOWARD COLLINS in 'Nature,' November 3, 1898.

the mothers' fertility is far too low. This again emphasises the point we have already referred to. The peculiar character of the selection, which leads to the female record in the Landed Gentry, is not one such as we have considered in our theory, where the record of any family is likely to appear in proportion to its size. Such a distribution is a *chance* distribution, but a selection of women inheriting land has not this character, and a woman who is the mother of co-heiresses is hardly doubly as likely to appear as the woman who is mother of one. A marriage in either case is likely to be arranged, and if we take only one daughter from each family the record will not already have weighted—at any rate to the full extent—every mother with her fertility. If the reader will compare the variation columns for both daughters and mothers in Table III. with the corresponding columns in Table II. or Table IV., he will at once see how anomalous is the selection of women given in the Landed Gentry.

(iv.) Table IV. gives the results for 1000 cases taken from the Peerage and Baronetage under the following limitations: one daughter only was taken for each mother, and in the case of both mother and daughter the marriage must have lasted at least 15 years. We found:

$$M_d = 4\cdot335, \qquad M_m = 5\cdot898,$$
$$\sigma_d = 2\cdot967, \qquad \sigma_m = 2\cdot830,$$
$$r_{md} = \cdot2130.$$

The probable error of $r_{md} = \cdot0204$. Thus, as it is now hardly necessary to repeat, fertility is certainly and markedly inherited. The regression coefficient is now as high as ·2233, the closest limit we have yet reached to the theoretical ·3 of the law of ancestral heredity.

Owing to the limitation to marriages of 15 or more years, the means of the fertilities of both mothers and daughters have risen, in the latter case more, however, than the former. It might have been expected that the fertility of mothers would have risen more, but it must be remembered that M_m is the apparent and not the *real* fertility of mothers; and further, since the record largely weights the more fertile women, the bulk of the mothers are already those with large families, *i.e.*, those whose marriages have lasted at least 15 years.

Assuming that there is no sensible secular change in unweighted fertility, *i.e.*, $\sigma_1 = \sigma_2$, we have from the formula on p. 284:

$$\sigma_1 = 2\cdot973.$$

From (v.) with $M_1 = M_2$ we find:

$$M_1 = 3\cdot845$$

for the real fertility of mothers. This is a sensible increase on the value 3·463 given in Case (ii.), in which there was no minimum duration to the length of the mother's marriage.

Applying formula (i.) we find:

$$M'_1 = 6{\cdot}144,$$

which is somewhat more than the observed value 5·898. The reason for this lies, we think, in the difficulty already referred to on p. 263. If we start extracting mothers, it is often difficult to follow the daughter's history; starting with the daughter it is much easier, although still laborious, to trace back her ancestry, and find the number of her brothers and sisters. Even in this case the search may be lengthy. But as daughters when married change their name, it requires great care in extracting large quantities to be sure that a mother is not repeated, *i.e.*, some approach made to weighting her with her fertility. Every care was made in extracting the records, but we cannot hope to have always avoided weighting to some extent a mother, and if this be done we shall have a transition from formulæ (xi.), (v.), and (i.) towards formulæ (xv.), (vii.), and (viii.), which would well account for the difference found between theory and observation.

If we sum up for inheritance of fertility in the female line on the basis of these four cases, we draw from each one of them the unquestionable result that fertility in woman is an inherited character. Further, the more we remove causes of fictitious values for the fertility in either generation, the closer does the value approach that required by the law of ancestral heredity. The two chief disturbing factors which we have not been able to eliminate are (*a.*) the age at which marriage is entered upon, (*b.*) restraint giving a fictitious value to the fertility. Both these causes must give a lessened value to the correlation of fertility between mother and daughter, and the first, judging from the great influence of age at marriage on fertility, cannot fail to give a serious diminution. Hence if we find the regression coefficient as high as ·2233, when we neglect these factors, it is no stretching of facts to conclude that it would in all probability rise to ·3 could we take them into account.

Our conclusion, therefore, is that fertility in woman is certainly inherited through the female line, and most probably according to the law of ancestral heredity. Reproductive selection is actually a *vera causa* of progressive change, but its influence is largely, if not entirely screened by the numerous factors tending to make the apparent fertility of women differ from their real or potential fertility.

(10.) *On the Inheritance of Fertility in Man.*

(i.) While many of the difficulties involved in the extraction of data for women still exist for man, a new and important feature tending to screen the full influence of the law of ancestral heredity arises in his case. The full fertility of the husband is not in the average case at all approached in the case of monogamic marriage. Hence, in considering the size of a man's family as a measure of his fertility we are measuring a character which differs largely from the character of fertility in woman. It is only in the case of sterile or even very sterile men that there is likely to be a correlation shown between the sizes of the families of fathers and sons.

The intensity or duration of fecundity in the husband must, one or other, be less than that of the wife,—and this will hardly be so in the great run of cases—if his family is to be in any way a measure of his fertility, or, as it might be better to call it in this case, his sterility. We are seeking to find a correlation between two characters, one in father and one in the son, neither of which we can measure unless they fall short of a certain limit. The result is that our correlated material is weakened down by the admixture of a mass of uncorrelated material in the manner indicated in Proposition V. of the theoretical part of this investigation. Within the family we cannot hope to get a correlation which will approach that indicated by the law of ancestral heredity. We may still, however, hope to ascertain whether fertility, respectively sterility, is an inherited character in man as well as woman.

(ii.) Our first attempt was to collect as much material as possible, so that our limitations were few. The Peerage, Baronetage, Landed Gentry, Family Histories, private pedigrees, and collected data provided the 6,070 cases arranged in Table V. Here large families were weighted because several, where available, were taken from one family. The son's marriage must either have lasted till the death of one partner or at least 15 years; there was no condition as to the duration of the father's marriage.

We have spoken of the correlation between fertility of father and son, but since only a single marriage of the father is taken, it may be equally well termed a correlation between the fertility of the mother and son, which may, perhaps, to some extent explain the relatively high values reached.

Let M_s, σ_s be the mean and standard-deviation of the son, M_p, σ_p of the parent, and r_{sp} the correlation; then we found:

$$M_s = 3{\cdot}871, \qquad M_p = 5{\cdot}831,$$
$$\sigma_s = 3{\cdot}003, \qquad \sigma_p = 3{\cdot}190,$$
$$r_{sp} = {\cdot}0514.$$

The probable error of $r_{sp} = {\cdot}0087$. Thus the correlation is nearly six times the probable error, or fertility in man is certainly inherited.

(iii.) Table VI. contains the result of extracting 1,000 cases from the Peerage, only one son being taken from each family, and his marriage having lasted at least 15 years. No attention was paid to the length of parents' marriage.

We found:

$$M_s = 5{\cdot}070, \qquad M_p = 5{\cdot}827,$$
$$\sigma_s = 2{\cdot}910, \qquad \sigma_p = 3{\cdot}142,$$
$$r_{sp} = {\cdot}0656.$$

The probable error of $r_{sp} = {\cdot}0212$. This case closely confirms the previous case; M_p and σ_p remain sensibly the same, M_s has risen owing to the longer period of duration of the son's marriage, and since there is a longer period for the possible exhaustion of the male fertility, we find r_{sp} is slightly larger. Although the numbers are smaller than in Case (i.), the probable error is not so large but that we can still assert an inheritance of fertility in man.

(iv.) Lastly, to compare with Case (iii.) for women, 1000 cases were extracted from the Landed Gentry, and are given in Table VII. Here no marriage of the son or parents was taken under a minimum of 15 years' duration, and only one son taken from each family. We found:

$$M_s = 5\cdot304, \qquad M_p = 6\cdot272,$$
$$\sigma_s = 2\cdot951, \qquad \sigma_p = 2\cdot911,$$
$$r_{sp} = \cdot1161.$$

Thus the longer duration of the marriage, which gives a greater chance for the exhaustion of the fertility of a partially sterile father, leads to an increased correlation. The probable error here is ·0210, and the correlation is thus unquestionable.

It would be idle to apply the theory before developed to these male cases, for the simple reason that we must certainly look upon them as containing a large proportion of uncorrelated material. But they suffice to show that male fertility is an inherited character, and although the results are widely different from those indicated by the law of ancestral heredity, they are large when we consider how little male fertility appears measurable by the results of monogamic marriage. Were an approximately close measure of male fertility available, there is certainly in the above results no reason to induce us to believe that it would not be found to obey the law of ancestral heredity.

(11.) *On the Inheritance of Fertility in Woman through the Male Line.*

Although we are not able to measure the potential fertility of the male, we are able to determine whether he transfers fertility from his mother to his daughter. This may be simply done by correlating the fertility of a woman and that of her paternal grandmother. This problem belongs to an important class—namely, questions as to the extent to which a sexual character is inherited through the opposite sex. DARWIN has touched upon this "transmission without development" in Chapter viii. of the 'Descent of Man,'* and we shall find his views amply verified.

The problem before us is: Does a woman have as close correlation with her paternal as with her maternal grandmother in the matter of fertility?

To solve this problem 1000 cases were taken out of the Peerage for the fertility of a woman and of her paternal grandmother. The marriages of the woman and of her grandmother were both taken with a minimum duration of fifteen years. Every care was taken that no weight should be given to fertile families by taking only one out

* Second Edition, p. 227, 'Laws of Inheritance.'

of each family, but, of course, the difficulty of avoiding this is increased when a pedigree must be traced through three instead of two generations.

If d denotes granddaughter, g grandmother, the following results were obtained (Table VIII.) :—

$$M_d = 4\cdot411, \qquad M_g = 5\cdot657,$$
$$\sigma_d = 2\cdot897, \qquad \sigma_g = 3\cdot056,$$
$$r_{dg} = \cdot1123.$$

The coefficient of regression of daughter's fertility on grandmother's fertility $= \cdot1065$.

The probable error of $r_{dg} = \cdot0211$.

According to the law of ancestral heredity* we should expect the grandparental correlation and regression to be half the parental and equal to ·15. Comparing the present result with Case (iv.), we see that ·1123 and ·1065 have to be compared with $\frac{1}{2}$(·2130) and $\frac{1}{2}$(·2233), or with ·1065 and ·1116. These are differences well within the probable error of our results, or we may conclude that the correlation of a woman with her paternal grandmother is exactly what from Case (iv.) of Section (9) we should expect to find for her correlation with her maternal grandmother. The reduction from ·15 to ·1123 is just what we might have predicted after the maternal reduction from ·3 to ·2130. We, therefore, conclude that the fertility of woman is inherited through the male line with the same intensity as through the female, and this intensity is most probably that which would be indicated by the law of ancestral heredity.

(12.) We do not stay to consider many points which flow from our tables, such, for example, as the amount of restraint indicated by the hump at the start of our various frequency distributions for size of families, partly because such consideration would lead us beyond our present scope, the inheritance of fertility, and partly because this point has been already dealt with by one of us in a paper on 'Reproductive Selection.' We consider that we have shown fertility in mankind to be an inherited character in both lines, and probably obeying the law of ancestral heredity.† By aid of our theoretical investigations it is clear that the average size of a family (M_1), as deduced from our record data (M'_1 or M''_1), is about 3·5 children, if the marriage lasts till the death of one partner, or at least till 15 years; it is about 3·9 to 4 children if the duration of the marriage is at least 15 years. Reproductive selection would increase this average by about ·5 child per generation were its influence not counteracted

* "Mathematical Contributions to the Theory of Heredity, on the Law of Ancestral Heredity," 'Roy. Soc. Proc.,' vol. 62, p. 397.

† In the paper on "The Law of Ancestral Heredity" ('Roy. Soc. Proc.,' vol. 62, p. 412) it is stated that fertility is probably inherited, but the amount falls below that which would be indicated by the law of ancestral heredity. At that time only Case (i.) of Section (9) and Case (i) of Section (10) had been worked out in detail. It is the rise of correlation with more stringent limitation of opposing influences, which suggests that after all that law is true for fertility as for other characters.

by a variety of other factors of evolution. These factors are so active that the influence is reduced to ·12 of a child per generation if we take Case (i.) of Section (9), and, we have little doubt, would be practically insensible did we take all marriages without any limitation whatever. Reproductive selection must, therefore, be looked upon as always tending to increase the fertility of a race; races are not only ever tending to increase, but tending to increase the rate at which they increase—a feature not recognised by MALTHUS, but which strengthens certain of his arguments. So soon, therefore, as environment, or other circumstance, relieves the pressure of opposing factors, a race will not only increase in numbers, but also in fertility. It is this inherited character of fertility, and its constant tendency to change unless held in check by natural selection or other factor of evolution, which seems to us the source of the immense diversity in fertility to be observed not only in different species, but in local races of the same species.

III. *On the Inheritance of Fecundity in Thoroughbred Racehorses.* By KARL PEARSON, F.R.S., with the assistance of LESLIE BRAMLEY-MOORE.*

(13.) The data provided for the fertility of thoroughbred racehorses by the *stud-books*, are of a kind which cannot be hoped for except in the cases of pedigree animals kept for breeding purposes, and of specially-arranged experiments on insects, &c. We have a practically complete record of the stud-life of every brood-mare. The sire by whom she has been covered in each year is stated, and the result, barren, dead foal, living filly or colt, twins, &c., can be ascertained. It is also possible to find out whether the foal dies young, say as a yearling. By examining the whole series of stud-books the complete pedigree of any mare or sire can nearly always be found, and the correlation theoretically worked out for almost any degree of relationship.

In starting an investigation of this kind on such a great mass of raw material, it is necessary to draw up certain rules for the extraction and arrangement of data. These rules must be prepared without any definite knowledge of the character of the material in *bulk*, for this can only be found after, perhaps, some 1000 cases have been extracted and worked out. Hence the rules originally adopted are often not such as an investigator would have arranged had he known beforehand the general character of the conclusions he would reach. But the statistician cannot, like the experimental physicist, modify without immense labour his methods and repeat his experiment. The collection of his data has frequently been far too laborious a task for repetition. His raw material has been prepared in a certain manner; he may

* During the three years in which this investigation has been in progress, a considerable number of friends have given me substantial aid in the arithmetical work, or in the preparation of the 6,000 pedigree cards on which the results are based. Mr. BRAMLEY-MOORE has latterly been my chief helper, but I am also much indebted to Miss ALICE LEE and Mr. G. U. YULE. Miss MARGARET SHAEN and Miss LINA ECKENSTEIN have also contributed to the labour of extracting the raw data from the stud-books.

sort and rearrange his data cards in a variety of ways, but to prepare new cards on a different system is practically beyond his powers.

These remarks are made in order to meet criticism of the method in which my data cards were prepared. I could *now* possibly extract more convenient data, but that is only because of the knowledge gained in the process of examining the fecundity of several thousand horses. I did not even know, *ab initio*, the extent of variability in equine fertility; I did not even know the immense preponderance which would have to be given to certain sires, at any rate I had no numerical estimate of it. Nor had I any percentage of the number of cases in which a pedigree might end abruptly with an alternative sire.*

I saw at once that the apparent fertility of racehorses was even less close to their potential fertility (which I presume to be the inherited character) than in the case of man. Mares go at different ages to the stud, they remain—for reasons not stated—uncovered for occasional years, or periods of years; they return to the training stable for a time; they are sold abroad; they are converted into hunters, put into harness, or, as is occasionally recorded, sold to cab proprietors. This by no means invariably denotes that their fertility is exhausted; their offspring may be bad racers, or their stock unfashionable. Very frequently also we find the mare put to a cart-horse stallion for a year, a few years, or for the remainder of her career, and then no record at all is given of the result. Thus the total fertility recorded can have but small correlation with the potential fertility, and I was compelled to deal with fecundity. The insufficiency of the *apparent* fertilities, as recorded in my mare index, to solve the problem, may be illustrated in the following manner: 1100 cases of the apparent fertilities of mares and dams having had at least four coverings were tabulated (Table IX.). The following results were calculated from this table, the subscript m referring to mare and d to dam :—

$$M_d = 7\cdot6655, \qquad M_m = 6\cdot1391,$$
$$\sigma_d = 3\cdot3652, \qquad \sigma_m = 3\cdot1617,$$
$$r_{md} = -\cdot0868.$$

The probable error of $r_{md} = \cdot0202$, and thus we might argue that a fertile dam has, on the average, infertile offspring. But an examination of the above numbers shows us that the dams are more variable than the mares,† and yet the dams have been theoretically subjected to the greater selection, for they must all be granddams, or the fertility of the mares could not have been recorded. We are forced to conclude that the mares have been in some manner selected, and the form of the selection is fairly obvious on examining the table. There appears a great defect of

* Even the pedigree of such a famous racehorse as Gladiateur is soon checked by the occurrence of alternative sires. His sire, Monarque, was the son of either The Baron, or Sting, or the Emperor.

† The variability of mares, as a whole, not separated into mares and dams, is (see Art. 16) 3·2775.

mares in the third quadrant,* *i.e.*, of mares and dams of large fertility, the frequency is cut off abruptly in this quadrant. The reason for this is fairly clear. We have dealt with a limited number of years, about 30, of horse-breeding; hence, when the dam has a long record, her later offspring at any rate cannot possibly have a long one; when she has a short one, it is possible for them to have a long one. Accordingly, there has been a process of unconscious selection, which has led to a negative correlation of the apparent fertilities.

To illustrate the point further, two more correlation coefficients were obtained. In Table X. are given the apparent fertilities of mares and their dams with a minimum of eight coverings. We find:

$$M_d = 8{\cdot}6191, \qquad M_m = 7{\cdot}6309,$$
$$\sigma_d = 3{\cdot}1656, \qquad \sigma_m = 2{\cdot}8149,$$
$$r_{md} = -{\cdot}0876.$$

The probable error is again about ·0202. While the mares now form a group with their mean fertility almost equal to that of the dams in the previous result, their variability is markedly less. Relatively to the dams its reduction is even greater. The correlation is sensibly the same. It would thus seem that the anomalous selection of mares which thus reduces their variability so markedly below that of the dams is not in the *low* fertilities.

I now removed from the Table IX. all parts of it concerning mares with a fertility greater than 8; 867 mares and dams remained with a minimum limit of four coverings, the mares not having a greater fertility than 8 offspring. I found:

$$M_d = 7{\cdot}7636, \qquad M_m = 4{\cdot}8558,$$
$$\sigma_d = 3{\cdot}3983, \qquad \sigma_m = 1{\cdot}9887,$$
$$r = -{\cdot}0190.$$

The probable error of $r = {\cdot}0229$. Now the line of regression for dams on mares ought to be the same, whether we obtain it from this result or from the first results in which mares with more than 8 offspring are included. Yet, in this case, there is no sensible correlation at all. In other words, if we exclude the data for large fertilities, we should have to conclude that there was no correlation between the apparent fertilities recorded for mares and their dams. We are thus forced to conclude that apparent fertility is a character depending on the manner in which the record is formed, and must be useless for the investigation of inheritance. This investigation strengthens my *à priori* reasons for selecting fecundity, not apparent fertility, as the character to be investigated. I took the fecundity of a brood-mare to be the number of her living offspring divided by the potential number of her offspring under the given circumstances. Of both numerator and denominator of this ratio I must say a few words.

* The portion of the table cut off by vertical and horizontal lines through the means of dams and mares.

In considering the inheritance of fertility I had two different problems in my mind: (i.) Is fertility pure and simple inherited? *i.e.*, Does a very fertile mare have offspring more fertile than the average? And (ii.) What effect does reproductive selection actually have on the population? *i.e.*, To what extent is it screened by other factors of evolution; does the very fertile mare actually have more offspring than the less fertile? Is, for example, her stock weedy and likely to die early? In the case of mankind, the fertility of a woman is, as a rule, effectively brought to its limit with the end of her marriage, and accordingly I started with completed marriages. In the case of a brood-mare her effective fertility depends not on the offspring she has but on the number of these which survive foaldom. It would doubtless have been better to have treated these two problems of fertility separately, but being fairly confident from Proposition I., p. 260, that fertility must be inherited, I was more interested to test the actual effect of reproductive selection. Accordingly I selected as the numerator of my fecundity ratio, not the number of foals born, but those who survived to the yearling sales. The difference is not very great, but quite sensible. For example, the mean fecundity of 3909 brood-mares, measured in my way, = ·6343, *i.e.*, 63 surviving offspring on the average of 100 coverings.

The following table gives the result of reckoning merely barren mares and those slipping foals or giving birth to dead foals in a twenty-year period:—

AVERAGE Fecundity of Brood-mares.

Year.	Average fecundity.	Year.	Average fecundity.
1873	·712	1883	·693
1874	·703	1884	·678
1875	·707	1885	·702
1876	·697	1886	·700
1877	·692	1887	·682
1878	·680	1888	·695
1879	·683	1889	·685
1880	·666	1890	·686
1881	·680	1891	·679
1882	·667	1892	·675

The averages of five-year periods are:

·702, ·675, ·691, ·684,

and of the whole period, ·688.

There does not appear to be sufficient evidence for any secular change here, and we may take ·688 to represent the average fecundity of the brood-mare, reckoning viable offspring to the number of coverings. The difference of ·688 and ·634 gives a death-rate of 5·4 foals in 68·8, or a death-rate of 7·85 per cent. of foals before maturity. If a considerable part of this death-rate be differential, we have room for natural selection influencing the drift of reproductive selection. The standard deviation in the fecundity is, however, about ·191, or about 19 foals in the 63, or about 30 per cent.—a very great variation, so that if fecundity be inherited, a differential death-rate of the immature will hardly suffice to check it.

So much then of the numerator of my ratio. I have spoken immediately above of the denominator as if it were the number of times the mare had been covered. It is generally this, but in the relatively few cases where the mare has given birth to twins, I have counted that covering *twice*. Had this not been done the fecundity might have been greater than unity, for example even in some exceptional cases have risen to two. On the other hand, a loss of twins would have been marked by no greater change in fecundity than a loss of one foal, or the survival of *one* twin would not have been different in its effect on fecundity to the birth of a foal. In order, therefore, to avoid these difficulties—especially that of isolated individuals lying far beyond the fecundity range of 0 to 1—when twins were born the potentiality of the covering was reckoned in the denominator as two. The relative infrequency of twins causes, however, this modification of the denominator to have small influence on the result.

My next step was to form some estimate of the extent to which fecundity thus measured was the same for different periods in a mare's breeding career. I expected fecundity to diminish with age as in the case of mankind, but taking out a fairly large test number of mares, I found that their fecundity for the periods covered by two successive stud-books was in the majority of cases closely the same. With larger experience I should now lay more weight on the decrease of fecundity with age; and I also think fecundity is smaller when the mare first goes to the stud. But even thus much of the reduced fecundity of old mares seems to arise from breeders sending famous mares to the sire long after their breeding days are passed. I have several records of old mares being covered seven or eight times without offspring. This custom of breeders was much more rife in the early days of breeding than it appears to be now, when some breeders discard or sell a fairly old mare, even if she is barren two or three successive years. Clearly the custom gives the mare a fictitious fecundity, far below her real value, and probably accounts for granddams having a somewhat less fecundity than their granddaughters.

The next problem to be answered was the effect the method of forming my fecundity ratios might have on the relative numbers which would be found in different element-groups. For example, supposing the element of fecundity to be 1/10, or the element-groups 0–1/20, 1/20–3/20, 3/20–5/20, . . . 17/20–19/20, 19/20–1, would the fact that the fecundity ratio is a ratio of *whole* numbers cause, *à priori*, a greater probability of frequency in one of these element-groups than another?

To begin with, all estimation of fecundity based on less than *four* coverings was discarded. Three coverings give too rough an appreciation of a mare's fecundity, it can only fall into one of the values 0, 1/3, 2/3, and 1. The question then arises, if all the fecundities:

0/4, 1/4, 2/4, 3/4, 4/4,

0/5, 1/5, 2/5, 3/5, 4/5, 5/5,

0/6, 1/6, 2/6, 3/6, 4/6, 5/6, 6/6,

. . . .

0/26, 1/26, 2/26, ... 25/26, 26/26,

were equally likely, how would the frequency depend on the grouping?*

Taking 26 coverings as the probable maximum—it actually occurs—we have for the total number of fecundities given above: $5 + 6 + 7 + \ldots + 27 = 368$ separate fecundities. Let us see how they would be divided in one or two cases.

Case (i.) Let the elements be based on 1/8, or be 0–1/16, 1/16–3/16, 3/16–5/16, 5/16–7/16, 7/16–9/16, 9/16–11/16, 11/16–13/16, 13/16–15/16, 15/16–1.

The half-groups at the ends are taken so that zero and perfect fecundity should really be plotted at the middle of a 1/8 element. We find, adding up the numbers of the above fecundities which fall into the nine groups, the following frequencies:—

33·5, 42, 43·5, 44, 42, 44, 43·5, 42, 33·5.

There is thus a somewhat deficient frequency in the terminal groups, and this would probably to some extent bias the distribution.

Case (ii.) Let the elements be based on 1/15, or be

0–1/30, 1/30–3/30, 3/30–5/30, . . . 25/30–27/30, 27/30–29/30, 29/30–1.

We have the following distribution:

23, 22, 23, 23·5, 22·5, 23·5, 23·5, 22, 22, 23·5, 23·5, 22·5, 23·5, 23, 22, 23.

The bias here is only slight and the distribution is on the whole very satisfactory.

Case (iii.) Let the elements be based on 1/20, or be

0–1/40, 1/40–3/40, 3/40–5/40, . . . 35/40–37/40, 37/40–39/40, 39/40–1.

We find for the groups:

23, 13, 17·5, 17·5, 17, 18, 17, 18, 17·5, 17, 18, 17, 17·5, 18, 17, 18, 17, 17·5, 17·5, 13, 23.

Here the terminal groups have too great a frequency, and the adjacent groups too little. It is clear that the division into 1/15 elements is better than those of 1/8 of 1/20, so far as these results go. But unfortunately the different coverings do not occur in anything like the same proportions. Their exact frequencies could only be found *à posteriori*, and I was desirous of having some idea of grouping before start-

* Such problems are really not infrequent in statistical investigations, and seem to be of some interest for the theory of fractional numbers. Mr. FILON worked out for me the details of the cases given below.

ing the labour of extraction I therefore weighted the different coverings on the basis of a small preliminary investigation as follows:

Case (iv.) Number of coverings,	4 to 5	inclusive,	loaded with	2.
,, ,,	6 to 9	,,	,,	3.
,, ,,	10 to 15	,,	,,	4.
,, ,,	16 to 18	,,	,,	2.
,, ,,	19 to 26	,,	,,	1.

The resulting system of frequencies was:

54, 42·5, 45·5, 47, 45·5, 46·5, 46, 45, 45, 46, 46·5, 45·5, 47, 45·5, 42·5, 54.

This system is not so uniform as in Case (ii.). I had hoped that the 744 frequencies would have been fairly closely the double system of Case (ii.). The main irregularity occurs at the terminal groups, or those having fecundities nearly zero and nearly perfect. These I considered would be relatively infrequent, when we started with as many as four coverings, and had an average failure of about 37 in 100. The sequel showed that the assumption was legitimate, so far as regards zero fecundity, but that perfect fecundity was sufficiently frequent to cause a hump in the frequency curve for fecundity, corresponding to the group-element 29/30 to 1. The frequency of this group is greater than that of the group 27/30 to 29/30, when we start from at least four coverings. This hump entirely disappears, however, if we start with at least eight coverings. Thus I take the hump to be purely "spurious," *i.e.*, a result of the arithmetical processes employed, and not an organic character in fecundity. It depends upon our definition of fecundity, which is not a truly continuous quantity.

As the theory of correlation applied is not in any way dependent on the form of the correlation surface, beyond the assumption of nearly linear regression, the hump cannot, I think, sensibly affect our conclusions. Had I known, however, *à priori*, what the frequency of different coverings and the nature of the fecundity frequency curve would be, I should have attempted to choose such a group-element, that, with proper weighting of the coverings, there would have been no *arithmetical* bias to the terminal groups. As it was, it seemed to me that the group-element of 1/15 gave fairly little arithmetical bias—at any rate where the bulk of the frequency would occur—and it was accordingly adopted as a basis for classifying fecundities.

The dfficulty illustrates the point I have referred to, namely, that in statistical investigations the best classification can only be found *à posteriori*, but the classification adopted has usually to be selected *à priori*.

The 1/15 element being selected, the letters *a*, *b*, *c*, *d*, *e*, *f*, *g*, *h*, *i*, *j*, *k*, *l*, *m*, *n*, *p*, *q* were given to the 16 groups of fecundities from 0 to 1, as cited under Case (ii.).*

* A table was formed of the 368 actually-occurring fecundities, from which it was possible to at once read off the group (or it might be two groups, *e.g.*, ·5 falls half into *h* and half into *i*) into which they each fell.

Thus the fecundity of a mare was described by one of these 16 letters. Here the centre of the *j* group, for example, is ·6, and it covers all fecundities from ·56 to ·63. Thus midway between *j* and *k* we are at about the mean fecundity.

The more recent Stud-Books, vols. 12 to 17, were taken as containing more complete details and, what is more important, less in-and-in breeding, although as we shall see, this is still an important factor. These volumes cover 30 and more years of English* stud life. From these 30 years' records upwards of 5000 mares, who had been covered upwards of four times, had their fecundity ascertained. The process was a very laborious one, as each mare had generally to be sought for in several volumes, and the records in each volume are not continuous, but overlap by quite arbitrary numbers of years. Further, great care had to be taken to identify each mare properly, as the same name is very frequently repeated, and the like difficulty occurs, though to a lesser extent, in the case of sires. A card was then written, giving the name of the mare and those of her sire, her dam, and her dam's sire. Upon this card the letter indicating her fecundity was placed. A card alphabet of mares was thus formed, consisting, in the first place, of about 3000 entries. This alphabet was again gone through and the fecundity of the dams of the mares inserted on the cards till there were about 2500 cases known of mare and dam. The dams were partly found from the existing series, but it was also largely necessary to work out fresh cases. Lastly, the cards were gone through and the fecundity of the grand-dams entered in upwards of 1000 cases. This forms the first series of cards.

In the next place a card index was formed of all the sires serving during these 30 years. This contained upwards of 1000 cards. On these cards the sire's sire was entered, and the fecundity of all the mares contained in the first or mare alphabet was now taken off and placed on the card of the mare's sire. Thus the card of each sire had the letters *a*, *b*, *c*, *d*, *e*, &c., upon it, and a frequency distribution was formed on the card of each sire for the fecundity of his daughters.

The same thing was done for the sires' sires; only here recourse had again to be had to the stud-books to obtain the fecundity of the daughters of the more ancient sires. Finally, a sire-alphabet was obtained which gave the average fertility of the daughters of a sire and of the daughters of his sire, or his half-sisters. On these cards was also placed the number of mares upon which each average was based.

These two card-alphabets, the mare and sire alphabets, form the "dressed" material upon which all the subsequent calculations were based.

(14.) At this point it seems desirable to insist somewhat on the many causes which tend to make the fecundity of mares, as thus determined, to a considerable extent fictitious. Many of these were only apparent to me as I became more and more familiar with the material.

* Irish mares were excluded except where, for pedigree purposes, it was necessary to deal with them. Many Irish mares were further included when it came to the valuation of the fertility of mares due to a given sire.

(*a.*) Mares appear to be less fecund at the beginning and end of their breeding career. Hence, when the fecundity is based on a part only of their career, as it often must be, we do not really get a fair appreciation.

(*b.*) A more fertile mare is likely to have more daughters go to the stud than a less fertile one, and hence we get a better appreciation of the fertility of the offspring of the former than of the latter.

(*c.*) Fashion among breeders interferes largely with the exhibition of the natural fecundity of a mare. She may be a famous mare and is sent to a famous sire, even though produce is not so likely as if she were put to a sire of a different class. This appears to be practically recognised when apparently barren mares are sent in one season to two, or even three sires, or again to half-bred horses or cart-horses.

(*d.*) Brood-mares which have produced performers are kept much longer at the stud, and we have the fecundity lowered by coverings after the mare is sensibly sterile. Less important mares are removed sooner from the stud.

(*e.*) Good racing mares are often put late to the stud.

(*f.*) In a certain number of cases we are simply told that the mare had no produce for a period of years, but whether she was covered or not is unrecorded.

(*g.*) Second-rate mares, or mares thought to be near the end of their fecundity, are often sold abroad. In the latter case the fecundity is fictitiously increased; in the former we have only a short period to base it on.

(*h.*) There is no record kept of the half-bred foals, which for our purpose are as important as the thoroughbred foals. "Put to a hunter" is a not uncommon record, with no statement of the result.

(*i.*) Comparatively infertile mares, unless of very valuable stock or famous as racers, are not kept long enough at the stud to get a reliable measure of their fecundity.

(*j.*) The smaller breeders will often put mares to inferior sires, already nearly worn out, either because they own them, or because their fee is low; and thus again a full chance is not given to the fecundity of the mare to exhibit itself.

(*k.*) We have excluded in our determination of the fecundity foals dying young. This is often due to the fault of the mare, but is often again due to the environment.

(*l.*) Lastly, thoroughbred mares are highly artificial creatures, and many must suffer from their environment,* either in the matter of barrenness or slipping foal, in a manner from which the wild horse or a more robust domesticated animal would be entirely free.

These considerations may suffice to show that our values of the fecundity will only roughly represent what may be termed the natural fecundity, and we ought not for

* I am told that there are like difficulties with cows. Cows are very liable to slip their calves, and one cow doing so, several others in the herd will or may follow her example. There is a strong folk-belief in Wiltshire—I give it merely as evidence of what a slight change in the environment is supposed to achieve—that the habitual presence of a donkey with the herd in some way soothes the cows, and renders them less ready to slip their calves.

a moment to expect inheritance in the full intensity of the Galtonian law to be exhibited by such material.

(15.) But there is another point of very considerable importance for the weakening of correlation, namely the effect of in-and-in breeding. To get correlation we must have a diversity of parents producing a diversity of offspring, but when the parents become more and more identical, we get larger and larger arrays between which and the parents the correlation is weakened. For example: suppose the correlation found between *all* parents and offspring in the general population, and now select only all the brothers in a large array and find the correlation between them and their offspring, we shall find that the correlation is lower than in the previous case.* It would be impossible to apply theory to the present case, however, because we can only roughly appreciate the extent of such in-and-in breeding. That it is great the following statistics will show.

Of the more than 1000 sires in my sire alphabet, only 760 were sires of mares which had been covered at least four times. These 760 sires had upwards of 5000 offspring, of whom I had the fecundity recorded, but when mares with alternative sires were excluded, there remained only 4677 available mares.† These mares were distributed as follows:—

Daughters	1	2	3	4	5	6	7	8	9	10	11	12
Sires	280	113	78	43	29	22	20	21	22	14	10	10
Daughters	13	14	15	16	17	18	19	20	Above 20			
Sires	11	11	8	6	2	6	4	4	46			

Here the second line gives the number of sires having the number of daughters in the first line in the 4677 cases, which I take to be a fair sample.

Thus over a third of the sires had only one mare. Two-thirds of the sires had together only one-fifth of the mares. *Seventy-six* of the sires were fathers of about half the mares, and 46 sires alone produced 1801 mares, almost as many as 642 sires did. We are here dealing with the fairly long period of 30 years, but even making due allowance for young stallions commencing and old stallions concluding their stud career, it will be manifest that our sample shows that the great bulk of mares for the period in question were the offspring of comparatively few sires.

But let us look at the problem from the standpoint of the sires. My 760 sires

* The theory of such cases is fully developed in a memoir on the influence of selection on correlation not yet published.

† Some other cases were also excluded for diverse reasons.

were *all* fathered among themselves except in 49 cases. In other words, they were the sons or grandsons of only 49 sires. Of these 49 sires, there were 12 whose pedigree I could not trace,* but they were very probably sons of sires already on my list or among the remaining 37. In the majority of cases they appeared only as the sire of one stallion. The remaining 37, whose pedigree I could trace, were descended at once or in very few generations from 9 sires.† Thus both from the standpoint of the mares and of the sires we are dealing with a closely in-bred stock, and this is one and probably a very important factor in the weakening of the fecundity correlation.

Having regard to these difficulties, if we can succeed in showing that fecundity in thoroughbred racehorses is inherited, we can be fairly confident that we have only reached a lower limit of the correlation coefficient.

(16.) *On the Inheritance of Fecundity in the Female Line.*

(i.) A preliminary investigation must here be made, in order to determine the ρ of the formulæ given in Proposition III. (p. 269) we want the correlation of fecundity with fertility. If ϕ be the fecundity, f the apparent fertility, and c the number of coverings, twins counting as a double covering, we have:

$$\phi = f/c,$$

whence if we determine the correlation between ϕ and f, numerous constants will follow. Table XI. gives the correlation between fertility and fecundity for 1000 brood-mares. We found:

$$M_\phi = \cdot 6375, \qquad M_f = 6\cdot 515,$$
$$\sigma_\phi = \cdot 1810, \qquad \sigma_f = 3\cdot 2775,$$
$$\rho = r_{\phi f} = \cdot 5152,$$
$$v_\phi = 100\sigma_\phi/M_\phi = 28\cdot 39, \qquad v_f = 100\sigma_f/M_f = 50\cdot 31,$$

where v_ϕ and v_f are the "coefficients of variation."‡ Here by YULE's Theorem § $r_{\phi f}\sigma_\phi/\sigma_f$ is the slope of the line which most closely fits the curve of regression for fecundity on fertility. If we supposed this curve to be straight, then the line must coincide with it. Now since fecundity vanishes with fertility, the curve passes through the origin, and hence, if the regression be linear, the line must also pass through the origin. In this case, as is shown on p. 270, $r_{\phi f} = v_\phi/v = \cdot 5644$. The difference between ·5644 and ·5152 may be taken, as it is several times the probable error, to indicate that the regression curve between fecundity and fertility is only approximately linear.

The variations in both fertility and fecundity are here large. Accordingly we

* Stockmar, Sovereign, Andover, Phaeton, Prince Caradoc, Robert Houdin, Pylades, King of Kent, Garry Owen, Calaban, Homily and Taurus.

† Tramp, Sir Peter Teazle, Catton, Buzzard, Orville, Diomed, Sorcerer, Dr. Syntax, Marske.

‡ 'Phil. Trans.,' A, vol. 187, p. 276.

§ 'Roy. Soc. Proc.,' vol. 60, p. 477.

must use the formula (i.) for the mean value of an index given in my memoir on spurious correlation.* We shall then obtain an approximate value to the mean number of coverings of each mare. Formulæ (iii.) of the same paper will then give the standard deviation for the number of coverings. In our present notation:

$$c = f/\phi,$$

and therefore:

$$M_c = \frac{M_f}{M_\phi}\left(1 + \left(\frac{v_\phi}{100}\right)^2 - r_{f\phi}\left(\frac{v_f v_\phi}{10{,}000}\right)\right),$$

$$\sigma_c = M_c \sqrt{\left(\frac{v_f}{100}\right)^2 + \left(\frac{v_\phi}{100}\right)^2 - 2r_{f\phi}\frac{v_f v_\phi}{10{,}000}}.$$

We find:

$$M_f/M_\phi = 10{\cdot}2196,$$

and:

$$M_c = 10{\cdot}2196 \times 1{\cdot}007 = 10{\cdot}2911,$$
$$\sigma_c = 4{\cdot}4455,$$
$$v_c = 43{\cdot}20.$$

To the same degree of approximation we can further ascertain the correlations between the number of coverings and the apparent fertility and fecundity, *i.e.*, r_{cf}, and $r_{c\phi}$. A short investigation similar to those in the memoir on spurious correlation just cited shows us that:

$$r_{cf} = (v_f - r_{f\phi}v_\phi)/v_c,$$
$$r_{c\phi} = (r_{f\phi}v_f - v_\phi)/v_c.$$

These lead to the numerical results:

$$r_{cf} = {\cdot}8259, \qquad r_{c\phi} = -{\cdot}0572.$$

The conclusions to be drawn from these results are all of some interest. In the first place we may ask: How does M_ϕ agree with its value found from other and more complete series? For 4677 mares—my complete series without mares with alternative sires—the average fecundity was ·6373. A better agreement could not have been hoped for. In a group of 1509 mares dealt with for variation only and entered as "daughters" on the cards—so that they had not been selected by the fact that their daughters must have recorded offspring, as is the case with "dam" entries—I found the following results:—

VARIATION in Fecundity of 1509 Brood-mares (Four Coverings).

Fecundity.	*a.*	*b.*	*c.*	*d.*	*e.*	*f.*	*g.*	*h.*	*i.*	*j.*	*k.*	*l.*	*m.*	*n.*	*p.*	*q.*
Frequency	9	3	11	26	46	43·5	85	122·5	154·5	232·5	194	223	146	100	23	90
Total 1509. $M_\phi = {\cdot}6345$. $\sigma_\phi = {\cdot}1965$.																

* 'Roy. Soc. Proc.,' vol. 60, p. 492.

Now this is precisely what we might expect; the mares belonged to a class, of which we are not certain whether their daughters have or have not recorded fecundity. The mean fecundity is therefore decreased and the variability increased. Add to this group 2400 mares, all of which had had their daughters' fertility recorded, and we find for 3909 mares, $M_\phi = \cdot 6345$ and $\sigma_\phi = \cdot 1910$, *i.e.*, the mean fecundity ascends and the variability falls. Illustration of this law will be found in the following two groups:—

	M_ϕ.	σ_ϕ.
1200 mares	·6337	·1888
1200 dams	·6525	·1643

Thus we send up the mean fertility and lower the variability by separating into two groups the pedigree of one which has a longer record. This is precisely in accordance with the theory already developed. Our mean fecundity and variability for brood-mares may be considered as constant characters, and variations in their values beyond their probable errors due to conscious or unconscious selection in the record itself, or in our extracting from it.

The reader will notice at once, if he turns to the diagram of the above frequency, (i.) that there is a small hump at (a) of no practical importance, and a larger one at (q), perfect fertility being fairly frequent with only four coverings, and there being from the arithmetical processes involved a bias towards (q) as compared with (p). (ii.) The distribution of frequency, although somewhat ragged, is quite clearly not normal, but of the character which in other papers I have called *skew*. Were there any occasion, it would be easy to fit it with one of my skew curves. To mark how (i.) will disappear and (ii.) become still more apparent, I have placed on the diagram the frequency distribution for 2000 mares reduced to the same scale.

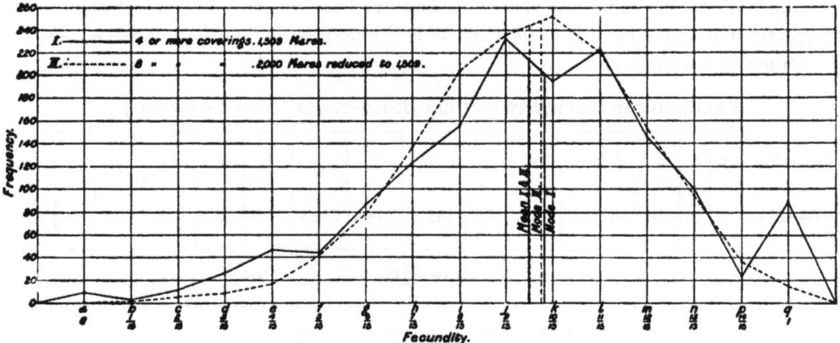

VARIATION in Fecundity of 2000 Brood-mares (Eight Coverings).

Fecundity	a.	b.	c.	d.	e.	f.	g.	h.	i.	j.	k.	l.	m.	n.	p.	q.
Frequency	0	2	7·5	11·5	21·5	55	104·5	182	271·5	315	337	293·5	204	127	49	19

Total, 2000. $M_4 = {\cdot}6330$. $\sigma_4 = {\cdot}1568$.

Thus, making the minimum number of coverings 8 instead of 4, has removed the terminal humps, zero fecundity is now unknown, and perfect fecundity very rare. We have reached a smooth skew frequency distribution ; we see fecundity as a continuous character obeying the usual laws of variation.* The mean fecundity in the two cases is sensibly the same, ·633, but owing to the fact that we have made a selection of a limited group in the second case, the variability is considerably decreased.

The average apparent fertility of brood-mares, 6·515, must not be confused with their average real fertility, for, as we have seen, we have in many cases not a complete record of their stud-life, or such a full record has not been used (*e.g.*, in case of mares still at the stud, but having been already covered four or more times). Its 50 per cent. variation shows that an apparent fertility of 9 to 12 is not infrequent. The average number of coverings being 10 and more, it will be seen that the records of between 50,000 and 60,000 coverings have been dealt with to form our mare and sire alphabets. The large variability in the number of coverings shows that 15 to 20 coverings will not be infrequent, and cases of 26 actually occurred. Lastly, we have the correlation between fertility and the number of coverings, high as might be supposed, for a high apparent fertility could only be exhibited by many coverings. Although a low apparent fertility might correspond to any number of coverings, still, in practice a sterile mare will not be sent indefinitely to the sire. The correlation between the number of coverings and the fecundity is small and *negative* (− ·0572). This follows from the principle that, fertility being the same, a high number of coverings reduces the fecundity, and this factor is more potent than the high correlation of fertility and the number of coverings.

(ii.) Table XII. exhibits the correlation of 1200 mares and their dams with regard to fecundity. Here the more fertile dams are weighted with their fertility, and at least four coverings were required of each mare. If the subscript m refers to mare, and d to dam, we find :

* The actual equation to the curve referred to the mode ·6531 as origin, the axis of x being positive towards perfect fecundity, and the unit of x being 1/15 is:

$$y = 342{\cdot}187\,(1 + x/47{\cdot}1358)^{82{\cdot}6261}\,(1 - x/12{\cdot}1106)^{21{\cdot}2291}.$$

The fit will be found to be very satisfactory.

$$M_m = \cdot 6337, \qquad M_d = \cdot 6525,$$
$$\sigma_m = \cdot 1888, \qquad \sigma_d = \cdot 1643,$$
$$r_{md} = \cdot 0831.$$

The coefficient of regression = ·0945.

The probable error of the correlation is ·0193 and of the regression* ·0195. Thus these quantities are four to five times their probable errors, and we conclude that fecundity is certainly inherited.

The intensity is far below that suggested by the law of ancestral heredity, but it nevertheless exists. Its lowness is probably due to the fictitious character of the fecundity owing to the causes indicated on pp. 298–9. An attempt must now be made to eliminate some of the factors disguising the fecundity, but to do so is by no means so easy as in the case of fertility in man.

(iii.) My first idea was that by taking a higher limit to the number of coverings a closer approach might be obtained to the true, *i.e.*, the inherited fecundity. Accordingly Table XIII. was formed for the correlation of 1000 mares and their dams, when the minimum number of coverings was eight. But I did not recognise that this would give far greater weight in the Table to the older mares, and that accordingly causes (*d*) and (*i*) of p. 298 would now play a much larger part in disguising the true fecundity than before. There appears to be no limit to the number of times a famous old mare may go to the stallion when there is very small hope of any offspring.

Table XIII. gives us the following results:

$$M_m = \cdot 6300, \qquad M_d = \cdot 6360,$$
$$\sigma_m = \cdot 1633, \qquad \sigma_d = \cdot 1500,$$
$$r_{md} = \cdot 0652.$$

The coefficient of regression = ·0708.

The probable error of the correlation is ·0212, and of the regression ·0213, both less than a third of the observed values. We should again conclude from this result that fecundity is inherited, although it offers less strong evidence than the previous case. The influence of selection† is at once apparent in the great reduction of the variabilities. The fact that we are throwing the determination of fecundity more on to the old age period of life appears from the reduced mean fecundities. I attribute the reduction in the fecundity-correlation to this source, *i.e.*, the very diverse treatment which old mares receive at the hands of different breeders.

(iv.) I made another attempt to remove screening causes by taking 1200 more

* PEARSON and FILON: 'Phil. Trans.,' A, vol. 191, p. 214.

† The effect of such a selection as the above in reducing correlation is dealt with in my paper on the influence of selection on correlation.

mares, not identical with the series in (ii.)* and working out their dams' records most carefully, rejecting any cases in which the breeder was clearly sending the mare to the stallion long after it was obvious (*post facto*) that she was sterile. In this case four coverings were retained as a minimum, and the results are given in Table XIV. We find:

$$M_m = \cdot 6369, \qquad M_d = \cdot 6616,$$
$$\sigma_m = \cdot 1885, \qquad \sigma_d = \cdot 1604,$$
$$r_{md} = \cdot 0995.$$

The coefficient of regression $= \cdot 1169$.

The probable error of the correlation is ·0193, and of the regression ·0194; the correlation is accordingly more than five, and the regression more than six times its probable error. We conclude that fecundity is most certainly inherited. The regression found is, however, only about two-fifths of what is required by the law of ancestral heredity,

(v.) It has been suggested that fertility or fecundity might alternate in *two* generations; when the offspring are numerous their offspring might have less fertile or fecund offspring. I do not see how this would be possible without its exercising an influence on the correlation of two generations, for we must come to one fertile followed by an infertile generation. But I had made preparations in my alphabet of mares for testing the correlation between mares and their granddams, and I went on to the construction of a table, although the results for mares and their dams showed me that whatever result might be reached, it would be within the probable error of the observations. I reached this conclusion in the following manner: If we go back one generation we introduce, owing to the nature of the record, so much fictitious correlation and so much in-and-in breeding that the coefficient of inheritance is reduced to two-fifths or less of what its value should be according to the law of ancestral heredity. In going back two generations we come to fewer mares, to more in-and-in breeding, and to just the type of famous old mare, whose breeder kept her at the stud long after she was sterile. I expected accordingly a great and artificial fall in the fecundity of granddams and a double drop, something like $\frac{2}{5} \times \frac{2}{5}$, in the value of the regression as indicated by the law of ancestral heredity. This would reduce the apparent regression to about $\frac{2}{5} \times \frac{2}{5}$ of ·15, or to about ·025, say, a value about equal to the probable error of the table. The results actually reached are given in Table XV., and we find, if the subscript g refer to granddam:

$$M_m = \cdot 6345, \qquad M_g = \cdot 6232,$$
$$\sigma_m = \cdot 2040, \qquad \sigma_g = \cdot 1687,$$
$$r_{mg} = \cdot 0169.$$

The coefficient of regression $= \cdot 0204$.

* In the first series the mares' names run from A to G; in the second from G to M, with 300 additions made to the A to G series, while I was completing my alphabet.

The probable error of the correlation is ·0213, and of the regression ·0213. Thus these results are not significant in themselves, but they are exactly what we might expect on the above hypothesis. Taken with the other five tables which we have worked out for the inheritance of fecundity, they are significant, for every one of them gives a *positive* correlation, however small it be, and thus adds to the accumulated evidence that fecundity is a heritable character.

(vi.) It remains to test our results by the theory developed on pp. 269 *et. seq.* But a difficulty comes in here. Turning to (xviii.) and (xix.) on p. 268, we cannot feel justified in putting $M_1 = M_2$, for there is a secular difference in the fecundity of mares and dams, owing to the fecundity of the older brood-mares being based on a longer period and liable to the disturbing causes so markedly manifest in the correlation of mares and granddams (see my remarks, p. 305). If we combine (xviii.) and (xix.) we find

$$M''_2 - M'_2 = r\frac{\sigma_2}{\sigma_1}(M''_1 - M'_1).$$

Now r is small, and it will accordingly be legitimate to put $M'_1 = M'_2$ and $\sigma_1 = \sigma_2$ on the right, we have then

$$(M''_2 - rM''_1)/(1 - r) = M'_2.$$

From this we deduce for the results in (ii.) on p. 304

$$M'_2 = \cdot 6321.$$

Turning now to (xix.), it may be written

$$M''_1 = M'_1\left\{1 + \frac{\sigma_1^2}{M_1^2}\left(\frac{\sigma'^2_1/\sigma_1^2}{1 + \sigma_1^2/(\rho^2 M_1^2)}\right)\right\}.$$

The second term in the curled brackets is small, and in it we may put to a first approximation $\sigma'_1 = \sigma_1 = \sigma_2$ and $M_1 = M'_2$. We then have

$$M''_1 = M'_1\left\{1 + \left(\frac{\cdot 1888}{\cdot 6321}\right)^2\left(\frac{1}{1 + \left(\frac{\cdot 1888}{\cdot 6321}\right)^2\frac{1}{(\cdot 5132)^2}}\right)\right\},$$

or,

$$M''_1 = M'_1 \times 1\cdot 0666.$$

Substituting the value of M''_1 we find

$$M'_1 = \cdot 6118.$$

We thus see a difference in the fecundities of the unweighted dams and unweighted mares of ·6118 and ·6321, or about 2 foals more in the hundred appear to survive in the later generation. This is very probably due to the causes already indicated as affecting the apparent fecundity of the older mares (see p. 298). The influence of reproductive selection changes these quantities to ·6337 in the case of the daughters, and to the *apparent* high fecundity of ·6525 in the case of the dams.

We can now find σ_2 to a second approximation by aid of (xxi.). In the small term multiplied by r, we put $\sigma_1 = \sigma_2 = \sigma''_2$. Hence we find

$$\sigma_2^2 = \sigma''^2_2 + r^2(\sigma''^2_2 - \sigma''^2_1),$$

and deduce, on substituting the numerical values,

$$\sigma_2 = \cdot 1896,$$

or is scarcely different from σ''_2. We accordingly conclude that we may quite reasonably assume the variability of the mares to represent the variability of the mares without reproductive selection, but the effect of weighting the dams with their fertility is to reduce the variability of the dams from about ·1896, if there be no secular change, to an apparent value as low as ·1643.

The same formulæ applied to the slightly better results in (iv.) on p. 305 give us:

$$M'_1 = \cdot 6205, \text{ and } M'_2 = \cdot 6342.$$

If we pass back from M'_1 and M'_2 to M_1 and M_2 we find:

	First case.	Second case.
M_1	·5460	·5567
M_2	·6266	·6278

If these results be considered as valid, we notice a remarkable difference between the fecundity of the younger and elder generation. While the crude results on pp. 304 and 305 might lead us on first examination to suppose the elder generation more fecund than the younger, these results show us that it is distinctly less so. The greater part of the difference, however, is due, not to a secular change, but to the causes we have so often referred to as weakening the fecundity recorded for the older mares. At the same time the whole system of breeding is so artificial that we may well doubt whether our equations (i.) and (v.) can be legitimately applied. For the chance of a mare getting into the stud-book as a dam, *i.e.*, having daughters at the stud, depends less on her fertility than on the degree of fashion in her stock. Thus the record weighting with fertility is hardly a probable hypothesis, and the values just given for M_1 are, I suspect, much below what they should be. For the above reason I have not proceeded to consider the changes in variability connoted by (ii.) and (xxii.). As I have made no attempt to form a correlation table for mares and dams in which the dam would have only one daughter to her record, I cannot make any plausible guess at the real magnitude of the cubic summation term in (xxii.). Apart, however, from the numerical application of these variation formulæ to a somewhat doubtful case, we see in these formulæ the theoretical basis for the observed fact that the fecundity of mothers is far less variable than that of daughters. It is really only an apparent divergence, due to the fact that the mothers have been weighted with their fertility; this, while it increases the apparent mean of their fecundity, reduces its apparent variability.

(17.) *On the Inheritance of Fecundity in the Brood-mare through the Male Line.*

For the thoroughbred horse this problem is fairly easily answered by investigating whether mares related to the same stallion have any correlation between their fecundities. The two cases I have selected are: (i.) "Sisters," daughters of the same sire, but in general not of the same mare; and then (ii.) "Nieces" and "Aunts," or daughters of a sire and the daughters of his sire. As we have only 760 sires and nearly 5000 mares, the daughters or aunts fall into rather large arrays, and we are compelled to use the methods discussed in Proposition IV., A and B. Even so the arithmetical work for a correlation based on the index of sires was far more laborious than for one based on the index of mares.

(i.) *To find the Correlation between Half-Sisters, Daughters of the same Sire.*

Here we have to use formulæ (xxiii.), (xxv.), and (xxvi.) of pp. 272–273. In order to do this a table was formed of the mean fecundity M of the array of sisters due to each sire, and of $\frac{1}{2}n(n-1)$, the number of pairs of sisters in each array. Then the products $\frac{1}{2}n(n-1)$ M and $\frac{1}{2}n(n-1)M^2$ were formed, and the numerator of (xxiii.), or σ_a^2, calculated by adding up for all the 760 sires. The result gave:

$$\sigma_a^2 = \cdot 6655167,$$

where the unit is the fecundity group element of 1/15. The number of pairs of sisters dealt with was 54,305. The denominator $\sigma_0^2(1-\rho^2)+\sigma_a^2$ is not so easily ascertained. σ_0 is the standard deviation of all the series of mares who are sisters without weighting; $\sigma_0\sqrt{(1-\rho^2)}$ is the standard deviation of an array of sisters, or if the regression be not linear, the mean of such standard deviations for all arrays, or rather its square is the mean of the squares of such standard deviations; ρ is the correlation between a patent character in the daughter and a purely latent character in the sire, and cannot therefore be found directly.

In order to get an appreciation of the standard deviation of an array of sisters—it being practically impossible to work out these quantities for 760 arrays—I selected twenty sires having fairly large arrays of daughters, and reached the following results:

TABLE of Arrays of Mares, which are Half-Sisters.

Sire.	No. of mares.	Mean fertility.	S.D. of array.
Speculum	76	9·697	2·989
Sterling	52	10·750	2·545
Scottish Chief	67	9·201	3·176
Newminster	64	8·875	2·497
Parmesan	37	9·708	3·076
Macaroni	81	10·210	2·770
King Tom	53	9·689	2·748
Lord Clifden	41	9·878	2·086
Hermit	79	9·437	3·003
Blair Athol	87	9·057	2·752
Lord Lyon	32	9·125	3·314
The Duke	35	9·186	2·474
Doncaster	37	9·297	3·021
Adventurer	58	10·466	2·621
Cathedral	43	9·267	2·847
Rosicrucian	59	10·932	3·094
Stockwell	80	9·131	2·093
Rataplan	40	8·222	2·201
Y. Melbourne	55	9·064	2·301
Thormanby	41	9·951	1·651
Totals	1117	191·143	53·259
Mean	55·85	9·55715	2·66295
Ditto in actual units*	..	·6371	·1775

I next took the mean and standard deviation of the 1117 mares to obtain σ_0. The mean fecundity was now found to be 9·5685 and $\sigma_0 = 2·7824$, or in actual units ·6379 and ·1855. Clearly only about $\frac{1}{8}$ per cent. difference is made whether we take the mean fecundity of the 1117 mares, or the mean of the unweighted means of the twenty arrays. Knowing σ_0 and $\sigma_0\sqrt{1-\rho^2}$ we can now find ρ. We have almost at once

$$\rho = ·2900.$$

This is probably the first determination of a coefficient of inheritance between a latent character in one sex and a patent character in the other sex. We see that it has almost exactly the value required (·3) by the law of ancestral heredity, or we conclude, *mares inherit from their sires a fecundity governed closely by the law of ancestral heredity.*

If the reader asks why is not the intensity reduced in this case in the same manner that we find it reduced in the case of the inheritance from the dam, the reply is :

(i.) In the case of the dam and mare, *both* quantities to be correlated are liable to fictitious values. In the case of sire and mare, we deal with only one.

* A fecundity unit is taken to be 1/15, for this is the unit of grouping.

(ii.) The influence of fictitious values has been shown on pp. 276–277 to chiefly affect the coefficient of correlation and not the standard deviation.

Now the present result is based solely on the calculation of standard deviations, or on the variability of fecundity as a whole and in arrays. It is accordingly not influenced to nearly the same extent by the existence of fictitious values. Could we calculate the variability of the arrays of daughters due to individual mares, we should probably get a better result for inheritance in the female line.*

The above result is so satisfactory that I have little doubt that we have determined a very good value for $\sigma_0\sqrt{1-\rho^2}$. Substituting it we find for the correlation between half-sisters:

$$r = \frac{\cdot 66552}{7\cdot 09130 + \cdot 66552} = \cdot 0858.$$

The law of ancestral heredity gives for *half*-sisters $r = \cdot 2$, and $\frac{2}{5}$ of this $= \cdot 08$.

Thus we see that the collateral heredity between half-sisters, daughters of the same sire, is quite sensible, and is almost what we might have predicted would be the result, if we supposed correlation to be weakened, as in the previous cases, to $\frac{2}{5}$ of its value by fictitious records.

It is worth while to consider the amount of fictitious fecundity suggested by the reduction factor $\frac{2}{5}$. We have only to suppose the n_1/N of our p. 277 to be $\frac{2}{5}$. Now we may well assume the chance of a fictitious fecundity being recorded to be the same for either one of a pair of sisters; hence we shall have $p = q$, and therefore, from the result on p. 276, we find $(p-1)^2/p^2 = \frac{2}{5}$. This gives us $(p-1)/p = \sqrt{\cdot 4}$, and $(n_1 + n_3)/N$ the fraction without fictitious values $= (p-1)/p = \cdot 6325$. Thus in order to introduce the reduction factor of $\frac{2}{5}$ by the occurrence of fictitious values of the fecundity, we should have to suppose about 37 per cent. of fictitious values to occur. This is, of course, a sort of average; many values will probably be only partially fictitious, *i.e.*, will to some extent approximate to their real values. Considering the very artificial character of the thoroughbred brood-mare, and the uncertainty of her treatment by breeders, this does not seem such an immense percentage that it would force us to the conclusion that the law of ancestral heredity cannot be true for the inheritance of fecundity.

(ii.) *To find the Correlation in Fecundity between the Sisters of a Sire and his Daughters.*

What we want is really the correlation between aunts and nieces, but they

* The standard deviations for the arrays of mares in Table XII. were indeed worked out for the twelve cases of dams from *e* to *q*. The mean of these cases was sensibly the same whether the simple mean, or the mean weighted with the numbers in the array was taken, and equalled 2·8091 or ·1823. This is $\sigma_0(1-r^2)^{1/2}$. But by p. 48, $\sigma_0 = \cdot 1888$, whence we deduce $r = \cdot 1375$, and the regression equals ·1581. Thus we have found a substantially larger value for r than that on p. 304 by dealing with variabilities, and not direct correlations. This gives additional evidence, if any were needed, of the inheritance of fecundity.

TABLE XVI.—Correlation Table for Weighted M)11.)

		c		*d*		*e*		*f*		*q*
		3–3·5.	3·5–4.	4–4·5.	4·5–5.	5–5·5.	5·5–6.	6–6·5.	15·5–16.	16–16·5.
Stallion's sire. *e*	4·5–5	..	..	..	..	..	..	..	..	..
	5–5·5	..	..	..	..	..	..	..		
f	5·5–6	..	..	..	..	..	..	..	..	..
	6–6·5	..	..	..	..	..	..	..	..	..
g	6·5–7	..	..	..	..	..	·5	·5	..	..
	7–7·5	..	..	..	..	..	·5	·5	..	..
h	7·5–8	..	..	..	..	..	..	..	2	2
	8–8·5	..	..	..	..	..	..	..	2	2
i	8·5–9	..	..	..	..	..	..	..	..	..
	9–9·5	9	9	..	20	49	37·5	8·5	20	20
j	9·5–10	3	3	..	66	66	80·5	64·5	32	32
	10–10·5	..	..	..	107	107	74·5	56	114	114
k	10·5–11	..	3	3	19·5	40	243	10·5	73·5	73·5
	11–11·5	29·5	29·5	..	6·75	35·75	29·75	·75	5·5	5·5
l	11·5–12	3·5	3·5	..	·75	6·25	13·25	7·75	10	10
	12–12·5	..	..	..	..	..	..	..	1	1
m	12·5–13	..	..	..	..	..	..	..	..	
	13–13·5	..	..	..	..	..	..	..	3	3
n	13·5–14	..	..	..	·25	·25	..	..	..	..
	14–14·5	..	..	..	·25	·25	..	..	..	..
p	14·5–15	..	..	..	..	..	..	..	..	..
	15–15·5	..	..	..	..	..	..	..		
q	15·5–16	..	..	..	..	..	..	..	..	..
	16–16·5	..	..	..	..	..	..	..	..	..
Totals		45	48	3	220·5	304·5	479·5	149	263	263

are not "aunts" and "nieces" in the human sense, for the aunts are only half-sisters of the sire. By a process similar to that on pp. 408 and 409 of my paper on the "Law of Ancestral Heredity,"* I deduce that the correlation between a sire's sisters and daughters ought to be ·05, and not ·15 as in the case of Man. If this be weakened down to the $\frac{2}{5}$ of previous results, we should not expect a result differing much from ·02. As the variability of the elder generation is always less than that of the younger, we ought to expect a coefficient of regression of about this value. The theory used will be that of p. 273 of the theoretical part of this paper. The weighted mean fecundity found for the arrays of aunts and nieces was as follows :—

	Without grouping.	With grouping.
Arrays of aunts	·6195	·6199
Arrays of nieces	·6346	·6338

The grouping was done in fecundity units of $\frac{1}{3}$, *i.e.*, 1/30 change in fecundity. The agreement may accordingly be considered very good. The "aunts" are the daughters of the older sires, who owing to in-and-in breeding form a comparatively small group, and are the sires of mares belonging to the older period, whose fecundity is much weakened by causes already referred to. Their mean fecundity is slightly less than that of granddams, given on p. 305, while the mean fecundity of their nieces agrees well with that for the corresponding group of mares.

The method of grouping being adopted, a correlation table was formed for the mean fecundities of arrays of mares, daughters of a sire, and of arrays of mares, daughters of his sire. This is Table XVI. Here each mean is weighted with the number of pairs of aunts and nieces in the two arrays, *i.e.*, the extent of the data on which it is based. It represents accordingly 138,424 pairs of aunts and nieces. The following results were obtained, corresponding to 687 pairs of sires :—

Sire's Sire.	Sire.
$M_a = \cdot 6199.$	$M_{a'} = \cdot 6338.$
$\bar{\sigma}_a = \cdot 04344.$	$\bar{\sigma}_{a'} = \cdot 07609.$

$$R = \cdot 1174.$$

It will be at once noticed how much more variable are the array-means for the sire than for the sire's sire. The means of many of the sire's arrays are based upon small numbers, which would have been selected out, if we had gone to another generation as in the case of the sire's sire.

It will clearly not be legitimate in this case to put $\bar{\sigma}'_a = \bar{\sigma}_a$ as suggested on p. 274. There is probably no secular change of importance here, but the sire's sire requiring

* 'Roy. Soc. Proc.,' vol. 62.

three generations from the record is really more stringently selected than the sire with only two. We can now form $\bar{\sigma}$ and $\bar{\sigma}'$ by (xxviii.) and (xxix.), if we adopt suitable values of $\bar{\sigma}_0$ and $\bar{\sigma}'_0$, ρ, as we have seen, may with high probability be put equal to ·3 (p. 309). σ_0 for groups of daughters, on p. 309, is given as ·1855, but since this certainly included a fair number of what are now aunts, it must be somewhat too low for $\bar{\sigma}'_0$. We can well put $\bar{\sigma}'_0$ equal to the ·1888 of the mares on p. 304. $\bar{\sigma}_0$ for aunts cannot be as low as the standard-deviation of dams on that page, as many of the aunts may never appear in the record as granddams,* *i.e.*, they are less stringently selected. The mean of the two results for mares and dams may, perhaps, be taken as a close enough approximation for our present purpose, or $\bar{\sigma}_0 = ·1765$. We then deduce

$$\bar{\sigma} = ·1739, \qquad \bar{\sigma}' = ·1955.$$

If we compare the results now found with those for sisters cited on pp. 308 and 309, we find :—

	"Aunts."	"Sisters."	"Nieces."
M	·6199	·6371	·6338
σ_a	·0434	·0544	·0761
σ'	·1739	·1855	·1955

The accordances and divergences are much what we might expect, except in the case of σ_a. We should, *à priori*, have expected "sisters" to have approached nieces more nearly than aunts. The work has been gone carefully through, but I have not succeeded in finding any error. In the "nieces," of course, the weighting of an outlying fecundity-mean due to a sire with but few daughters, may still be large, if his sire have numerous daughters; this cannot occur in the case of "sisters," as the weighting depends only on the number in the array. The like heavy weighting cannot usually occur in the case of "aunts," for they are, as a rule (owing to selection to the third generation) daughters of old and famous sires, with plenty of material for basing averages upon. We do not get *many* "nieces" attached to "aunts," who are not daughters of famous sires. Such is probably the source of divergence in σ_a between nieces and sisters.

Using formula (xxviii.), on p. 274, we find

$$r' = ·0114,$$

and for the regression coefficient ·0128.

This value is much below the ·05 of the law of ancestral heredity, and below the reduced value ·02, which we might have expected to reach. Still, it again shows

* Every dam appears as a granddam, otherwise the fecundity of the daughter could not have been found.

positive correlation, and we may conclude that the patent character in the daughter is inherited latently through the male line.*

But there is another and far more significant method of looking at this result, namely, by considering the meaning of R on p. 274. We may treat the fecundity of daughters as really a character of the sire, and their mean fecundity as a measure of a latent character in him. R is then the correlation between a latent character in both a stallion and his sire, and we see that it is sensibly inherited for R = ·1174. To compare with the law of ancestral heredity, we must use the coefficient of regression, for the stallions are much more variable than their sires. We find

Regression of stallion on sire = ·2056,

which carries us a long way in the direction indicated by that law. Thus it is extremely probable that this law of inheritance applies not only to the inheritance of a patent character, or of a character latent in one sex and transmitted to a second, but also to the inheritance of a character latent both in the transmitter and receiver. The present method accordingly seems applicable to the inheritance of a character latent in two individuals, if we take the mean of the character, when patent in the offspring, as a measure of its strength in the individual in whom it is latent. If l_1 be the measure of a latent character in a parent, then the offspring will have a mean value $ql_1 + c_1$ of this character, where q is the coefficient of parental regression and c_1 a constant. If l_2 be the measure of the same latent character in a relative, then the offspring in this case will have $ql_2 + c_2$ of the character. But the correlation of l_1 and l_2 will be identical with that of $ql_1 + c_1$ and $ql_2 + c_2$, as I have shown elsewhere.† Thus the mean of the patent character in the offspring may be used to measure the correlation between latent characters in their parents.

To sum up our results for thoroughbred mares, we conclude that their fecundity, notwithstanding the imperfections and difficulties of the record, has been demonstrated to be inherited, and this, both through the male and female line, so far as we can judge, with an equal intensity. The *apparent* value of this intensity, except in the case of latent characters, is much below that required by the law of ancestral heredity, roughly, perhaps, 2/5 of that value; but there is considerable reason to think that this reduction may take place owing to the presence of fictitious values in the record arising from the peculiar circumstances under which thoroughbred horses are reared and bred. These fictitious values would hardly influence the means and variability of arrays like they must do the relationship between pairs of individuals. Hence, when we deal with such means and variabilities as in the cases on pp. 309 and 313, we find a much closer approach to the law of ancestral heredity. Fecundity is certainly inherited; that it is inherited according to the Galtonian law

* As a matter of fact, this conclusion is stronger than it appears here, for the correlation between nieces and aunts was worked out, without grouping, for fourteen distinct series, and in *thirteen* of them was found to be sensibly positive; in the fourteenth it was found to have an insignificant negative value.

† "On the Reconstruction of the Stature of Prehistoric Races," 'Phil. Trans.,' A, vol. 192, p. 183.

is not demonstrated, but may be treated as probable until the results of further investigations—preferably by breeding experiments instituted for this very purpose—are available.

(18.) *Conclusion.*—The investigations of this memoir have been to some extent obscure and difficult, but the general result is beyond question.

Fertility and fecundity, as shown by investigations on mankind and on the thoroughbred horse, are inherited characters.

The laws of inheritance of these characters are with considerable probability those already developed in my memoir on the Law of Ancestral Heredity for the inheritance of directly measurable organic characters.

In the course of the work it has been shown how a numerical measure may be obtained for the inheritance of a character by one sex from the other, when it is patent in the former and latent in the latter. Fertility and fecundity purely latent in the male (in the sense here used) are shown to be transferred by him from his mother to his daughter. Thus DARWIN'S views with regard to the transmission through one sex of a character peculiar to the other are given a quantitative corroboration.*

When we turn from these points to their weight and importance for the theory of evolution, we are at once encountered by all the wide-reaching principles which flow from the demonstration that genetic (reproductive) selection is a true factor of development. Let us look at these a little more closely.

If natural selection were to be absolutely suspended, *i.e.*, if there were no differential death-rate at all, then development would not for a moment cease. Not only is fertility inherited, but there can be small doubt that it is closely correlated with all sorts of organic characters; thus the inheritance of fertility marks, the moment natural selection is suspended, a progressive change in a great variety of organic characters. Without a differential death-rate the most fertile will form in every generation a larger and larger percentage of the whole population. There are very few characters which may not be supposed to be more or less directly correlated with fertility, and in reproductive selection we see a cause of progressive change continuously at work.† There is, so to speak, in every species an innate tendency to progressive change, quantitatively measurable by determining the correlation coefficients between fertility and organic characters, and between fertility in the parents and in the offspring. This "innate tendency" is no mysterious "force" causing evolution to take place in a pre-ordained direction; it is simply a part of the physical organisation of the individual, which does not leave fertility independent of

* The method is perfectly general, and a value can always be found for the intensity of transmission of a sexual character through the opposite sex. We could obtain, for example, a numerical measure of the manner in which a bull transmits good milking qualities to its offspring.

† I have endeavoured to show ('Roy. Soc. Proc.,' vol. 59, p. 303), that fertility is correlated with stature in woman. I hope later to return to the correlation of fertility and physique.

physique and organic relationship, or leave these characters uncontrolled by the principle of heredity. It seems to me, therefore, that the results of this memoir force on us some modification of current views of evolution. The suspension of natural selection does not denote either the regression of a race to past types, as the supporters of panmixia suggest, or the permanence of the existing type, as others have believed. It really denotes full play to genetic or reproductive selection, which will progressively develop the race in a manner which can be quantitatively predicted when once we know the numerical constants which define the characters of a race and their relation to racial fertility. In other words, natural selection must not be looked upon as moulding an otherwise permanent or stable type; it is occupied with checking, guiding, and otherwise controlling a progressive tendency to change.

So soon as a species is placed under a novel environment, either artificially or naturally, the equilibrium is disturbed, and it will begin to progress in the manner indicated by genetic (reproductive) selection, until this progress is checked by the development of characters in a manner or to an extent which is inconsistent with fitness to survive in the new surroundings. Within a very few generations a novel environment, sympathetic so to speak to the progressive tendency indicated by reproductive selection, produces the suitable variations without the assistance of natural selection. It seems to me that this principle ought to be borne in mind when, in laboratory experiments or in artificial breeding, natural selection is wholly or largely suspended, or again is altered in type; the species dealt with is unlikely to remain constant for several generations, but will develop in the direction indicated by genetic selection. Further, when stable types of life like the English sparrow are taken to America, or the English rabbit to Australia, where initially they fill a more or less vacant field among living forms, and natural selection is in part suspended, we should expect in a few generations a considerable divergence in type.* The converse aspect of the problem is also of great importance; namely, the natural selection of physical characters must tend to indirectly modify fertility and fecundity, if these be correlated with those characters. Variations in the fertility of local races need not be looked upon as due directly to environment, but may arise from the selection of characters correlated with fertility, combined with the law that fertility is itself an inherited character.

Lastly, the inheritance of fertility involves the "acceleration" of fertility; a race, natural selection being suspended, tends not only to increase but to increase at an increasing rate. This principle is again full of meaning, not only for the study of the manner in which lower types of life rapidly expand under changed environment, but also for the problems set to those philosophers who may desire that the most social and not the most fertile type of citizen may predominate in our modern civilised communities, where the state and public opinion to a greater or less extent hinder natural selection from playing the great part it does in wild life.

* It would be interesting to know whether the size or frequency of the litter of the Australian rabbit is greater than that of the English.

Table I.—Correlation of Fertility of a Woman and of her Daughter. Marriages completed, or having lasted at least 15 years in Daughter's case only. 4418 cases. Fertility of Mother weighted. (See p. 281.)

Daughter's fertility.	Mother's fertility.																			Totals.
	1	2	3	4	5	6	7	8	9	10	11	12	13	14	15	16	17	18	19	
0	28	42	70	98	109	90	71	81	67	52	29	22	3	3	4	3	1	..	2	775
1	22	56	78	104	80	70	64	51	39	31	20	19	1	4	4	4	..	..	..	647
2	21	39	63	65	78	73	47	47	40	29	20	11	2	2	..	2	..	..	2	541
3	20	35	44	53	53	75	57	57	33	37	21	13	10	5	1	2	..	..	..	516
4	17	27	58	52	69	62	53	46	41	24	16	12	11	6	1	..	..	..	..	495
5	16	28	26	37	52	55	43	32	40	25	12	14	5	2	1	..	..	..	..	388
6	13	20	26	42	41	39	30	38	28	19	5	8	3	2	1	..	..	..	1	316
7	3	14	33	23	33	31	24	32	26	14	10	9	5	1	1	1	..	..	..	260
8	6	9	15	16	25	16	22	15	20	11	6	6	6	..	1	3	..	..	..	177
9	4	7	13	13	13	10	14	12	10	8	5	8	3	1	..	1	..	..	..	122
10	..	13	2	8	10	7	13	7	8	6	1	3	1	1	..	..	..	..	..	80
11	7	2	6	9	7	1	4	4	4	3	2	1	1	1	..	..	..	..	..	52
12	..	1	2	1	4	4	..	2	1	.	2	1	2	..	..	..	..	..	..	20
13	..	..	2	1	1	2	4	2	1	2	1	1	1	..	..	..	..	..	..	18
14	..	1	..	1	..	2	1	..	..	1	..	..	..	..	..	..	..	..	..	6
15	..	..	..	.	..	.	.	..	..	1	..	..	..	..	..	..	..	..	..	1
16	1	2	1	..	..	..	..	..	..	..	..	..	..	.	..	..	..	..	..	4
Totals ..	158	296	439	523	575	537	447	426	358	263	150	128	54	28	14	16	1	0	5	4418

Table II.—Correlation between the Fertility of a Woman and that of her Daughter. No weight to the Fertility of the Mother. Marriage completed or having lasted at least 15 years in the Daughter's case only. 1000 cases from the Peerage. (See p. 282.)

Daughter's fertility.	Mother's fertility.															Totals.
	1	2	3	4	5	6	7	8	9	10	11	12	13	14	15	
0	3	6	18	15	15	12	11	5	6	3	1	1	..	..	1	97
1	6	13	16	23	17	16	10	11	6	7	2	..	..	..	..	127
2	6	7	17	22	17	15	9	14	7	4	2	..	..	..	..	120
3	8	12	18	20	13	22	12	12	10	6	1	3	..	..	..	137
4	5	6	16	15	21	19	22	8	12	5	3	4	1	..	..	137
5	4	7	4	12	16	16	17	11	13	7	6	3	..	..	..	116
6	1	9	8	12	13	9	13	9	6	6	2	4	1	..	..	93
7	1	2	5	6	9	9	5	7	5	6	2	2	2	..	1	62
8	..	1	6	2	8	6	5	5	5	3	2	.	2	..	..	45
9	1	1	1	5	4	4	6	1	1	1	2	.	.	..	..	27
10	..	3	1	2	2	2	..	4	3	2	..	1	.	1	..	21
11	..	..	1	2	1	1	2	..	..	..	1	1	.	..	..	9
12	..	..	.	.	2	.	..	..	..	..	..	.	2	..	..	4
13	..	..	..	..	..	1	2	..	..	..	..	..	1	..	1	5
Totals ..	35	67	111	136	138	132	114	87	74	50	24	19	9	1	3	1000

TABLE III.—Correlation between the Fertility of a Woman and that of her Daughter. No weight to Fertility of the Mother. Marriage completed or having lasted at least 15 years in Daughter's case only. 1000 cases, largely "Heiresses" from the Landed Gentry. (See p. 283.)

Daughter's fertility.	Mother's fertility.																					Totals.
	1	2	3	4	5	6	7	8	9	10	11	12	13	14	15	16	17	18	19	20	21	
0	20	12	9	6	18	16	14	15	7	8	3	5	1	..	..	..	..	..	..	..	..	134
1	16	10	17	23	12	11	11	7	3	7	4	..	..	..	..	..	..	..	..	..	..	121
2	6	11	11	9	18	12	18	9	2	2	1	2	..	1	..	..	..	..	..	..	..	102
3	15	9	8	13	19	10	6	10	4	4	1	..	..	1	..	..	..	..	..	..	..	100
4	23	8	7	17	11	9	9	11	7	3	2	3	2	..	..	..	..	..	..	..	..	112
5	17	8	8	16	13	10	12	10	6	6	2	..	3	..	..	..	..	..	..	..	..	111
6	12	3	2	15	9	8	8	7	3	4	5	1	..	2	1	..	..	..	..	..	..	80
7	14	3	11	3	7	7	7	9	4	3	3	..	3	1	1	..	..	..	..	..	..	76
8	9	5	4	6	6	6	5	2	3	2	2	2	..	1	..	2	..	..	..	..	..	55
9	10	3	1	2	1	2	5	6	..	1	1	2	1	2	..	..	..	..	..	..	..	37
10	2	..	3	3	4	2	5	6	1	2	..	2	..	..	1	..	..	..	..	..	..	31
11	4	..	2	..	3	2	1	3	1	..	..	..	..	..	..	..	..	..	..	..	..	16
12	1	1	1	1	2	1	..	..	2	1	1	1	1	..	..	..	..	..	..	..	..	13
13	..	..	..	..	2	..	1	2	..	..	..	..	1	..	..	..	..	..	..	..	1	7
14	..	..	..	..	1	1	..	..	..	..	..	..	..	..	..	..	..	..	..	..	..	2
15	..	..	..	..	..	..	..	..	..	..	1	..	..	..	..	..	..	..	..	..	..	1
16	..	..	..	..	..	..	..	..	..	..	..	..	..	..	..	..	..	..	..	..	..	0
17	..	..	..	..	..	..	..	..	..	..	..	..	..	..	..	..	..	..	..	..	..	0
18	..	..	..	..	..	..	..	..	..	..	..	..	..	..	..	..	..	..	..	..	..	0
19	..	..	..	..	..	..	..	..	..	..	..	..	..	..	..	..	..	..	..	..	..	0
20	..	..	..	..	..	..	..	..	..	..	..	..	..	..	..	..	..	..	..	..	..	0
21	..	..	..	..	..	1	..	..	..	..	..	..	..	..	..	..	..	..	..	..	..	1
22	..	..	..	..	..	..	1	..	..	..	..	..	..	..	..	..	..	..	..	..	..	1
Totals ..	149	73	84	114	126	98	103	97	43	43	26	18	12	8	3	2	0	0	0	0	1	1000

TABLE IV.—Correlation between the Fertility of a Woman and that of her Mother, when the Marriages of both have lasted at least 15 years and no weight is given to the Fertility of the Mother. 1000 cases from the Peerage. (See p. 285.)

Daughter's fertility.	Mother's fertility.																	Totals.
	0	1	2	3	4	5	6	7	8	9	10	11	12	13	14	15	16	
0	..	5	9	11	18	21	15	8	9	6	3	2	3	..	..	..	..	110
1	..	12	5	14	15	10	13	9	8	5	3	2	2	..	..	..	..	98
2	..	9	9	10	15	18	15	9	3	2	4	2	.	..	..	..	1	97
3	..	5	10	16	11	9	14	13	10	4	8	2	3	..	..	..	..	105
4	..	5	5	19	17	21	15	18	10	14	2	1	5	1	..	..	..	133
5	..	7	6	7	17	23	9	12	13	14	8	3	2	2	..	..	..	123
6	..	4	5	8	11	15	12	15	14	7	5	3	3	1	..	..	..	103
7	..	5	4	3	8	4	13	9	8	5	10	2	1	1	..	..	..	73
8	..	1	2	4	12	9	9	8	5	12	3	4	1	2	1	..	..	73
9	..	..	..	4	3	3	4	7	5	3	2	2	1	.	..	..	..	34
10	..	..	..	1	2	1	3	4	6	3	2	.	1	.	1	..	..	24
11	..	..	..	2	1	1	1	..	..	1	2	.	..	.	..	..	..	8
12	..	..	2	1	2	3	..	1	1	..	..	1	..	1	..	1	..	13
13	..	..	..	..	..	2	1	..	..	..	..	1	..	2	..	..	..	6
14	..	..	..	..	..	.	..	..	..	..	..	.	..	.	..	..	..	0
15	..	..	..	..	..	.	..	..	..	..	..	.	..	.	..	..	..	0
16	..	..	..	..	..	.	..	..	..	..	..	..	..	..	..	..	..	0
Totals ..	0	53	57	100	132	140	124	113	92	76	52	25	22	10	2	1	1	1000

TABLE V.—Correlation of the Fertility of a Man and of his Son as shown in Marriage. The Marriage of the Son being completed or having lasted at least 15 years. Weight given to the Fertility of Father. 6070 cases. (See p. 287.)

Number of son's children.	Number of father's children.																								Totals.
	0	1	2	3	4	5	6	7	8	9	10	11	12	13	14	15	16	17	18	19	20	21	22	23	
0	..	36	67	79	78	97	86	70	72	36	35	19	8	8	5	10	2	..	..	1	..	..	1	..	710
1	..	62	89	114	104	97	77	81	69	50	25	24	11	7	1	6	4	..	1	..	..	..	..	1	823
2	..	58	75	87	106	103	85	75	68	41	33	23	16	15	2	2	5	..	..	..	..	1	..	..	795
3	..	39	84	89	98	87	87	83	65	54	40	28	10	8	7	6	1	1	1	2	..	..	..	..	790
4	..	38	66	84	90	89	76	73	68	51	27	22	15	14	5	2	3	1	..	1	..	..	..	..	725
5	..	37	53	70	81	68	58	66	61	41	31	29	17	5	3	4	1	..	..	..	..	..	..	..	625
6	..	22	43	55	59	67	41	41	49	40	17	22	9	5	1	..	3	1	..	..	..	1	..	..	476
7	..	18	34	35	30	34	51	39	42	33	24	15	5	5	1	2	..	..	..	1	..	..	..	..	369
8	..	18	20	26	36	36	28	30	25	17	12	12	7	3	1	2	2	..	..	1	..	..	..	..	276
9	..	10	12	22	24	27	21	24	14	11	9	4	5	1	2	..	1	..	..	2	..	..	1	..	190
10	..	4	11	17	13	18	14	10	12	14	6	4	2	2	..	..	..	..	..	..	..	..	..	..	127
11	..	4	7	6	7	18	7	3	3	3	6	3	3	..	1	1	..	..	..	1	..	..	..	..	73
12	..	4	..	3	5	6	2	2	3	1	5	2	1	..	1	..	..	..	..	..	..	..	..	..	35
13	..	..	..	3	1	4	2	1	3	2	..	3	..	..	..	..	..	..	..	..	..	..	..	..	19
14	..	..	1	..	2	2	2	4	..	..	1	1	..	..	..	..	..	1	..	..	..	..	..	..	14
15	..	..	..	..	2	..	1	..	1	..	..	2	..	..	2	..	..	..	..	..	..	..	..	..	8
16	..	..	1	..	1	..	..	1	1	..	..	..	1	..	..	..	..	..	..	..	..	..	..	..	5
17	..	..	..	..	..	1	1	..	1	..	..	..	..	..	..	..	..	..	..	..	..	..	..	..	3
18	..	..	..	..	..	..	..	..	..	..	1	..	..	..	..	..	..	..	..	..	..	..	..	..	1
19	..	..	..	..	..	..	..	1	..	..	..	1	..	..	..	..	..	..	..	..	..	..	..	..	2
20	..	..	..	..	..	..	..	..	..	..	..	..	..	..	..	..	..	..	..	..	..	..	..	..	0
21	..	..	..	..	1	..	1	..	..	..	..	..	..	..	..	..	..	..	..	..	..	..	..	..	2
22	..	..	..	..	1	..	..	..	..	..	..	..	..	..	..	..	..	..	..	..	..	..	..	..	1
23	..	..	..	..	..	..	..	..	..	..	..	..	..	..	..	..	..	..	..	..	..	..	..	..	0
24	..	1	..	..	..	..	..	..	..	..	..	..	..	..	..	..	..	..	..	..	..	..	..	..	1
Totals ..	0	351	563	690	739	754	640	604	557	394	272	214	110	73	32	35	22	4	2	9	0	2	2	1	6070

TABLE VI.—Correlation of the Fertility in Marriage of a Man and his Son; no weight being given to the Fertility of the Father, and the Marriage of the Son being completed or having lasted at least 15 years. 1000 cases from the Peerage. (See p. 287.)

	Fertility of the father.																	Totals.
Fertility of sons.	1	2	3	4	5	6	7	8	9	10	11	12	13	14	15	16	17	
0	5	8	7	14	18	2	2	3	8	3	4	4	..	..	..	..	..	78
1	3	3	6	5	8	8	6	5	4	..	2	..	1	..	..	..	..	51
2	7	5	6	13	12	12	12	6	5	4	2	1	1	..	..	..	..	86
3	5	10	13	11	17	13	13	12	10	4	1	2	1	..	..	..	..	112
4	4	16	18	24	23	5	18	10	7	8	1	5	2	1	..	..	1	143
5	9	8	11	14	16	12	16	12	2	5	8	6	3	..	2	..	..	124
6	3	4	10	16	13	11	11	10	10	1	2	2	1	..	..	..	..	94
7	5	6	8	7	10	14	11	12	4	7	1	1	..	..	..	..	..	86
8	3	5	4	15	19	7	10	8	4	2	2	1	..	..	1	..	..	81
9	1	6	5	9	5	5	8	5	4	3	3	1	..	..	..	..	..	55
10	2	3	9	3	2	5	4	6	4	3	1	1	1	..	..	..	..	44
11	1	1	1	4	2	1	1	2	1	1	.	.	..	..	..	..	..	15
12	..	..	2	2	2	..	1	1	1	2	.	1	..	..	..	..	..	13
13	1	..	..	1	3	..	1	1	1	1	.	1	..	1	..	..	..	10
14	..	1	..	..	..	1	..	..	..	..	1	..	..	..	..	..	..	3
15	..	..	1	..	..	..	..	1	..	..	1	..	..	..	..	..	..	3
16	..	..	..	..	..	..	..	..	1	1	.	..	..	..	..	..	..	2
Totals ..	49	76	101	138	150	96	114	94	66	45	29	26	10	2	3	0	1	1000

TABLE VII.—Correlation of the Fertility in Marriage of a Man and his Son; no weight being given to the Fertility of the Father, and the Marriages of both Father and Son having lasted at least 15 years. 1000 cases from the Landed Gentry. (See p. 288.)

Fertility of the son.	Fertility of father. 1	2	3	4	5	6	7	8	9	10	11	12	13	14	15	16	Totals.
0	1	1	..	11	9	6	4	5	6	2	1	..	2	..	..	..	48
1	..	5	7	11	5	8	7	2	7	1	..	2	..	1	..	..	56
2	2	6	9	12	8	7	8	4	1	4	3	1	1	..	..	..	66
3	5	8	15	14	11	10	10	10	7	5	3	4	..	..	..	..	103
4	4	9	14	22	16	21	10	12	11	7	1	8	..	2	1	..	138
5	5	7	12	19	30	12	17	13	12	10	3	3	2	..	..	..	145
6	5	4	9	12	17	16	20	14	8	8	5	4	..	3	..	..	125
7	..	4	5	23	13	12	16	8	9	6	3	2	..	2	1	..	104
8	3	2	5	7	8	12	7	11	8	5	4	3	..	1	..	1	77
9	1	..	6	8	3	7	5	6	3	1	2	1	..	..	..	1	44
10	..	1	5	7	4	3	4	7	1	4	2	1	..	1	1	..	41
11	..	1	1	3	3	4	5	2	4	2	2	..	..	..	..	..	27
12	..	..	1	..	2	1	..	4	..	3	1	..	.	2	..	..	14
13	..	..	2	1	1	1	1	..	..	..	1	..	..	1	..	..	8
14	..	..	..	1	1	..	..	..	..	..	..	..	..	..	..	..	2
15	..	..	..	..	.	..	..	..	..	..	..	..	..	..	..	..	0
16	..	..	..	..	.	..	..	..	1	..	..	..	..	..	..	..	1
17	..	..	..	..	1	..	..	..	..	..	..	..	..	..	..	..	1
Totals ..	26	48	91	151	132	120	114	98	78	58	31	29	5	13	8	3	1000

TABLE VIII.—Correlation of the Fertility of a Woman and her Paternal Grandmother. No weight being given to the Fertility of the Grandmother, and the Marriages of both having lasted for at least 15 years. 1000 cases from the Peerage and Baronetage. (See p. 289.)

Fertility of granddaughter.	Fertility of paternal grandmother.																					Totals.
	1	2	3	4	5	6	7	8	9	10	11	12	13	14	15	16	17	18	19	20	21	
0	6	12	14	11	15	15	8	5	6	2	2	2	1	1	..	..	..	..	..	..	1	101
1	10	10	7	13	13	5	7	4	2	7	1	1	..	..	..	..	..	..	..	..	..	80
2	6	7	20	14	11	10	11	6	6	4	2	1	..	..	..	1	..	..	..	..	..	99
3	9	15	12	20	13	9	15	14	10	1	.	5	..	..	..	1	..	..	1	..	..	125
4	7	13	16	19	19	11	11	13	8	2	5	4	..	..	..	1	..	..	..	..	..	129
5	4	9	10	14	20	15	16	12	9	9	5	2	..	..	..	..	..	..	..	..	..	125
6	5	10	17	9	16	10	8	8	6	5	5	1	2	1	..	..	..	..	..	..	..	103
7	7	4	8	7	8	9	10	15	10	5	2	1	2	..	..	..	..	..	..	..	..	88
8	6	7	5	5	3	6	9	12	6	1	2	1	..	..	..	1	..	..	..	..	..	64
9	3	3	3	4	4	4	1	2	6	..	1	.	2	..	..	..	..	..	1	..	..	34
10	2	1	4	2	3	1	1	6	1	5	1	.	..	..	..	..	..	..	..	..	..	27
11	1	..	2	2	..	2	1	..	.	..	1	.	..	..	..	..	..	..	..	..	..	9
12	..	..	.	2	..	5	1	1	1	..	1	1	..	..	..	..	..	..	..	..	..	12
13	1	..	..	1	..	..	..	..	1	..	..	1	..	..	..	..	..	..	..	..	..	4
Totals ..	67	91	118	123	125	102	99	98	72	41	28	20	7	2	0	4	0	0	2	0	1	1000

Table IX.—Correlation between the Apparent Fertilities of Mares and Dams, with a minimum of at least four coverings. 1100 cases. (See p. 291.)

Dam's foals.	Mare's foals.																			Totals.
	0	1	2	3	4	5	6	7	8	9	10	11	12	13	14	15	16	17	18	
1	..	..	..	..	..	1	..	1	..	..	1	..	..	2	..	..	..	..	..	5
2	..	1	8	7	4	4	6	5	3	2	1	2	3	..	..	..	..	..	..	46
3	..	2	5	5	3	5	10	7	5	3	4	1	2	3	..	1	..	..	..	56
4	..	1	5	7	12	15	13	9	13	8	7	4	4	4	2	3	1	1	..	109
5	..	2	6	10	7	13	18	7	11	8	6	10	2	1	3	1	..	1	..	106
6	..	7	10	17	16	14	6	9	8	10	7	2	5	1	4	..	1	..	..	117
7	1	2	15	17	22	18	10	10	14	8	6	2	2	..	1	..	1	..	1	130
8	1	1	7	17	20	19	11	14	16	6	4	9	2	2	..	..	..	..	..	129
9	1	1	8	15	6	6	7	8	6	1	6	3	2	2	1	..	..	..	..	73
10	..	2	4	9	15	16	12	9	4	5	6	2	2	3	1	..	..	..	..	90
11	..	2	7	14	11	7	10	13	8	4	.	3	2	..	1	..	..	..	..	82
12	..	2	6	7	6	11	8	8	5	..	1	2	..	1	..	..	..	..	..	57
13	..	..	2	4	9	3	3	5	3	4	1	..	..	..	..	..	..	..	..	34
14	..	1	4	3	6	5	5	5	2	2	4	..	..	..	..	..	..	..	..	37
15	..	2	..	1	1	1	4	1	1	3	..	..	1	..	1	..	..	..	..	16
16	..	..	..	..	..	..	1	1	3	..	..	1	..	..	1	..	..	..	..	7
17	..	..	..	..	..	2	.	1	1	1	..	1	..	..	..	..	..	..	..	6
18	..	..	..	..	..	..	..	..	..	..	..	..	..	..	..	..	..	..	..	0
Totals ..	3	26	27	133	138	140	124	113	103	65	54	42	27	19	15	5	3	2	1	1100

TABLE X.—Correlation between the Apparent Fertility of Mares and Dams, with a minimum of at least eight coverings. 1100 cases. (See p. 292.)

Dam's foals.	Mare's foals.																		Totals.
	1	2	3	4	5	6	7	8	9	10	11	12	13	14	15	16	17	18	
1	..	..	..	..	1	..	..	..	..	..	..	..	1	..	..	..	..	..	2
2	..	..	1	1	2	..	2	1	1	..	..	..	..	..	..	..	..	..	8
3	1	..	1	..	1	2	4	..	2	..	2	..	2	.	..	1	..	..	16
4	..	1	1	1	7	7	6	10	5	3	8	2	5	..	2	1	1	..	60
5	..	1	1	5	7	15	9	13	12	1	11	4	3	3	..	2	1	..	88
6	..	1	3	9	17	16	14	15	12	16	5	5	4	5	..	1	..	..	123
7	2	2	7	16	18	17	24	22	19	9	9	1	3	3	1	1	..	1	155
8	2	1	6	6	18	21	25	26	15	7	15	5	6	2	1	1	..	..	157
9	..	1	6	4	9	10	20	9	4	9	7	3	3	5	..	..	..	..	90
10	..	..	3	8	14	17	16	8	9	9	7	2	4	1	1	..	..	..	99
11	..	1	3	4	10	19	18	14	7	1	6	2	1	1	..	..	..	..	87
12	..	1	2	6	11	11	14	12	6	6	4	1	1	1	..	..	..	..	76
13	..	..	1	4	7	11	8	9	9	1	3	1	..	..	..	..	..	..	54
14	..	2	2	1	4	7	6	3	3	4	2	..	..	..	..	..	..	..	34
15	..	..	..	1	2	5	2	4	6	1	..	2	..	1	..	..	..	..	24
16	..	..	..	1	..	3	2	6	..	..	1	..	..	1	..	..	..	..	14
17	..	..	..	.	2	..	2	1	1	..	2	..	1	..	..	..	..	..	9
18	..	..	..	..	..	2	..	1	..	..	..	..	..	1	..	..	..	..	4
Totals. . ..	5	11	37	67	130	163	172	154	111	67	82	28	34	24	5	7	2	1	1100

Table XI.—Correlation between the Fecundity and the *Apparent* Fertility of Brood-mares. 1000 cases. (See p. 300.)

Fecundity.	Apparent fertility. 0	1	2	3	4	5	6	7	8	9	10	11	12	13	14	15	16	17	18	Totals.
a	2	..	..	..	..	..	..	..	..	..	..	..	..	..	..	..	..	..	..	2
b	..	1	..	..	..	..	..	..	..	..	..	..	..	..	..	..	..	..	..	1
c	..	2·5	1	..	..	..	..	..	..	..	..	..	..	..	..	..	..	..	..	3·5
d	..	5·5	2	..	1	..	..	..	..	..	..	..	..	..	..	..	..	..	..	8·5
e	..	13	5	3·5	2	..	1	..	..	..	..	..	..	..	..	..	..	..	..	24·5
f	..	..	10	4·5	9	4	3	..	..	..	..	..	..	..	..	..	..	..	..	30·5
g	..	..	16	13	6	13	9	5	1	1	..	..	..	..	..	..	..	..	..	64
h	..	..	17·5	5	13	12·5	11	10	6·5	5·5	1·5	..	..	..	..	..	..	..	..	82·5
i	..	..	16·5	5	6	19·5	23	14	17·5	13·5	6·5	3	1	..	..	..	..	..	..	125·5
j	..	..	..	21	21	21	11	17	18	4	21	6	2	3	..	..	..	..	..	145
k	..	..	4	..	22	..	9	21·5	12	18	5	15	9	2	3·5	..	..	..	..	121
l	..	1	1	45	..	13	19	10·5	11	12	6	5	8	9	3·5	2	2	..	..	148
m	..	..	..	..	31	13	..	10	17	10	12·5	7	3	4	6	3·5	..	1	..	118
n	..	..	..	..	..	13	9	10	10	1·5	3·5	2	3	.	1	2·5	2	1	·5	59
p	..	..	..	..	..	..	..	..	..	1·5	4	5	2	2	1	..	1	1	·5	18
q	..	..	..	..	14	9	5	10	4	2	..	..	2	2	1	..	..	..	..	49
Totals ..	2	23	73	97	125	118	100	108	97	69	60	43	30	22	16	8	5	3	1	1000

TABLE XII.—Correlation Table for 1200 Mares and their Dams. Fecundity with a minimum of four coverings. (See p. 303.)

Dams.	Mares.																Totals.
	a	*b*	*c*	*d*	*e*	*f*	*g*	*h*	*i*	*j*	*k*	*l*	*m*	*n*	*p*	*q*	
a	..	..	..	..	..	..	..	..	..	..	..	..	..	..	..	..	0
b	..	..	..	..	..	..	..	..	..	..	..	1	..	..	..	..	1
c	..	..	..	..	..	·5	..	..	1	·5	..	..	..	..	..	..	2
d	..	..	..	..	..	·5	..	1	1	2·5	..	3	1	..	..	..	9
e	..	..	..	..	2·5	1	1·5	1	3	2·5	..	2·5	1	·5	·5	·5	16·5
f	..	..	1	..	·5	1	4·5	2·5	2·5	·5	2	3·5	3	1	1	1·5	24·5
g	..	..	..	1	2	3	5	2·5	10·5	10	3	6·5	1	1	1	4	50·5
h	·5	..	·5	·5	3	5·5	9	10·25	10·25	14·5	14·25	15·75	13·5	4	3	6·5	111
i	·5	..	..	1	4·5	3	12	14·25	19·25	21·5	17·25	18·25	10	5·5	1·5	4·5	133
j	1	..	..	5	3	3	10	11	18·5	24	34	26	21·5	8	3·5	12	180·5
k	..	..	1	1·5	5·5	4	14	17	21	19	28·75	31·75	23	13·25	3·75	8·5	192
l	1	1	1·5	3	6·5	7	11	13·5	17·5	32	25·75	27·75	15·5	11·25	2·25	9·5	186
m	..	..	·5	3	3	7·5	5·5	9·25	13·75	19·5	18·5	24	14	8	3	10	139·5
n	..	..	·5	..	1·5	4·5	5·5	7	6·5	10·5	9·5	13	11·5	9·5	1	7·5	88
p	..	..	..	..	1	·5	..	1·25	2·25	6	2·5	5·5	4	..	.	2·5	25·5
q	..	..	.	1	1	..	4	3·5	4·5	9	5	4	3·5	2·5	1	2	41
Totals ..	3	1	5	16	34	41	82	94	131·5	172	160·5	182·5	122·5	64·5	21·5	69	1200

TABLE XIII.—Correlation between Brood-mares and their Dams with regard to Fecundity with a minimum of eight coverings. 1000 cases. (See p. 304.)

Fecundity of dams.	Fecundity of mares.																Totals.
	a	*b*	*c*	*d*	*e*	*f*	*g*	*h*	*i*	*j*	*k*	*l*	*m*	*n*	*p*	*q*	
a	..	..	..	..	..	..	..	..	..	..	..	..	..	..	..	..	0
b	..	..	..	..	..	..	..	..	..	..	..	1	..	..	..	..	1
c	..	..	..	..	..	·5	..	..	1	·5	..	..	..	..	..	..	2
d	..	..	..	..	..	·5	..	..	1	·5	..	1	..	..	..	..	3
e	..	..	..	..	..	1	1	1	..	1	·5	2	..	..	..	..	6·5
f	..	..	1	..	..	1	4	3·5	2·5	1	3·5	2	2	..	1	..	21·5
g	..	..	..	..	1·5	2·5	3	2	11	10	7·5	4·5	4	4	2	..	52
h	..	..	·5	..	2·5	5	8	10	9·5	17	18·25	10·25	10·5	5	3·5	..	100
i	..	..	1·5	2	2	2·5	9	12·5	22	22	24·75	16·75	15·5	7	2·5	2	142
j	..	..	..	1	1·5	4·5	9	8	17	25	34	22	20·5	6	4·5	4	157
k	..	..	·5	·5	2·5	3·5	5·5	14·25	22·25	19·5	31·5	31	19	9·5	6	2	167·5
l	..	1	·5	1·5	5	4	5	14·25	19·25	25·5	26	20	17	5·5	3	2	149·5
m	..	..	·5	3·5	..	2	3	9·25	13·25	13·5	12	17·5	8·5	5·25	3·75	1	93
n	..	..	1	..	..	6	5	6·5	8	13	8·5	12	11·5	3·75	2·25	2	79·5
p	..	..	..	..	..	·5	.	2·25	2·25	7·5	2	4	2·5	1·5	..	..	20·5
q	..	..	..	..	..	..	..	·5	·5	2	1	..	..	..	..	1	5
Totals ..	0	1	5·5	8·5	15	33·5	52·5	82	129·5	158	169·5	144	111	47·5	28·5	14	1000

TABLE XIV.—Correlation of the Fecundity of Brood-mares and their Dams with a minimum of four coverings. Second series. 1200 cases. (See p. 305.)

	Fecundity of mares.																Totals.
Fecundity of dams.	*a*	*b*	*c*	*d*	*e*	*f*	*g*	*h*	*i*	*j*	*k*	*l*	*m*	*n*	*p*	*q*	
a	..	..	..	..	..	..	..	..	..	..	..	..	..	..	..	..	0
b	..	..	..	..	..	..	..	..	..	..	..	..	..	..	..	..	0
c	..	..	..	..	..	..	..	1·25	·25	..	1	..	..	..	..	..	2·5
d	..	..	..	..	..	1	..	·75	·75	..	..	..	..	..	..	..	2·5
e	..	..	..	..	..	1	1	·25	·25	..	·5	1·5	·5	1·5	..	..	6·5
f	..	..	..	..	1	..	5	3·25	2·25	5	2·5	1·5	1·5	·5	1	..	23·5
g	..	..	..	..	2·5	1·5	6	5	8	9	7	7·5	6	7	1	..	60·5
h	·5	..	1·25	1·25	4	3	6	8·25	9·75	19	11	9·5	13·25	7·5	·25	3	97·5
i	1·5	..	2·25	3·25	3	3	6	9·25	14·75	30	15	15·5	16·75	6	2·25	5	133·5
j	..	..	·5	3·5	3·5	4·5	16	12·5	20·5	27	27·5	28·5	28	11	2	6	191
k	2	..	..	·5	5	4	13	18·25	28·75	23·5	20·25	25·75	23	12·75	2·75	13·5	193
l	1	..	..	2·5	6	3	4	22·25	23·75	24	27·75	18·25	22·5	13·25	4·75	13·5	186·5
m	..	..	·5	2	4·5	5	10	7·75	10·25	18	15	31	12	9·5	2	5·5	133
n	1	..	1·5	1	2·5	5	7	5·5	7	14·5	11·5	10·5	9·25	4·75	1·5	8	90·5
p	..	..	..	1	1	..	..	·25	1·25	6	2·5	5·5	3·25	3·25	..	2·5	26·5
q	..	..	..	..	..	1	2	2·5	5·5	6	9	10	7·5	2	2·5	5	53
Totals ..	6	0	6	15	33	32	76	97	133	182	150·5	165	143·5	79	20	62	1200

TABLE XV.—Correlation of the Fecundity of Brood-mares and of their Granddams, with a minimum of four coverings. 1000 cases. (See p. 305.)

Fecundity of granddams.	Fecundity of mares.																Totals.
	a	*b*	*c*	*d*	*e*	*f*	*g*	*h*	*i*	*j*	*k*	*l*	*m*	*n*	*p*	*q*	
a	..	..	..	..	..	..	..	..	..	..	..	..	..	..	..	..	0
	..	..	..	..	..	..	..	..	..	..	1	..	..	..	..	..	1
c	..	..	..	..	·5	..	·5	·75	1·25	·5	..	1	·5	..	..	..	5
d	..	..	..	..	·5	..	·5	·25	·75	1·5	1	1	·5	..	..	..	6
e	..	1	..	..	..	·5	..	·5	1·5	1·5	..	3	..	·5	..	..	8·5
	..	..	..	..	..	1·5	..	2·5	3·5	5·5	4·5	3·5	6·5	2	..	1	30·5
g	1	..	..	4	3	4	4	5	7	7	7·5	15·5	12·5	3·5	..	6	80
h	..	..	1	1·5	4·5	4	6	7·5	12·5	11	11·5	19	10·25	5·25	..	5·5	99·5
i	..	..	2	2·5	7	5·5	6	13·5	14·5	18·5	15·5	29	13·75	8·75	1	8·5	146
j	1	..	2	3	7	6	8	10·5	14·5	16	18	22	19	11·5	·5	17	156
k	1	..	1	3	4	7	11·5	11·25	15·25	25	10·75	21·75	10·5	14	1	13	150
l	..	..	..	2	2	2	7·5	8·75	12·75	25·5	10·75	20·75	16	5	1·5	9	123·5
m	..	..	..	..	7	4	3·5	8	13	5	11	11	14	5·5	·5	6·5	89
n	1	..	..	·5	2	3	4·5	3·5	6·5	8·5	4·5	6	4	2·5	..	3·5	50
p	..	..	..	·5	..	..	1	1·5	1·5	1·5	·5	4	1·5	1	..	4	17
q	1	..	1	3	..	..	4	1	3	4	1·5	8·5	4·5	4·5	1	1	38
Totals ..	5	1	7	20	37·5	37·5	57	74·5	107·5	131	98	166	113·5	64	5·5	75	1000

INDEX SLIP.

---

in the University of

F.R.S.

98.

the laws of FARADAY,
ned up by the formula

),

ium,
r $1/\eta$ is the charge per

olyte per unit volume of

ll), called in the sequel

ould travel if, all other
and the potential slope

cities.

e taken as representing
at any given moment in

20.4.99

TABLE XV.—Correlation of the Fecundity of Brood-mares and of their Granddams, with a minimum of four coverings. 1000 cases. (See p. 305.)

	Fecundity of mares.																Totals.
	a	b	c	d	e	f	g	h	i	j	k	l	m	n	p	q	
a	..	..	..	..	..	..	..	..	..	..	..	..	..	..	..	..	0
	..	..	..	..	..	..	..	..	..	..	1	..	..	..	..	..	1
																	5
																	6
																	8·5
																	30·5
																	80
																	99·5
																	146
																	156
																	150
																	123·5
																	89
																	50
																	17
																	38
																	1000

II.—*Ionic Velocities.*

By Orme Masson, *M.A., D.Sc., Professor of Chemistry in the University of Melbourne.*

Communicated by Professor W. Ramsay, *F.R.S.*

Received December 12,—Read December 15, 1898.

Introduction.

The general theory of electrolytic conduction, involving the laws of Faraday, Hittorf, Kohlrausch, and Arrhenius, may be briefly summed up by the formula

$$C = A \frac{n}{\eta} (U + V) = A \frac{n}{\eta} \pi x (u + v),$$

where

C is the current,

A is the area of cross-section of the conducting medium,

η is the electro-chemical equivalent of hydrogen or $1/\eta$ is the charge per monad ion,

n is the number of monad equivalents of the electrolyte per unit volume of solution,

U is the average working velocity of the cations,

V is that of the anions,

π is the fall of potential per unit of length (dP/dl), called in the sequel potential slope,

x is the coefficient of ionization,

u is the velocity with which the same cations would travel if, all other things being equal, ionization were complete, and the potential slope had unit value, and

v is the corresponding value for the anions.

u and v are referred to in the sequel as *specific velocities.*

In further explanation it may be pointed out that x may be taken as representing either that fraction of the total dissolved molecules which is at any given moment in the ionized state, or (which is the same thing) that fraction of the total time during which, on the average, any given dissolved molecule is ionized;* and that the relation of the *working velocity* U (or V) to the *running velocity* πu (or πv) is therefore similar to that which holds between the average speed of a train for its whole journey, including stoppages, and its actual average speed between stations. Briefly, $U = \pi x u$ and $V = \pi x v$.

The values of u and v are not necessarily quite the same for the same ions in different strengths of solution, for the running speed, apart from stoppages, may be, and almost certainly is, affected by the concentration. Nor can it be assumed that all ions are equally affected in this manner: more probably each has what may be called its own frictional coefficient. In other words, the value of the ratio u/v for any given electrolyte may be expected to show some variation according to the strength of the solution, though in dilute solutions these variations may practically vanish. At extreme dilution the maximum values u_∞ and v_∞ are attained. Here, also, x attains its maximum value 1; so that

$$U_\infty = \pi u_\infty \quad \text{and} \quad V_\infty = \pi v_\infty;$$

or the working and running velocities are identical.

The history of the study of ionic velocities divides itself naturally into three chapters. The first may be called the Hittorfian chapter, the second the Kohlrauschian, and the third may be associated with the names of LODGE and WHETHAM.

HITTORF, and those who have since adopted his well-known method, studied the changes of concentration in the neighbourhood of the electrodes and deduced from these the ratios $\frac{U}{U+V}$ and $\frac{V}{U+V}$, or (which is the same thing) the ratios $\frac{u}{u+v}$ and $\frac{v}{u+v}$. These ratios, generally called the transport numbers of the cation and anion, may be conveniently represented in the sequel by the symbols $1-p$ and p. They represent respectively the cation share and the anion share of the current.

The classical work of KOHLRAUSCH consists essentially in the measurement of current and potential difference in an electrolytic cell of known dimensions, and containing a uniform solution of an electrolyte of known concentration. Thus all the values in the general equation, as given above, can be observed except $x(u+v)$, and this can be calculated if the truth of the equations be assumed. From this value of $x(u+v)$ and the value of $\frac{v}{u+v}$, as determined by Hittorfian methods, the separate values xu ($= U/\pi$) and xv ($= V/\pi$) for any given concentration may also be calculated. Further, by working with various strengths of solution up to extreme dilution, x is eliminated and $u_\infty + v_\infty$ obtained. But here it is obviously impracticable to determine the Hittorfian ratio by experiment; so that a certain assumption is necessary in calculating the separate values of u_∞ and v_∞. This assumption is that

* Compare WHETHAM, 'Phil. Trans.,' A, 1893, p. 340.

the transport numbers determined in dilute solutions are approximately correct for extreme dilution, and that they may be legitimately so far modified as to lead to constant specific velocities, u_∞ and v_∞, in all combinations of the corresponding ions. KOHLRAUSCH's values are thus obtained.

In general, electrolytes composed of monad ions, such as the chlorides of the alkali metals, are those which yield the best and most consistent results, while compounds containing divalent and polyvalent ions, such as Zn or SO_4, do not behave so conformably with the law represented by the equation. This is shown both by the great alteration in the value of the transport numbers produced by change of concentration in many such cases, and by the fact that polyvalent ions do not appear to afford constant values of u_∞ and v_∞ in their different combinations, on which account they are excluded from KOHLRAUSCH's tables of specific ionic velocities. Whatever may be the full explanation of these apparent contradictions of theory, it is probable that in these cases the ionization is not of such a character that the nature and number of ions of either kind can be deduced direct from the known composition of the solution; in other words, the number of active ions of either kind is not related in the usual simple manner to n.

The third chapter in the history of the study of ionic velocities is that which deals with their direct observation and measurement, and was begun by LODGE in 1885 ('Brit. Ass. Reports,' 1886, p. 389). The great value of his work lay in the ingenious conception of the possibility of actually watching the advance of ions whose colour renders their progress through an otherwise colourless solution visible, and of ions which, though themselves colourless, may be detected in progress by their interaction with indicators; and, further, in the introduction for these purposes of solid jellies in place of ordinary aqueous solutions, and the avoidance by this means of various sources of error, such as convection currents due to gravity and to temperature changes. The actual experiments were, however, of a pioneer character; and the interpretation of them seems to have been vitiated by a misunderstanding of the mechanism of the process on which they depended. It is necessary to point this out; first, because the author's experiments cannot be properly discussed unless this be done, and secondly, because most of the recent text-books dealing with electrolysis quote LODGE's experiment on the velocity of the hydrogen ion as affording the first and chief direct experimental verification of KOHLRAUSCH's theory, and do not direct attention to the difficulty in question.

From the general equation and the explanations already given it is obvious that mere direct measurement of U and V, or of both, cannot by itself give results of exact value for comparison with calculated velocities (xu and xv). At best it can afford only an indication of whether something like the right order of magnitude has been arrived at by such calculation. To obtain data for exact comparison not only U or V, or both, but also the potential slope π, causing U or V, must be correctly measured. But the work of LODGE does not show that this latter was determined in any of his experiments; or rather it shows that the value of π was incorrectly assumed. Thus he assumed, in the hydrogen experiment, that the potential slope causing the observed hydrogen velocity was 1 volt per centim., because the tube was 40 centims. long, and there was a difference of potential of 40 volts between the electrodes. It is obvious, however, that this may be a very misleading assumption where the value of π is required in one particular part of a tube which contains quite different electrolytic solutions in different portions of its length. There is, therefore, no exact information to be obtained by comparing LODGE's observed velocity of from ·0024 to ·0029 centim. per second with KOHLRAUSCH's calculated value ·0032· Considerations of temperature and of concentration, though important, are less so than that of correct potential slope, and therefore may be passed over.

Perhaps the most striking of all LODGE's experiments were those in which he observed the velocities of Cl, Br, or I entering and travelling through a jelly tube from the cathode end, while Sr or Ba travelled in the opposite direction. The original jelly was charged with, among other things, a small proportion of Ag ions to act by partial precipitation as an indicator of the progress of the halogen, and with SO_4 ions to play a similar part towards the new cations. The observed velocities of the former were in all cases approximately double those of the latter; whence LODGE concluded that Cl, Br, and I are, as ions, naturally twice as fast as Sr and Ba. Here again, however, the fact that the observed velocities were caused by unknown, and presumably different, potential slopes necessarily vitiates the conclusion drawn. It will be shown in the sequel that what really determined the interesting and simple velocity ratios observed in this set of experiments was not the specific character of the ions under inspection, but the composition of the intermediate solution into which the Cl and Ba, or similar ions, had not yet penetrated. As, however, this was a mixture, and the indications given of its composition are rather qualitative than quantitative, no results of theoretical value can be deduced.

WHETHAM's method (*loc. cit.*) rendered the use of gelatine unnecessary, as he avoided gravity currents, at all events, by employing a vertical tube in which to observe the rate of migration of the boundary between a coloured solution and a colourless one during the passage of a current, the lighter solution lying above. He also avoided the occurrence of different and unknown potential slopes in different parts of the column by selecting for each experiment a pair of solutions of, as nearly as possible, equal specific resistance. The results so obtained were in very good accord with the calculated velocities (xu or xv) of the same ions in similar solutions of the same concentration, and thus afforded the first exact confirmation of the KOHLRAUSCH theory. But, from the very nature of the method, its application was restricted to a very few cases, as it is obviously not easy to find solutions suitable in all respects.*

* WHETHAM's determination of the velocity of the copper ion, and his comparison of the results with the calculated number, are open to the objection that what he observed was not the copper ion at all, but

New Method of Observing Ionic Velocities.

This method resembles LODGE's, in so far as it makes use of electrolytic jellies and of visible moving boundaries, but differs from his in its essential principle, as will be explained, and also in the fact that it seeks to avoid such sources of error as change of temperature, the use of mixtures of unknown composition, and the introduction of indicators that react with the ions under observation. The use of gelatine necessarily introduces a small amount of electrolytic impurity which must have some disturbing effect; but the solid gelatine used in making the jellies contains less than ·5 per cent. of its weight of mineral matter, and a plain 12 per cent. jelly was found to have a conductivity so small as to be practically negligible in comparison with those of the salt jellies used for experiment. That this is so is shown by the results obtained; but the fact that the best available gelatine has some conductivity of its own would introduce a real difficulty in any attempt to apply the method to solutions of small concentration.*

Fig. 1.

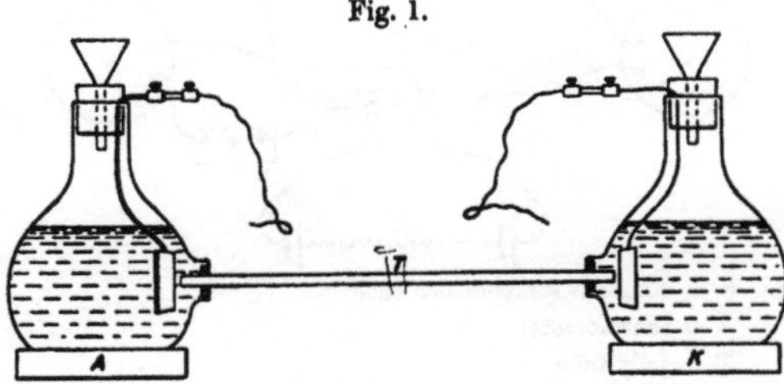

A sketch of the apparatus is shown in fig. 1. A straight tube of convenient length and uniform narrow bore, the dimensions of which are known, is graduated by

a complex cuprammonium ion. He says (*loc. cit.*, p. 344): "The first solutions used were those of copper and ammonium chlorides with just enough ammonia added to each to bring out the deep blue colour of the copper." This is certainly not the colour of the copper ion. In support of this statement it may be mentioned that the deep blue ion of FEHLING's solution, which has as much right to be called copper as has the deep blue ion of WHETHAM's experiment, can be proved by direct observation to be a negative ion, which travels towards the anode while its associated K ions carry the current towards the cathode. This observation led the author, in conjunction with B. D. STEELE, to an investigation of the cupro-tartrates, which they propose to communicate to the Chemical Society. Attention is there directed to earlier evidence of the same fact adduced by others.

* Some trials were made with agar-agar in place of gelatine. It proved inferior, however, in respect to freedom from electrolytic impurity; and, though it affords jellies of high melting-point, and otherwise admirable, they have the fatal habit, after setting firm in the tube, of contracting away from its walls and exuding an aqueous solution. It is then easy to blow the whole cylinder of jelly out of the tube by the application of slight pressure at one end. The gelatine jellies used by the author showed no such tendency; nor did any extension from the tube occur in the course of the experiments in the manner described by LODGE. The difference may be due to the use of stiffer jellies, and particularly to the use of a constant temperature bath to prevent heating of the jelly by the current.

an etched scale from end to end. The two ends of the tube can be fitted, water-tight, into the short side necks of two cells of relatively large capacity, so that the tube forms a horizontal connection between them. For an experiment the tube is filled (by means of tubular elbow-pieces and rubber connections) with a molten jelly, which is then allowed to set at the experimental temperature. This jelly contains a known quantity per cub. centim. of the salt, whose ionic velocities it is desired to observe, *e.g.* KCl. The ends of the jelly, after removal of the elbow pieces, are shaped true with a knife; and the tube is then connected with the empty cells, and the whole apparatus is placed in a large bath of water, so that only the mouths of the cells

Fig. 2.

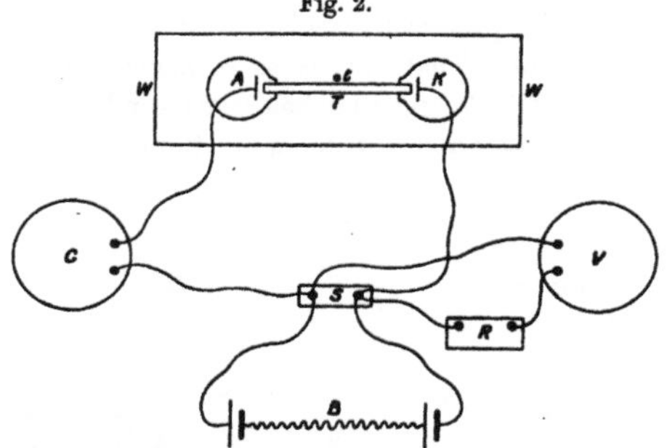

$W\ W$ = constant temperature bath.
t = thermometer.
T = jelly-tube.
A = anode cell.
K = cathode cell.
C = low resistance galvanometer for current measurement.
V = high resistance galvanometer for voltage measurement.
R = added resistance in voltmeter circuit.
S = mercury connections.
B = battery.

remain above the surface. The temperature of the bath is kept constant at any desired point below the melting-point of the jelly, *e.g.* 18° C. The graduations of the tube are easily read through the water in a good light, and parallax is avoided by having the tube marked both back and front. The electrodes, which have surfaces very large in comparison with the bore of the tube, are placed in the cells and fixed so as to be close to the ends of the tube without touching it. They are connected with a battery of sufficient voltage, as constant as possible. A low-resistance galvanometer serves to register the current, and one of high resistance is arranged in parallel circuit to indicate the voltage. Fig. 2 shows the whole arrangement diagrammatically. The main circuit is completed and the experiment started by filling the cells with aqueous solutions, and a stop-watch is started simultaneously.

The nature of these cell solutions is all-important to the theory and practice of the method. They may be distinguished as the anode solution and the cathode solution respectively. They must fulfil four conditions, as follows:—In the first place, each must possess a strong and characteristic colour; but the anode solution must owe its colour to its cation, and the cathode solution must owe its colour to its anion. In the second place, the coloured ions must not be such as to act chemically on the salt-jelly, so as to form a precipitate in the tube through which they are to travel. In the third place, the cell solutions themselves must not, during an experiment, undergo such chemical change as to lead to the production of a new sort of cathion (*e.g.*, H ions) in the anode cell, or of a new sort of anion (*e.g.*, OH ions) in the cathode cell. The fourth condition, which will be explained more fully later, is that the coloured ions must be specifically slower than the corresponding ions of the salt-jelly. A suitable anode solution in most cases is made with copper sulphate, provided that the anode be made of copper, to prevent, or at least minimise, the production of free acid, *i.e.*, of H ions. A generally suitable cathode solution is made with potassium chromate and sufficient bichromate to prevent the formation of free alkali, *i.e.* of OH ions.* The cathode should be of platinum. The strength of these solutions should be known, but need not be proportioned to that of the salt in the jelly tube. All that is necessary is that there shall be plenty of coloured ions in the neighbourhood of the electrode and tube for the carriage of the current into the latter.

During the experiment, the procession of the original cations (say K) of the jelly is followed through the tube by a corresponding procession of blue Cu ions, while the opposite procession of original anions (say Cl) is followed by a corresponding procession of yellow CrO_4, or of mixed CrO_4 and Cr_2O_7. Thus the tube is soon seen to contain blue ($CuCl_2$) jelly at one end, colourless (KCl) jelly in the middle, and yellow (K_2CrO_4) jelly at the other end, of which the first and third continually grow in length at the expense of the second, intermediate, part. The ratio of the lengths of blue and yellow is constant, and these eventually meet, to the extinction of the colourless portion. There is no mixing of K with Cu, nor of Cl with CrO_4. The blue and yellow boundaries remain quite clear cut, and may be sharply located at any points in their course, the former marching steadily through the solid jelly towards the cathode, the latter towards the anode, till they meet. It may be mentioned that the blue boundary is always slightly convex in the direction of its migration, while the yellow boundary is always slightly concave, so that each presents a meniscus with its convexity towards the cathode.

At intervals throughout the experiment observations are taken of the positions of the blue and yellow boundaries, of the time, of the current, and of the voltage; also of

* The passage of CrO_4 and Cr_2O_7 ions into the jelly-tube, where they meet with K or other cations, does not chemically affect the gelatine. In fact, a transparent orange half-normal bichromate-jelly, containing 12 per cent. of gelatine, may be prepared, solidified, and remelted, without any precipitation of the gelatine; but precipitation occurs on the addition of free mineral acid.

the bath temperature, which should keep constant. With constant E.M.F. the current and the velocities steadily diminish in a manner determined by the nature and strength of the salt-jelly, but they remain in direct proportion to one another. The experiment is terminated when the two boundaries meet, which they do at a point that can always be predicted with very considerable accuracy from the first readings.

The essential principle of the method may now be explained. The visible moving boundaries mark respectively, not only the rates of advance of the foremost Cu and CrO_4, but also those of the rearmost K and the rearmost Cl. These are themselves invisible, but the immediately following coloured ions may be taken as their indicators. Now the intermediate colourless part of the jelly is at the start of uniform composition and remains so throughout the experiment, however much its length may be curtailed by the progress of the blue and the yellow; so that (to use the same symbols as before) n and x have constant values from start to finish and in all parts of the colourless jelly, while π, which diminishes as the experiment proceeds, has yet always the same value there, whether the part near the blue boundary or the part near the yellow boundary be considered. Therefore the rearmost K at the one end and the rearmost Cl at the other are comparable in all respects; and a comparison of their working velocities U and V, made visible by the indicators, gives at once the ratio u/v for the particular concentration employed. Obviously also the result may be put in the form of HITTORF's transport number $\frac{v}{u+v}$ (or p), and may be compared with the values obtained by the indirect Hittorfian method. Of course, the observed U is also that of the copper ions and the observed V that of the CrO_4 ions; but the experiment affords no indication of the corresponding values of π and x, which are certainly very different in the two cases. Hence there is nothing gained by regarding the observations from this point of view, as has already been pointed out in connection with LODGE's experiments.

While the first result is the determination of u/v for the original salt, a second is the testing of the general equation $C = A\frac{n}{\eta}(U + V)$, or, if its truth be assumed, of the efficacy of the method itself. For each quantity is independently determined, and, since all may be expressed in the same (C.G.S.) units, the value 1 should be obtained by dividing the left-hand side by the right.

The experiment, as carried out, also affords data for the determination of the working velocities per unit potential slope, viz., of $U/\pi = xu$ and $V/\pi = xv$. For the total resistance for each position of the boundaries is given by the readings of voltage and current; and, as the increase of total resistance is directly proportional to the diminution of the length of the colourless jelly, and as a constant correction can be introduced for the resistance of the galvanometer, and approximately also for that of the solutions between the electrodes and the ends of the tubes, the resistance of the tube full of the colourless jelly, and therefore of $x(u+v)$, follows. Obviously the same result can be obtained by separate measurement of resistance by KOHLRAUSCH'S method, and this has certain advantages. Arrangements are also possible for the direct measurement of π in the various parts of the tube during the progress of an experiment by means of wires sealed through the walls of the tube, but this would introduce considerable complication. The velocities per unit potential slope, obtained as above, are not dealt with in the present paper; but it may be mentioned that they show (in accordance with conductivity results obtained by ARRHENIUS and others) a considerable percentage reduction in solid jelly as compared with aqueous solutions, but a reduction which is, at all events approximately, the same for different salts. The values of the relative velocities of the different ions should, therefore, be fairly comparable with those found by the older methods.

It is clear that any value which the method may have must depend on the justice of the assumption that the observed velocities of the boundaries are determined by, and may be taken as indicative of, those of the intermediate colourless ions, or that no mixing of these with their coloured pursuers occurs. In support of this there are both experimental facts and theory.

At the end of an experiment in which the tube was originally full of a strong chloride jelly, the author has frequently melted out the yellow part (without disturbing the blue, which consists of $CuCl_2$), and tested it for chlorine without finding more than, at most, a barely perceptible trace. After one experiment with KCl, in which a wide tube was employed, so that the quantities were considerable, only a doubtful trace of potassium could be found in the blue part, after separation of the copper by H_2S, evaporation with sulphuric acid, and ignition to destroy organic matter.

Less direct, but no less striking, evidence to the same effect is afforded by the fact, established by preliminary experiments, that the ratio of velocities of the blue and yellow boundaries is practically the same in different experiments with the same concentration of the same salt, no matter how the dimensions of the tube be varied. There is also the fact, already stated, that this ratio in any one experiment remains practically constant from first to last. Very slight variations, it is true, do occur, attributable most probably to the presence of impurities in the gelatine or in the salts employed, or to slight heating by the current in spite of the constant temperature bath. But such small deviations from perfect constancy need not be considered at present.

A third line of evidence is found in the fact that, while it is possible to greatly vary the relative velocities of the same indicators by using different salt-jellies, it is found that the relative rates of advance of the two boundaries remain the same with different indicators and the same salt-jelly. The latter part of this statement has, however, so far been tested only by the substitution of potassium ferrocyanide, and by that of a tartrate solution, for the usual chromate in experiments with potassium sulphate and sodium chloride. In the tartrate experiment the indicator was, of course, quite invisible in itself, but it met the copper and formed a visible precipitate across the tube at the exact point calculated from previous experiments made in the usual way. The relative velocities of Na and Cl were thus given the same value, whether Cu and CrO_4 or Cu and $C_4H_4O_6$ were used as indicators.*

Besides these experimental indications that there is no commingling of the ions at the blue and yellow boundaries, there are theoretical reasons in favour of the same conclusions, *provided that the coloured ion is specifically slower than the one it is following*. This has been already mentioned as one of the requisite characters of a satisfactory indicator. Imagine naturally slow Cu ions travelling behind naturally faster K ions, with Cl ions travelling past both in the opposite direction. If Cu lag behind K, or K run away from Cu, a region will be established where cations are deficient, a state of affairs that must immediately correct itself by reason of the consequent E.M.F. If, on the other hand, they keep pace with one another, as in fact they do, it must be by virtue of a steeper potential slope in the blue. If now some K ion accidentally lags behind its fellows, it will find itself in this region and be at once hurried forward again; while any ambitious Cu ion, trying to penetrate

* The precipitates formed across the tube by the meeting of Cu ions with CrO_4 ions and $Fe(CN)_6$ ions are of the nature of the semi-permeable membranes used by TRAUBE and PFEFFER in the study of osmotic pressure. An interesting fact has been observed with both these membranes. They are first seen as fine transverse films across the tube, but, if the experiment be not stopped, they rapidly thicken up till they form discs about half a millim. wide. Simultaneously the galvanometer shows a rapid fall of current, which becomes almost nil within a very few minutes of the first meeting. If now, or later, the current be reversed, the galvanometer deflection rapidly goes up almost to its previous value, though the membrane remains apparently quite unaffected even when the reversed current is maintained for hours. By again reversing it the phenomena may be repeated, though the current does not now fall off immediately. The explanation suggests itself that the membrane is impervious to the ions (Cu and CrO_4 or $Fe(CN)_6$) which have produced it, but not to other ions such as K and Cl. Before the first reversal of the current, only Cu ions can reach the membrane from the anode side and only CrO_4 (or $Fe(CN)_6$) from the cathode side, and these cannot pass. After reversal, K and Cl, or other corresponding ions respectively, carry their charges to and through the membrane. These are now on the wrong side; so that when the current is again reversed, it is some time before the original state of affairs is restored and the current again cut down. If, after the membrane is first completely formed, the circuit be broken and everything left *in situ* for 24 hours or so, it is found that, on re-connecting without reversal, a very fair current will pass; but it does not last long. In all probability this is due to simple diffusion through the membrane, by which a little K and a little Cl find their way across to the parts of the tube previously free from them.

It is a curious fact, no certain explanation of which has yet been arrived at, that when the intermediate salt is a sulphate instead of a chloride, the copper and chrome ions do *not* form a membrane on meeting, but simply intermix with production of a greenish colour; nor is the current cut down. This difference of behaviour has been consistently manifested in all the experiments recorded in this paper. Yet copper sulphate and potassium chromate solutions precipitate copper chromate when mixed. The usual membrane was, however, obtained with a K_2SO_4 jelly and Cu and $Fe(CN)_6$ indicators.

The making of osmotic pressure apparatus might be improved by depositing the membrane electrolytically in the walls of the pot, previously charged with a potassium chloride jelly. Current readings would give a sure indication of the condition of the membrane during and after its formation.

the K region, will find itself forced to drop back. This explanation of the sharpness of certain margins was advanced by WHETHAM (*loc. cit.*), but bears repetition here. It comes to this—that the boundary will possess stability if the necessary condition be fulfilled, but not otherwise. As a matter of fact it has been found by experiment that the colourless SO_4 ion following the yellow CrO_4 ion through a jelly tube overtakes it continuously, so that there is *no* boundary visible, but only a gradual fading out of colour. A similar result was got also by following the blue Cu ions with colourless Zn ions through the tube: again there was *no* boundary, but the colour gradually faded out. These cases probably illustrate the non-fulfilment of the condition that the foremost ion must be by nature the faster.

Theory of the Moving Boundary.

It is easy to deduce the behaviour of the ions on each side of a moving boundary from the fundamental equation given at the beginning of this paper, and from the fact that the visible (coloured) ion keeps pace with the invisible (colourless) ion in front of it.

Let the boundary in question be that between visible and invisible cations travelling with the current and matched by invisible anions, all of one kind, travelling against it. Let the symbols be used with the same meaning as before, but let those applying to the coloured part of the jelly be marked with dashes to distinguish them from those applying to the colourless jelly on the other side of the boundary.

Since equal currents cross all sections of the tube at the same time,

$$n'(U' + V') = n(U + V).$$

But

$$U' = U.$$

Hence

$$\frac{n}{n'} = \frac{U'}{U' + V'} \times \frac{U + V}{V} = \frac{1 - p'}{1 - p}.$$

Also

$$\frac{V'}{V} = \frac{n}{n'} \times \frac{V'}{U' + V'} \times \frac{U + V}{V} = \frac{p'(1 - p)}{p(1 - p')}.$$

Thus the concentrations of the salts in the two portions of the jelly are directly as the corresponding cations' transport numbers; and the working velocities of the common anion on the two sides of the boundary are directly as its own transport numbers, and inversely as those of the corresponding cations. Since $\frac{p'(1-p)}{p(1-p')} = \frac{uv'}{u'v'}$, and since v and v' may be considered for practical purposes as of equal value, the second of these rules may be put more simply, though not quite so correctly, in the form $V'/V = u/u'$; or the working velocities of the common anion are inversely as the specific velocities of the corresponding cations.

It is possible to test the foregoing conclusions by analytical experiments conducted as follows:—The exact ratio of the velocities of the K and Cl ions having been first found by experiments with two coloured indicators in the usual manner, a tube of suitable length and considerable capacity is marked with a file at the place where these (or other) indicators should meet. It is then filled with the KCl jelly and connected with the cells, in *both* of which is placed copper sulphate solution. A copper anode and a platinum cathode are used. During the progress of the experiment, two boundaries travel as usual, though only one is visible, viz., that between KCl and $CuCl_2$. It will, however, meet the other, *viz.*, that between KCl and K_2SO_4, at the mark previously made; and the circuit must be broken when the blue boundary reaches this point. The cathode will have gained in weight by as many milligramme equivalents of copper as there were milligramme molecules of KCl originally present in the whole tube; and this may be taken as a test of the correctness of the experiment. That it is so, is evident from the fact that there will have travelled across that section of the tube, where the meeting of the margins occurred, all the K ions originally present on one side of it, and all the Cl ions originally present on the other, but nothing else. These are together equal in number to the total K, or to the total original KCl molecules. The tube is cut in two at the mark as soon as the experiment is over, and the contents of each part are then analysed separately, Cl and Cu being estimated in the part nearest the anode, and SO_4 and K in the other. The results may be checked by estimating the Cl that has escaped into the anode cell, and the K that has escaped into the cathode cell. The results give the value of n' directly for each end of the tube, with which the value of the original n may be compared. The relative specific velocities (u/v) for Cu and Cl, and for K and SO_4 follow from consideration of the exchanges at the anode end and the cathode end respectively; and, as those of K and Cl are already known, as determining the meeting point, the specific velocities of all four ions may be compared with one another.

One such experiment has been carried out, but the accuracy of the analytical results was, to some extent, spoilt by unforeseen difficulties that cropped up in the course of the work, due in part to the presence of gelatine in the solutions of the tube-contents, and in part to the formation of cup*rous* chloride in the *anode* cell and on the anode. This formation of cuprous chloride has since been observed in other experiments. It may be seen on the anode as a white crust after it has been washed with water, alcohol, and ether; and its formation there causes the anode to lose *less* than the calculated weight instead of more, as is usual. It is intended to repeat the experiment described above, taking all precautions to ensure accuracy. In the meantime it may be said that the results, though rough, tended entirely towards the confirmation of the theory. Thus they showed, as might be expected, that very little change of concentration or of K velocity occurs across the KCl/K_2SO_4 boundary, but that at the blue boundary the Cl approximately doubles its velocity and reduces its concentration to about two-thirds. These figures (which are only rough approximations) agree, according to the formulæ already given, with a chlorine transport number (p') in $CuCl_2$ of ·67, taking its value (p) in KCl as ·5. The author is not aware of any Hittorfian experiments with $CuCl_2$; but, to judge from those made with other chlorides and with other copper salts, this value for p' can not be far wrong.

Experimental Details.

The following facts, in addition to those already stated, may be recorded concerning the series of experiments, the results of which are tabulated in this paper.

Gelatine.—The small conductivity of this has been already referred to. As additional evidence that its impurities are unimportant, it may be mentioned that practically no difference is observed in the value of u/v got for the same concentration of the same salt, whether the jelly contain, as usual, 12 per cent. of gelatine or only half that amount.

Salts.—The salts used were all re-crystallized, and were what is generally called pure; but no very special purification was attempted, as the present object was rather to test the method in a preliminary manner than to get the most accurate quantitative results attainable.

Jellies.—These were made as follows. The required quantity of the salt was weighed into a beaker, dissolved in water, and washed into a stoppered 50 cub. centim. flask containing 6 grammes of gelatine in small pieces. The flask was warmed till the gelatine had dissolved, water was added to the mark, and the contents mixed and cooled. The exact volume was then made up with water and, after re-warming and thoroughly mixing, the jelly was ready for use. It was always made fresh for each experiment. For the experiments with lithium chloride, which is very deliquescent, the calculated quantity of carbonate was dissolved, with all precautions, in slight excess of hydrochloric acid, and the chloride was obtained neutral in reaction by evaporation and drying at about 150° C. The jellies were all clear when solid and were transparent in the tube; except the twice-normal lithium sulphate one, which was, in bulk, very slightly opalescent. The melting-points were by no means all the same, but all were completely solid at the temperature employed.

Temperature of the Bath.—This was in all cases very close to 18° C., the average in each experiment lying below rather than above that point. The extreme variation during any experiment did not exceed ·5° on either side of 18°, and was generally less.

Cell-solutions.—These were always of normal strength, *i.e.*, the anode cell contained $\frac{1}{2}$ $CuSO_4$ gramme per litre, and the cathode cell contained $\frac{8}{18}$ K_2CrO_4 + $\frac{1}{18}$ $K_2Cr_2O_7$ gramme per litre. Each cell contained 100 cub. centims.

The Tube.—This was the same in all the experiments tabulated. Its ends were ground smooth, and its bore uniform. Its length was 15 centims. and its area of cross-section (A), which was determined carefully by the weight of its mercury contents, was ·0378 sq. centim. It was divided into half centims. on both sides to avoid parallax, and the scale divisions were picked out with red. They were easily seen through the water against the white porcelain bottom of the bath. It was quite possible to divide by eye to less than half a millim.; but only the readings made when the boundary was crossing a scale division have been used in drawing the curves. The walls of this tube were rather thicker than is advisable, since it is important to maintain free thermal communication between the jelly and the bath. Nearly 1½ centims. at each end of the tube were hidden in the cell neck, so that readings, after the start, could not be made till the boundaries reached that mark.

The Battery employed consists of 48 cells of zinc | potash | copper oxide, arranged in one series and giving about 40 volts. It is very constant, whether in use or not, even many months after setting up. The current, in the experiments quoted, varied from about 2 to about 13 milliamperes.

The Time for an experiment, from the starting of the current till the meeting of the boundaries, varied from nearly three hours to over seven. The tube was under constant observation. Time readings were made to the nearest quarter-minute.

TABLE I.—Experimental Results.

Salt.	n	$\frac{u}{v}$	$\frac{v}{u+v}$	p	$\frac{C\eta}{An(U+V)}$
NH_4Cl	1	1·041	·490	·508 – ·517 (H)	1·005
KCl	·5 1 3	1·021 1·041 1·069	·495 ·490 ·483	·503 – ·516 (H)	1·017 1·032 1·014
NaCl	·5 1 2	·671 ·681 ·703	·598 ·595 ·587	·622 – ·648 (H)	1·023 1·013 1·022
LiCl	·5 1	·456 ·471	·687 ·680	·674 – ·773 (K)	1·004 1·013
½ K_2SO_4	·5 1	1·143 1·143	·467 ·467	·498 – ·499 (H)	1.002 ·951
½ Na_2SO_4	·5 1	·765 ·765	·567 ·567	·634 – ·641 (H)	1·005 ·950
½ Li_2SO_4	·5 1 2	·538 ·508 ·515	·650 ·663 ·660	·595 – ·649 (K)	1·021 1·006 ·907
½ $MgSO_4$	·5 1 2	·463 ·422 ·442	·684 ·703 ·693	·656 – ·762 (H)	·942 ·861 ·807

Experimental Results.

In Table I., the first column gives the salt used, the formula specifying the quantity required, in grammes per litre, for a normal solution, *i.e.*, for one in which $n = 1$. The values of n given in the second column refer to the number of gramme formula weights per litre. In the third column are given the ratios, u/v, these being the values calculated from the observed meeting points of the blue and yellow boundaries. In one case, however, viz., that of half normal KCl, this final reading was rejected and the ratio calculated from the readings immediately preceding it; for the experiment, otherwise one of the best as regards constancy of u/v, was spoilt just at the end by a disturbance that had a very visible effect on the meeting point, spoiling the boundaries and shifting the film formation 1 millim. This was the only case of the kind. In the fourth column the results are expressed as $\frac{v}{u+v}$, the transport numbers of the anions; so that they may be compared with the corresponding values obtained for aqueous solutions by the indirect Hittorfian method. These latter are given, under p, in the fifth column, where the experimenter's name (HITTORF or KUSCHEL) is signified by its initial letter. To be strictly comparable, these values should be for the same concentrations; but, as such data are not in most cases available, the extreme values of p are given, the most dilute value being placed first. For most of the salts these extreme values of p refer to concentrations outside those employed in the jelly-tube experiments, so that they may be taken as including the values for corresponding concentrations. Two points may be noticed. The first is that, while the author's results for lithium and magnesium salts fall within the p limits, those found by him for sodium, potassium, and ammonium salts give smaller anion transport numbers than the smallest value obtained by HITTORF. The second point is that the new values for the chlorides tend to decrease slightly as the concentration increases, whereas the opposite tendency is manifested by the Hittorfian values. There is, however, a general similarity that is evident on comparing the two columns. The last column of the table gives the values of $\frac{C\eta}{An(U+V)}$, for the calculation of which all the quantities were expressed in C.G.S. units. As already explained, the approximation of this value to unity may be taken as a test of the method, and it is evident that the result is, on the whole, satisfactory. Only the stronger sulphate solutions give too small values; and it is notable that magnesium sulphate, which contains two divalent ions, is abnormal even in the most dilute solution employed, though it would evidently behave normally with greater dilution. It has already been pointed out that it is salts of divalent and polyvalent ions that have always proved least consistent with theory.

For the calculation of the numbers in this last column it was necessary to obtain strictly comparable current and velocity values. Now those directly obtained in an experiment are not comparable, those of C referring to current at given instants and those of U and V to average velocities for a long period of time; and it has been already told how both current and velocity steadily diminish at a rate determined by the nature and strength of the jelly. Either, then, an average current value must be obtained, or velocity values at definite points. Both methods have been used. For the former, since the quantities involved were too small to admit of the use of a voltameter, the average current was calculated from the area of the time-current curve. This was done for all the experiments except those with lithium and magnesium sulphates; and it was found in all cases that the average current value was attained at a point very close to that at which the indicators had travelled half their distance. With lithium and magnesium sulphates the current and velocities were so nearly constant that an appreciable error could hardly occur by assuming this as the correct point. The numbers given in the table were calculated in this manner.*

The other mode of calculation, which is a more rigid test of the experimental method, was applied to only two cases, but with satisfactory results. The values of U + V at intervals throughout the experiment were deduced by drawing tangents to a curve got by plotting the added lengths of the coloured portions of the tube against the time. The current-time curve gave the corresponding current values. Numerous values of the ratio $\frac{C}{U + V}$ were thus obtained, which in the case of half-normal KCl were constant to within 1 per cent. of their mean value, and in the case of twice-normal KCl were nearly, but not quite, so consistent. These mean values corresponded

* A rather frequent occurrence in these experiments may be mentioned here. It is one which, at first sight, looks as if it must be fatal to them, but which has been proved to be really unimportant. Flaws, having the appearance of small bubbles or cracks, are apt to appear in the jelly towards the end of an experiment. They seldom occur till the boundaries have done most of their journey, and they generally appear near the centre of the tube. In many experiments they do not appear at all, but when a flaw does come it is apt to extend in a rather curious manner. This occurrence of flaws at once causes a reduction of current, and it is easy to locate their first appearance in the time-current curve, even were it not noted at the time, as was always done. It has, however, very little, if any, effect on the velocities of the boundaries; and the ratio of these (u/v) is absolutely unaffected. It is indeed striking to watch the calm indifference with which these flaws are treated by the travelling colour boundaries, and to contrast the behaviour of the latter with that of the galvanometer needle. The facts are explicable on the assumption that the flaws reduce the current by reducing the value of A, leaving the current density (C/A), and therefore the actual velocities, as well as their ratio, unaffected. It is obvious that, in such a case, the experiment must be regarded as finished as soon as flaws appear, where the object is to test the truth of the equation involving the original, and only known, value of A; but that, for the determination of u/v (or of p), the experiment may be continued as usual till the boundaries meet. These rules have been observed in all cases. The occurrence of flaws of the kind described was noticed by LODGE (*loc. cit.*). Their cause has not been ascertained. Whether they will occur or not in any given experiment almost seems to be decided by caprice.

respectively to the figures 1·015 and 1·002 for $\frac{C\eta}{An(U+V)}$, in place of 1·017 and 1·014, found by the other method and given in the table.

TABLE II.—Relative Velocities of the Ions.

	Chlorides.			Sulphates.			KOHLRAUSCH.		
	n = ·5.	n = 1.	n = 2.	n = ·5.	n = 1.	n = 2.	1879.	1885.	1893.
K	100	100	100	100	100	...	100	100	100
Na	65·7	65·4	65·8	66·9	66·9	...	65	62	68
Li	44·7	45·2	...	47·1	44·4	45·2	44	46	55
NH_4	...	100	...	...	...	...	98	96	100
½Mg	...	...	...	40·5	36·9	38·7	...	...	...
Cl	97·9	96·1	93·6	...	...	...	102	104	105
½SO_4	...	...	...	87·7	87·7	87·7	...	...	...

In Table II., the results of the same experiments are given in such a form that the specific velocities of the different ions may be compared with one another and with those calculated for the same ions by KOHLRAUSCH. From the found values of $\frac{u\,(K)}{v\,(Cl)}$ and $\frac{u\,(Na)}{v\,(Cl)}$, the value of $\frac{u\,(Na)}{u\,(K)}$ may evidently be deduced, and so with the others. By making u (K) at each concentration equal to 100, comparable values are obtained. These are given for the chlorides and sulphates employed. Under KOHLRAUSCH's name are given, in parallel columns, the relative values of his specific velocities for the same ions, also reduced to the basis $u\,(K) = 100$. The figures in the first of these columns correspond to the specific velocities calculated by him in 1879 ('Ann. Phys. Chem.,' vol. 6, p. 172) by extrapolation to extreme dilution from experiments with solutions the strength of which was not less than half normal. The figures in the second column correspond to the velocities given by him in 1885 ('Ann. Phys. Chem.,' vol. 26, p. 214) for solutions one-tenth normal, while the figures in the last column correspond to his well-known specific velocities at infinite dilution, calculated in 1893 ('Ann. Phys. Chem.,' vol. 50, p. 408) by extrapolation from his later work with exceedingly dilute solutions.

The theoretical considerations advanced in the earlier part of this paper show that no exact agreement need be expected between the values for the same ion in the various columns of this Table. A general agreement might, however, be expected; and it is

Fig. 3.

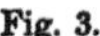

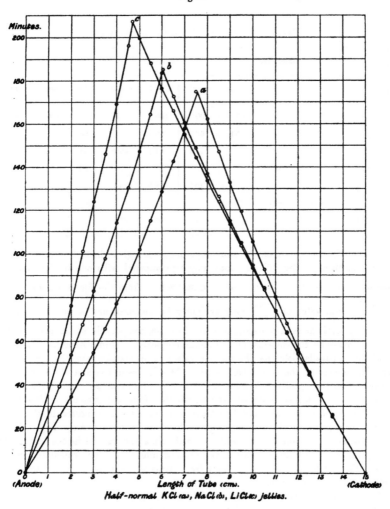

Half-normal KCl (a), NaCl (b), LiCl (c) jellies.

Fig. 4.

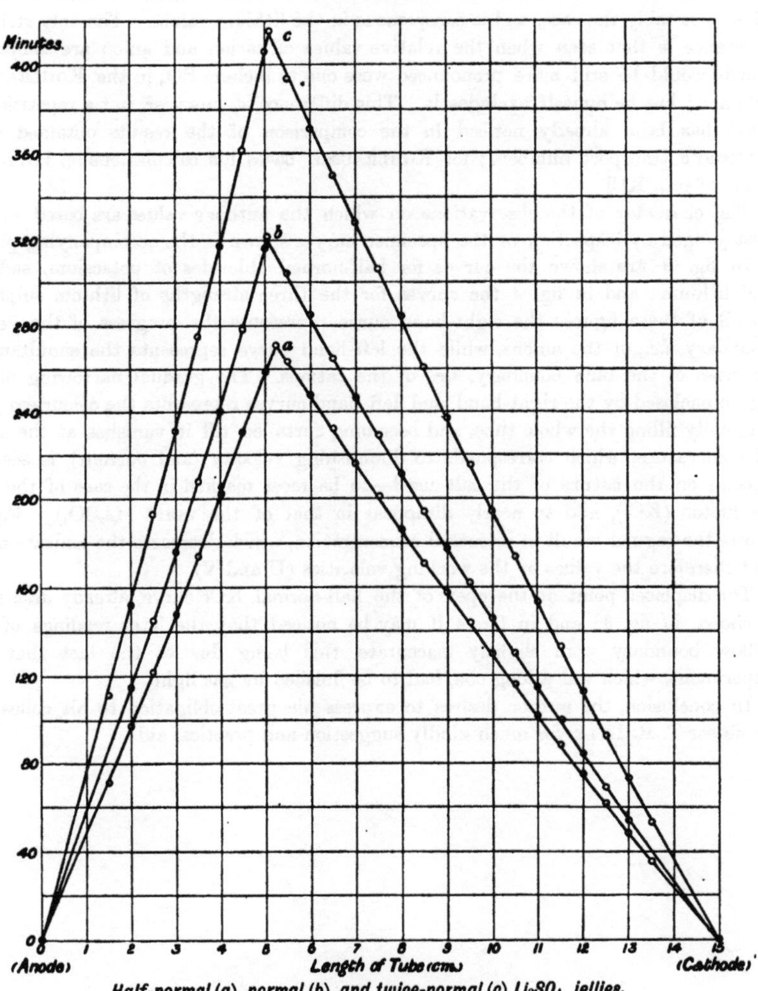

Half-normal (a), normal (b), and twice-normal (c) Li_2SO_4 jellies.

certainly seen to exist. Moreover, one would naturally expect the author's results to show better agreement with KOHLRAUSCH'S 1879 values than with his later ones, as the former were deduced from the conductivities of normal and half-normal solutions. This is notably the case, and is conspicuous in the lithium values. The only striking difference is that seen when the relative values of cation and anion are compared, which would be still more pronounced were one to include SO_4 in the KOHLRAUSCH columns; but he himself excludes it. This difference is, however, but a repetition of what has been already noticed in the comparison of the results obtained with HITTORF'S transport numbers; for KOHLRAUSCH bases his calculations on HITTORF'S value of p in KCl.

The character of the observations on which the author's values are based will be best judged by inspection of the specimen curves shown in the accompanying plates.

In fig. 3 are shown the curves for half-normal chlorides of potassium, sodium, and lithium; and in fig. 4 the curves for the three strengths of lithium sulphate. In all of these figures the right-hand curve represents the progress of the yellow boundary, *i.e.*, of the anions, while the left-hand curve represents the simultaneous progress of the blue boundary, *i.e.*, of the cations. The gradual narrowing of the figure enclosed by the right-hand and left-hand curves represents the colourless jelly originally filling the whole tube, and becoming curtailed till it vanishes at the apex. The curvature, which corresponds to diminishing velocity (and current), is seen to depend on the nature of the salt used—to be most marked in the case of the best conductor (KCl), and to nearly disappear in that of the worst (Li_2SO_4). Fig. 4 shows the regular result of increased concentration, which decreases the ionization (x), and therefore the values of the working velocities (U and V).

The displaced point at the apex of the half-normal KCl curve, already described, is shown in fig. 3; and in fig. 4 it may be noticed that the later readings of the yellow boundary were slightly inaccurate, this being due to the fact that the experiment, which was a long one, had to be finished by gas light.

In conclusion, the author desires to express his great obligation to his colleague, Professor T. R. LYLE, for much kindly suggestion and practical aid.

rs.—Part II.

.S.

INDEX SLIP.

---

proper development,
The subject-matter
ecessary consequence
into prominence.

t all the integers are
ast-written condition

essive parts, and we

brought under view,
he seven symbols

5.5.99

We thus obtain 7^{s-1} different sets of conditions that may be assigned; these are not all essentially different and in many cases they overlap.

Art. 65. For the moment I concentrate attention upon the symbol

$$\geqq,$$

and remark that the $s-1$ conditions, which involve this symbol, set forth above, constitute one set of a large class of sets which involve the symbol. We may have the single condition

$$A_1^{(1)}\alpha_1 + A_2^{(1)}\alpha_2 + A_3^{(1)}\alpha_3 + \ldots + A_s^{(1)}\alpha_s \geqq 0,$$

wherein $A_1, A_2, A_3 \ldots A_s$ are integers $+$, zero or $-$, of which at least one must be positive, or we may have the set of conditions

$$\left.\begin{array}{l} A_1^{(1)}\alpha_1 + A_2^{(1)}\alpha_2 + A_3^{(1)}\alpha_3 + \ldots + A_s^{(1)}\alpha_s \geqq 0 \\ A_1^{(2)}\alpha_1 + A_2^{(2)}\alpha_2 + A_3^{(2)}\alpha_3 + \ldots + A_s^{(2)}\alpha_s \geqq 0 \\ A_1^{(3)}\alpha_1 + A_2^{(3)}\alpha_2 + A_3^{(3)}\alpha_3 + \ldots + A_s^{(3)}\alpha_s \geqq 0 \\ \cdot \quad \cdot \quad \cdot \quad \cdot \quad \cdot \quad \cdot \quad \cdot \quad \cdot \quad \cdot \quad \cdot \quad \cdot \quad \cdot \\ A_1^{(r)}\alpha_1 + A_2^{(r)}\alpha_2 + A_3^{(r)}\alpha_3 + \ldots + A_s^{(r)}\alpha_s \geqq 0 \end{array}\right\}$$

as the definition of the partitions considered. If the symbol be $=$ instead of $\geqq$ the solution of the equations falls into the province of linear Diophantine analysis. The problem before us may be regarded as being one of linear partition analysis. There is much in common between the two theories; the problems may be treated by somewhat similar methods.

The partition analysis of degree higher than the first, like the Diophantine, is of a more recondite nature, and is left for the present out of consideration.

I treat the partition conditions by the method of generating functions. I seek the summation

$$\Sigma X_1^{\alpha_1} X_2^{\alpha_2} X_3^{\alpha_3} \ldots X_s^{\alpha_s}$$

for every set of values (integers)

$$\alpha_1, \alpha_2, \alpha_3, \ldots \alpha_s$$

which satisfy the assigned conditions.

It appears that there are, in every case, a finite number of ground or fundamental solutions of the conditions, viz.:—

$$\begin{array}{l} \alpha_1^{(1)}, \alpha_2^{(1)}, \alpha_3^{(1)} \ldots \alpha_s^{(1)} \\ \alpha_1^{(2)}, \alpha_2^{(2)}, \alpha_3^{(2)} \ldots \alpha_s^{(2)} \\ \cdot \quad \cdot \quad \cdot \quad \cdot \quad \cdot \quad \cdot \quad \cdot \quad \cdot \\ \alpha_1^{(m)}, \alpha_2^{(m)}, \alpha_3^{(m)} \ldots \alpha_s^{(m)} \end{array}$$

such that every solution

$$\alpha_1,\ \alpha_2,\ \alpha_3,\ \ldots\ \alpha_s$$

is such that

$$\alpha_1 = \lambda_1\alpha_1^{(1)} + \lambda_2\alpha_1^{(2)} \ldots + \lambda_m\alpha_1^{(m)}$$

$$\alpha_2 = \lambda_1\alpha_2^{(1)} + \lambda_2\alpha_2^{(2)} \ldots + \lambda_m\alpha_2^{(m)}$$

$$\alpha_3 = \lambda_1\alpha_3^{(1)} + \lambda_2\alpha_3^{(2)} \ldots + \lambda_m\alpha_3^{(m)}$$

$$\cdot\quad\cdot\quad\cdot\quad\cdot\quad\cdot\quad\cdot\quad\cdot\quad\cdot\quad\cdot\quad\cdot\quad\cdot\quad\cdot$$

$$\alpha_s = \lambda_1\alpha_s^{(1)} + \lambda_2\alpha_s^{(2)} \ldots + \lambda_m\alpha_s^{(m)}$$

$\lambda_1, \lambda_2, \ldots \lambda_m$ being positive integers.

This arises from the fact that every term

$$X_1^{\alpha_1}X_2^{\alpha_2}X_3^{\alpha_3}\ldots X_s^{\alpha_s}$$

of the summation is found to be expressible as a product

$$\begin{aligned}&\{X_1^{\alpha_1^{(1)}}X_2^{\alpha_2^{(1)}}X_3^{\alpha_3^{(1)}}\ldots X_s^{\alpha_s^{(1)}}\}^{\lambda_1}\\ \times&\{X_1^{\alpha_1^{(2)}}X_2^{\alpha_2^{(2)}}X_3^{\alpha_3^{(2)}}\ldots X_s^{\alpha_s^{(2)}}\}^{\lambda_2}\\ \times&\ \cdot\quad\cdot\quad\cdot\quad\cdot\quad\cdot\quad\cdot\quad\cdot\quad\cdot\quad\cdot\\ \times&\{X_1^{\alpha_1^{(m)}}X_2^{\alpha_2^{(m)}}X_3^{\alpha_3^{(m)}}\ldots X_s^{\alpha_s^{(m)}}\}^{\lambda_m}\end{aligned}$$

Denoting this product by

$$P_1^{\lambda_1}P_2^{\lambda_2}\ldots P_m^{\lambda_m}$$

the generating function assumes the form

$$\frac{1-(Q_1^{(1)}+Q_1^{(2)}+Q_1^{(3)}+\ldots)+(Q_2^{(1)}+Q_2^{(2)}+Q_2^{(3)}+\ldots)-(Q_3^{(1)}+\ldots)+\ldots}{(1-P_1)(1-P_2)\ldots(1-P_m)}$$

wherein the denominator indicates the ground solutions and the numerator the simple and compound syzygies which unite them.

The terms

$Q_1^{(1)}$, $Q_1^{(2)}$, $Q_1^{(3)}$ . . . denote first syzygies
$Q_2^{(1)}$, $Q_2^{(2)}$, $Q_2^{(3)}$ . . . „ second „
$Q_3^{(1)}$, $Q_3^{(2)}$, $Q_3^{(3)}$ . . . „ third „
. . . .

The reader will note the striking analogy with the generating functions of the theory of invariants.

Similar results are obtained as solutions of linear Diophantine equations.

The generating functions under view are *real* in the sense of CAYLEY and SYLVESTER. Enumerating generating functions of various kinds are obtained by assigning equalities between the suffixed capitals

$$X_1, X_2, \ldots X_s.$$

Putting, *e.g.*,

$$X_1 = X_2 = \ldots = X_s = x,$$

we obtain the function which enumerates by the coefficient of x^n, in the ascending expansion, the numbers of solutions for which

$$\alpha_1 + \alpha_2 + \ldots + \alpha_s = n.$$

It will be gathered that the note of the following investigation is the importation of the idea that the solution of any system of equations of the form

$$A_1\alpha_1 + A_2\alpha_2 + A_3\alpha_3 + \ldots + A_s\alpha_s \geqq 0$$

(all the quantities involved being integers) is a problem of partition analysis, and that the theory proceeds *pari passu* with that of the linear Diophantine analysis.

Section 5.

Art. 66. I propose to lead up to the general theory of partition analysis by considering certain simple particular cases in full detail.

Suppose we have a function

$$F(x, a)$$

which can be expanded in ascending powers of x. Such expansion being either finite or infinite, the coefficients of the various powers of x are functions of a which in general involve both positive and negative powers of a. We may reject all terms containing negative powers of a and subsequently put a equal to unity. We thus arrive at a function of x only, which may be represented after CAYLEY (modified by the association with the symbol $\geqq$) by

$$\underset{\geqq}{\Omega} F(x, a),$$

the symbol $\geqq$ denoting that the terms retained are those in which the power of a is $\geqq 0$.

Similarly we may indicate by the operation

$$\underset{=}{\Omega}$$

that the only terms retained are those in which a occurs to the power zero and the meaning of the operations

$$\underset{>}{\Omega}, \quad \underset{<}{\Omega}, \quad \underset{=}{\Omega}, \quad \underset{\geqq}{\Omega}$$

will be understood without further explanation. To generalise the notion we may consider

$$\underset{\geqq}{\Omega} F(X_1, X_2, \ldots X_s, a_1, a_2, \ldots a_t)$$

to mean that the function is to be expanded in ascending powers of $X_1, X_2, \ldots X_s$, the terms involving any negative powers of $a_1, a_2, \ldots a_t$ are to be rejected, and that subsequently we are to put

$$a_1 = a_2 = \ldots = a_t = 1.$$

In this case the operation Ω has reference to each of the letters $a_1, a_2, \ldots a_t$ and a term involving any negative power of either of these quantities is rejected.

If the quantities $a_1, a_2, \ldots a_t$ be not all subjected to the same operation we may denote the whole operation by

$$\overset{a_1}{\underset{\geqq}{\Omega}} \overset{a_2}{\underset{>}{\Omega}} \overset{a_3}{\underset{\geqq}{\Omega}} \ldots \overset{a_t}{\underset{=}{\Omega}} F(X_1, X_2, \ldots X_s, a_1, a_2, a_3 \ldots a_t)$$

wherein $\overset{a_r}{\underset{\sigma_r}{\Omega}}$ operates upon a_r according to the law of the symbol σ_r.

The operation, *quâ* a single quantity and the symbol $\geqq$, have been studied by CAYLEY.* *Quâ* more than one quantity it has presented itself in a memoir on partitions by the present author.†

These Ω functions are of moment in all questions of partition and linear Diophantine analysis.

Art. 67. I will construct Ω functions to serve as generators of well-known solutions and enumerations in the theory of unipartite partition.

Problem I. To determine the number of partitions of w into i or fewer parts.

Graphically considered we have i rows of nodes

$$\begin{array}{llllllll} a_1 & \cdot & \cdot & \cdot & \cdot & \cdot & \cdot & \cdot \\ a_2 & \cdot & & & & & & \\ a_3 & \cdot & & & & & & \\ \vdots & & & & & & & \\ a_i & \cdot & & & & & & \end{array}$$

* "On an Algebraical Operation," 'Collected Papers,' vol. 9, p. 537.

† "Memoir on the Theory of the Partitions of Numbers," Part I., 'Phil. Trans.,' A, vol. 187, pp. 619–673, 1896.

$\alpha_1, \alpha_2, \ldots$ denoting the numbers of nodes in the first, second, &c., rows,

$$\begin{gathered} \alpha_1 \geqq \alpha_2 \\ \alpha_2 \geqq \alpha_3 \\ \vdots \\ \alpha_{i-1} \geqq \alpha_i \end{gathered}$$

To find

$$\Sigma X_1^{\alpha_1} X_2^{\alpha_2} \ldots X_i^{\alpha_i}$$

for all sets of integers satisfying the conditions take

$$\underset{\geqq}{\Omega} \frac{1}{1 - a_1 X_1 \,.\, 1 - \frac{a_2}{a_1} X_2 \,.\, 1 - \frac{a_3}{a_2} X_3 \ldots 1 - \frac{a_i}{a_{i-1}} X_i}$$

where observe that the factors $\frac{1}{1 - a_1 X_1}, \frac{1}{1-(a_2/a_1)X_2}, \ldots$ generate the successive rows of nodes and that the method of placing the letters $a_1, a_2, \ldots$ ensures the satisfaction of the first, second, &c., conditions.

Continued application of the simple theorem

$$\underset{\geqq}{\Omega} \frac{1}{1 - ax \,.\, 1 - \frac{1}{a} y} = \frac{1}{1 - x \,.\, 1 - xy},$$

applied in respect of the quantities $a_1, a_2, \ldots$ in succession, reduces the Ω function to the form

$$\frac{1}{1 - X_1 \,.\, 1 - X_1 X_2 \,.\, 1 - X_1 X_2 X_3 \ldots 1 - X_1 X_2 X_3 \ldots X_i}$$

the *real* generating function.

The ground solutions or fundamental partitions are, as shown by the denominator factors,

$$(\alpha_1, \alpha_2, \alpha_3, \ldots \alpha_i)$$

$$= \left\{ \begin{array}{l} (1,\ 0,\ 0,\ \ldots 0) \\ (1,\ 1,\ 0,\ \ldots 0) \\ (1,\ 1,\ 1,\ \ldots 0) \\ \ \cdot \quad \cdot \quad \cdot \quad \cdot \quad \cdot \\ (1,\ 1,\ 1,\ \ldots 1) \end{array} \right.$$

and, as might have been anticipated, the graphical representation is in evidence.

Art. 68. By choosing to sum the expression

$$\Sigma X_1^{\alpha_1} X_2^{\alpha_2} \dots X_i^{\alpha_i},$$

every solution of the given conditions has been generated. The same result might have been achieved by other summations such as

$$\Sigma X_1^{\lambda_1\alpha_1} X_2^{\mu_2\alpha_2} \dots X_i^{\eta_i\alpha_i},$$

$\lambda_1, \lambda_2, \dots \lambda_i$ being given positive integers, or as

$$\Sigma X_1^{\alpha_1-\alpha_2} X_2^{\alpha_2-\alpha_3} \dots X_{i-1}^{\alpha_{i-1}-\alpha_i} X_i^{\alpha_i}.$$

We, in fact, may take as indices of $X_1, X_2, \dots X_i$ any given linear functions of $\alpha_1, \alpha_2, \dots \alpha_i$, and form the corresponding generating function.

For the two cases specified, the Ω functions are

$$\underset{\geq}{\Omega} \frac{1}{1 - a_1 X_1^{\lambda_1} . 1 - \frac{a_2}{a_1} X_2^{\mu_2} \dots 1 - \frac{1}{a_{i-1}} X_i^{\eta_i}},$$

$$\underset{\geq}{\Omega} \frac{1}{1 - a_1 X_1 . 1 - \frac{a_2}{a_1}\frac{X_2}{X_1} . 1 - \frac{a_3}{a_2}\frac{X_3}{X_2} \dots 1 - \frac{1}{a_{i-1}}\frac{X_i}{X_{i-1}}},$$

and the reduced functions

$$\frac{1}{1 - X_1^{\lambda_1} . 1 - X_1^{\lambda_1} X_2^{\mu_2} \dots 1 - X_1^{\lambda_1} X_2^{\mu_2} \dots X_i^{\eta_i}},$$

$$\frac{1}{1 - X_1 . 1 - X_2 . 1 - X_3 \dots 1 - X_i}$$

respectively.

Generally for the sum

$$\Sigma X_1^{\lambda_1\alpha_1 + \mu_1\alpha_2 + \dots} X_2^{\lambda_2\alpha_1 + \mu_2\alpha_2 + \dots} \dots X_i^{\lambda_i\alpha_1 + \mu_i\alpha_2 + \dots + \eta_i\alpha_i}$$

the two functions are

$$\underset{\geq}{\Omega} \frac{1}{1 - a_1 X_1^{\lambda_1} X_2^{\lambda_2} \dots X_i^{\lambda_i} . 1 - \frac{a_2}{a_1} X_1^{\mu_1} X_2^{\mu_2} \dots X_i^{\mu_i} \dots 1 - \frac{1}{a_{i-1}} X_1^{\eta_1} X_2^{\eta_2} \dots X_i^{\eta_i}}$$

and

$$\frac{1}{1 - X_1^{\lambda_1} X_2^{\lambda_2} \dots X_i^{\lambda_i} . 1 - X_1^{\lambda_1+\mu_1} X_2^{\lambda_2+\mu_2} \dots X_i^{\lambda_i+\mu_i} \dots 1 - X_1^{\lambda_1+\dots+\eta_1} X_2^{\lambda_2+\dots+\eta_2} \dots X_i^{\lambda_i+\dots+\eta_i}}.$$

Art. 69. In any of these instances we have i quantities at disposal, viz.:

$$X_1, X_2, \dots X_i,$$

in order to derive enumerating generating functions corresponding to certain problems. In the last-written general case, the quantities $\lambda, \mu, \ldots \eta$ being given integers, put as a particular case,

$$X_1 = X_2 = \ldots = X_i = x.$$

The reduced function is

$$\frac{1}{1 - x^{\Sigma\lambda} \,.\, 1 - x^{\Sigma\lambda + \Sigma\mu} \,.\, - \ldots 1 - x^{\Sigma\lambda + \Sigma\mu + \ldots + \Sigma\eta}},$$

and herein the coefficients of x^w, in the expansion, give the number of partitions

$$\alpha_1, \alpha_2, \alpha_3, \ldots \alpha_i$$

of all numbers which satisfy the equation

$$\Sigma\lambda \,.\, \alpha_1 + \Sigma\mu \,.\, \alpha_2 + \ldots + \Sigma\eta \,.\, \alpha_i = w,$$

$\alpha_1, \alpha_2, \ldots \alpha_i$ being in descending order.

For the three particular cases considered above this equation takes the forms

$$\alpha_1 + \alpha_2 + \ldots + \alpha_i = w,$$
$$\lambda_1\alpha_1 + \mu_2\alpha_2 + \ldots + \eta_i\alpha_i = w,$$
$$\alpha_1 = w,$$

connected with the reduced generators,

$$\frac{1}{1 - x \,.\, 1 - x^2 \,.\, 1 - x^3 \ldots 1 - x^i},$$

$$\frac{1}{1 - x^{\lambda_1} \,.\, 1 - x^{\lambda_1 + \mu_2} \ldots 1 - x^{\lambda_1 + \mu_2 + \ldots + \eta_i}},$$

$$\frac{1}{(1 - x)^i},$$

respectively.

Further, we may separate $X_1, X_2, \ldots X_i$ in any manner into k sets and put those which are in the first set equal to x_1, those in the second equal to x_2, and so on, and so reach an enumerating function involving k quantities, $x_1, x_2, x_3, \ldots x_k$.

Ex. gr. Put

$$X_1 = X_3 = X_5 = \ldots = x_1,$$
$$X_2 = X_4 = X_6 = \ldots = x_2,$$

and suppose i even. We obtain

$$\frac{1}{1 - x_1 \,.\, 1 - x_1x_2 \,.\, 1 - x_1^2x_2 \,.\, 1 - x_1^2x_2^2 \ldots 1 - x_1^{\frac{1}{2}i}x_2^{\frac{1}{2}i}},$$

to enumerate by the coefficient of $x_1^{w_1}x_2^{w_2}$ those partitions of $w_1 + w_2$ for which

$$\alpha_1 + \alpha_3 + \alpha_5 + \ldots = w_1$$
$$\alpha_2 + \alpha_4 + \alpha_6 + \ldots = w_2.$$

This enumerating function, since it involves x_1 and x_2, is one connected also with the partitions of bipartite numbers. In general when k sets are taken, we have a theorem of k-partite partitions. When $k = i$, we have at once a *real* generating function for unipartites and an enumerating function for i-partites, for, from the latter point of view, the number unity which appears as the coefficient of $X_1^{\alpha_1}X_2^{\alpha_2}\ldots X_i^{\alpha_i}$ shows that the multipartite number

$$\overline{\alpha_1\alpha_2\ldots\alpha_i}$$

can be partitioned in one way only into the parts

$$\overline{1\,0\,.\,.\,.\,.\,.\,.\,.}$$
$$\overline{1\,1\,0\,.\,.\,.\,.\,.\,.}$$
$$\overline{1\,1\,1\,.\,.\,.\,.\,.\,.}$$
$$.\;.\;.\;.\;.\;.\;.\;.\;.$$
$$\overline{1\,1\,1\,.\,.\,.\,.\,.\,1}$$

there being i figures in each part.

Art. 70. We may now enquire into the partitions of all numbers

$$\alpha_1, \quad \alpha_2, \quad \alpha_3, \ldots \alpha_i,$$

subject to the given conditional relations and also to the linear equations

$$\lambda_1\alpha_1 + \mu_2\alpha_2 + \ldots + \eta_i\alpha_i = w$$
$$\lambda'_1\alpha_1 + \mu'_2\alpha_2 + \ldots + \eta'_i\alpha_i = w'$$
$$.\;.\;.\;.\;.\;.\;.\;.\;.\;.\;.\;.$$
$$\lambda_1^{(s)}\alpha_1 + \mu_2^{(s)}\alpha_2 + \ldots + \eta_i^{(s)}\alpha_i = w^{(s)}.$$

To illustrate the method, it suffices to take $s = 2$, and then we have to perform the summation

$$\Sigma X_1^{\lambda_1\alpha_1}X_2^{\mu_2\alpha_2}\ldots X_i^{\eta_i\alpha_i}Y_1^{\lambda'_1\alpha_1}Y_2^{\mu'_2\alpha_2}\ldots Y_i^{\eta'_i\alpha_i}.$$

The Ω function reduced is

$$\frac{1}{1 - X_1^{\lambda_1}Y_1^{\lambda'_1}\,.\,1 - X_1^{\lambda_1}X_2^{\mu_2}Y_1^{\lambda'_1}Y_2^{\mu'_2}\ldots 1 - X_1^{\lambda_1}X_2^{\mu_2}\ldots X_i^{\eta_i}Y_1^{\lambda'_1}Y_2^{\mu'_2}\ldots Y_i^{\eta'_i}},$$

wherein putting

$$X_1 = X_2 = \ldots = X_i = x,$$
$$Y_1 = Y_2 = \ldots = Y_i = y,$$

we obtain the enumerating function

$$\frac{1}{1 - x^{\lambda_1}y^{\lambda'_1} . 1 - x^{\lambda_1+\mu_2}y^{\lambda'_1+\mu'_2} \ldots 1 - x^{\lambda_1+\mu_2+\ldots+\nu_i}y^{\lambda'_1+\mu'_2+\ldots+\nu'_i}},$$

in which we seek the coefficient of $x^w y^{w'}$.

Art. 71. *Ex. gr.* Consider the particular case

$$\alpha_1 + \alpha_2 + \ldots + \alpha_i = w,$$
$$\alpha_1 + 2\alpha_2 + \ldots + i\alpha_i = w',$$

$\alpha_1, \alpha_2, \ldots \alpha_i$ being, as usual, subject to the conditional relations.

The enumerating function is

$$\frac{1}{1 - xy . 1 - x^2y^3 . 1 - x^3y^6 \ldots 1 - x^iy^{\frac{1}{2}i(i+1)}},$$

and it is obvious also that the partitions of the bipartite $\overline{ww'}$ which satisfy the conditions may be composed by the biparts

$$\overline{11},\ \overline{23},\ \overline{36}, \ldots \overline{i, \tfrac{1}{2}i(i+1)}.$$

The corresponding graphical representation is not by superposition of lines of nodes, but by angles of nodes, of the natures

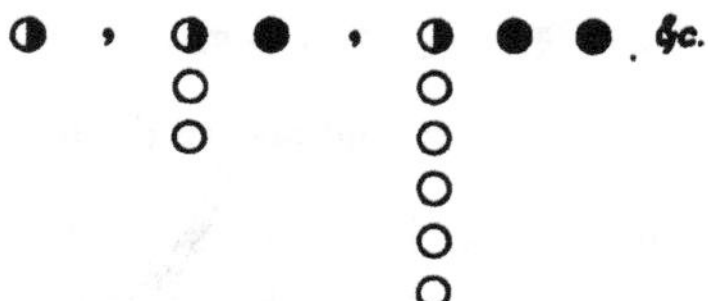

Art. 72. It is convenient, at this place, to give some elementary theorems concerning the Ω function which will be useful in what follows.

$$\underset{\geq}{\Omega} \frac{1}{1 - ax . 1 - \frac{1}{a}y} = \frac{1}{1 - x . 1 - xy},$$

$$\underset{>}{\Omega} \frac{1}{1 - ax . 1 - ay . 1 - \frac{1}{a}z} = \frac{1 - xyz}{1 - x . 1 - y . 1 - xz . 1 - yz},$$

$$\underset{\geqq}{\Omega}\frac{1}{1-ax.1-\frac{1}{a}y.1-\frac{1}{a}z}=\frac{1}{1-x.1-xy.1-xz},$$

$$\underset{\geqq}{\Omega}\frac{1}{1-a^2x.1-\frac{1}{a}y}=\frac{1+xy}{1-x.1-xy^2},$$

$$\underset{\geqq}{\Omega}\frac{1}{1-ax.1-\frac{1}{a^2}y}=\frac{1}{1-x.1-x^2y},$$

$$\underset{\geqq}{\Omega}\frac{1}{1-a^3x.1-\frac{1}{a}y}=\frac{1+xy+xy^2}{1-x.1-xy^3},$$

$$\underset{\geqq}{\Omega}\frac{1}{1-ax.1-\frac{1}{a^3}y}=\frac{1}{1-x.1-x^3y},$$

$$\underset{\geqq}{\Omega}\frac{1}{1-a^2x.1-ay.1-\frac{1}{a}z}=\frac{1+xz-xyz-xyz^2}{1-x.1-y.1-yz.1-xz^2},$$

$$\underset{\geqq}{\Omega}\frac{1}{1-a^2x.1-\frac{1}{a}y.1-\frac{1}{a}z}=\frac{1+xy+xz+xyz}{1-x.1-xy^2.1-xz^2},$$

$$\underset{\geqq}{\Omega}\frac{1}{1-ax.1-ay.1-az.1-\frac{1}{a}w}=\frac{1-xyw-xzw-yzw+xyzw+xyzw^2}{1-x.1-y.1-z.1-xw.1-yw.1-zw},$$

$$\underset{\geqq}{\Omega}\frac{1}{1-ax.1-ay.1-\frac{1}{a}z.1-\frac{1}{a}}=\frac{1-xyz-xyw-xyzw+xy^2zw+x^2yzw}{1-x.1-y.1-xz.1-xw.1-yz.1-yw}.$$

Art. 73. I pass on to consider the partitions of numbers into parts limited not to exceed i in magnitude.

The Ω function is clearly

$$\underset{\geqq}{\Omega}\frac{1-(a_1X_1)^{i+1}}{1-a_1X_1}\cdot\frac{1-\left(\frac{a_2}{a_1}X_2\right)^{i+1}}{1-\frac{a_2}{a_1}X_2}\cdot\frac{1-\left(\frac{a_3}{a_2}X_3\right)^{i+1}}{1-\frac{a_3}{a_2}X_3}\ldots \textit{ad inf.}$$

In this form I have not succeeded in effecting the reduction, but if we put at once

$$X_1=X_2=X_3=\ldots=x,$$

the reduced form is

$$\frac{1}{1-x.1-x^2.1-x^3\ldots1-x^i}$$

If the parts be limited to i in number and to j in magnitude, we find

$$\underset{\geq}{\Omega}\frac{1-(a_1x)^{j+1}}{1-a_1x}\cdot\frac{1-\left(\frac{a_2}{a_1}x\right)^{j+1}}{1-\frac{a_2}{a_1}x}\cdots\frac{1-\left(\frac{1}{a_{i-1}}x\right)^{j+1}}{1-\frac{1}{a_{i-1}}x}=\frac{1-x^{j+1}\,.\,1-x^{j+2}\,.\,1-x^{j+3}\ldots 1-x^{j+i}}{1-x\,.\,1-x^2\,.\,1-x^3\ldots 1-x^i}$$

the well-known result.

Art. 74. It is to be remarked that the generating function in question may also be written

$$\underset{\geq}{\Omega}\frac{\frac{1}{1-g}}{1-a_1gx\,.\,1-\frac{a_2}{a_1}x\,.\,1-\frac{a_3}{a_2}x^2\ldots 1-\frac{1}{a_{i-1}}x},$$

in which we have to seek the coefficient of g^j. This function reduces to

$$\frac{1}{1-g\,.\,1-gx\,.\,1-gx^2\,.\,1-gx^3\ldots 1-gx^i}$$

the well-known form.

In general, when a generating function reduces to the product of factors

$$\frac{1}{1-x^s},$$

the part-magnitude being unrestricted, we obtain a product of factors

$$\frac{1}{1-gx^s}$$

for the restricted case, and this is frequently exhibitable, as regards the coefficients of g^i, as a product of factors

$$\frac{1-x^{i+s}}{1-x^s}.$$

The Ω function is not altered by the interchange of the letters i, j.

Art. 75. If the successive parts of the partition are limited in magnitude by

$$j_1, j_2, \ldots j_i,$$

numbers necessarily in descending order, the generating function is

$$\underset{\geq}{\Omega}\frac{1-(a_1x)^{j_1+1}}{1-a_1x}\cdot\frac{1-\left(\frac{a_2}{a_1}x\right)^{j_2+1}}{1-\frac{a_2}{a_1}x}\cdots\frac{1-\left(\frac{1}{a_{i-1}}x\right)^{j_i+1}}{1-\frac{1}{a_{i-1}}x}.$$

For $i = 2$, this may be shown to be equal to

$$\frac{(1 - x^{j_2+1})(1 - x^{j_1+2})}{(1 - x)(1 - x^2)} + x^{j_2+1} \frac{(1 - x^{j_2+1})(1 - x^{j_1-j_2})}{(1 - x)(1 - x^2)},$$

but for $i > 2$, the functions are obtained with increasing labour, and are of increasing complexity.

Many cases present themselves, similar to the one before us, where the Ω function is written down with facility, but no serviceable reduced function appears to exist. On the other hand, we meet with astonishing instances of compact reduced functions which involve valuable theorems.

Art. 76. From the reduced function we can frequently proceed to an $\underset{\geqq}{\Omega}$ function, thus inverting the usual process. If, for example, we require an Ω equivalent to

$$\frac{1}{1 - x^{P_1} . 1 - x^{P_2} . 1 - x^{P_3} \ldots 1 - x^{P_i}},$$

a little consideration leads us to

$$\underset{\geqq}{\Omega} \frac{1}{1 - a_1 x^{P_1} . 1 - \frac{a_2}{a_1} x^{P_2 - P_1} . 1 - \frac{a_3}{a_2} x^{P_3 - P_2} \ldots 1 - \frac{1}{a_{i-1}} x^{P_i - P_{i-1}}}.$$

This indicates that a unipartite partition into the parts $P_1, P_2, \ldots P_i$ may be represented by a two-dimensional partition of another kiud which involves the parts

$$P_1, P_2 - P_1, P_3 - P_2, \ldots P_i - P_{i-1}.$$

Ex. gr., the numbers P_1, P_2, P_3 being in ascending order, the line partition

$$P_3 P_3 P_3 P_2 P_2 P_1$$

can be thrown into the plane partition

$$\begin{array}{cccccc} P_1 & P_1 & P_1 & P_1 & P_1 & P_1 \\ P_2 - P_1 & P_2 - P_1 & P_2 - P_1 & P_2 - P_1 & P_2 - P_1 & \\ P_3 - P_2 & P_3 - P_2 & P_3 - P_2. & & & \end{array}$$

of the nature of a regularised graph in the elements $P_1, P_2 - P_1, P_3 - P_2$, though these quantities are not necessarily in any specified order of magnitude. We obtain, in fact, a mixed numerical and graphical representation of a partition of a new kind. If

$$(P_1, P_2, P_3) = (1, 3, 4),$$

the partition 4 3 3 3 1 has the mixed graph

1 1 1 1 1
2 2 2 2
1

as well as its ordinary unit-graph.

In one case the mixed graph is composed entirely of units, and is, moreover, the graph conjugate to the unit graph.

This happens when

$$(P_1, P_2, P_3, \ldots) = (1, 2, 3, \ldots).$$

Thus, *quâ* these elements,

4 3 3 3 1

has the mixed (here the conjugate) graph

1 1 1 1 1
1 1 1 1
1 1 1 1 .
1

Art. 77. Observe that a partition may be such *quâ* the parts which actually appear in it, *or* it may be *quâ*, in addition, certain parts which might appear, but which happen to be absent. A mixed graph corresponds to each such supposition.

Ex. gr. :—

Partition.	*Quâ* elements.	Graph.
4 3	4, 3	3 3 1
4 3	4, 3, 1	1 1 2 2 1
4 3	4, 3, 2	2 2 1 1 1
4 3	4, 3, 2, 1	1 1 1 1 1 1 1

We thus arrive at a generalization of the notion of a conjugate partition, and are convinced that the proper representation of a Ferrers-graph is not by nodes or points, but by units.

When the mixed elements

$$P_1, P_2 - P_1, P_3 - P_2, \ldots$$

are in descending order of magnitude we have a correspondence between unipartite partitions and multipartite partitions of a certain class.

Art. 78. It is usual to consider the parts of a partition arranged in descending order. The Ω function enables us to assign any desired order of magnitude between the successive parts.

In the case of three parts we have already considered the system

$$\alpha_1 \geqq \alpha_2, \quad \alpha_2 \geqq \alpha_3.$$

For the system

$$\alpha_1 \geqq \alpha_2, \quad \alpha_3 \geqq \alpha_2,$$

we have the solution

$$\underset{\geqq}{\Omega} \frac{1}{1 - a_1X_1 \, . \, 1 - \dfrac{1}{a_1a_2}X_2 \, . \, 1 - a_2X_3},$$

and thence the *real* reduced generator

$$\frac{1}{1 - X_1 \, . \, 1 - X_1X_2X_3 \, . \, 1 - X_3},$$

and the enumerating function

$$\frac{1 + x}{(1 - x)(1 - x^2)(1 - x^3)}.$$

On the other hand, for the system

$$\alpha_2 \geqq \alpha_1, \quad \alpha_2 \geqq \alpha_3,$$

we construct

$$\underset{\geqq}{\Omega} \frac{1}{1 - \dfrac{X_1}{a_1} \, . \, 1 - a_1a_2X_2 \, . \, 1 - \dfrac{X_3}{a_2}},$$

leading to the real and enumerating functions

$$\frac{1 - X_1X_2^2X_3}{1 - X_2 \, . \, 1 - X_1X_2 \, . \, 1 - X_2X_3 \, . \, 1 - X_1X_2X_3},$$

$$\frac{1 + x^2}{(1 - x)(1 - x^2)(1 - x^3)};$$

of the former, the denominator shows the ground solutions, *id est*, fundamental partitions,

$$(\alpha_1, \alpha_2, \alpha_3) = (0, 1, 0); \ (110); \ (011); \ (111);$$

and the enumerator points to the syzygy

$$X_2 \, . \, X_1X_2X_3 - X_1X_2 \, . \, X_2X_3 = 0.$$

Art. 79. If the partition be into i parts, we can assign 2^{i-1} different orders depending upon the symbols $\geqq$, $\leqq$, and these can all be expressed by conditional relations affecting $\alpha_1, \alpha_2, \ldots \alpha_i$, involving the symbol $\geqq$ only. These are not all *essentially* different, as one order does or does not give rise to a different order by inversions of parts. Denoting $\geqq$, $\leqq$ by the letters d, a, we have for $i = 3$ the orders dd, da, ad, aa; the orders dd, aa are not essentially different, because interchange of a and d combined with inversion converts the one into the other; da, ad are essentially different, because this two-fold operation leaves each of these unchanged. Hence there are three orders to be considered, and the results have been obtained above.

For $i = 4$ we have the essentially different orders ddd, dda, dad, add. The first of these has been obtained; the other three are solved by the Ω functions:

$$\underset{\geqq}{\Omega} \frac{1}{1 - a_1X_1 . 1 - \frac{a_2}{a_1}X_2 . 1 - \frac{X}{a_2a} . 1 - a_3X_4};$$

$$\underset{\geqq}{\Omega} \frac{1}{1 - a_1X_1 . 1 - \frac{X_2}{a_1a_2} . 1 - a_2a_3X_3 . 1 - \frac{X_4}{a_3}};$$

$$\underset{\geqq}{\Omega} \frac{1}{1 - \frac{X_1}{a_1} . 1 - a_1a_2X_2 . 1 - \frac{a_3}{a_3}X_3 . 1 - \frac{X_4}{a_3}};$$

which reduce to the three expressions:

$$\frac{1}{1 - X_1 . 1 - X_4 . 1 - X_1X_2 . 1 - X_1X_2X_3X_4}$$

$$\frac{1 - X_1X_2X_3^2X_4}{1 - X_1 . 1 - X_3 . 1 - X_3X_4 . 1 - X_1X_2X_3 . 1 - X_1X_2X_3X}$$

$$\frac{1 - X_1X_2^2X_3 - X_1X_2^2X_3X_4 - X_1X_2^2X_3^2X_4 + X_1X_2^2X_3^2X_4 + X_1^2X_2^2X_3^2X}{1 - X_2 . 1 - X_1X_2 . 1 - X_2X_3 . 1 - X_1X_2X_3 . 1 - X_2X_3X_4 . 1 - X_1X_2X_3X};$$

and to the three enumerating functions:

$$\frac{1 + x + x^2}{1 - x . 1 - x^2 . 1 - x^3 . 1 - x^4};$$

$$\frac{1 + x + x^2 + x^3 + x^4}{1 - x . 1 - x^2 . 1 - x^3 . 1 - x^4};$$

$$\frac{1 + x^2 + x^3}{1 - x . 1 - x^2 . 1 - x^3 . 1 - x^4}.$$

The last real generating function that has been written down gives the solution of the system of conditions

$$\alpha_2 \geqq \alpha_1, \quad \alpha_2 \geqq \alpha_3, \quad \alpha_3 \geqq \alpha_4;$$

the ground solutions are

$$(\alpha_1, \alpha_2, \alpha_3, \alpha_4) = (0, 1, 0, 0), (1, 1, 0, 0), (0, 1, 1, 0), (1, 1, 1, 0), (0, 1, 1, 1), (1, 1, 1, 1);$$

the three simple syzygies are given by

$$\begin{aligned} X_2 . X_1X_2X_3 - X_1X_2 . X_2X_3 &= S_1 = 0, \\ X_2 . X_1X_2X_3X_4 - X_1X_2 . X_2X_3X_4 &= S_2 = 0, \\ X_2X_3 . X_1X_2X_3X_4 - X_1X_2X_3 . X_2X_3X_4 &= S_3 = 0, \end{aligned}$$

and the two compound syzygies by

$$\begin{aligned} X_2X_3X_4 . S_1 - X_2X_3 . S_2 &= 0, \\ X_1X_2X_3 . S_2 - X_1X_2 . S_3 &= 0. \end{aligned}$$

Art. 80. In general, when the number of parts is i, we have k_i orders which are altered by interchange of d and a, combined with inversion, and l_i which are unaltered where

$$2k_i + l_i = 2^{i-1}.$$

Hence the number of essentially different orders is

$$k_i + l_i = 2^{i-2} + \tfrac{1}{2}l_i.$$

To determine l_i observe that an order

$$d^{\lambda_1}a^{\mu_1}\,d^{\lambda_2}a^{\mu_2}\ldots d^{\lambda_{s-1}}a^{\mu_{s-1}}\,d^{\lambda_s}a^{\mu_s}$$

will be unaltered by the operations spoken of when

$$\lambda_1 - \mu_s = \mu_1 - \lambda_s = \lambda_2 - \mu_{s-1} = \mu_2 - \lambda_{s-1} = \ldots = 0;$$

so that $i - 1$ must be even and there will be two such unaltered orders for each partition of $i - 1$ into even parts.

Hence the generating function for $k_i + l_i$ is

$$\frac{x^2}{1-2x} + \frac{x}{(1-x^2)(1-x^4)(1-x^6)\ldots ad.\ inf.},$$

giving for

$$i = 2, 3, 4, 5, 6, 7, \ldots$$

the numbers

$$1, 3, 4, 10, 16, 35, \ldots$$

Section 6.

Art. 81. The theory, so far, has been concerned with partitions upon a line. The parts were supposed

$$\alpha_1 \quad \alpha_2 \quad \alpha_3 \quad \alpha_4 \quad \alpha_5 \ldots \alpha_{i-1} \quad \alpha_i$$

to be placed at the points upon a line with one of the symbols $\geqq$, $\leqq$ placed between every pair of consecutive points.

When the symbol was invariably $\geqq$ the enumerating function found was

$$\frac{(j+1)}{(1)} \cdot \frac{(j+2)}{(2)} \cdot \frac{(j+3)}{(3)} \ldots \frac{(j+i)}{(i)}$$

wherein (s) denotes $1 - x^s$. If we place these factors at the successive points of the line we obtain a diagrammatic exhibition of the generating function, viz. :—

$$\frac{(j+1)}{(1)} \quad \frac{(j+2)}{(2)} \quad \frac{(j+3)}{(3)} \quad \frac{(j+4)}{(4)} \ldots \frac{(j+i-1)}{(i-1)} \quad \frac{(j+i)}{(i)}$$

a simple fact that the following investigation shows to be fundamental in idea.

Art. 82. I pass on to consider partitions into parts placed at the points of a two-dimensional lattice.

For clearness take the elementary case of four parts placed at the points of a square.

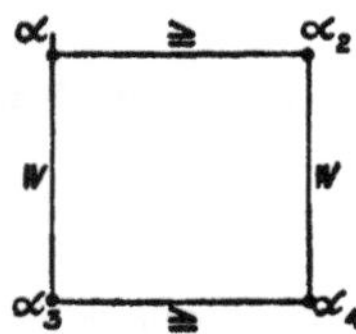

with symbols $\geqq$ placed as shown. We have to solve the conditional relations

$$\alpha_1 \geqq \alpha_2, \qquad \alpha_2 \geqq \alpha_4$$
$$\alpha_1 \geqq \alpha_3, \qquad \alpha_3 \geqq \alpha_4.$$

The four parts are subject to two descending orders. For the sum

$$\Sigma X_1^{\alpha_1} X_2^{\alpha_2} X_3^{\alpha_3} X_4^{\alpha_4}$$

we have the Ω function

$$\underset{\geqq}{\Omega} \frac{1}{1 - abX_1 \,.\, 1 - \frac{d}{a}X_2 \,.\, 1 - \frac{c}{b}X_3 \,.\, 1 - \frac{1}{cd}X_4}$$

which reduces to

$$\frac{1 - X_1^2X_2X_3}{1 - X_1 \,.\, 1 - X_1X_2 \,.\, 1 - X_1X_3 \,.\, 1 - X_1X_2X_3 \,.\, 1 - X_1X_2X_3X_4},$$

establishing the ground solutions

$$(\alpha_1, \alpha_2, \alpha_3, \alpha_4) = (1, 0, 0, 0);\quad (1, 1, 0, 0);\quad (1, 0, 1, 0);\quad (1, 1, 1, 0);\quad (1, 1, 1, 1).$$

connected by the syzygy indicated by

$$X_1 \,.\, X_1X_2X_3 - X_1X_2 \,.\, X_1X_3 = 0,$$

and leading to the enumerating function

$$\frac{1}{(1-x)(1-x^2)^2(1-x^3)}.$$

Art. 83. If the parts be restricted not to exceed j in magnitude, we may take as Ω function

$$\underset{\geqq}{\Omega} \frac{1 - (abX_1)^{j+1}}{1 - abX_1 \,.\, 1 - \frac{d}{a}X_2 \,.\, 1 - \frac{c}{b}X_3 \,.\, 1 - \frac{1}{cd}X_4},$$

and herein putting $X_1 = X_2 = X_3 = X_4 = x$, and reducing, we get

$$\frac{1 - x^{j+1}}{1 - x} \,.\, \left(\frac{1 - x^{j+2}}{1 - x^2}\right)^2 \,.\, \frac{1 - x^{j+3}}{1 - x^3},$$

and we notice that we may represent this diagrammatically on the points of the original lattice, viz. :—

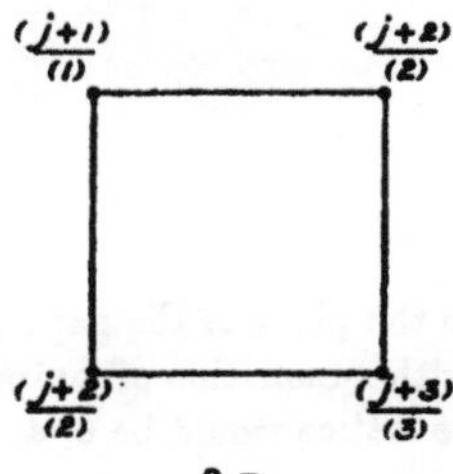

Art. 84. We next have to observe the identity

$$\underset{\geqq}{\Omega} \frac{1}{1 - abX_1 \,.\, 1 - \frac{d}{a}X_2 \,.\, 1 - \frac{c}{b}X_3 \,.\, 1 - \frac{1}{cd}X_4} = \underset{\geqq}{\Omega} \frac{1}{1 - aX_1 \,.\, 1 - abX_1X_2 \,.\, 1 - \frac{1}{a}X_3 \,.\, 1 - \frac{1}{ab}X_3X_4},$$

and to note that the dexter leads to the enumerating function

$$\underset{\geqq}{\Omega} \frac{1}{1 - ax \,.\, 1 - abx^2 \,.\, 1 - \frac{1}{a}x \,.\, 1 - \frac{1}{ab}x^2},$$

corresponding to the problem of two superposable layers of units, each of two rows;

```
1 1 1 1 1 1        1 1 1 1
1 1 1 1            1 1     ,
```

in the case indicated superposition yields

```
2 2 2 2 1 1
2 2 1 1      ;
```

the first row contains a combined number of two's and units $\geqq$ the combined numbers in the second row, and further, the number of two's in first row, $\geqq$ the number of two's in second row. In the Ω function these conditions are secured by the auxiliaries a, b, respectively, and it is established that the problem of partition at the points of the elementary (*i.e.*, simple square) lattice is identical with that of two superposable unit-graphs, each of at most two rows.

In fact, the graph

```
2 2 2 2 1 1 . . . . . . . x|
2 2 1 1                    |
                           |
|                          |
|                          |
y
```

the axis of z being perpendicular to the plane of the paper, is immediately convertible to the lattice form by projection, with summation of units, upon the plane $y\,z$. The numbers at the points of the square lattice would be 6, 4, 4, 2 respectively.

Art. 85. Observe too that the partition is also one upon another kind of lattice in which the part-magnitude is limited not to exceed 2.

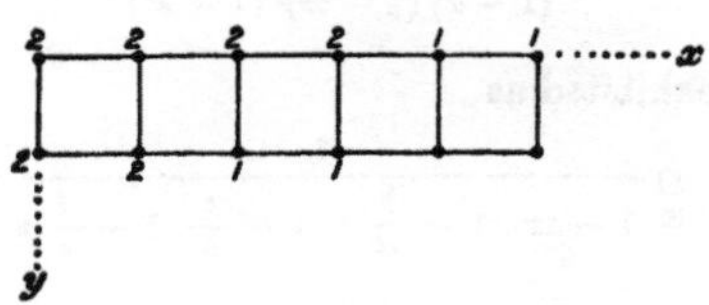

Here, starting from the origin, we may proceed to the opposite point of the lattice along any line of route which proceeds in the positive direction along either axis, and the condition is that along each line of route (here there are six) the numbers must be in descending order and limited in magnitude to 2.

Art. 86. We have, therefore, solved the system of conditions :

$$\begin{array}{ccccccc} \alpha_1 & \geq & \alpha_2 & \geq & \alpha_3 \geq \ldots\ldots & \geq & \alpha_j \\ \text{IV} & & \text{IV} & & \text{IV} & & \text{IV} \\ \beta_1 & \geq & \beta_2 & \geq & \beta_3 \geq \ldots\ldots & \geq & \beta_j \end{array}$$

$$2 \geq \alpha_1 \geq 0,$$

which is seen to possess the same solution as the system

$$\begin{array}{ccc} \alpha_1 & \geq & \alpha_2 \\ \text{IV} & & \text{IV} \\ \alpha_3 & \geq & \alpha_4 \end{array}$$

$$j \geq \alpha_1 \geq 0;$$

and we remark the diagrammatic representation

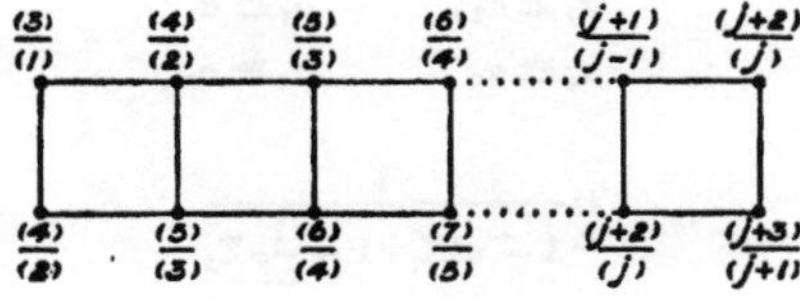

the product of all the factors being

$$\frac{(j+1)(j+2)^2(j+3)}{(1)\quad(2)^2\quad(3)}.$$

Art. 87. I return to the enumerating function

$$\frac{1}{(1-x)(1-x^2)^2(1-x^3)},$$

to note that it may be exhibited as

$$\underset{\geq}{\Omega}\frac{1}{1-ax\,.\,1-\frac{b}{a}x\,.\,1-\frac{c}{b}\,.\,1-\frac{1}{c}x};$$

the interpretation of which is that the coefficient of x^w in the development gives the number of instances in which

$$\alpha_1+\alpha_2+\alpha_4=w_1,$$

$\alpha_1, \alpha_2, \alpha_3, \alpha_4$ being integers satisfying the conditions

$$\alpha_1 \geq \alpha_2 \geq \alpha_3 \geq \alpha_4.$$

We arrive at the form in question if for these conditions we construct

$$\Sigma X_1^{\alpha_1} X_2^{\alpha_2} X^{\alpha_4}$$

and then put $X_1 = X_2 = X_4 = x$.
The graphical representation is of the form

```
1 1 1 1 1 1 1 1 . . .
1 1 1 1 1 . . .
0 0 0 0 . . .
1 1 . . .
```

the numbers of figures in the rows being in descending order and the third row of figures zeros.

Art. 88. As another instance of the elementary lattice take the system

$$\alpha_1 \geq \alpha_2, \qquad \alpha_1 \geq \alpha_3$$
$$\alpha_4 \geq \alpha_2, \qquad \alpha_4 \geq \alpha_3,$$

leading to

$$\underset{\geq}{\Omega}\frac{1}{\begin{array}{c}1-abX_1\,.\,1-\frac{1}{ad}X_2\\ 1-\frac{1}{bc}X_3\,.\,1-cdX_4\end{array}},$$

reducing to

$$\frac{1-X_1^2X_2X_3X_4^2}{1-X_1\,.\,1-X_1X_2X_4\,.\,1-X_1X_3X_4\,.\,1-X_1X_2X_3X_4\,.\,1-X_4},$$

establishing the fundamental solutions

$(\alpha_1, \alpha_2, \alpha_3, \alpha_4) = (1, 0, 0, 0); \quad (1, 1, 0, 1); \quad (1, 0, 1, 1); \quad (1, 1, 1, 1,); \quad (0, 0, 0, 1);$

connected by the syzygy indicated by

$$X_1 . X_1X_2X_3X_4 . X_4 - X_1X_2X_4 . X_1X_3X_4 = 0.$$

Art. 89. A more general generating function connected with the elementary lattice and descending orders is

$$\underset{\geqq}{\Omega} \frac{\begin{array}{c}1 - (abX_1)^{j_1+1} . 1 - \left(\frac{d}{a} X_2\right)^{j_2+1} \\ 1 - \left(\frac{c}{b} X_3\right)^{j_3+1} . 1 - \left(\frac{1}{cd} X_4\right)^{j_4+1}\end{array}}{\begin{array}{c}1 - abX_1 . 1 - \frac{d}{a} X_2 \\ 1 - \frac{c}{b} X_3 . 1 - \frac{1}{cd} X_4\end{array}},$$

where now $\alpha_1, \alpha_2, \alpha_3, \alpha_4$ are restricted not to exceed j_1, j_2, j_3, j_4 respectively, and of course

$$\begin{array}{ccc} j_1 & \geqq & j_2 \\ \text{IV} & & \text{IV} \\ j_3 & \geqq & j_4 \end{array}$$

are conditions.

It should be remarked that we examine the case of bipartite partitions with regular graphs by putting $X_2 = X_1$, $X_4 = X_3$.

Part-magnitude being unlimited, the reduced function is

$$\frac{1 - X_1^2 X_3}{1 - X_1 . 1 - X_1^2 . 1 - X_1X_3 . 1 - X_1^2X_3 . 1 - X_1^2X_3^2},$$

and is *real*.

Art. 90. Leaving the particular case, I pass on to consider the general theory of partitions at the points of a lattice in two dimensions. It can be shown immediately that it is coincident with the theory of those partitions of all multipartite numbers which can be represented by regular graphs in three dimensions. For consider the superposition of any number of unit graphs, adding into single numbers the units in the same vertical line. We obtain a scheme of numbers

$$\begin{array}{cccccccc} a_{11} & a_{12} & a_{13} & a_{14} & . & . & . & x \\ a_{21} & a_{22} & a_{23} & . & . & . & & \\ a_{31} & a_{32} & . & . & . & & & \\ a_{41} & & & & & & & \\ . & & & & & & & \\ . & & & & & & & \\ y. & & & & & & & \end{array}$$

in which all the rows and all the columns taken in the positive directions along the axes of x and y are in descending order. We may consider these numbers to be placed at the points of a lattice of which the sides involve m and l points along the sides parallel to the axes of x and y respectively; m will then be a limit to the number of units in any row of a unit graph, and l will be the limit to the number of rows.

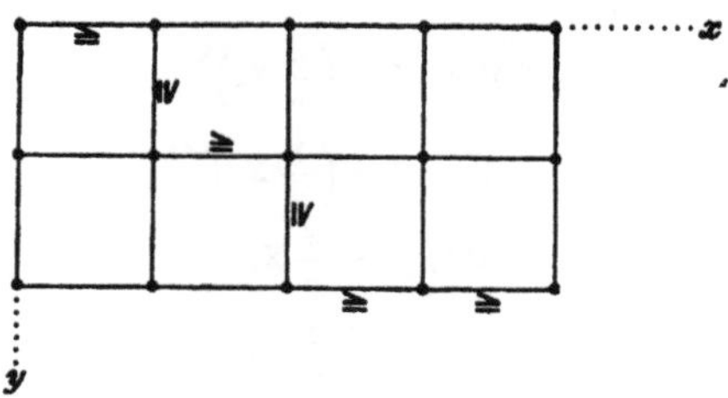

There is a descending order along each line of route from the origin to the opposite corner of the lattice, and there are altogether

$$\binom{l+m-2}{l-1} \text{ such lines of route.}$$

Art. 91. The theory of the regular partitions of multipartite numbers is thus reduced to a lattice partition into $l\,m$ parts *in plano*. The conditional relations may be written

$$\begin{array}{ccccccccc}
\alpha_{11} & \geq & \alpha_{12} & \geq & \alpha_{13} & \ldots & \alpha_{1.m-1} & \geq & \alpha_{1m} \\
\text{IV} & & \text{IV} & & \text{IV} & & \text{IV} & & \text{IV} \\
\alpha_{21} & \geq & \alpha_{22} & \geq & \alpha_{23} & \ldots & \alpha_{2.m-1} & \geq & \alpha_{2.m} \\
\cdot & & \cdot & & \cdot & \ldots & & & \\
\alpha_{l-1.1} & \geq & \alpha_{l-1.2} & \geq & \alpha_{l-1.3} & \ldots & \alpha_{l-1.m-1} & \geq & \alpha_{l-1.m} \\
\text{IV} & & \text{IV} & & \text{IV} & & \text{IV} & & \text{IV} \\
\alpha_{l.1} & \geq & \alpha_{l.2} & \geq & \alpha_{l.3} & \ldots & \alpha_{l.m-1} & \geq & \alpha_{l.m}
\end{array}$$

and for the sum

$$\Sigma \prod_{s=1}^{s=l} \prod_{t=1}^{t=m} X_{st}^{\alpha_{st}} .$$

we at once write down the Ω generating function, viz. :—

$$\underset{\geqq}{\Omega}\ \frac{1}{\begin{array}{llll} 1-a_1\alpha_1 X_{11} & .\ 1-\frac{a_2}{a_1}\beta_1 X_{12} & .\ 1-\frac{a_3}{a_2}\gamma_1 X_{13} & \ldots \text{ to } m \text{ factors} \\ 1-b_1\frac{\alpha_2}{\alpha_1}X_{21} & .\ 1-\frac{b_2}{b_1}\frac{\beta_2}{\beta_1}X_{22} & .\ 1-\frac{b_3}{b_2}\frac{\gamma_2}{\gamma_1}X_{23} & \ldots \text{ to } m \text{ factors} \\ 1-c_1\frac{\alpha_3}{\alpha_2}X_{31} & .\ 1-\frac{c_2}{c_1}\frac{\beta_3}{\beta_2}X_{32} & .\ 1-\frac{c_3}{c_2}\frac{\gamma_3}{\gamma_2}X_{33} & \ldots \text{ to } m \text{ factors} \\ \vdots & \vdots & \vdots & \&\text{c.} \\ \text{to } l \text{ factors} & \text{to } l \text{ factors} & \text{to } l \text{ factors} & \end{array}}$$

If the part-magnitude be limited to n, we must place as numerator in the function

$$\begin{array}{lll} 1-\left(a_1\alpha_1X_{11}\right)^{n+1} & .\ 1-\left(\frac{a_2}{a_1}\beta_1X_{12}\right)^{n+1} & \ldots \text{ to } m \text{ factors} \\ 1-\left(b_1\frac{\alpha_2}{\alpha_1}X_{21}\right)^{n+1} & .\ 1-\left(\frac{b_2}{b_1}\frac{\beta_2}{\beta_1}X_{22}\right)^{n+1} & \ldots \text{ to } m \text{ factors} \\ \vdots & \vdots & \&\text{c.} \\ \text{to } l \text{ factors} & \text{to } l \text{ factors} & \end{array}$$

and if we please we may reject all the numerator factors except

$$1-(a_1\alpha_1X_{11})^{n+1}.$$

Art. 92. The existence of the three-dimensional graph shows that this function remains unaltered, when X_{st} is put equal to x, for every substitution impressed upon the numbers

$$l,\ m,\ n,$$

but there is a still more refined theorem of reciprocity connected with a more general generating function.

Suppose that the number of layers which involve 1, 2, 3, &c. rows be restricted to

$$l_1,\ l_2,\ l_3,\ \ldots;$$

that the successive layers are restricted to involve at most

$$m_1,\ m_2,\ m_3,\ \ldots \text{ rows};$$

and that the successive rows of the layers are restricted to contain at most

$$n_1,\ n_2,\ n_3,\ \ldots \text{ units.}$$

We have then the comprehensive Ω function :—

$$\underset{\geqq}{\Omega}\ \frac{\begin{array}{llll} 1-(a_1\alpha_1X_{11})^{n_1+1} & .\ 1-\left(\frac{a_2}{a_1}\beta_1X_{12}\right)^{n_2+1} & .\ 1-\left(\frac{a_3}{a_2}\gamma_1X_{13}\right)^{n_3+1} & \text{to } m_1 \text{ factors} \\ 1-\left(b_1\frac{\alpha_2}{\alpha_1}X_{21}\right)^{n_1+1} & .\ 1-\left(\frac{b_2}{b_1}\frac{\beta_2}{\beta_1}X_{22}\right)^{n_2+1} & .\ 1-\left(\frac{b_3}{b_2}\frac{\gamma_2}{\gamma_1}X_{23}\right)^{n_3+1} & \text{to } m_2 \text{ factors} \\ 1-\left(c_1\frac{\alpha_3}{\alpha_2}X_{31}\right)^{n_1+1} & .\ 1-\left(\frac{c_2}{c_1}\frac{\beta_3}{\beta_2}X_{32}\right)^{n_2+1} & .\ 1-\left(\frac{c_3}{c_2}\frac{\gamma_3}{\gamma_2}X_{33}\right)^{n_3+1} & \text{to } m_3 \text{ factors} \\ \vdots & \vdots & \vdots & \&c. \\ \text{to } l_1 \text{ factors} & \text{to } l_2 \text{ factors} & \text{to } l_3 \text{ factors} & \end{array}}{\begin{array}{llll} 1-a_1\alpha_1X_{11} & .\ 1-\frac{a_2}{a_1}\beta_1X_{12} & .\ 1-\frac{a_3}{a_2}\gamma_1X_{13} & \ldots \text{ to } m_1 \text{ factors} \\ 1-b_1\frac{\alpha_2}{\alpha_1}X_{21} & .\ 1-\frac{b_2}{b_1}\frac{\beta_2}{\beta_1}X_{22} & .\ 1-\frac{b_3}{b_2}\frac{\gamma_2}{\gamma_1}X_{23} & \ldots \text{ to } m_2 \text{ factors} \\ 1-c_1\frac{\alpha_3}{\alpha_2}X_{31} & .\ 1-\frac{c_2}{c_1}\frac{\beta_3}{\beta_2}X_{32} & .\ 1-\frac{c_3}{c_2}\frac{\gamma_3}{\gamma_2}X_{33} & \ldots \text{ to } m_3 \text{ factors} \\ \vdots & \vdots & \vdots & \&c. \\ \text{to } l_1 \text{ factors} & \text{to } l_2 \text{ factors} & \text{to } l_3 \text{ factors} & \end{array}},$$

wherein, naturally, each of the series

$$\begin{array}{l} l_1,\ l_2,\ l_3,\ \ldots \\ m_1,\ m_2,\ m_3,\ \ldots \\ n_1,\ n_2,\ n_3,\ \ldots \end{array}$$

is in descending order, and the theorem of reciprocity involved in the fact of the existence of the graph consists in the circumstance that the function remains unaltered, when X_{st} is put equal to x, for any substitution impressed upon the unsuffixed symbols l, m, n.

In the corresponding lattice the conditions are :—

(i.) The first, second, &c., rows do not contain more than n_1, n_2, &c. numbers respectively ;

(ii.) The first, second, &c., rows do not contain higher numbers than l_1, l_2, &c. . . . ;

(iii.) No number so great as s occurs below row m_s for all values of s; $m_1, m_2, \ldots m_s, \ldots$ being of course in descending order of magnitude.

Art. 93. The reduction of this Ω function presents great difficulties, and I propose to restrict consideration to the case

$$\begin{array}{l} l_1 = l_2 = l_3 = \ldots = l \\ m_1 = m_2 = m_3 = \ldots = m \\ n_1 = n_2 = n_3 = \ldots = n. \end{array}$$

To adapt the function to enumerate the partitions into at most m parts of l-partite numbers, such partitions being such as possess regular graphs *in solido*, put

$$X_{11} = X_{12} = X_{13} = \ldots = X_{1m} = x_1$$
$$X_{21} = X_{22} = X_{23} = \ldots = X_{2m} = x_2$$
$$\cdot \quad \cdot \quad \cdot \quad \cdot \quad \cdot \quad \cdot \quad \cdot \quad \cdot$$
$$X_{l_1} = X_{l_2} = X_{l_3} = \ldots = X_{l_m} = x_l,$$

and the resulting function enumerates by the coefficients of

$$x_1^{p_1} x_2^{p_2} \ldots x_l^{p_l},$$

the number of partitions of the l-partite

$$\overline{(p_1 p_2 \ldots p_l)}$$

into at most m parts.

Art. 94. Further putting

$$x_1 = x_2 = x_3 = \ldots = x_l = x,$$

the coefficients of x^w gives the number of graphs *in solido or* unipartite partitions upon a two-dimensional lattice, limited, as indicated above, by the numbers l, m, n.

This function appears to be reducible to the product of factors shown in the tableau below:—

$$\frac{(n+1)}{(1)} \cdot \frac{(n+2)}{(2)} \cdot \frac{(n+3)}{(3)} \cdots \frac{(n+m)}{(m)};$$

$$\frac{(n+2)}{(2)} \cdot \frac{(n+3)}{(3)} \cdot \frac{(n+4)}{(4)} \cdots \frac{(n+m+1)}{(m+1)};$$

$$\frac{(n+3)}{(3)} \cdot \frac{(n+4)}{(4)} \cdot \frac{(n+5)}{(5)} \cdots \frac{(n+m+2)}{(m+2)};$$

$$\vdots$$

$$\frac{(n+l)}{(l)} \cdot \frac{(n+l+1)}{(l+1)} \cdot \frac{(n+l+2)}{(l+2)} \cdots \frac{(n+m+l-1)}{(l+m-1)}.$$

This result, verified in a multitude of particular cases, awaits demonstration. For $l = 2$ it has been proved independently by Professor FORSYTH and by the present author. The diagrammatic exhibition of the result at the points of the lattice is clear, and since the product is an invariant for any substitution impressed upon the letters l, m, n, it appears that such exhibition is six-fold. Taking a lattice whose sides contain l m points respectively, so that l m points in all are involved, we mark a corner point, regarding it as an origin of rectangular axes *one*, and proceed to the opposite corner, along any line of route, such that progression along any branch or section of the lattice is in the positive direction, marking the successive points reached *two*, *three*, &c.

For every point, marked s, we have a factor,

$$\frac{(n+s)}{(s)},$$

and express the generating function as a product of lm such factors. If n be ∞, each factor is of the form

$$\frac{1}{(s)},$$

and if the number s appears σ times on the lattice, we have a factor $(s)^{-\sigma}$, and the complete result may be written

$$\frac{1}{(s_1)^{\sigma_1}(s_2)^{\sigma_2}(s_3)^{\sigma_3}\dots}.$$

Art. 95. Hence the enumeration is identical with that of the partitions of a unipartite number into an unlimited number of parts of $\sigma_1+\sigma_2+\sigma_3+\dots$ different kinds, viz. :—

σ_1 of the numerical value s_1 but differently coloured.
σ_2 ,, ,, s_2 ,, ,,
σ_3 s_3
. . . .

The number of distinct lines of route in a lattice of lm points is

$$\binom{l+m-2}{l-1},$$

so that, in general, on the lattice we have partitions of a number into lm parts subject to $\binom{l+m-2}{l-1}$ descending orders.

Such a partition is transformable ($l \geqq m$) into one composed of the parts

1	of	1	colour
2	,,	2	,,
$\vdots$		$\vdots$	
m	,,	m	,,
$\vdots$		$\vdots$	
l	,,	m	,,
$l+1$	,,	$m-1$	,,
$\vdots$		$\vdots$	
$l+m-1$	,,	1	,,

a theorem of reciprocity analogous to and including the well-known theorem connected with the partitions of a number on a line. There is also a lattice theory connected with unipartite partitions on a line, for the unit-graph of such a partition is nothing more than a number of units and zeros placed at the points of a two-dimensional lattice, such numbers being subject to the $\binom{l+m-2}{l-1}$ descending orders.

Art. 96. The fact is that the theory of the two-dimensional lattice, the part-magnitude being restricted to unity, is co-extensive with the whole theory of partitions upon a line. Hence for such partitions we may represent the generating function, diagrammatically, in two ways upon a lattice as well as in two ways upon a line.

The two representations upon a line are

$$\frac{(l+1)}{(1)}\quad\frac{(l+2)}{(2)}\quad\frac{(l+3)}{(3)}\quad\frac{(l+4)}{(4)}\quad\cdots\quad\frac{(l+m-1)}{(m-1)}\quad\frac{(l+m)}{(m)}.$$

$$\frac{(m+1)}{(1)}\quad\frac{(m+2)}{(2)}\quad\frac{(m+3)}{(3)}\quad\frac{(m+4)}{(4)}\quad\cdots\quad\frac{(l+m-1)}{(l-1)}\quad\frac{(l)}{(l)}.$$

Upon a lattice we have

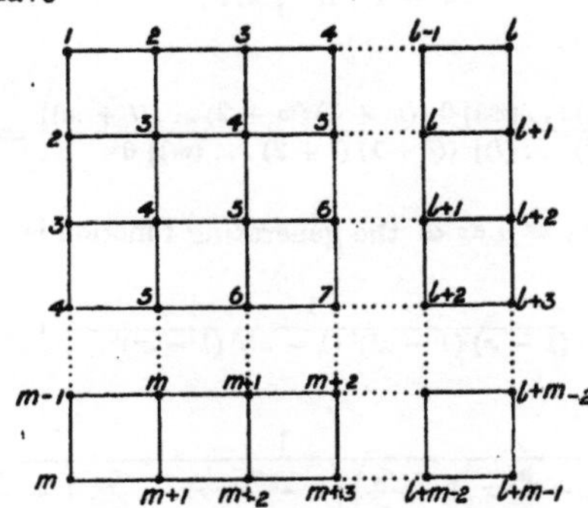

and at the point marked s we place the factor

$$\frac{(s+1)}{(s)}.$$

The second lattice is obtained by interchange of l and m.

The product thus obtained is

$$\prod_{s=1}^{s=l+m-1}\left\{\frac{(s+1)}{(s)}\right\}^{b_s-b_{s-l}-b_{s-m}}$$

b_s denoting the sth figurate number of the second order, and $b_s - b_{s-l} - b_{s-m}$ is easily shown to be equal to the number of points of the lattice marked s. We have to show that this is equal to

$$\prod_{s=1}^{s=m} \frac{(l+s)}{(s)}.$$

Taking $l \geqq m$, observe that $(l+s)$ occurs in the former to the power

$$\begin{gathered} b_{l+s-1} - b_{s-1} - b_{l+s-1-m} \\ - b_{l+s} + b_s + b_{l+s-m} \end{gathered}$$

which

$$\begin{aligned} &= 1 \text{ if } l + s > m \\ &= 0 \text{ if } l + s \leqq m; \end{aligned}$$

whilst (s) occurs to the power

$$\begin{gathered} b_{s-1} - b_{s-1-l} - b_{s-1-m} \\ - b_s + b_{s-l} + b_{s-m} \end{gathered}$$

which

$$\begin{aligned} &= 1 \quad \text{if} \quad s > m \\ &= 0 \quad \text{if} \quad s > l \text{ and } \leqq m \\ &= -1 \quad \text{if} \quad s \leqq l; \end{aligned}$$

the product is, therefore,

$$\frac{\{(l+1)(l+2)\ldots(m)\}\, 0\, \{(m+1)(m+2)\ldots(l+m)\}}{\{(1)(2)\ldots(l)\}\, \{(l+1)(l+2)\ldots(m)\}\, 0} = \prod_{s=1}^{s=m} \frac{(l+s)}{(s)}.$$

Art. 97. When $l = m = n = \infty$ the generating function is

$$\frac{1}{(1-x)(1-x^2)^2(1-x^3)^3(1-x^4)^4\ldots},$$

which may be written

$$\underset{\geqq}{\Omega} \frac{1}{(1-a_1x)\left(1-\frac{a_2}{a_1}x \,.\, 1-\frac{a_3}{a_2}\right)\left(1-\frac{a_4}{a_3}x \,.\, 1-\frac{a_5}{a_4} \,.\, 1-\frac{a_6}{a_5}\right)(\ldots)\ldots}$$

from which is deduced a graphical representation in two dimensions involving units and zeros.

The graph is regular, and the successive rows involve the numbers

$$1;\ 1, 0;\ 1, 0, 0;\ 1, 0, 0, 0;\ \ldots$$

respectively. In the general case there is a similar representation, proper restrictions being placed upon the numbers of figures in the rows.

Section 7.

Art. 98. It might have been conjectured that the lattice *in solido* would have afforded results of equal interest, but this on investigation does not appear to be the case. The simplest of such lattices is that in which the points are the summits of a cube and the branches the edges of the cube.

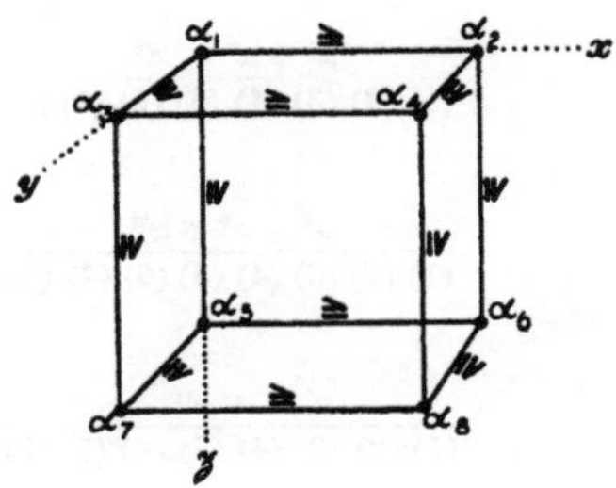

$\alpha_1\ \alpha_2\ \alpha_3\ \alpha_4\ \alpha_5\ \alpha_6\ \alpha_7\ \alpha_8$ is a partition of a number into eight parts, satisfying the conditional relations indicated by the symbols $\geqq$ as shown. The descending order is in the positive direction parallel to each axis. The Ω function

$$\underset{\geqq}{\Omega}\ \frac{1}{\begin{array}{c}1-a_1a_2a_3X_1\,.\,1-\frac{a_4a_5}{a_1}X_2\,.\,1-\frac{a_6a_7}{a_2}X_3\,.\,1-\frac{a_8a_9}{a_3}X_4\\ 1-\frac{a_{10}}{a_4a_6}X_5\,.\,1-\frac{a_{11}}{a_7a}X_6\,.\,1-\frac{a_{12}}{a_5a_9}X_7\,.\,1-\frac{1}{a_{10}a_{11}a_{12}}X_8\end{array}}$$

is difficult to deal with, and the result which I have obtained too complicated to be worth preserving. I therefore put at once

$$X_1 = X_2 = X_3 = X_4 = X_5 = X_6 = X_7 = X_8 = x,$$

and seek the sum $\Sigma x^{\alpha_1+\alpha_2+\alpha_3+\alpha_4+\alpha_5+\alpha_6+\alpha_7+\alpha_8}$. I divide the calculation into eighteen parts as follows :—

Conditions.	Result.
$\alpha_8 \geqq \alpha_7 \geqq \alpha_4$ $\alpha_5 \geqq \alpha_2,\quad \alpha_5 \geqq \alpha_3$	$\dfrac{1+x^3+x^4}{(1)(2)(3)(4)(5)(6)(7)(8)}$
$\alpha_8 \geqq \alpha_7 \geqq \alpha_4$ $\alpha_2 \geqq \alpha_3,\quad \alpha_2 > \alpha_5$	$\dfrac{x^2+x^3+x^6}{(1)(2)(3)(4)(5)(6)(7)(8)}$

Conditions.	Result.
$\alpha_6 \geqq \alpha_7 \geqq \alpha_4$ $\alpha_3 > \alpha_2, \quad \alpha_3 > \alpha_5$	$\frac{x^2 + x^5}{(1)(2)(3)(4)(5)(6)(7)(8)}$
$\alpha_6 \geqq \alpha_4, \quad \alpha_4 > \alpha_7$ $\alpha_5 \geqq \alpha_2, \quad \alpha_5 \geqq \alpha_3$	$\frac{x^6 + x^9 + x^{10}}{(1)(2)(3)(4)(5)(6)(7)(8)}$
$\alpha_6 \geqq \alpha_4, \quad \alpha_4 > \alpha_7$ $\alpha_2 \geqq \alpha_3, \quad \alpha_2 > \alpha_5$	$\frac{x^8 + x^9 + x^{12}}{(1)(2)(3)(4)(5)(6)(7)(8)}$
$\alpha_6 \geqq \alpha_4, \quad \alpha_4 > \alpha_7$ $\alpha_3 > \alpha_2, \quad \alpha_3 > \alpha_5$	$\frac{x^8 + x^{11}}{(1)(2)(3)(4)(5)(6)(7)(8)}$
$\alpha_4 > \alpha_6, \quad \alpha_6 \geqq \alpha_7$ $\alpha_5 \geqq \alpha_2, \quad \alpha_5 \geqq \alpha_3$	$\frac{x^5 + x^8}{(1)(2)(3)(4)(5)(6)(7)(8)}$
$\alpha_4 > \alpha_6, \quad \alpha_6 \geqq \alpha_7$ $\alpha_2 \geqq \alpha_3, \quad \alpha_2 > \alpha_5$	$\frac{x^4 + x^7 + x^8}{(1)(2)(3)(4)(5)(6)(7)(8)}$
$\alpha_4 > \alpha_6, \quad \alpha_6 \geqq \alpha_7$ $\alpha_3 > \alpha_2, \quad \alpha_3 > \alpha_5$	$\frac{x^6 + x^7 + x^{10}}{(1)(2)(3)(4)(5)(6)(7)(8)}$
$\alpha_4 > \alpha_7, \quad \alpha_7 > \alpha_6$ $\alpha_5 \geqq \alpha_2, \quad \alpha_5 \geqq \alpha_3$	$\frac{x^{11} + x^{14}}{(1)(2)(3)(4)(5)(6)(7)(8)}$
$\alpha_4 > \alpha_7, \quad \alpha_7 > \alpha_6$ $\alpha_2 \geqq \alpha_3, \quad \alpha_2 > \alpha_5$	$\frac{x^{10} + x^{13} + x^{14}}{(1)(2)(3)(4)(5)(6)(7)(8)}$
$\alpha_4 > \alpha_7, \quad \alpha_7 > \alpha_6$ $\alpha_3 > \alpha_2, \quad \alpha_3 > \alpha_5$	$\frac{x^{12} + x^{13} + x^{16}}{(1)(2)(3)(4)(5)(6)(7)(8)}$
$\alpha_7 \geqq \alpha_4 > \alpha_6$ $\alpha_5 \geqq \alpha_2, \quad \alpha_5 \geqq \alpha_3$	$\frac{x^6 + x^9 + x^{10}}{(1)(2)(3)(4)(5)(6)(7)(8)}$
$\alpha_7 \geqq \alpha_4 > \alpha_6$ $\alpha_2 \geqq \alpha_3, \quad \alpha_2 > \alpha_5$	$\frac{x^8 + x^9}{(1)(2)(3)(4)(5)(6)(7)(8)}$

Conditions.	Result.
$\alpha_7 \geqq \alpha_4 > \alpha_6$ $\alpha_3 > \alpha_2, \quad \alpha_3 > \alpha_5$	$\dfrac{x^8 + x^{11} + x^{12}}{(1)(2)(3)(4)(5)(6)(7)(8)}$
$\alpha_7 > \alpha_6 \geqq \alpha_4$ $\alpha_5 \geqq \alpha_2, \quad \alpha_5 \geqq \alpha_3$	$\dfrac{x^4 + x^5 + x^8}{(1)(2)(3)(4)(5)(6)(7)(8)}$
$\alpha_7 > \alpha_6 \geqq \alpha_4$ $\alpha_2 \geqq \alpha_3, \quad \alpha_2 > \alpha_5$	$\dfrac{x^7 + x^8}{(1)(2)(3)(4)(5)(6)(7)(8)}$
$\alpha_7 > \alpha_6 \geqq \alpha_4$ $\alpha_3 > \alpha_2, \quad \alpha_3 > \alpha_5$	$\dfrac{x^6 + x^7 + x^{10}}{(1)(2)(3)(4)(5)(6)(7)(8)}$

and by addition the resulting generating function* is

$$\frac{1 + 2x^2 + 2x^3 + 3x^4 + 3x^5 + 5x^6 + 4x^7 + 8x^8 + 4x^9 + 5x^{10} + 3x^{11} + 3x^{12} + 2x^{13} + 2x^{14} + x^{16}}{(1)(2)(3)(4)(5)(6)(7)(8)}.$$

Art. 99. By analogy with the lattice *in plano* one might have conjectured that the result would have been

$$\frac{1}{(1)(2)^3(3)^3(4)};$$

but this is not so, although the two functions do coincide as far as the coefficient of x^6 inclusive. In fact, the two expansions yield respectively

$$1 + x + 4x^2 + 7x^3 + 14x^4 + 23x^5 + 41x^6 + 63x^7 + \ldots,$$
$$1 + x + 4x^2 + 7x^3 + 14x^4 + 23x^5 + 42x^6 + 63x^7 + \ldots,$$

the succeeding coefficients becoming widely divergent. This at first seemed surprising, but observe that analogy might also lead us to expect that, if the part-magnitude be limited to i, the result would be

$$\frac{(i+1)(i+2)^3(i+3)^3(i+4)}{(1)(2)^3(3)^3(4)};$$

but this does not happen to be expressible in a finite integral form for all values of i, a fact which necessitates the immediate rejection of the conjecture. The expression in question is only finite and integral when i is of the form $3p$ or $3p+1$. We have,

* Mr. A. B. KEMPE, Treas. R.S., has verified this conclusion by a different and most ingenious method of summation, which also readily yields the result for any desired restriction on the part-magnitude.

further, the fact that the expression does give the enumeration when $i = 1$, for then the generating function is easily ascertainable to be

$$1 + x + 3x^2 + 3x^3 + 4x^4 + 3x^5 + 3x^6 + x^7 + x^8,$$

which may be exhibited in the forms

$$\frac{(4)^2 (5)}{(1)(2)^2} \equiv \frac{(3)(4)^2(5)}{(1)(2)^2(3)} \equiv \frac{(2)(3)^2(4)^2(5)}{(1)(2)^2(3)^2(4)}.$$

Art. 100. The second of these forms immediately arrests the attention, for, *in plano*, it denotes the number of partitions on a lattice of four points (in fact, a square), the part-magnitude being limited not to exceed 2. The reason of this is as follows :—

Taking the cube with any distribution of units at the summits, we may project the summits upon the plane of $y\,z$, adding up the units on the cube edges at right

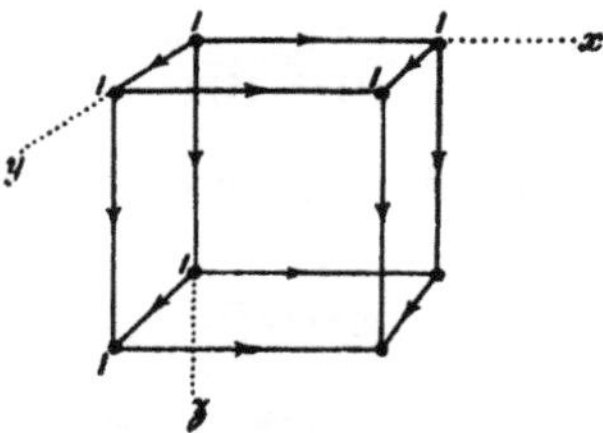

angles to that plane, and thus obtain a distribution, on the points of the cube face in that plane, of numbers limited in magnitude to 2.

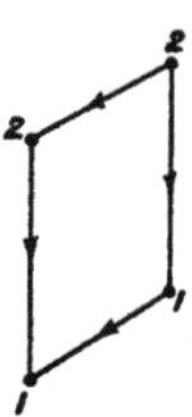

This projection establishes the theorem, which may now be generalized. Conceive a lattice *in solido* having l, m, n points along the axes of x, y, z respectively, and a distribution of units at the points of the lattice which form an unbroken succession along each line of route through the lattice from the origin to the opposite corner, a line of route always proceeding parallel to the axes in a positive sense. Now project and sum units on the plane of $y\,z$.

The result is a partition of the number at the points of a lattice *in plano* whose sides contain m and n points respectively, the part-magnitude being limited not to exceed l. The descending order in this lattice is clearly from the origin to the opposite corner in the plane yz along each of its lines of route.

The enumerating generating function is

$$\begin{array}{l} \frac{(l+1)}{(1)} \cdot \frac{(l+2)}{(2)} \cdot \frac{(l+3)}{(3)} \cdots \frac{(l+m)}{(m)} \\ \times \frac{(l+2)}{(2)} \cdot \frac{(l+3)}{(3)} \cdot \frac{(l+4)}{(4)} \cdots \frac{(l+m+1)}{(m+1)} \\ \times \frac{(l+3)}{(3)} \cdot \frac{(l+4)}{(4)} \cdot \frac{(l+5)}{(5)} \cdots \frac{(l+m+2)}{(m+2)} \\ \quad \vdots \qquad\quad \vdots \qquad\quad \vdots \qquad\qquad \vdots \\ \times \frac{(l+n)}{(n)} \cdot \frac{(l+n+1)}{(n+1)} \cdot \frac{(l+n+2)}{(n+2)} \cdots \frac{(l+m+n)}{(m+n)} . \end{array}$$

Each factor may be supposed at a point of the corresponding lattice ; if any point is the sth along a line of route the factor is

$$\frac{(l+s)}{(s)}.$$

The number of points at which we place

$$\frac{(l+s)}{(s)}$$

is equal to the coefficient of x^s in the expansion of

$$x(1 + x + x^2 + \ldots + x^{m-1})(1 + x + x^2 + \ldots + x^{n-1})$$

that is of

$$\frac{x}{(1+x)^2}(1 - x^m)(1 - x^n).$$

If m, n be in ascending order and b_s denote the sth figurate number of the second order, this coefficient is

$$b_s - b_{(s-m)} - b_{(s-n)}$$

the term $+ b_{s-m-n}$ being omitted because s is at most $m + n - 1$.

Hence the generating function may be written

$$\prod_{s=1}^{s=m+n-1} \left\{\frac{(l+s)}{(s)}\right\}^{b_s - b_{s-m} - b_{s-n}}.$$

Art. 101. It is now important to show the connexion between this result and the original lattice *in solido*.

I say that this generating function may be exhibited by factors placed at the points of the lattice *in solido*. These factors are of form

$$\frac{(s+1)}{(s)},$$

and such a factor must be placed at every point which is the sth occurring along a line of route in the cubic reticulation.

I take l, m, n in ascending order, and remark that the number of points possessing this property is the coefficient of x^s in the product

$x(1+x+x^2+\ldots+x^{l-1})(1+x+x^2+\ldots+x^{m-1})(1+x+x^2+\ldots+x^{n-1})$, which is

$$\frac{x}{(1-x)^3}(1-x^l)(1-x^m)(1-x^n),$$

and that, if c_s denote the sth of the third order of figurate numbers, this coefficient is

$$c_s - c_{s-l} - c_{s-m} - c_{s-n} + c_{s-l-m} + c_{s-l-n} + c_{s-m-n},$$

the term $-c_{s-l-m-n}$ being omitted, because s is at most $l+m+n-2$.

I propose, therefore, to prove the identity

$$\prod_{s=1}^{s=m+n-1}\left\{\frac{(l+s)}{(s)}\right\}^{b_s-b_{s-m}-b_{s-n}} = \prod_{s=1}^{s=l+m+n-2}\left\{\frac{(s+1)}{(s)}\right\}^{c_s-c_{s-l}-c_{s-m}-c_{s-n}+c_{s-l-m}+c_{s-l-n}+c_{s-m-n}}.$$

The factor $(l+s)$ occurs to the power

$$-b_{l+s}+b_s-b_{s-m}-b_{s-n}+b_{l+s-m}+b_{l+s-n}$$

on the sinister side, and to the power

$$-(c_{l+s}-c_{l+s-1})+(c_s-c_{s-1})-(c_{s-m}-c_{s-m-1})$$
$$-(c_{s-n}-c_{s-n-1})+(c_{l+s-m}-c_{l+s-m-1})+(c_{l+s-n}-c_{l+s-n-1})$$

on the dexter. But

$$c_k - c_{k-1} = b_k = k.$$

Hence, under all circumstances, the two powers must be equal.

Again the factor (s) occurs to the power,

$$-b_s+b_{s-l}+b_{s-m}+b_{s-n}-b_{s-l-m}-b_{s-l-n}$$

on the sinister side, and to the power

$$-(c_s - c_{s-1}) + (c_{s-l} + c_{s-l-1}) + (c_{s-m} - c_{s-m-1})$$
$$+ (c_{s-n} - c_{s-n-1}) - (c_{s-l-m} - c_{s-l-m-1}) - (c_{s-l-n} - c_{s-l-n-1})$$

on the dexter, and again the two powers are equal.

Hence the identity under consideration is established, and this carries with it the proof of the diagrammatic representation of the generating function on the points o the solid reticulation.

Art. 102. I resume the general theory of the partitions on the summits of a cube. When the parts are unrestricted in magnitude the generating function has been found. A process similar to that employed leads to the theorem that when the parts are restricted not to exceed t in magnitude the generating function is the quotient of

$$\begin{aligned} 1 &+ a(2x^2 + 2x^3 + 3x^4 + 2x^5 + 2x^6) \\ &+ a^2(x^5 + 3x^6 + 4x^7 + 8x^8 + 4x^9 + 3x^{10} + x^{11}) \\ &+ a^3(2x^{10} + 2x^{11} + 3x^{12} + 2x^{13} + 2x^{14}) \\ &+ a^4 . x^{16} \end{aligned}$$

by

$$(1-a)(1-ax)(1-ax^2)(1-ax^3)(1-ax^4)(1-ax^5)(1-ax^6)(1-ax^7)(1-ax^8),$$

the required number being given by the coefficient of $a^t x^w$. Denoting the numerator by $1 + a\mathrm{P}(x) + a^2\mathrm{Q}(x) + a^3\mathrm{R}(x) + a^4 . x^{16}$, the whole coefficient of a^t is

$$\frac{(9)(10)\ldots(t+8)}{(1)(2)\ldots(t)} + \mathrm{P}(x)\frac{(9)(10)\ldots(t+7)}{(1)(2)\ldots(t-1)} + \mathrm{Q}(x)\frac{(9)(10)\ldots(t+6)}{(1)(2)\ldots(t-2)}$$
$$+ \mathrm{R}(x)\frac{(9)(10)\ldots(t+5)}{(1)(2)\ldots(t-3)} + x^{16} . \frac{(9)(10)\ldots(t+4)}{(1)(2)\ldots(t-4)} .$$

Denoting this generating function by $\mathrm{F}_t(x)$, I find

$$\mathrm{P}(x) = \mathrm{F}_1(x) - \frac{(9)}{(1)},$$

$$\mathrm{Q}(x) = \mathrm{F}_2(x) - \frac{(9)}{(1)}\mathrm{F}_1(x) + x\frac{(8)(9)}{(1)(2)},$$

$$\mathrm{R}(x) = \mathrm{F}_3(x) - \frac{(9)}{(1)}\mathrm{F}_2(x) + x\frac{(8)(9)}{(1)(2)}\mathrm{F}_1(x) - x^3\frac{(7)(8)(9)}{(1)(2)(3)},$$

$$x_{16} = \mathrm{F}_4(x) - \frac{(9)}{(1)}\mathrm{F}_3(x) + x\frac{(8)(9)}{(1)(2)}\mathrm{F}_2(x) - x^3\frac{(7)(8)(9)}{(1)(2)(3)}\mathrm{F}_1(x) + x^6\frac{(6)(7)(8)(9)}{(1)(2)(3)(4)},$$

whence

$$\mathrm{F}_5(x) = \frac{(9)}{(1)}\mathrm{F}_4(x) - x\frac{(8)(9)}{(1)(2)}\mathrm{F}_3(x) + x^3\frac{(7)(8)(9)}{(1)(2)(3)}\mathrm{F}_2(x)$$
$$- x^6\frac{(6)(7)(8)(9)}{(1)(2)(3)(4)}\mathrm{F}_1(x) + x^{10} . \frac{(5)(6)(7)(8)(9)}{(1)(2)(3)(4)(5)},$$

and in general

$$\mathrm{F}_t(x) = \frac{(9)(10)\ldots(t+4)}{(1)(2)\ldots(t-4)}\mathrm{F}_4(x) - x\frac{(8)(9)\ldots(t+4)}{(1)(2)\ldots(t-3)}\cdot\frac{(t-4)}{(1)}\mathrm{F}_3(x)$$

$$+ x^3\frac{(7)(8)\ldots(t+4)}{(1)(2)\ldots(t-2)}\cdot\frac{(t-4)(t-3)}{(1)(2)}\cdot\mathrm{F}_2(x)$$

$$- x^6\frac{(6)(7)\ldots(t+4)}{(1)(2)\ldots(t-1)}\cdot\frac{(t-4)(t-3)(t-2)}{(1)(2)(3)}\mathrm{F}_1(x)$$

$$+ x^{10}\frac{(5)(6)\ldots(t+4)}{(1)(2)\ldots(t)}\cdot\frac{(t-4)(t-3)(t-2)(t-1)}{(1)(2)(3)(4)}.$$

Art. 103. This appears to be the most symmetrical form in which the generating function can be exhibited, and it may be assumed that the like function for the solid reticulation in general will be of complicated nature. The argument that has been given shows that the theory of the n-dimensional lattice (easily realizable *in plano*), the part-magnitude being limited so as not to exceed unity, is co-extensive with the whole theory of partitions on the lattice of $n-1$ dimensions.

Section 8.

Art. 104. The enumerating generating functions that are met with at the outset in the theory of the partitions of numbers are such as are formed by factors of the forms

$$\frac{1-x^{n+s}}{1-x^s},$$

written for brevity $\frac{(n+s)}{(s)}$. All those which appear in connection with regular graphs in two and three dimensions are so expressible, and the mere fact of such expression proves beyond question that the numerator of the generating function is exactly divisible by the denominator; in other words, it proves that the function can be put into a finite integral form. It is quite natural therefore to seek the general expression of functions of this form, which possesses this property of competency to generate a finite number of terms. Moreover, it is conceivable that such a determination will indicate the paths of future research in these matters: will be in fact a sign-post at the cross-ways. This is the reason why I undertook the investigation; but, as frequently happens in similar cases, the problem proves *à posteriori* to be *per se* of great interest and to involve in itself a notable theorem in partitions.

Art. 105. I consider the function

$$\frac{(n+1)^{a_1}(n+2)^{a_2}(n+3)^{a_3}\ldots(n+s)^{a_s}}{(1)^{a_1}\quad(2)^{a_2}\quad(3)^{a_3}\quad\ldots\quad(s)^{a_s}},$$

which I also write

$$\mathrm{X}_1^{a_1}\mathrm{X}_2^{a_2}\mathrm{X}_3^{a_3}\ldots\mathrm{X}_s^{a_s},$$

and investigate the sum

$$\Sigma X_1^{\alpha_1} X_2^{\alpha_2} X_3^{\alpha_3} \ldots X_s^{\alpha_s}$$

for all values of $\alpha_1, \alpha_2, \alpha_3, \ldots \alpha_s$, which render the expression under the sign of summation expressible in a finite integral form *for all values of the integer n.*

Art. 106. Let ξ_t be that factor of $1 - x^t$ which, when equated to zero, yields all the primitive roots of the equation

$$1 - x^t = 0.$$

Then $1 - x^t = \xi_1 \xi_{d_1} \xi_{d_2} \ldots \xi_t$ where $1, d_1, d_2, \ldots t$ are all the divisors of t. We must find the circumstances under which every expression ξ_t will occur at least as often in the numerator as in the denominator. We need not attend to ξ_1, since it occurs with equal frequency in numerator and denominator. In regard to ξ_2, we have equal frequency if $n + 1$ be uneven, but if $n + 1$ be even we must have

$$\alpha_1 + \alpha_3 + \alpha_5 + \ldots \geqq \alpha_2 + \alpha_4 + \alpha_6 + \ldots$$

For ξ_3 if $n + 1 \equiv 0 \bmod 3$,

$$\alpha_1 + \alpha_4 + \alpha_7 + \ldots \geqq \alpha_3 + \alpha_6 + \alpha_9 + \ldots,$$

and if $n + 1 \equiv 1 \bmod 3$,

$$\alpha_2 + \alpha_5 + \alpha_8 + \ldots \geqq \alpha_3 + \alpha_6 + \alpha_9 + \ldots,$$

while the case of $n + 1 \equiv 2 \bmod 3$ need not be attended to.

Proceeding in this manner we find the following conditions :—

$$\alpha_1 + \alpha_3 + \alpha_5 + \ldots \geqq \alpha_2 + \alpha_4 + \alpha_6 + \ldots$$

$$\left\{\begin{array}{l} \alpha_1 + \alpha_4 + \alpha_7 + \ldots \geqq \alpha_3 + \alpha_6 + \alpha_9 + \ldots \\ \alpha_2 + \alpha_5 + \alpha_8 + \ldots \geqq \alpha_3 + \alpha_6 + \alpha_9 + \ldots \end{array}\right.$$

$$\left\{\begin{array}{l} \alpha_1 + \alpha_5 + \alpha_9 + \ldots \geqq \alpha_4 + \alpha_8 + \alpha_{12} + \ldots \\ \alpha_2 + \alpha_6 + \alpha_{10} + \ldots \geqq \alpha_4 + \alpha_8 + \alpha_{12} + \ldots \\ \alpha_3 + \alpha_7 + \alpha_{11} + \ldots \geqq \alpha_4 + \alpha_8 + \alpha_{12} + \ldots \end{array}\right.$$

$$\ldots\ldots\ldots\ldots\ldots\ldots$$

$$\left\{\begin{array}{ll} \alpha_1 + \alpha_s + \ldots\ldots & \geqq \alpha_{s-1} \\ \alpha_2 & \geqq \alpha_{s-1} \\ \vdots & \vdots \\ \alpha_{s-2} & \geqq \alpha_{s-1} \end{array}\right.$$

$$\left\{\begin{array}{ll} \alpha_1 & \geqq \alpha_s \\ \alpha_2 & \geqq \alpha_s \\ \vdots & \vdots \\ \alpha_{s-1} & \geqq \alpha_s \end{array}\right.$$

$\frac{1}{2} s(s-1)$ in number.

The next step is to construct an Ω function which shall express these conditions and lead practically to the desired summation.

Art. 107. First take $s = 2$; there is but one condition

$$\alpha_1 \geqq \alpha_2,$$

and the function is

$$\underset{\geqq}{\Omega} \frac{1}{1 - a_1 X_1 \,.\, 1 - \dfrac{1}{a_1} X_2} = \frac{1}{1 - X_1 \,.\, 1 - X_1 X_2},$$

and every term in the ascending expansion of this function is of the required form, and no other forms exist. The general term being

$$X_1^{\alpha_1 - \alpha_2} (X_1 X_2)^{\alpha_2} \qquad \alpha_1 \geqq \alpha_2,$$

we may call X_1 and X_1X_2 the ground forms from which all other forms are derived.

Art. 108. Next take $s = 3$. The conditions are

$$\left.\begin{array}{l} \alpha_1 + \alpha_3 \geqq \alpha_2 \\ \alpha_1 \geqq \alpha_3 \\ \alpha_2 \geqq \alpha_3 \end{array}\right\},$$

leading to the summation formula

$$\underset{\geqq}{\Omega} \frac{1}{1 - a_1 a_2 X_1 \,.\, 1 - \dfrac{a_3}{a_1} X_2 \,.\, 1 - \dfrac{a_1}{a_2 a_3} X_3},$$

the auxiliaries a_1, a_2, a_3 determining the first, second and third conditions respectively. The function is equal to

$$\underset{\geqq}{\Omega} \frac{1}{1 - a_1 a_2 X_1 \,.\, 1 - \dfrac{1}{a_1} X_2 \,.\, 1 - \dfrac{1}{a_2} X_2 X_3}$$

$$= \underset{\geqq}{\Omega} \frac{1}{1 - a_1 X_1 \,.\, 1 - \dfrac{1}{a_1} X_2 \,.\, 1 - a_1 X_1 X_2 X_3}$$

$$= \underset{\geqq}{\Omega} \frac{1}{1 - a_1 X_1} \left\{ \frac{1}{1 - a_1 X_1 X_2 X_3 \,.\, 1 - X_1 X_2^2 X_3} + \frac{\dfrac{X_2}{a_1}}{1 - \dfrac{X_2}{a_1} \,.\, 1 - X_1 X_2^2 X_3} \right\}$$

$$= \frac{1}{1 - X_1 \,.\, 1 - X_1 X_2 X_3 \,.\, 1 - X_1 X_2^2 X_3} + \frac{X_1 X_2}{1 - X_1 \,.\, 1 - X_1 X_2 \,.\, 1 - X_1 X_2^2 X_3}$$

$$= \frac{1 - X_1^2 X_2^2 X_3}{1 - X_1 \,.\, 1 - X_1 X_2 \,.\, 1 - X_1 X_2 X_3 \,.\, 1 - X_1 X_2^2 X_3},$$

representing the complete solution.

The denominator factors yield the ground forms

$$X_1X_2X_3, \qquad X_1X_2^2X_3$$

in addition to those previously met with, whilst the numerator factor indicates the ground form syzygy

$$X_1 \cdot X_1X_2^2X_3 - X_1X_2 \,.\, X_1X_2X_3 = 0.$$

Observe that

$$X_1X_2X_3 = \frac{1 - x^{n+1} . 1 - x^{n+2} . 1 - x^{n+3}}{1 - x . 1 - x^2 . 1 - x^3}$$

$$X_1X_2^2X_3 = \frac{(1 - x^{n+1})(1 - x^{n+2})^2(1 - x^{n+3})}{(1 - x)(1 - x^2)^2(1 - x^3)}$$

are those with which we are familiar in the theories of simple and compound partition respectively.

Art. 109. I pass on to the case $s = 4$; the conditions are

$$\begin{aligned} \alpha_1 + \alpha_3 &\geqq \alpha_2 + \alpha_4 \\ \alpha_1 + \alpha_4 &\geqq \alpha_3 \\ \alpha_2 &\geqq \alpha_3 \\ \alpha_1 &\geqq \alpha_4 \\ \alpha_2 &\geqq \alpha_4 \\ \alpha_3 &\geqq \alpha_4 \end{aligned}$$

We neglect the fifth of these as being implied by the remainder and from the function

$$\underset{\geqq}{\Omega} \frac{1}{1 - a_1a_2a_4X_1 . 1 - \frac{a_3}{a_1}X_2 . 1 - \frac{a_1a_5}{a_2a_3}X_3 . 1 - \frac{a_2}{a_1a_4a_5}X_4}$$

which, when reduced, is

$$\frac{1}{1 - X_1 . 1 - X_1X_2 . 1 - X_1X_2X_3X_4 . 1 - X_1X_2^2X_3^2X_4}$$

$$+ \frac{X_1X_2^2X_3}{1 - X_1 . 1 - X_1X_2 . 1 - X_1X_2^2X_3 . 1 - X_1X_2^2X_3^2X_4}$$

$$+ \frac{X_1X_2X_3}{1 - X_1 . 1 - X_1X_2X_3 . 1 - X_1X_2^2X_3 . 1 - X_1X_2^2X_3^2X_4}$$

showing that the new ground forms are $X_1X_2X_3X_4$ and $X_1X_2^2X_3^2X_4$, both of which have presented themselves before.

The result may be written

$$\frac{1 - X_1^2X_2^2X_3 - X_1^2X_2^2X_3^2X_4 - X_1^2X_2^3X_3^3X_4^1 + X_1^3X_2^3X_3^2X_4 + X_1^3X_2^4X_3^3X_4}{1 - X_1 . 1 - X_1X_2 . 1 - X_1X_2X_3 . 1 - X_1X_2^2X_3 . 1 - X_1X_2X_3X_4 . 1 - X_1X_2^2X_3^2X_4}.$$

and the numerator now indicates the existence of first and second syzygies between the ground forms.

We have the first syzygies

$$\begin{aligned}(\mathrm{A}) &= \mathrm{X}_1\mathrm{X}_2 \,.\, \mathrm{X}_1\mathrm{X}_2\mathrm{X}_3 - \mathrm{X}_1 \,.\, \mathrm{X}_1\mathrm{X}_2^2\mathrm{X}_3 = 0,\\ (\mathrm{B}_1) &= \mathrm{X}_1\mathrm{X}_2\mathrm{X}_3 \,.\, \mathrm{X}_1\mathrm{X}_2\mathrm{X}_3\mathrm{X}_4 - \mathrm{X}_1 \,.\, \mathrm{X}_1\mathrm{X}_2^2\mathrm{X}_3^2\mathrm{X}_4 = 0,\\ (\mathrm{B}_2) &= \mathrm{X}_1\mathrm{X}_2^2\mathrm{X}_3 \,.\, \mathrm{X}_1\mathrm{X}_2\mathrm{X}_3\mathrm{X}_4 - \mathrm{X}_1\mathrm{X}_2 \,.\, \mathrm{X}_1\mathrm{X}_2^2\mathrm{X}_3^2\mathrm{X}_4 = 0,\end{aligned}$$

and the second syzygies

$$\begin{aligned}\mathrm{X}_1(\mathrm{B}_2) - \mathrm{X}_1\mathrm{X}_2(\mathrm{B}_1) &= 0,\\ \mathrm{X}_1\mathrm{X}_2\mathrm{X}_3(\mathrm{B}_2) - \mathrm{X}_1\mathrm{X}_2^2\mathrm{X}_3(\mathrm{B}_1) &= 0.\end{aligned}$$

Art. 110. For $s = 5$, the generating function is

$$\underset{\geqq}{\Omega} \frac{1}{1 - a_1a_2a_3a_4\mathrm{X}_1 \,.\, 1 - \frac{b_2b_3b_4}{a_1}\mathrm{X}_2 \,.\, 1 - \frac{a_1c_3c_4}{a_2b_2}\mathrm{X}_3 \,.\, 1 - \frac{a_2d_4}{a_1a_3b_3c}\mathrm{X}_4 \,.\, 1 - \frac{a_1b_2a_3}{a_4b_4c_4d_4}\mathrm{X}_5}$$

and there is no difficulty in continuing the series. The obtaining, however, of the reduced forms soon becomes laborious.

Art. 111. There is another method of investigation. Guided by the results obtained let us restrict consideration to the forms

$$\mathrm{X}_1^{\alpha_1}\mathrm{X}_2^{\alpha_2} \ldots \mathrm{X}_s^{\alpha_s}$$

which are such that

$$\alpha_m = \alpha_{s+1-m} \,.^*$$

This is of great importance, because we are thus able, for any given order, to generate the functions of that order alone.

$$\text{Put } \mathrm{X}_m\mathrm{X}_{s+1-m} = \mathrm{Y}_m \text{ and seek } \Sigma\mathrm{Y}_1^{\alpha_1}\mathrm{Y}_2^{\alpha_2} \ldots$$

Art. 112. For $s = 2$, the generating function is simply

$$\frac{1}{1 - \mathrm{Y}_1} = \frac{1}{1 - \mathrm{X}_1\mathrm{X}_2}.$$

Art. 113. For $s = 3$, the conditions

$$2\alpha_1 \quad a \geqq a_1$$

lead to

$$\underset{\geqq}{\Omega} \frac{1}{1 - \frac{a^2}{b}\mathrm{Y}_1 \,.\, 1 - \frac{b}{a}\mathrm{Y}_2},$$

the letters a, b determining the first and second conditions respectively

* The validity of this assumption will be considered later.

This is on reduction

$$\frac{1}{1 - Y_1Y_2 \,.\, 1 - Y_1Y_2^2} = \frac{1}{1 - X_1X_2X_3 \,.\, 1 - X_1X_2^2X_3}$$

a real generating function.

Art. 114. For $s = 4$, the conditions are the same, viz. :—

$$2\alpha_1 \geqq \alpha_2 \geqq \alpha_1$$

and the Ω function, where now

$$Y_1 = X_1X_4, \;\; Y_2 = X_2X_3,$$

is

$$\underset{\geqq}{\Omega} \frac{1}{1 - \frac{a^2}{b} Y_1 \,.\, 1 - \frac{b}{a} Y_2} = \frac{1}{1 - Y_1Y_2 \,.\, 1 - Y_1Y_2^2}$$

$$= \frac{1}{1 - X_1X_2X_3X_4 \,.\, 1 - X_1X_2^2X_3^2X_4}$$

yielding the ground forms already found by the first method.

Art. 115. For $s = 5$, the conditions are

$$\alpha_1 + \alpha_2 \geqq \alpha_3 \geqq \alpha_2,$$

$$2\alpha_1 \geqq \alpha_2 \geqq \alpha_1 ;$$

leading to

$$\underset{\geqq}{\Omega} \frac{1}{1 - \frac{ab^2}{d} Y_1 \,.\, 1 - \frac{ad}{bc} Y_2 \,.\, 1 - \frac{c}{a} Y},$$

where

$$Y_1 = X_1X_5, \;\; Y_2 = X_2X_4, \;\; Y_3 = X_3,$$

and this is

$$\underset{\geqq}{\Omega} \frac{1}{1 - \frac{ab^2}{d} Y_1 \,.\, 1 - \frac{d}{b} Y_2Y_3 \,.\, 1 - \frac{1}{a} Y}$$

$$= \underset{\geqq}{\Omega} \frac{1}{1 - abY_1Y_2Y_3 \,.\, 1 - \frac{1}{b} Y_2Y_3 \,.\, 1 - \frac{1}{a} Y_3}$$

$$= \underset{\geqq}{\Omega} \frac{1}{1 - bY_1Y_2Y_3 \,.\, 1 - \frac{1}{b} Y_2Y_3 \,.\, 1 - bY_1Y_2Y_3^2}$$

$$= \frac{1 - Y_1^2Y_2^3Y_3^4}{1 - Y_1Y_2Y_3 \,.\, 1 - Y_1Y_2Y_3^2 \,.\, 1 - Y_1Y_2^2Y_3^2 \,.\, 1 - Y_1Y_2^2Y_3^3}$$

$$= \frac{1 - X_1X_2X_3X_4X_5}{1 - X_1X_2X_3X_4X_5 \,.\, 1 - X_1X_2X_3^2X_4X_5 \,.\, 1 - X_1X_2^2X_3^2X_4^2X_5 \,.\, 1 - X_1X_2^2X_3^3X_4^2X_5},$$

establishing the ground forms

$$X_1X_2X_3X_4X_5, \;\; X_1X_2X_3^2X_4X_5$$

$$X_1X_2^2X_3^2X_4^2X_5, \;\; X_1X_2^2X_3^3X_4^2X_5$$

connected by the simple syzygy

$$(X_1X_2X_3X_4X_5)(X_1X_2^2X_3^3X_4^2X_5) - (X_1X_2X_3^2X_4X_5)\,X_1X_2^2X_3^2X_4^2X_5) = 0.$$

Art. 116. I stop to remark that one of these ground forms, viz.:—

$$X_1X_2X_3^2X_4X_5$$

is new, not having so far presented itself in a partition theorem. It is one of an infinite system which merits, and will receive, separate consideration later on. The one before us is associated with partitions at the points of the dislocated lattice.

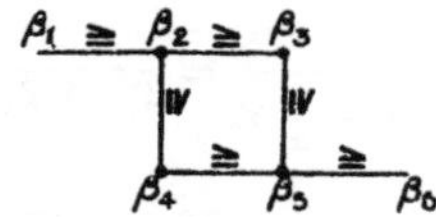

Art. 117 For $s = 6$, the conditions are:

$$2\alpha_2 \geq \alpha_1 + \alpha_3$$
$$2\alpha_1 \geq \alpha_2$$
$$\alpha_3 \geq \alpha_2,$$

leading to

$$\underset{\geq}{\Omega}\frac{1}{1-\frac{b^2}{a}Y_1 \,.\, 1-\frac{a^2}{bc}Y_2 \,.\, 1-\frac{c}{a}Y_3},$$

where

$$Y_1 = X_1X_6, \quad Y_2 = X_2X_5, \quad Y_3 = X_3X_4.$$

This is

$$\underset{\geq}{\Omega}\frac{1}{1-\frac{b^2}{a}Y_1 \,.\, 1-\frac{a}{b}Y_2Y_3 \,.\, 1-\frac{1}{a}Y}$$

$$= \underset{\geq}{\Omega}\frac{1}{1-\frac{1}{a}Y_3 \,.\, 1-\frac{a}{b}Y_2Y_3 \,.\, 1-bY_1Y_2Y_3}$$

$$= \frac{1}{1-Y_1Y_2Y_3 \,.\, 1-Y_1Y_2^2Y_3^2 \,.\, 1-Y_1Y_2^2Y_3^3},$$

establishing the ground forms:

$$X_1X_2X_3X_4X_5X_6$$
$$X_1X_2^2X_3^2X_4^2X_5^2X_6$$
$$X_1X_2^2X_3^3X_4^3X_5^2X_6,$$

unconnected by any syzygy.

Art. 118. For $s = 7$, the independent conditions are:

$$
\begin{aligned}
2\alpha_1 + 2\alpha_3 &\geqq 2\alpha_2 + \alpha_4 \\
2\alpha_2 &\geqq \alpha_4 \\
\alpha_1 + \alpha_2 &\geqq \alpha_3 \\
\alpha_4 &\geqq \alpha_3 \\
2\alpha_1 &\geqq \alpha_2 \\
\alpha_3 &\geqq \alpha_2 \\
\alpha_2 &\geqq \alpha_1,
\end{aligned}
$$

and these lead to

$$\underset{\geqq}{\Omega} \frac{}{1 - \frac{a^2ce^2}{g}\mathrm{Y}_1 . 1 - \frac{b^2cg}{a^2ef}\mathrm{Y}_2 . 1 - \frac{a^2f}{cd}\mathrm{Y}_3 . 1 - \frac{d}{ab}\mathrm{Y}_4}$$

and eliminating d, f, g in succession

$$= \underset{\geqq}{\Omega} \frac{1}{1 - \frac{a^2ce^2}{g}\mathrm{Y}_1 . 1 - \frac{b^2cg}{a^2ef}\mathrm{Y}_2 . 1 - \frac{af}{bc}\mathrm{Y}_3\mathrm{Y}_4 . 1 - \frac{1}{ab}\mathrm{Y}_4}$$

$$= \underset{\geqq}{\Omega} \frac{1}{1 - \frac{a^2ce^2}{g}\mathrm{Y}_1 . 1 - \frac{bg}{ae}\mathrm{Y}_2\mathrm{Y}_3\mathrm{Y}_4 . 1 - \frac{a}{bc}\mathrm{Y}_3\mathrm{Y}_4 . 1 - \frac{1}{ab}\mathrm{Y}_4},$$

$$= \underset{\geqq}{\Omega} \frac{1}{1 - abce\mathrm{Y}_1\mathrm{Y}_2\mathrm{Y}_3\mathrm{Y}_4 . 1 - \frac{b}{ae}\mathrm{Y}_2\mathrm{Y}_3\mathrm{Y}_4 . 1 - \frac{a}{bc}\mathrm{Y}_3\mathrm{Y}_4 . 1 - \frac{1}{ab}\mathrm{Y}_4}.$$

and eliminating e

$$= \underset{\geqq}{\Omega} \frac{1}{1 - abc\mathrm{Y}_1\mathrm{Y}_2\mathrm{Y}_3\mathrm{Y}_4 . 1 - b^2c\mathrm{Y}_1\mathrm{Y}_2^2\mathrm{Y}_3^2\mathrm{Y}_4^2 . 1 - \frac{a}{bc}\mathrm{Y}_3\mathrm{Y}_4 . 1 - \frac{1}{ab}\mathrm{Y}}$$

and eliminating c

$$= \underset{\geqq}{\Omega} \frac{1 - a^2b^2\mathrm{Y}_1^2\mathrm{Y}_2^3\mathrm{Y}_3^4\mathrm{Y}_4^4}{\begin{array}{c}1 - ab\mathrm{Y}_1\mathrm{Y}_2\mathrm{Y}_3\mathrm{Y}_4 . 1 - a^2\mathrm{Y}_1\mathrm{Y}_2\mathrm{Y}_3^2\mathrm{Y}_4^2 . 1 - b^2\mathrm{Y}_1\mathrm{Y}_2^2\mathrm{Y}_3^2\mathrm{Y}_4^2, \\ 1 - ab . \mathrm{Y}_1\mathrm{Y}_2^2\mathrm{Y}_3^3\mathrm{Y}_4^3 . 1 - \frac{1}{ab}\mathrm{Y}_4\end{array}}.$$

And on further reduction it is finally

$$\frac{1 - \mathrm{Y}_1^2\mathrm{Y}_2^3\mathrm{Y}_3^4\mathrm{Y}_4^4 . 1 - \mathrm{Y}_1^2\mathrm{Y}_2^3\mathrm{Y}_3^4\mathrm{Y}^3}{\begin{array}{c}1 - \mathrm{Y}_1\mathrm{Y}_2\mathrm{Y}_3\mathrm{Y}_4 . 1 - \mathrm{Y}_1\mathrm{Y}_2\mathrm{Y}_3\mathrm{Y}_4^2 . 1 - \mathrm{Y}_1\mathrm{Y}_2\mathrm{Y}_3^2\mathrm{Y}_4^2, \\ 1 - \mathrm{Y}_1\mathrm{Y}_2^2\mathrm{Y}_3^2\mathrm{Y}_4^2 . 1 - \mathrm{Y}_1\mathrm{Y}_2^2\mathrm{Y}_3^3\mathrm{Y}_4^3 . 1 - \mathrm{Y}_1\mathrm{Y}_2^2\mathrm{Y}_3^3\mathrm{Y}_4^4\end{array}}$$

establishing the ground forms

$$Y_1Y_2Y_3Y_4 \equiv X_1X_2X_3X_4X_5X_6X_7$$
$$Y_1Y_2Y_3Y_4^2 \equiv X_1X_2X_3X_4^2X_5X_6X_7$$
$$Y_1Y_2Y_3^2Y_4^2 \equiv X_1X_2X_3^2X_4^2X_5^2X_6X_7$$
$$Y_1Y_2^2Y_3^2Y_4^2 \equiv X_1X_2^2X_3^2X_4^2X_5^2X_6^2X_7$$
$$Y_1Y_2^2Y_3^3Y_4^3 \equiv X_1X_2^2X_3^3X_4^3X_5^3X_6^2X_7$$
$$Y_1Y_2^2Y_3^3Y_4^4 \equiv X_1X_2^2X_3^3X_4^4X_5^3X_6^2X_7$$

connected by the simple syzygies

$$(Y_1Y_2Y_3Y_4)(Y_1Y_2^2Y_3^3Y_4^3) - (Y_1Y_2Y_3^2Y_4^2)(Y_1Y_2^2Y_3^2Y_4^2) = 0,$$
$$(Y_1Y_2Y_3Y_4)(Y_1Y_2^2Y_3^3Y_4^4) - (Y_1Y_2Y_3Y_4^2)(Y_1Y_2^2Y_3^3Y_4^3) = 0,$$

and, denoting these respectively by A and B, the numerator term $+ Y_1^4Y_2^6Y_3^8Y_4^9$ indicates the second, or compound, syzygy :—

$$(Y_1Y_2Y_3Y_4)(Y_1Y_2^2Y_3^3Y_4^4)(A) - (Y_1Y_2Y_3^2Y_4^2)(Y_1Y_2^2Y_3^2Y_4^2)(B) = 0.$$

Art. 119. I remark that the forms

$$Y_1Y_2Y_3Y_4^2, \qquad Y_1Y_2Y_3^2Y_4^2$$

are new to partition theory.

Art. 120. For $s = 8$, the reduced conditions are

$$\alpha_2 + \alpha_3 \geq \alpha_1 + \alpha_4$$
$$\alpha_1 + \alpha_2 \geq \alpha_3$$
$$\alpha_4 \geq \alpha_3$$
$$2\alpha_1 \geq \alpha_2$$
$$\alpha_3 \geq \alpha_2$$

leading to

$$\underset{\geq}{\Omega} \frac{1}{1 - \frac{bd^2}{a}Y_1 \, . \, 1 - \frac{ab}{de}Y_2 \, . \, 1 - \frac{ae}{bc}Y_3 \, . \, 1 - \frac{c}{a}Y_4}$$

$$= \underset{\geq}{\Omega} \frac{1}{1 - \frac{bd^2}{a}Y_1 \, . \, 1 - \frac{a}{d}Y_2Y_3Y_4 \, . \, 1 - \frac{1}{b}Y_3Y_4 \, . \, 1 - \frac{1}{a}Y_4}$$

$$= \underset{\geq}{\Omega} \frac{1}{1 - bdY_1Y_2Y_3Y_4 \, . \, 1 - \frac{1}{d}Y_2Y_3Y_4 \, . \, 1 - \frac{1}{b}Y_3Y_4 \, . \, 1 - \frac{1}{d}Y_2Y_3Y_4^2}$$

$$= \underset{\geq}{\Omega} \frac{1}{1 - bY_1Y_2Y_3Y_4 \, . \, 1 - bY_1Y_2^2Y_3^2Y_4^2 \, . \, 1 - \frac{1}{b}Y_3Y_4 \, . \, 1 - bY_1Y_2^2Y_3^2Y_4^3}$$

$$= \frac{1 - Y_1^2Y_2^3Y_3^4Y_4^4 - Y_1^2Y_2^3Y_3^4Y_4^5 - Y_1^2Y_2^4Y_3^5Y_4^6 + Y_1^3Y_2^5Y_3^6Y_4^7 + Y_1^3Y_2^5Y_3^7Y_4^8}{\begin{array}{c}1 - Y_1Y_2Y_3Y_4 \, . \, 1 - Y_1Y_2Y_3^2Y_4^2 \, . \, 1 - Y_1Y_2^2Y_3^2Y_4^2 \\ 1 - Y_1Y_2^2Y_3^2Y_4^3 \, . \, 1 - Y_1Y_2^2Y_3^3Y_4^3 \, . \, 1 - Y_1Y_2^2Y_3^3Y_4^4\end{array}}$$

indicating the ground forms

$$Y_1Y_2Y_3Y_4 \equiv X_1X_2X_3X_4X_5X_6X_7X_8$$
$$Y_1Y_2Y_3^2Y_4^2 \equiv X_1X_2X_3^2X_4^2X_5^2X_6^2X_7X_8$$
$$Y_1Y_2^2Y_3^2Y_4^2 \equiv X_1X_2^2X_3^2X_4^2X_5^2X_6^2X_7^2X_8$$
$$Y_1Y_2^2Y_3^2Y_4^3 \equiv X_1X_2^2X_3^2X_4^3X_5^3X_6^2X_7^2X_8$$
$$Y_1Y_2^2Y_3^3Y_4^3 \equiv X_1X_2^2X_3^3X_4^3X_5^3X_6^3X_7^2X_8$$
$$Y_1Y_2^2Y_3^3Y_4^4 \equiv X_1X_2^2X_3^3X_4^4X_5^4X_6^3X_7^2X_8$$

Art. 121. So far it appears that all products which can be placed in the form of a rectangle

$$\begin{array}{llll} X_1 & X_2 & X_3 & \ldots X_l \\ X_2 & X_3 & X_4 & \ldots X_{l+1} \\ \vdots & \vdots & \vdots & \quad\vdots \\ X_m & X_{m+1} & X_{m+2} & \ldots X_{l+m-1} \end{array}$$

are ground forms for all values of l and m.

I have established this independently, and thus proved that the conjectured result for the general lattice *in plano* is, at any rate, finite and integral, as it should be.

It is desirable to obtain information concerning the ground forms which are not within the rectangular tableau.

The forms

$$X_1^{\alpha_1}X_2^{\alpha_2}\ldots X_\nu^{\alpha_\nu},$$

which appear in the tableau, may be eliminated from consideration, with the exception of the form

$$X_1X_2\ldots X_\nu,$$

by ascribing additional conditions such as

$$\alpha_1 = \alpha_2,$$

which are not true in the tableau.

The condition of this tableau is that if $\alpha_p = \alpha_{p+1}$, no index α_{p+q} is greater than α_p; after a repetition of index, no rise in index takes place. In the Y form, therefore, we may assign the conditions

$$\alpha_p = \alpha_{p+1} < \alpha_{p+2}$$

for any value of p, as one excluding the whole of the forms appertaining to the tableau.

We may impress the conditions

$$\alpha_1 = \alpha_2 < \alpha_3$$
$$\alpha_2 = \alpha_3 < \alpha_4$$
$$\alpha \;= \alpha_4 < \alpha_5$$

in succession, and we may combine any number of such conditions as are independent.

Art. 122. I postpone further investigation into this interesting theory, and will now give a formal proof that the product tableau is, in fact, finite and integral. The product in question for $m \geqq l$ is

$$X_1X_2^2 \ldots X_{l-1}^{l-1}(X_lX_{l+1} \ldots X_m)^l X_{m+1}^{l-1}X_{m+2}^{l-2} \ldots X_{m+l-2}^2 X_{m+l-1}$$

so that

$$\begin{array}{lll} \alpha_s = s & \text{for} & l \geqq s \\ \alpha_s = l & \text{for} & s > l \text{ and } < m+1 \\ \alpha_{m+s} = l - s & \text{for} & l - 1 \geqq s \end{array}$$

All the conditions may be resumed in the single formula

$$\alpha_s + \alpha_{2s+t} + \alpha_{3s+2t} + \ldots \geqq \alpha_{s+t} + \alpha_{2s+2t} + \alpha_{3s+3t} + \ldots$$

s and t being any integers.

Let the greatest integer in $\frac{l+t-1}{s+t}$ be denoted by $I_1\frac{l+t-1}{s+t}$ or by I_1 simply for brevity. Similarly let I_2 refer to $\frac{m+t}{s+t}$, I_3 to $\frac{l+m+t-1}{s+t}$, J_1 to $\frac{l-1}{s+t}$, J_2 to $\frac{m}{s+t}$, and J_3 to $\frac{l+m-1}{s+t}$. We derive

$$\begin{array}{rcl} I_1 = J_1 & \text{or} & J_1 + 1 \\ I_2 = J_2 & \text{or} & J_2 + 1 \\ I_3 = J_3 & \text{or} & J_3 + 1 \\ I_1 + I_2 = I_3 & \text{or} & I_3 + 1 \\ J_1 + J_2 = J_3 & \text{or} & J_3 - 1 \end{array}$$

and we have ten possible cases to consider, viz. :—

Case 1.

$$\begin{array}{ll} I_1 = J_1 & I_1 + I_2 = I_3 \\ I_2 = J_2 & J_1 + J_2 = J_3 \\ I_3 = J_3 & \end{array}$$

Case 2.

$$\begin{array}{ll} I_1 = J_1 + 1 & I_1 + I_2 = I_3 \\ I_2 = J_2 & J_1 + J_2 = J_3 - 1 \\ I_3 = J_3 & \end{array}$$

Case 3.

$$\begin{array}{ll} I_1 = J_1 + 1 & I_1 + I_2 = I_3 + 1 \\ I_2 = J_2 & J_1 + J_2 = J_3 \\ I_3 = J_3 & \end{array}$$

Case 4.

$$\mathrm{I}_1 = \mathrm{J}_1 \qquad \mathrm{I}_1 + \mathrm{I}_2 = \mathrm{I}_3$$
$$\mathrm{I}_2 = \mathrm{J}_2 + 1 \qquad \mathrm{J}_1 + \mathrm{J}_2 = \mathrm{J}_3 - 1$$
$$\mathrm{I}_3 = \mathrm{J}_3$$

Case 5.

$$\mathrm{I}_1 = \mathrm{J}_1 \qquad \mathrm{I}_1 + \mathrm{I}_2 = \mathrm{I}_3 + 1$$
$$\mathrm{I}_2 = \mathrm{J}_2 + 1 \qquad \mathrm{J}_1 + \mathrm{J}_2 = \mathrm{J}_3$$
$$\mathrm{I}_3 = \mathrm{J}_3$$

Case 6.

$$\mathrm{I}_1 = \mathrm{J}_1 + 1 \qquad \mathrm{I}_1 + \mathrm{I}_2 = \mathrm{I}_3 + 1$$
$$\mathrm{I}_2 = \mathrm{J}_2 + 1 \qquad \mathrm{J}_1 + \mathrm{J}_2 = \mathrm{J}_3 - 1$$
$$\mathrm{I}_3 = \mathrm{J}_3$$

Case 7.

$$\mathrm{I}_1 = \mathrm{J}_1 + 1 \qquad \mathrm{I}_1 + \mathrm{I}_2 = \mathrm{I}_3$$
$$\mathrm{I}_2 = \mathrm{J}_2 \qquad \mathrm{J}_1 + \mathrm{J}_2 = \mathrm{J}_3$$
$$\mathrm{I}_3 = \mathrm{J}_3 + 1$$

Case 8.

$$\mathrm{I}_1 = \mathrm{J}_1 \qquad \mathrm{I}_1 + \mathrm{I}_2 = \mathrm{I}_3$$
$$\mathrm{I}_2 = \mathrm{J}_2 + 1 \qquad \mathrm{J}_1 + \mathrm{J}_2 = \mathrm{J}_3$$
$$\mathrm{I}_3 = \mathrm{J}_3 + 1$$

Case 9.

$$\mathrm{I}_1 = \mathrm{J}_1 + 1 \qquad \mathrm{I}_1 + \mathrm{I}_2 = \mathrm{I}_3$$
$$\mathrm{I}_2 = \mathrm{J}_2 + 1 \qquad \mathrm{J}_1 + \mathrm{J}_2 = \mathrm{J}_3 - 1$$
$$\mathrm{I}_3 = \mathrm{J}_3 + 1$$

Case 10.

$$\mathrm{I}_1 = \mathrm{J}_1 + 1 \qquad \mathrm{I}_1 + \mathrm{I}_2 = \mathrm{I}_3 + 1$$
$$\mathrm{I}_2 = \mathrm{J}_2 + 1 \qquad \mathrm{J}_1 + \mathrm{J}_2 = \mathrm{J}_3$$
$$\mathrm{I}_3 = \mathrm{J}_3 + 1$$

For the series

$$\alpha_s + \alpha_{2s+t} + \alpha_{3s+2t} + \ldots,$$

we have, as far as α_{l-1}, I_1 terms; as far as α_m, I_2 terms; and, as far as α_{l+m-1}, I_3 terms.

Hence the summation gives :—

$$\begin{aligned}&\tfrac{1}{2}\mathrm{I}_1\{2s + (\mathrm{I}_1 - 1)(s + t)\} + l(\mathrm{I}_2 - \mathrm{I}_1)\\ &\qquad + \tfrac{1}{2}(\mathrm{I}_3 - \mathrm{I}_2)\{2l + 2m + 2t - 2(s + t)(\mathrm{I}_2 + 1) - (s + t)(\mathrm{I}_3 - \mathrm{I}_2 - 1)\}\\ &\quad = \tfrac{1}{2}(s + t)(\mathrm{I}_1^2 + \mathrm{I}_2^2 - \mathrm{I}_3^2) + (\tfrac{1}{2}s - \tfrac{1}{2}t - \tfrac{1}{2}l)\,\mathrm{I}_1\\ &\qquad + (\tfrac{1}{2}s - \tfrac{1}{2}t - m)\,\mathrm{I}_2 + (l + m - \tfrac{1}{2}s + \tfrac{1}{2}t)\,\mathrm{I}_3.\end{aligned}$$

Summing similarly the series

$$a_{s+t} + a_{2s+2t} + a_{3s+3t} \ldots$$

we find

$$\tfrac{1}{2}(s+t)(J_1^2 + J_2^2 - J_3^2) + (\tfrac{1}{2}s + \tfrac{1}{2}t - l)J_1$$
$$+ (\tfrac{1}{2}s + \tfrac{1}{2}t - m)J_2 + (l + m - \tfrac{1}{2}s - \tfrac{1}{2}t)J_3,$$

and we have in each of the ten cases to establish the relation

$$\tfrac{1}{2}(s+t)(I_1^2 + I_2^2 - I_3^2) + (\tfrac{1}{2}s - \tfrac{1}{2}t - l)I_1$$
$$+ (\tfrac{1}{2}s - \tfrac{1}{2}t - m)I_2 + (l + m - \tfrac{1}{2}s + \tfrac{1}{2}t)I_3,$$
$$\geq \tfrac{1}{2}(s+t)(J_1^2 + J_2^2 - J_3^2) + (\tfrac{1}{2}s + \tfrac{1}{2}t - l)J_1$$
$$+ (\tfrac{1}{2}s + \tfrac{1}{2}t - m)J_2 + (l + m - \tfrac{1}{2}s - \tfrac{1}{2}t)J_3$$

for all values of s and t.

For Case 1 it reduces to

$$I_1 + I_2 \geq I_3,$$

which is true, for here $I_1 + I_2 = I_3$.

For Case 2, making use of $J_1 + J_2 = J_3 - 1$, the reduction is to

$$J_1 \geq \frac{l - s - t}{s + t},$$

and J_1 being the greatest integer in $\frac{l-1}{s+t}$, and moreover $s + t$ being at least unity, the relation is obviously satisfied.

For Case 3, making use of $J_1 + J_2 = J_3$, we find

$$J_1 \geq \frac{l - s}{s + t},$$

and this is satisfied as $s \geq 1$.

For Case 4, reducing by $J_1 + J_2 = J_3 - 1$, we find

$$J_2 \geq \frac{m}{s+t} - 1$$

obviously true from the definition of J_2.

For Case 5, reducing by $J_1 + J_2 = J_3$, we find

$$J_2 \geq \frac{m - s}{s + t}$$

obviously satisfied.

For Case 6, reducing by $J_1 + J_2 = J_3 - 1$, we find

$$J_3 \geq \frac{l + m - s}{s + t}$$

clearly satisfied.

For Case 7, reducing by $J_1 + J_2 = J_3$, we find

$$J_2 \geqq \frac{m}{s+t},$$

which is right.

For Case 8, reducing by $J_1 + J_2 = J_3$, we find

$$J_1 \geqq \frac{l}{s+t},$$

which is satisfied.

For Case 9, reducing by $J_1 + J_2 = J_3 - 1$, the ratio is one of equality.

For Case 10, reducing by $J_1 + J_2 = J_3$, we find

$$s + t \geqq 0,$$

which is right.

Hence the relation is universally satisfied, and we have proved that the expression

$$X_1 X_2^2 \dots X_{l-1}^{l-1} (X_l X_{l+1} \dots X_m)^l X_{m+1}^{l-1} X_{m+2}^{l-2} \dots X_{l+m-1}$$

is in every case finite and integral.

Art. 123. In Part 3 of this Memoir I hope to treat of other systems of algebraical and arithmetical functions which fall within the domain of partition analysis and the theory of the linear composition of integers; also to take up the general theories of partition analysis and linear Diophantine analysis, with possible extensions to higher degrees.

ction of Röntgen Rays, Agents.

niversity of Cambridge.

F.R.S.

INDEX SLIP.

265, 1897) I described
ted with water vapour,
if the maximum degree
tain limit. Using v_2/v_1
the expansion, we may

xceeds 1·25; the drops
not exceeded. Beyond
th increasing expansion
even slightly exceeding

n (approximately eight-
ondenses independently
with which it is mixed.
v_2/v_1 lies between 1·25

12.5.99

IX. *On the Condensation Nuclei produced in Gases by the Action of Röntgen Rays, Uranium Rays, Ultra-violet Light, and other Agents.*

By C. T. R. WILSON, *M.A., Clerk-Maxwell Student in the University of Cambridge.*

Communicated by Professor J. J. THOMSON, *F.R.S.*

Received October 29,—Read November 24, 1898.

CONTENTS.

IN a previous communication ('Phil. Trans.,' A, vol. 189, p. 265, 1897) I described experiments proving that when dust-free air, initially saturated with water vapour, is allowed to expand adiabatically, condensation takes place, if the maximum degree of supersaturation resulting from the expansion exceeds a certain limit. Using v_2/v_1 the ratio of the final to the initial volume as a measure of the expansion, we may describe the phenomena briefly as follows :—

Condensation only takes place throughout the gas if v_2/v_1 exceeds 1·25 ; the drops are comparatively few, provided a second limit ($v_2/v_1 = 1·38$) is not exceeded. Beyond this second limit the rate of increase in the number of drops with increasing expansion is extremely rapid, very dense fogs resulting from expansions even slightly exceeding this limit.

The view was taken that when the degree of supersaturation (approximately eight-fold) corresponding to the second limit is reached, the vapour condenses independently of any nuclei other than its own molecules or those of the gas with which it is mixed.

The rain-like condensation which takes place in air when v_2/v_1 lies between 1·25 and 1·38 was taken as indicating the presence of some other kind of nuclei than the molecules of gas or vapour.

It was further found that when the gas was exposed to even weak Röntgen radiation, comparatively dense fogs were obtained when v_2/v_1 exceeded 1·25 (the supersaturation being then approximately four-fold), no condensation taking place with smaller expansions. Thus, exposure of the gas to Röntgen rays causes nuclei to be produced, requiring a definite degree of supersaturation in order that water may condense upon them. Later ('Proc. Camb. Phil. Soc.,' vol. 9, p. 333, 1897) it was found that nuclei, requiring exactly the same minimum expansion to catch them, are produced in air by the action of uranium rays.

In the paper just referred to the conclusion was drawn that the nuclei produced by X-rays and uranium rays are identical with one another, as well as with those always present in small numbers in moist air, and causing the rain-like condensation which results when v_2/v_1 lies between 1·25 and 1·38. It was also there suggested that these nuclei are to be identified with the "ions," to the presence of which the conducting power of gases exposed to X-rays or uranium rays is due.

The primary object of the experiments described in the present paper was the study, by comparison of their efficiency as nuclei of condensation, of the carriers of the electricity in gases, when these are made by any of the known methods to be capable of allowing the passage of electricity through them. In the course of the work certain other kinds of nuclei were unexpectedly met with, which appear not to be associated with any conducting power in the gas. The method by which nuclei carrying a charge of electricity were distinguished from such uncharged nuclei is described in § 10.

I must explain here the meaning to be attached to certain expressions frequently used throughout this paper to avoid circumlocution. I have spoken of the expansion required to "catch" nuclei, meaning the expansion required to cause water to condense on such nuclei. The expressions "larger" and "smaller" are often used of nuclei instead of "requiring a less degree of supersaturation," or "requiring a greater degree of supersaturation, in order that condensation may take place on them." Nuclei are often said to "grow" when they become larger in the sense just defined.

It is probable that the expressions "larger" and "smaller" may be taken literally, without error; for we may suppose such nuclei to be very small drops of water, which are able to persist in spite of their small size, because the effect of the curvature of the surface in raising the equilibrium vapour-pressure is balanced by the opposite effect produced by the drop either being charged with electricity or containing some substance in solution. An increase in the charge of electricity or of the quantity of dissolved substance, either of which would increase the efficiency of the drop as a condensation nucleus, would also result in an immediate increase in the size of the nucleus necessary for equilibrium.

2. Expansion Apparatus.

The apparatus that I have most frequently used for bringing about the sudden expansion required in these experiments is represented in fig. 1.

A is the cloud chamber, where any drops which may be formed on expansion are made visible by the light from a luminous gas flame, brought by means of a lens to

Fig. 1.

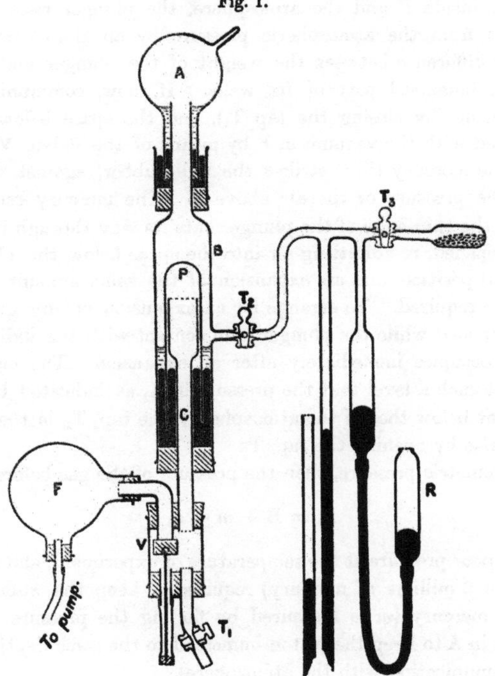

a focus within it. As the form of the cloud chamber had very frequently to be changed, it was convenient to arrange that it could readily be detached from the rest of the apparatus. An air-tight joint was made by means of the indiarubber stopper and mercury-cup arrangement shown in the diagram. The form of the cloud chamber there shown is the simplest that was used.

The cylindrical glass tube B (internal diameter = 2·7 centim.) is closed at its lower end by an indiarubber stopper, through the centre of which passes a glass tube, C, about 1 centim. in diameter, with a wider portion at its upper end serving as a guide to the light glass plunger, P, which slides freely over it. The plunger is made from a thin-walled test-tube, the open end of which has been cut perpendicular to the sides and ground smooth. Its lower edge is always immersed in the mercury which fills the lower part of B, and thus the gas in A and the upper part of B is completely cut off from the air inside P. The external diameter of the plunger is 2 millims. less than the internal diameter of the outer tube; there is thus a space of 1 millim. all round the tubes. When the tap, T_1, is open and there is thus free communication between the space inside P and the atmosphere, the plunger rises till the pressure in A only differs from the atmospheric pressure by an almost negligible amount, depending on the difference between the weight of the plunger and of the mercury displaced by the immersed part of its walls. If, now, communication with the atmosphere be cut off (by closing the tap T_1), and the space below the plunger be suddenly connected with the vacuum in F by means of the valve, V, the plunger is driven through the mercury till it strikes the indiarubber, against which it remains tightly held by the pressure of the air above it. The mercury remains practically stationary, while the thin edge of the plunger cuts its way through it.

If T_1 be again opened, re-admitting air into the space below the plunger, the latter rises to its original position, and an expansion of the same amount can be repeated as often as may be required. To arrange for an expansion of any given amount, the tap, T_2, must be opened while the plunger is in contact with the indiarubber, that is, in the position it occupies immediately after an expansion. The mercury reservoir, R, is then fixed at such a level that the pressure in A, as indicated by the gauge, is the desired amount below that of the atmosphere; the tap, T_2, is then closed and the plunger made to rise by opening the tap, T_1.

If B be the barometric pressure, then the pressure of the gas before expansion is

$$P_1 = B + m - \pi,$$

where π is the vapour pressure at the temperature of experiment, and m is the pressure (amounting to 1 or 2 millims. of mercury) required to keep the walls of the plunger immersed in the mercury (m is measured by finding the pressure which has to be applied to the air in A to keep the piston immersed to the same depth when the space below it is in communication with the atmosphere).

The pressure of the gas after expansion is

$$P_2 = B - p - \pi,$$

where p is the difference of pressure indicated by the open mercury gauge when put in connection with A before the previous contraction.

Then the ratio of the final to the initial volume of the gas is (if BOYLE's law holds)

$$\frac{v_2}{v_1} = \frac{P_1}{P_2} = \frac{B + m - \pi}{B - p - \pi}.$$

P_2, it will be noticed, is the pressure, not at the moment when the expansion is completed, but after the temperature has risen to its original value.

As the initial pressure, P_1, in these experiments is always approximately equal to the atmospheric pressure, it is sufficient for many purposes to take $P_1 - P_2$, or p, as a measure of the expansion without further reduction.

To make, for example, $v_2/v_1 = 1{\cdot}25$, $P_1 - P_2$ must be equal to 15 centims. of mercury if the barometric pressure $= 760$ millims. and the temperature $= 15°$ C.; while, as long as the atmospheric pressure lies between 740 and 780 millims., and the temperature lies between 10° and 25° C., $P_1 - P_2$ for the same expansion will always lie between 14·4 and 15·4 centims.

The gas with which the apparatus is to be charged is introduced through the stopcock, T_3, a side tube on the cloud chamber A being connected to a water air-pump, so that a stream of the gas at low pressure may pass through the apparatus. (At this stage a sufficiently low pressure must, of course, be maintained below the plunger, P, to prevent it rising out of the mercury in B.) The side tube is afterwards sealed off, and when sufficient gas has been generated to bring the pressure nearly up to that of the atmosphere, T_3 is closed.

The stopcocks, T_2, T_3, were lubricated with water only and protected by mercury cups. The mercury in B, as well as that covering the indiarubber stopper over which A is slipped, is prevented from coming in contact with the gas in the apparatus by a layer of distilled water.

In most of the experiments in which large expansions were required the expansion apparatus had the form described above. Many of the experiments with air, however, were performed with an expansion apparatus resembling that described in the 'Camb. Phil. Soc. Proc.' (*loc. cit.*). In it the plunger works in water instead of mercury, and is made to fit the outer tube like a piston, instead of working on an internal guide tube. The only advantage of the form with mercury is the absence of any risk of contamination of the gas in the apparatus by air which, when the plunger works in water, may gain entrance by solution and diffusion through the latter. The mercury apparatus is also suited for experiments with other liquids than water; such experiments, have, however, not yet been made.

Both these forms of apparatus give results almost identical with those obtained by means of the apparatus used in the earlier experiments ('Phil. Trans.,' *loc. cit.*). No considerable error appears to be produced by the yielding of the indiarubber when struck by the plunger, or by the momentum acquired by the air in the narrower part of the tube.

It will be convenient to mention here the methods used in preparing the gases required for the experiments. Oxygen was prepared by heating potassium permanganate. Hydrogen was obtained from palladium, which had previously been charged with the gas, obtained from the purest zinc and dilute sulphuric acid, the gas being passed through potassium permanganate solution before reaching the palladium.

Carbonic acid was prepared by heating potassium bicarbonate. In each case the tube in which the gas was produced was fused to the rest of the apparatus with the blow-pipe.

§ 3. Nuclei produced by X-rays.

In the experiments with X-rays described in the 'Phil. Trans.,' *loc. cit.*, the gas was only exposed to such radiation as was able to penetrate glass. It was of interest to know whether under the action of strong radiation condensation would take place with a less degree of supersaturation, or whether merely the number of nuclei would be increased. In a postscript to that paper the results of further experiments were given, showing that the latter alternative was the true one. Further experiments have confirmed this.

For the experiments with X-rays the expansion apparatus used differed from that represented in fig. 1, in being without the part A, the tube B being cut off square at the top instead of being prolonged into a narrower tube. With the help of an india-rubber washer, its upper end, which was ground smooth, was closed by a thin sheet of aluminium. This was held down by a brass diaphragm, which was screwed tight by three bolts, attached to a similar brass plate pressing against the lower surface of the indiarubber stopper which closes the lower end of B.

A "focus" tube giving out strong X-radiation was fixed a few centimetres above the aluminium plate closing the top of the tube B. In some of the experiments the expansion apparatus was wrapped in tinfoil (provided with the necessary apertures for observing the result of the expansions); this was found to be without effect on the appearance of the fogs.

The results of expansions in the immediate neighbourhood of the point where condensation first begins are given below.

I. Air exposed to X-rays.

B = barometer reading = 767 millims.; t = temperature = 18° C.; π = maximum vapour pressure at t° C. = 15 millims.; m = pressure required to sink plunger = 0 millim.

Gauge reading (in millims.) = p.	$v_2/v_1 = \frac{B - \pi + m}{B - \pi - p}$.	Result of expansion.
146	1·241	No drops
149	1·247	Very few drops
156	1·259	Fog

Least value of v_2/v_1, with which condensation was observed = 1·247.

When the expansion was made without exposure of the air to X-rays, only a very few scattered drops were seen even with v_2/v_1 as great as 1·279.

Similar results were obtained with oxygen.

II. Oxygen exposed to X-rays.

B = barometer reading = 767 millims.; t = temperature = 21·5° C.; π = maximum vapour pressure at t° C. = 19 millims.; m = pressure required to sink plunger = 0.

Focus bulb 3 centims. above aluminium window.

Duration of exposure before expansion.	Gauge reading (in millims.) = p.	$v_2/v_1 = \frac{B - \pi + m}{B - \pi - p}$.	Result of expansion.
1 min.	146	1·243	No drops
½ "	152	1·256	Fog
1 "	149	1·249	Shower
1 "	147	1·245	No drops
10 secs.	159	1·271	Dense fog

Least value of v_2/v_1, with which condensation was observed = 1·249.

The number of drops produced even with expansions exceeding any of those given in the table is exceedingly small in the absence of the rays.

The minimum expansion required, in order that condensation in the form of drops may take place, is, it will be noticed, clearly defined; the increase in the expansion corresponding to a change in the result from entire absence of drops to dense fog being very small. It is also independent of the strength of the radiation, as is seen from the identity of the results here given with those previously obtained with weak rays ('Phil. Trans.,' *loc. cit.*). The increase of the density of the fogs with increasing expansion continues till v_2/v_1 is about 1·31; beyond that point, as far as can be judged from the appearance of the fogs, the increase in the number of the drops is slight, till the second limit v_2/v_1 = 1·37 is reached, beyond which the region of dense fogs, due to great supersaturation alone, is entered. Thus expansions exceeding v_2/v_1 = 1·31 appear to be sufficient to catch nearly all the nuclei produced by the rays.

Prolonged exposure to the rays does not cause the nuclei to grow larger (or become otherwise more effective in helping the condensation) than the limit corresponding to the expansion v_2/v_1 = 1·25. The observations given in Table II. show this, an exposure of 10 seconds producing nuclei enough to give a dense fog with v_2/v_1 = 1·271, while even after 1 minute not one nucleus has grown sufficiently to be caught by an expansion v_2/v_1 = 1·245.

The nuclei introduced by the X-rays, as has already been pointed out ('Phil. Trans.,' *loc. cit.*), rapidly diminish in number after the radiation has been cut off, but several seconds are required for their complete disappearance; thus an expansion made 5 seconds after switching off the current from the induction coil will give a shower very much denser than would have resulted had there been no exposure of the gas to the rays. No trace of the nuclei can, however, be detected 30 seconds after cutting off the rays. The rapid diminution of the number of the nuclei is readily explained, if we regard them as consisting of positively and negatively charged ions which tend to recombine and neutralize one another.

Experiments were also made with carbonic acid.

CO_2 EXPOSED to X-rays.

B = barometer reading = 768 millims.; t = temperature = 21° C.; π = vapour pressure at t° C. = 18 millims.; m = pressure required to sink plunger = 0.

Gauge reading (in millims.) = p.	$v_2/v_1 = \frac{B - \pi + m}{B - \pi - p}$.	Result of expansion.
189	1·337	No drops
190	1·339	" "
190	1·339	Very few drops
192	1·344	Slight shower
235	1·45	Dense fog showing colours

Least value of v_2/v_1 with which condensation was observed = 1·339.

Even with an expansion the same as in the last observation given in the table, only a slight shower was obtained in the absence of the rays.

The expansion, found in the previous experiments ('Phil. Trans.,' *loc. cit.*) to be necessary to cause rain-like condensation to take place in the absence of X-rays, was v_2/v_1 = 1·36. Dense condensation began at the limit v_2/v_1 = 1·53. In the experiments now described, condensation was again found to begin in the absence of the rays when v_2/v_1 = 1·36.

The experiments with CO_2 were made with a thin glass cloud chamber, such as is shown in fig. 1.

4. Nuclei produced by Uranium Rays.

In the experiments on the action of the uranium rays on condensation, described in a previous paper ('Camb. Phil. Soc. Proc.,' vol. 9, p. 333), the air was contained in a glass vessel which the rays had to penetrate; by far the larger part of the radiation being thus absorbed before it reached the air. Experiments were, therefore, performed in which the uranium compound was inside the vessel and thus actually in contact with the air, so that the maximum intensity of the radiation was obtained.

For this purpose the apparatus used was that shown in fig. 1, the cloud chamber being a thin glass bulb. Inside it was fixed, by means of a copper wire wound round the top of the narrow prolongation of B, a small shallow glass cup containing some uranium oxide.

AIR in contact with Uranium Oxide.

B = barometer reading = 759 millims.; t = temperature = 13° C.; π = maximum vapour pressure at t° C. = 11 millims.; m = pressure required to sink plunger = 1 millim.

Gauge reading (in millims.) = p.	$v_2/v_1 = \frac{B - \pi + m}{B - \pi - p}$	Result of expansion.
147	1·246	0
148	1·249	1 or 2 drops
149	1·251	0
150	1·253	1 or 2 drops
154	1·261	Dense shower
174	1·305	Fog
196	1·357	Fog

Least value of v_2/v_1 with which condensation was observed = 1·249.

This is identical with the number obtained when the uranium was contained in a glass bulb outside the expansion apparatus. ('Camb. Phil. Soc. Proc.,' *loc. cit.*) The number of drops is, as was to be expected, very much greater.

Similar experiments were made with hydrogen.

HYDROGEN in contact with Uranium Oxide.

B = barometer reading = 759 millims.; t = temperature = 14·5° C.; π = maximum vapour pressure at t° C. = 12 millims.; m = pressure required to sink plunger = 1 millim.

Gauge reading (in millims.) = p.	$v_2/v_1 = \frac{B - \pi + m}{B - \pi - p}$	Result of expansion.
151	1·255	No drops
153	1·259	1 or 2 drops
161	1·277	Shower
170	1·296	Dense shower
198	1·362	Dense shower

Least value of v_2/v_1 with which condensation was observed = 1·259.

HYDROGEN exposed to Uranium Rays (through glass).

B = barometer reading = 767 millims.; t = temperature = 13° C.; π = maximum vapour pressure at t° C. = 11 millims.; m = pressure required to sink plunger = 1 millim.

Gauge reading (in millims.) = p.	$v_2/v_1 = \frac{B - \pi + m}{B - \pi - p}$	Result.
150	1·246	No drops
153	1·255	Very few drops

Least value of v_2/v_1 with which condensation was observed = 1·255.

It is remarkable that the minimum supersaturation required to cause condensation on the ions should be the same in hydrogen as in air. It must, however, be remembered that in all these experiments water vapour is necessarily present; and some of the ions may always be derived from it. It may be that those requiring the minimum expansion to make condensation take place on them are produced from this source, when hydrogen is the gas under investigation.

In hydrogen which is not exposed to Uranium rays or other nucleus-producing agent, no drops at all are produced even with v_2/v_1 as great as 1·3 (see 'Phil. Trans.,' *loc. cit.*). A small quantity of uranium oxide contained in a thin glass bulb, 16 centims. away from the glass bulb forming the cloud chamber, was found to give quite a noticeable shower with $v_2/v_1 = 1{\cdot}277$. An expansion apparatus filled with hydrogen thus forms a very sensitive detector of Uranium or X-rays. If we take the view already suggested, that these nuclei actually are identical with the ions, each individual ion being made visible on expansion by the formation of a visible drop around it, it is not surprising that the method should be more delicate than the photographic or electrical modes of detecting such rays.

§ 5. Nuclei produced by Ultra-violet Light.

In some experiments of LENARD and WOLFF ('Wied. Ann.,' vol. 37, p. 443, 1889), light, rich in ultra-violet rays, was admitted through a quartz window into a vessel containing moist dust-free air. They found that if the air was allowed to expand after being exposed for some minutes to the light, a fog was produced showing that nuclei of some kind had been produced by the action of the ultra-violet rays. Similar results were obtained in steam-jet experiments. They regarded their experiments as proving that the ultra-violet rays caused the posterior surface of the quartz to disintegrate, the small particles thrown off constituting the nuclei on which condensation took place.

For the purpose of measuring the expansion required to make water condense on the nuclei produced in this way, I used apparatus identical with that made for the experiments with Röntgen rays, a quartz plate being, however, substituted for the aluminium closing the top of the cylindrical tube B.

The quartz plate was attached in the same way as the aluminium plate, being screwed tightly against an indiarubber band placed on the top of B. As in LENARD and WOLFF's experiments, the source of the ultra-violet light was the spark between zinc terminals produced by an induction coil; a Leyden jar being inserted in the secondary circuit to brighten the spark. Short sparks of about 2 millims. in length were generally used. Cadmium terminals were substituted for zinc in many of the experiments, but with no great increase in the effect. The expansion apparatus was wrapped in tinfoil, provided with windows to enable the fogs to be seen, the quartz itself being covered with wire gauze, placed on the brass diaphragm which held the quartz in position.

The effects described below are certainly due to the ultra-violet rays. That they are due to light of some kind is easily shown by interposing a quartz lens, so that an image of the source is formed, which, by a slight displacement of the lens, may be made to fall either on the quartz window or just to one side of it. In the latter case all effect on the condensation ceases, while so long as the concentrated light from the quartz lens does enter the window, the effect is immensely increased by its presence. Even exceedingly thin glass or mica interposed anywhere between the source and the cloud-chamber prevents all action. It is, therefore, the ultra-violet rays alone which are active in producing nuclei.

The results of one series of experiments, extending over four consecutive days, are given in the tables which follow. Many experiments of the same kind were made with exactly similar results. Since, however, as will be seen in what follows, the absolute numbers are of no particular interest, the one series has been considered sufficient. For the same reason the value of v_2/v_1 has not been calculated, the gauge reading p (approximately equal to $P_1 - P_2$) being used as a measure of the amount of expansion. The time for which the air was exposed to the ultra-violet rays before the expansion was made is given, as well as the interval, if any, which elapsed between the cutting-off of the rays and the expansion. When the number under this last heading is zero, it is to be taken as indicating that the expansion was brought about while the sparks were still passing.

DISTANCE of Spark from Quartz Plate = 4·5 centims.

p = gauge reading (approx. = pressure fall).	Duration of exposure.	Interval after exposure.	Result of expansion.
(1) 10 millims.	1 min.	0	Fog
(2) 112 „	10 mins.	5 mins.	Dense coloured fog

DISTANCE of Spark from Quartz Plate = 17 centims.

p = gauge reading (approx. = pressure fall).	Duration of exposure.	Interval after exposure.	Result of expansion.
(3) 102 millims.	5 mins.	0	0
(4) 133 „	30 secs.	0	Slight shower
(5) 133 „	7 mins.	0	Slight shower
(6) 145 „	30 secs.	0	Shower
(7) 145 „	5 mins.	0	Shower
(8) 160 „	30 secs.	0	Fog
(9) 157 „	10 „	0	Very dense shower

DISTANCE of Spark from Quartz Plate = 11 centims.

p = gauge reading (approx. = pressure fall).	Duration of exposure.	Interval after exposure.	Result of expansion.
(10) 101 millims.	30 secs.	0	Dense fog
(11) 74 „	30 „	0	Slight shower
(12) 74 „	3 mins.	0	Slight shower
(13) 134 „	30 secs.	0	Fog
(14) 68 „	30 „	0	Slight shower
(15) 38 „	5 mins.	0	Few drops
(16) 140 „	5 secs.	0	0
(17) 140 „	20 „	0	Fog
(18) 140 „	5 „	1 sec. (approx.)	1 drop seen
(19) 140 „	10 „	„ „	Shower
(20) 140 „	20 „	„ „	Very dense shower
(21) 140 „	1 min.	„ „	Fog
(22) 163 „	Current switched on momentarily.	...	Shower

DISTANCE of Spark from Quartz Plate = 5·5 centims.

p = gauge reading (approx. = pressure fall).	Duration of exposure.	Interval after exposure.	Result of expansion.
(23) 85 millims.	10 secs.	0	Shower
(24) 85 „	5 „	0	0
(25) 85 „	10 „	0	Shower
(26) 85 „	15 „	about 1 sec.	Fog
(27) 41 „	30 „	„	Shower
(28) 41 „	20 „	„	0
(29) 41 „	40 „	„	Fog

DISTANCE of Spark from Quartz Plate = 32 centims.

p = gauge reading (approx. = pressure fall).	Duration of exposure.	Interval after exposure.	Result of expansion.
(30) 156 millims.	20 secs.	0	Shower
(31) 140 „	2 mins.	0	0
(32) 169 „	5 secs.	0	Fog
(33) 169 „	3 „	about 1 sec.	Dense shower
(34) 153 „	10 „	0	1 or 2 drops
(35) 185 „	20 „	10 secs.	Fog
(36) 185 „	20 „	20 „	Very dense shower
(37) 185 „	20 „	30 „	Slight shower

DISTANCE of Spark from Quartz Plate = 42 centims.

p = gauge reading (approx. = pressure fall).	Duration of exposure.	Interval after exposure.	Result of expansion.
(38) 148 millims.	60 secs.	0	0
(39) 152 „	60 „	0	Slight shower

DISTANCE of Spark from Quartz Plate = 21 centims.

p = gauge reading (approx. = pressure fall).	Duration of exposure.	Interval after exposure.	Result of expansion.
(40) 128 millims.	60 secs.	0	Very few drops
(41) 140 „	5 „	0	0
(42) 140 „	60 „	0	Few drops
(43) 155 „	3 „	0	Shower

The nuclei produced by the action of ultra-violet light differ in many ways from those produced by X-rays or Uranium rays.

The expansion required to make water condense upon them depends on the strength of the radiation, and when this is strong, only a very slight expansion is necessary, as is shown in Experiments (1) and (29) in the above table. The smallest expansion of which the apparatus admitted was, in fact, sufficient to form a fog with strong radiation, and indeed, in later experiments, fogs were obtained without any expansion. Under the action of the weakest rays used, however (Expts. 30–39), the expansion required to obtain condensation is as great (p = about 150 millims.) as that required to catch the nuclei produced by X-rays.

The expansion required to make condensation take place on these nuclei depends on the time during which the apparatus has been exposed to the action of the rays before the expansion, being less the longer the exposure. The number which can be caught by a given expansion also increases with the time of exposure. (See Expts. 16–21, 23–29.) The nuclei thus appear to grow under the action of the ultra-violet rays. The increase in the size of the nuclei, or in the number exceeding a given size, does not, however, continue indefinitely with increasing time of exposure, but after a time a steady state is reached, the result of a given expansion becoming independent of the time of exposure, if this be long enough. (Expts. 3–7, 10–12.)

The time for which the nuclei persist depends on the size to which they have attained. When the radiation is so weak that the nuclei are only caught if an expansion be made as great as would be required for X-ray nuclei, by far the larger number have disappeared in 30 seconds after cutting off the radiation. (Expts.

35–37.) They do, however, last longer than the nuclei produced by X-rays. When the nuclei are large enough to be caught by a very slight expansion they last for many minutes at least (Expt. 2). As will be seen later, those produced by very strong radiation last for many hours. The shorter life of the smaller nuclei is probably due mainly to their more rapid rate of diffusion.

The limit to the size attained by the nuclei for a given strength of radiation when the time of exposure is indefinitely prolonged is, perhaps, also to be explained by diffusion. For, if the radiation be weak, the nuclei may reach the walls by diffusion before any considerable growth has time to take place; whereas, with stronger radiation, not only will the drops grow more in a given time, but the slower rate of diffusion resulting from their increased size must increase the time for which they remain exposed. It is not surprising, according to this view, that a comparatively small increase in the intensity of the radiation may result in a very great diminution in the least expansion required to catch the nuclei.

Fig. 2.

Fig. 3.

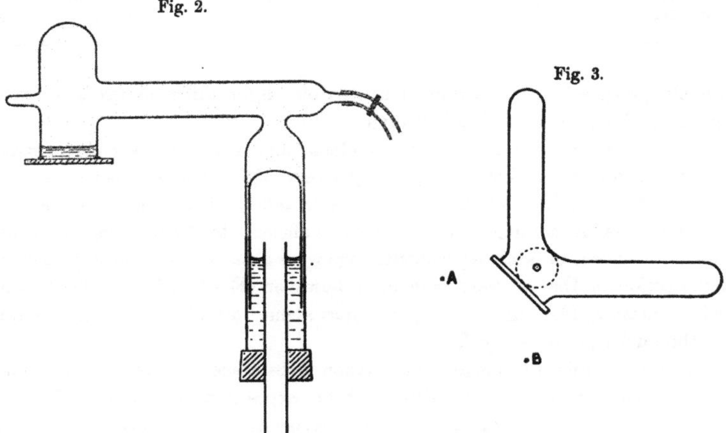

Experiments were now carried out with the object of deciding whether the nuclei are produced throughout the volume of the moist air, or, only at the surface of the quartz, as LENARD and WOLFF supposed (*loc. cit.*). For this purpose an expansion apparatus was made of the form shown in fig. 2. The ultra-violet light entered from below through a quartz plate, which was covered with water to a depth of about 5 millims. The quartz plate was held up against the ground edge of the cloud-chamber by means of two elastic bands, an indiarubber washer being used as before to make an air-tight joint. The apparatus was wrapped in tinfoil, provided with the necessary apertures.

Dense fogs were obtained under the action of the ultra-violet light (zinc-spark); the expansion being that corresponding to a pressure fall of 10 centims.

The nuclei produced under these conditions cannot have arisen from the disintegration of the quartz.

Another proof that the quartz is not the source of the nuclei is furnished by the results of experiments made with an expansion apparatus of the form shown in fig. 3. It consisted of three glass tubes meeting at right angles, two being horizontal and the third pointing downwards and containing the piston. A quartz window was fixed in a vertical plane making an angle of 45° with each of the horizontal arms. By placing the zinc points forming the source of the ultra-violet light at a position such as A, the rays may be made to pass along only one limb of the apparatus. If the source is transferred to B the air in the other limb is exposed to the rays and none of them traverse the first limb. The quartz being equally inclined to both limbs, any nuclei which are thrown off from it will find their way equally readily along either. The experiments show, however, that nuclei are only introduced into that limb along which the ultra-violet light passes.

The fogs were made visible by the light from a gas flame, which could be concentrated by a condensing lens at any part of either tube. The time for which the rays were allowed to act before the expansion was made was generally 5 seconds. In some of the experiments the rays were made approximately parallel by means of a quartz lens. The expansions used, measured by the pressure fall p, varied from 13 to 16 centims. of mercury.

In every case a shower of fog was produced from end to end of the tube traversed by the ultra-violet rays, while no effect could be detected in the other branch even at a point not more than 1 centim. from the junction of the tubes. The exposure could be made twice as long without any effect being obtained in the branch not exposed to the rays.

In all these experiments attention was confined to the small portion of one tube which was illuminated by the light from the luminous gas flame, which was brought to a focus at that point; observations being made alternately with corresponding portions of the two branches successively illuminated in this way. Finally, however, experiments were made, in which, owing to the use of stronger radiation this was unnecessary, the fogs produced being well seen without any condensing lens. The sparks were produced between cadmium terminals, and a more powerful induction coil was used than in the previous experiments. A parallel beam of ultra-violet light was not used, but the cadmium points were brought to within 1·5 centims. from the quartz plate. The experiments were made with a pressure fall of 14 centims., the time of exposure being 20 seconds. Under these conditions, the tube along which the rays were directed was filled with fog on expansion, the other tube remaining empty. The fogs were well seen by means of the light from the gas flame without any condensing lens, so that a general view of the result of expansion throughout both branches was obtained at once. On displacing the cadmium points so that the rays now passed along the other tube, and repeating the experiments, the tube which before was filled with fog now remained dark, and the other was filled with a white or coloured fog.

These experiments prove that the nuclei produced by the action of ultra-violet light do not have their origin at the surface of the quartz. It might still be supposed that they are produced at the surface of the glass, where this is exposed to the ultra-violet rays. LENARD and WOLFF, however, were able to detect no effect of this kind with glass. With the object of testing this point, a T-shaped expansion apparatus was now made (fig. 4). The length of the horizontal tube amounted to

Fig. 4.

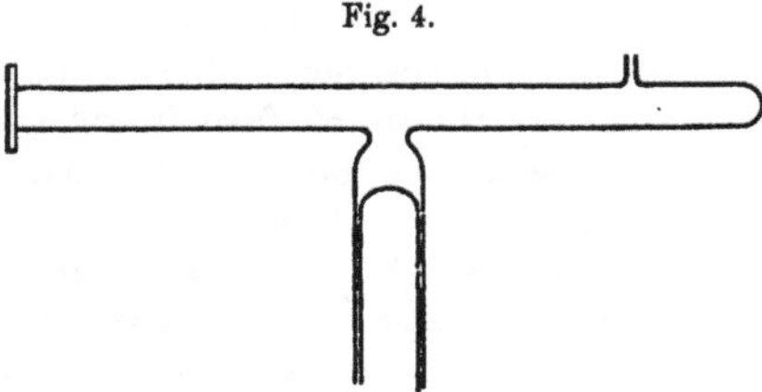

27 centims., and the internal diameter was 1·3 centim. One end of this tube was closed by a quartz plate cemented on with shellac.

The rays from a spark between cadmium terminals were sent axially along the tube, a quartz lens being inserted to make the rays converge to a point slightly beyond the far end of the tube. By observing the image of the cadmium points which was formed by the quartz lens, it was easy to test whether the light was passing axially, and also whether the points were sufficiently near together to give an image considerably smaller than the diameter of the tube. The length of the spark-gap was generally rather less than 1 millim.

In the earlier experiments made with this apparatus, the fogs, which were produced on expansion under the influence of the ultra-violet light, although very dense near the quartz plate, diminished rapidly in denseness with increasing distance from the quartz, and did not reach the far end of the tube at all. This was at first interpreted as indicating either that the nuclei arose at the surface of the quartz, or that the active rays were absorbed by a comparatively small thickness of moist air. The latter view was easily disproved by interposing a layer of moist air (in an open tube 17 centims. long) between the source and the expansion apparatus. This exercised no appreciable absorption. The whole effect was finally traced to a deposit of fine dew on the inside of the quartz plate. On removing this by gentle warming, uniform fogs from end to end of the tube were obtained on expansion. There was never any indication of any increase in the density of the fog close to the far end of the tube, where the rays strike the glass. To ensure that there should be no effect throughout the length of the tube due to rays grazing the walls, a tinfoil diaphragm with an aperture of 5 millims. was inserted in front of the quartz plate. The fogs still remained uniform from end to end. A want of uniformity in the density of the fog at once shows itself in the curvature of the upper surface of the fog, due to the more rapid settling where the drops are fewer and larger. In these experiments the time of exposure before expansion was from 3 to 5 seconds.

It is easy to understand the great effect produced by a slight dimness of the quartz plate, for scattering of the ultra-violet rays by the small droplets on the plate is likely to take place to a much greater degree than that of the luminous rays. The rays which escape scattering or deflection at the quartz may not be strong enough of themselves to produce, in the time for which the exposure lasts, nuclei large enough to be caught with the degree of expansion used, while together with the scattered rays they may be more than strong enough close to the quartz for this purpose. Now the scattered light (the quartz plate being small compared with the length of the tube) will fall off approximately inversely as the square of the distance from the quartz. It is thus readily understood why the fog extended only a short distance from the quartz when this was covered with a deposit of dew.

That the uniformity of the fog from end to end of the tube, when the contents are actually exposed to equally intense radiation throughout, is not due to diffusion of the nuclei before expansion, or mixing of the air in the tube in consequence of the expansion, is certain. For, in the experiments in which there was a deposit of dew on the quartz, no fog was produced at the far end, even with an exposure of 60 seconds; while a very dense fog was obtained near the quartz with an equal expansion, with an exposure of only 10 seconds. Similar results were obtained in experiments in which the Cadmium points were displaced to one side, so that only a small portion of the tube near the quartz was exposed to the rays. The fog obtained on expansion only extended a short distance beyond the part of the tube reached by the rays.

The experiments with this apparatus make the superficial origin of the nuclei very improbable; for, if they arose only where the rays fell on a surface, the fogs would have been confined to the ends of the tube. To account otherwise for the fact that whenever a fog was produced (with the light passing axially) it was uniform from end to end of the tube, we would have to suppose that on account of undetected scattering of the ultra-violet rays at the ends, the walls throughout the whole length of the tube received approximately uniform ultra-violet illumination.

Perhaps the most striking proof that the nuclei produced by ultra-violet light are formed throughout the volume of the moist air, and not at the surface of the vessel containing it, is furnished by experiments with very strong radiation. As already stated in a preliminary note on the subject ('Camb. Phil. Soc. Proc.,' vol. 9, p. 392), under the influence of very strong ultra-violet light fogs are produced without any expansion, even in unsaturated air. The nuclei which, when they are only exposed to very weak ultra-violet light, do not grow beyond the stage at which a four-fold supersaturation is required to make condensation take place upon them, grow under the influence of very strong radiation till they become large enough to scatter ordinary light.

TYNDALL, many years ago ('Phil. Trans.,' vol. 160, p. 333, 1870), showed that when the more refrangible rays from an arc lamp, such as were able to traverse blue glass but not red glass, were concentrated within a tube containing air mixed with amyl-nitrite or certain other vapours, dense clouds were produced. He was unable to obtain any such effect with pure air and water only. The experiments to be described differ in no essential respect from his, except in the fact that the rays from the arc lamp were allowed to traverse no material such as glass, which is opaque to the ultra-violet rays, before entering the tube containing the moist air. Under these conditions, air containing water vapour only, shows the phenomena that were observed by TYNDALL with other vapours.

I have found the apparatus shown in fig. 5 convenient for experiments on this

Fig. 5.

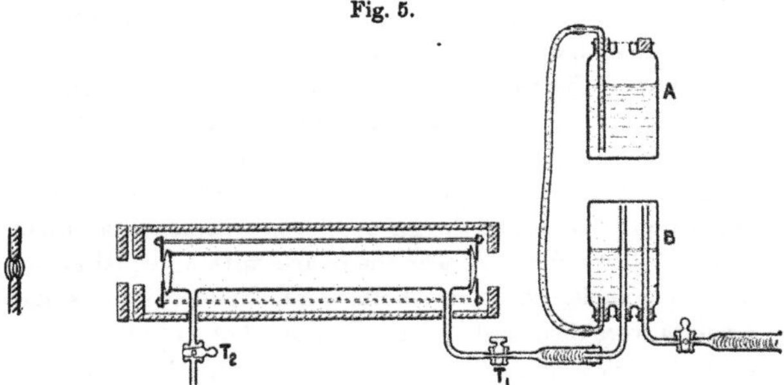

subject. It consists of a glass tube 34 centims. long, and 4 centims. in diameter, provided with a side tube near each end. The ends are closed by quartz lenses, which are fixed air-tight by means of indiarubber washers. They are pressed tightly against these indiarubber rings by means of two brass diaphragms screwed together by means of three bolts just outside the tube. By means of the two Wolff bottles, A, B, a current of filtered air can be driven through the apparatus. The air is filtered before entering B, and again on leaving it. If, while A was fixed at some height above B, the stop-cock T_2 remained closed while T_1 was open, the pressure in the apparatus was greater than the atmospheric pressure. On opening T_2 the pressure was suddenly reduced to that of the atmosphere, and the expansion produced in this way was sufficient to cause condensation on ordinary dust particles. A small quantity of water was contained in the tube to keep the air saturated. To enable even faint clouds to be seen, the tube was contained in a blackened box, open along one side, and with a hole somewhat smaller than the diameter of the tube at each end, so that a beam of light from an arc lamp might pass along the axis of the tube.

The quartz lens through which the light entered had a focal length of about 10 centims., and the arc was generally placed at such a distance that the light was brought to a focus about the middle of the tube. A strong arc is necessary in these experiments.

Filtered air was passed through the apparatus until all dust particles were removed. The presence or absence of dust particles was easily ascertained by allowing expansion to take place as described above, while the light from the arc traversed the tube, a sheet of mica being interposed at this stage to cut off the ultra-violet rays. When drops ceased to appear on opening the stopcock T_2, a few minutes were generally allowed to elapse before exposing to the ultra-violet rays, to enable the air inside the tube to come to rest. To start the exposure to ultra-violet rays the mica was removed.

Under the conditions described above, a bluish fog is seen in the tube in about two minutes, making its appearance first near the apex of the beam of light, and then extending both ways in the form of a double cone. That the fog when it first appears is confined to the path of the light is easily proved by displacing the tube slightly to one side; or better, by inserting the mica screen and moving the box with the tube fixed inside it nearer the arc, so that the luminous rays converge to a focus much nearer the far end of the tube; or, without moving the tube, by inserting a glass lens just in front of the tube after the fog has appeared, so as to bring the luminous rays to a new focus. In each of these ways it is easy to prove that the fog does actually arise, not near the quartz nor the glass walls of the tube, but along the axis of the tube in the neighbourhood of the point where the light is most concentrated. It was found that the shape of the cloud was specially well defined when the water in the tube contained two or three per cent. of caustic potash. This prevented any deposit of fine drops on the inner surface of the quartz by keeping the inside of the tube not quite saturated.

We thus obtain a further confirmation of the conclusion already drawn from the expansion experiments, that ultra-violet light produces nuclei throughout the volume of the moist air which it traverses, and not only at the surface of the quartz or the glass walls of the tube.

On allowing the air to expand after the fog has appeared, condensation takes place throughout the tube, showing that outside the part traversed by the strongest radiation, nuclei have been formed, but not large enough or numerous enough to form a visible fog. These may arise partly through some of the nuclei produced in the strongest part of the beam travelling into other parts of the tube; they may also have been produced by the action of ultra-violet rays, scattered by the cloud particles produced in the direct path of the light. These will scatter ultra-violet light even before they have grown large enough to make themselves visible by scattering the luminous rays. The scattering of the ultra-violet rays by the cloud particles is probably also the cause of the fog (without expansion) spreading further and reaching a greater density on the side of the focus next the source.

If the exposure to the rays of the arc be continued after the fog has become visible, this often assumes very remarkable shapes, resembling those obtained by TYNDALL with air which had been passed through a solution of hydrobromic or hydroiodic acid.* The fog may, for example, develop dark striæ which may be straight and vertical, or may have quite complicated forms. They sometimes produce a regular cone-in-cone structure (*cf.* TYNDALL, *loc. cit.*), or the fog may become divided up into rounded clouds often connected by a thread of fog along the axis of the tube. These complicated forms were noticed most frequently in some of the earlier experiments, in which a longer tube (60 centims.) of the same cross-section was used, with a quartz lens whose focal length was 25 centims. The light was thus less concentrated and the fog took much longer to become visible, generally about ten minutes.

There can be no doubt that all these complicated cloud forms owe their origin to air currents in the tube. A dark stria may always be produced at will at any part of the fog by warming the lower edge of the tube at that point, by holding one's finger against it. A stream of air, free from fog, rises at this point in a narrow layer, and a dark vertical stria is produced,† the brightness of the fog immediately on each side of it being also increased. Such a dark band persists for a long time after the exciting cause has been removed.

No condensation could be produced in pure steam even with prolonged exposure. The air was expelled from the tube by allowing a rapid current of steam to pass through the apparatus for an hour at low pressure. The tube was allowed to cool to the temperature of the room (15° C.) before exposing to the ultra-violet rays. The failure to produce any visible condensation in the steam was not due to the quartz becoming dimmed through drops of water condensing on it, for on letting in a small quantity of air the effect was readily obtained. Fogs without expansion were obtained without difficulty with an air pressure of 5 centims. of mercury.

To determine to what extent water vapour was necessary for the production of these fogs, experiments were made with a much smaller apparatus than that just described. This is shown in fig. 6. The tube was 16 centims. long and 4 centims. in diameter. A solution of potash, or of sulphuric acid, was placed in the bottom of the tube. The stop-cocks were lubricated with H_2SO_4. A slow current of filtered air was drawn through the apparatus. This was then allowed to stand for some time to enable the equilibrium vapour pressure to be attained before the exposure to the ultra-violet rays was begun.

* TYNDALL, 'Roy. Soc. Proc.,' vol. 17, p. 92, 1869.

† The production of dark bands of this kind in fogs was observed by TYNDALL ('Roy. Inst. Proc.,' vol. 6, p. 1, 1870), and further studied by Lord RAYLEIGH ('Roy. Soc. Proc.,' vol. 34, p. 414, 1882), LODGE ('Nature,' vol. 28, p. 297, 1883), and AITKEN ('Trans. Roy. Soc.,' Edin., vol. 32 (1), p. 239).

No effect could be obtained with solid KOH or strong H_2SO_4 in the tube. With aqueous H_2SO_4 containing 45 per cent. of H_2SO_4, corresponding to a relative humidity of about 50 per cent., no fog was obtained. Over 10 per cent. sulphuric acid, which had been all night in the tube, so that there can be no doubt that the equilibrium vapour pressure was reached, a fog very quickly appeared under the action of the ultra-violet light. Over aqueous caustic potash, containing about 17 per cent. of KOH, a fog was readily obtained. The relative humidity over such a

Fig. 6.

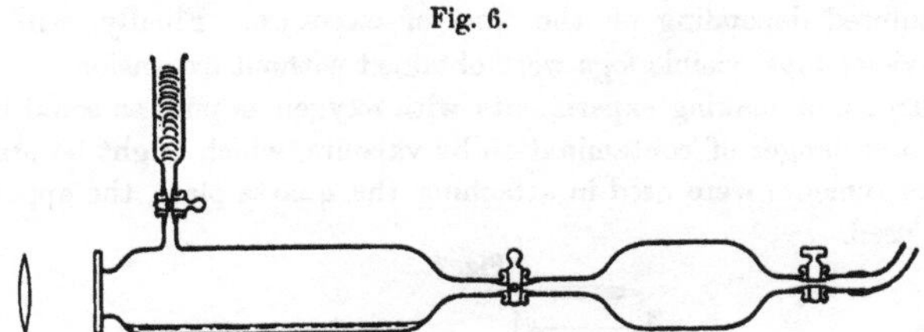

solution is less than 90 per cent. Experiments have not been tried with humidity between 50 and 90 per cent. These experiments then show that both air and water vapour are necessary for the production of the ultra-violet light fogs; it is not necessary that the air should be saturated.

The cloud particles produced by the action of ultra-violet light persist, for some hours at least, after the rays have been cut off. This is so, even when the air is unsaturated; for example, the fog produced over a 17 per cent. potash solution was found to be still visible three hours after the arc was stopped. The drops are therefore small enough to settle with extreme slowness; yet in spite of their small size there is no indication of any tendency for them to evaporate again. It is probable, therefore, that the drops do not consist of pure water. We might, it is true, account for their persistence by supposing each to have become charged with electricity under the influence of the ultra-violet rays. In the light of later experiments, however, the former view appears to be the more probable.

Any discussion of the nature of these fogs is, however, postponed till the experiments made with other gases than air have been described.

Before going on to describe the experiments made with oxygen, mention should be made of experiments in which no indiarubber or cement of any kind came in contact with the air exposed to the ultra-violet rays. For this purpose a test-tube with the open end ground smooth was closed by a plano-convex quartz lens, simply held in position by an indiarubber band, no indiarubber washer being inserted. The inside of the tube was moistened with distilled water; the air inside was at atmospheric pressure. The tube was fixed in a horizontal position and left for two days to allow the dust particles to settle. On exposing to the light of an arc lamp, placed so that its light was brought to a focus near the middle of the tube, a fog developed in less than two minutes.

Expansion experiments were made with oxygen with the same apparatus as was used in the experiments on air. The results obtained were identical with those obtained in the former experiments. The oxygen was obtained by heating potassium permanganate. With weak radiation nuclei were produced requiring, however long an exposure might be made, a pressure fall of 15 centims. or more; while, when the radiation was stronger, fogs were produced with comparatively slight expansions, the expansion required depending on the time of exposure. Finally, with sufficiently strong ultra-violet rays, visible fogs were obtained without expansion.

For the purpose of making experiments with oxygen as pure as could be obtained; and without any danger of contamination by vapours, which might be present if any indiarubber or cements were used in attaching the quartz plate, the apparatus shown in fig. 7 was used.

Fig. 7.

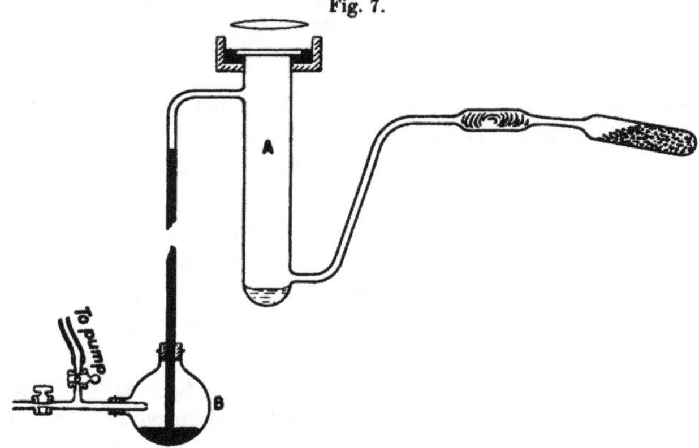

The vertical tube A, in which the gas was exposed to the action of the rays, was 16 centims. long and 3 centims. in diameter. Its open end was carefully ground flat and closed by a quartz plate, which was simply placed upon it, mercury being then poured into the wooden collar surrounding it to make a tight joint. The tube was exhausted before filling the collar with mercury.

A small quantity (less than 1 cub. centim.) of well boiled distilled water was placed in the tube immediately before closing it with the quartz plate. The water was drawn up into a pipette while still boiling and run into the apparatus after cooling slightly. The apparatus was then pumped out by connecting to a water pump.

The oxygen was prepared by heating potassium permanganate, which had been twice recrystallised, and then heated in an open dish till the greater part of it fell to powder. Between the tube which contained the permanganate and the rest of the apparatus a tube of glass wool was inserted, to prevent any of the small particles passing over with the oxygen.

The apparatus was kept at a low pressure by the pump, and a stream of oxygen made to pass through it by heating the permanganate tube. The gas passed through the mercury in B on its way to the pump. The connection to the pump could be closed at any time and the space above the mercury in B connected with the atmosphere. The mercury then rose in the long vertical tube, which dipped into it to a depth of about 1 centim., and now served to indicate the pressure. The pressure could now be raised to any desired amount, less than that of the atmosphere, by further heating of the permanganate. By closing the connection between B and the atmosphere and opening that leading to the pump, the apparatus could be again exhausted and a stream of oxygen allowed to pass through it at a pressure of a few centimetres of mercury.

A quartz lens, fixed above the quartz plate, served to bring the light from an arc to a focus a little below the middle of the tube. The arc was formed between two horizontally placed carbons contained in a box with an aperture below somewhat smaller than the diameter of the tube.

A current of oxygen was passed through the apparatus, while this was connected to the pump for 15 minutes, the apparatus was then left for one night and oxygen again allowed to stream through it for 30 minutes. After the pressure had been raised to 70 centims. the contents of the tube were exposed to the ultra-violet light of the arc. A fog appeared in a very few minutes. The apparatus was again left for a day and then pumped out, and a stream of oxygen allowed to pass for 10 minutes. Again, less than two minutes' exposure to the ultra-violet rays was sufficient to produce a fog. Again, after standing for three days, while repeatedly exposed to the ultra-violet rays till a fog appeared, the oxygen was pumped out and a vigorous stream passed for 30 minutes. The pressure was then brought up to 50 centims., and the gas exposed to the ultra-violet rays. A fog appeared after an exposure of about one minute.

The presence of nitrogen thus appears to be unnecessary for the production of cloud by the ultra-violet rays. There is no indication of any diminution in the density of the clouds or in the ease with which they are produced as the gas becomes purer.

The quantity of matter in the clouds which develop under the action of ultra-violet light is very small: as is seen from the fact that even isolated patches of the fog remain suspended in the tube. Since the mass of each drop, even if its diameter be as great as one mean wave-length of light, does not amount to 10^{-12} gram, a very large number of drops may be present although the total weight of the fog is very small.

The small quantity of matter in these clouds makes it very difficult to exclude the possibility of their formation being due to the presence of traces of some vapour, which might become oxidised under the influence of the ultra-violet rays. That it is not due to mercury vapour was proved by making the exposure immediately after the stream of oxygen had stopped before the mercury vapour in the gauge could have had time to diffuse into the rest of the apparatus. Before the final experiment, also, the tubes leading to the gauge and the permanganate tube were sealed off. After the cloud appeared the exposure was continued for three-quarters of an hour and the apparatus left till the next day. The fog was found to have disappeared by that time, as was seen on exposing to the light of the arc with a plate of mica interposed; in less than two minutes after the removal of the mica, however, a new fog was produced. The last experiment was performed with the object of determining whether the formation of the cloud depended on the presence of minute traces of some substance, which might all be used up to form fog if the exposure was sufficiently prolonged. After the fog had settled to the bottom of the vessel, it was thought that the complete or partial removal of the active substance would be made manifest in an increased difficulty in producing a second fog. No such effect was found; the exposure was perhaps, however, not sufficiently prolonged for any very great weight to be attached to the result.

The experiments described leave little room for doubt that pure oxygen and water vapour alone are sufficient to enable a cloud to be produced under the influence of ultra-violet light.

In hydrogen, fogs could not be obtained under the influence of ultra-violet light without expansion. Experiments were made with an expansion apparatus to see if nuclei of any kind were produced in this gas when exposed to ultra-violet light.

Fig. 8.

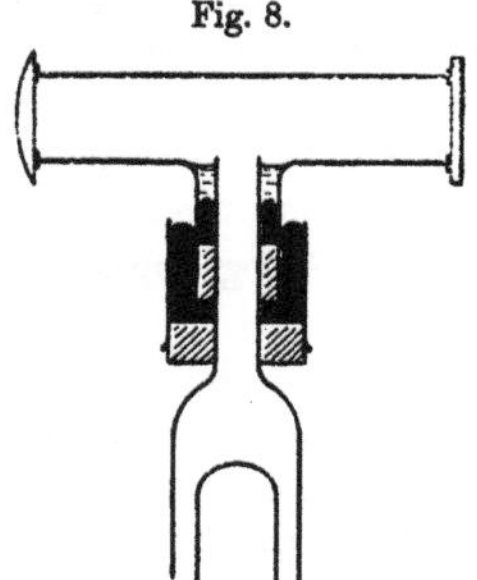

A T-shaped cloud-chamber (fig. 8) was used, the rest of the apparatus being that shown in fig. 1. The horizontal tube (fig. 8) was 9 centims. long and 1·8 centims. in diameter. The ends were closed by quartz plates, fixed like those in the large apparatus which was used for experiments on the clouds produced in moist air by ultra-violet light without expansion.

The arc was used as the source of ultra-violet rays. These were made to converge by means of a quartz lens to a point slightly beyond the middle of the tube. The hydrogen was obtained, as in previous experiments, by heating a tube containing palladium saturated with the gas.

In hydrogen prepared in this way nuclei were produced under the influence of the ultra-violet rays, but never in very large numbers, and always requiring great supersaturation to make water condense upon them, however long the exposure. On replacing the hydrogen by air, without making any other alteration in the apparatus, and again exposing to the rays, a fog was obtained without expansion in less than 1 minute.

HYDROGEN exposed to Ultra-violet Rays of Arc.

Gauge reading (in millims.) (approx. = pressure fall).	Result of expansion.
137	0
147	Slight shower
167	Dense shower
167 (mica interposed)	0

In all the above observations the time of exposure was thirty seconds.

FRESH hydrogen prepared.

Gauge reading (in millims.) (approx. = pressure fall).	Result of expansion.
147	Very few drops
142 (exposed for 60 seconds)	0
150	Dense shower

The expansion required to make condensation take place upon these nuclei is, it will be seen, approximately the same as is required in the case of the nuclei produced by X-rays or Uranium rays.

With CO_2 exposed to ultra-violet rays fogs were obtained with slight expansion and even without expansion, but stronger radiation was found to be necessary than in the case of air or oxygen. For these experiments an apparatus like that used for hydrogen was used. The CO_2 was prepared by heating potassium bicarbonate.

The nuclei produced by the action of ultra-violet light on moist air, oxygen or carbonic acid, are thus seen to be capable of growing under the action of the rays till they actually become large enough to scatter ordinary light. At least in the case of air these visible fogs may persist for hours, although the air be not saturated with water vapour. Later experiments make it very improbable that the growth of these nuclei is due to each one becoming charged with electricity under the action of the ultra-violet light. The most obvious way of accounting for the growth of the nuclei into visible particles, and their persistence even in unsaturated air, is to suppose that by the action of the ultra-violet light some compound is formed in solution in each drop. Were it not for the fact that the fogs are produced in pure oxygen as well as in air, one would naturally consider the combination of oxygen and nitrogen with the water of the incipient drop to form nitric acid as the most likely reaction which could account for the phenomena. Possibly when the clouds are produced in moist air this may be, in fact, the reaction which takes place. When the clouds are produced in oxygen, however, the only possible combination which can account for the phenomena is that of oxygen and water to form hydrogen peroxide. The formation of ozone would not enable us to explain the production of the clouds, and indeed although clouds are very easily produced in ozonised oxygen it is, as the experiments of MEISSNER and others show, only as a consequence of reactions, by which some of the ozone is destroyed.

The view here taken is then, that under the action of the ultra-violet light small drops of water combine with the oxygen in contact with them, and in consequence of the lowering of the equilibrium vapour pressure by the dissolved H_2O_2 they are able to grow, when similar drops of pure water would evaporate.

The time taken by the nuclei to grow to any given size depends simply on the time required for the quantity of dissolved substance produced in each drop by the action of the ultra-violet light to become sufficient to enable a drop of that size to be in equilibrium. That, for a drop containing a definite quantity of dissolved substance, there is a definite size necessary for equilibrium, is obvious from the fact that the lowering of vapour pressure due to the dissolved substance is proportional to the concentration, that is, inversely proportional to the volume, while the increase of vapour pressure due to the curvature of the surface is inversely proportional to the radius. By the growth of the drop, if initially the solution is too strong for equilibrium, the lowering of vapour pressure due to the dissolved substance will very quickly diminish till it ceases to exceed the rise of vapour pressure due to the curvature.

If it is only, as is in itself quite likely, at the surface of separation of the gas and liquid that the ultra-violet rays cause combination to take place, the maximum effect will be produced where, as in this case, the water is in the form of a cloud of minute particles; for it is only in very small drops that any considerable proportion of the molecules are situated in the surface layer.

The absence of any effect of this kind in moist hydrogen is in agreement with the view that the growth of the drops in air or oxygen is due to the formation of hydrogen peroxide.

§ 6. NUCLEI PRODUCED BY SUNLIGHT.

AITKEN* has shown that many vapours when exposed to sunlight in glass vessels

* AITKEN, 'Trans. Roy. Soc.,' Edin., vol. 39 (1), p. 15, 1897.

become charged with nuclei, on which condensation takes place when supersaturation is brought about by expansion. He was unable to detect any such effect when moist air was exposed in this way to the action of the sun's rays.

In experiments such as AITKEN'S, in which the sunlight has to traverse glass, any ultra-violet rays which may be present are cut off before reaching the gas under investigation. Now the light from an arc lamp, when deprived of its ultra-violet rays by passing through glass, was found to have no cloud-producing effect. An apparatus provided with a quartz window was therefore used.

This had the form shown in fig. 9. The quartz plate was fixed, as in the other

Fig 9.

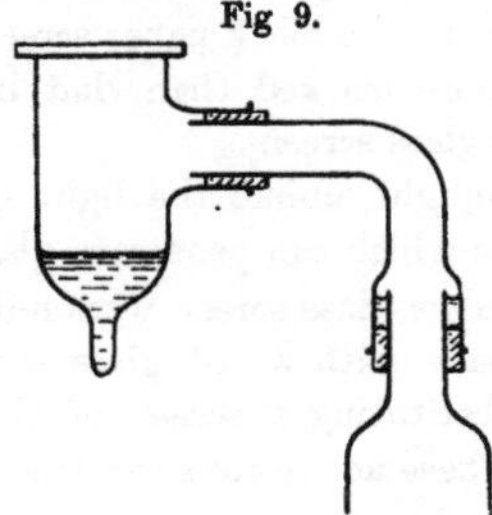

expansion experiments in which a quartz window was required, with the help of an indiarubber washer. There was a joint on the horizontal part of the tube, made by means of an indiarubber stopper, as shown in the figure; the connection between the expansion apparatus and the gauge also was made by means of an indiarubber tube, instead of glass tubing with the joints made by the blow-pipe, as in other experiments. The cloud-vessel could thus be placed so that the quartz plate was directly facing the sun.

The presence of indiarubber vapour is doubtless a disadvantage, but the experiments with ultra-violet light from other sources make it highly improbable that any complications are thereby introduced.

On account of the heating effect of the sunlight, accurate measurements of the expansion were not possible.

The experiments were made during the month of August, between the hours of 10 A.M. and noon. The apparatus was placed at an open window, which was closed when it was desired to make an experiment with the ultra-violet rays intercepted by glass. Even when a quartz lens was used to concentrate the sunlight no nuclei were produced which could be caught with slight expansions. Even with a pressure fall of 137 millims. no drops were seen under these conditions.

With pressure falls exceeding 15 centims., showers or fogs were obtained in which the drops were plainly more numerous when no glass was interposed than when a glass screen was used to cut off the ultra-violet rays.

Observations were now made without any glass screen, the quartz lens being still used to concentrate the light. The expansions were made alternately with the apparatus unscreened, and with a screen of black paper in front of the quartz plate. The black paper was removed immediately after the expansion, to enable the drops to be seen.

The pressure fall being 174 millims., the drops were very few when the black paper was interposed, while a fog resulted when the expansion was made without the screen.

Similar experiments, made with a glass screen interposed, the expansion being the same as before, showed again a very marked difference between the results of expansions made with and without the black paper screen.

The difference is, in fact, more marked than that between the showers or fogs obtained with and without the glass screen.

It is plain, therefore, that sunlight, unlike the light from the other sources tried, contains nucleus-producing rays which can penetrate glass.

The black paper and the window-glass screen were now removed, and expansions of the same amount as before made with a red glass screen interposed. Only a few drops were produced. On substituting a screen of blue glass, a fog was obtained under the same conditions. These active rays can thus penetrate blue glass, but not red glass.

In connection with the above results it is of interest to notice that ELSTER and GEITEL ('Wied. Ann.,' vol. 38, p. 497, 1889) found that the actino-electric effect of sunlight was not stopped by window-glass or blue glass (red glass being almost opaque to it), while glass is quite opaque to the active rays from a zinc-spark or arc.

To determine to what extent the unconcentrated light of the sun was effective in producing nuclei, the quartz lens was removed, and expansions again made with and without a black paper screen, which was removed immediately after the expansion. A glass lens was interposed immediately after the expansion to make the drops readily visible.

With expansions sufficient to give a few drops in the absence of sunlight, comparatively dense showers were obtained when the air had been exposed to the rays immediately before expansion.

In AITKEN'S experiments on the effect of sunlight the expansion was probably not sufficiently great to make condensation take place on the nuclei produced by it in moist air. Moreover, nuclei, which require such a large degree of supersaturation of water vapour before it can condense upon them, have never been found to persist for more than a few seconds; while in AITKEN'S experiments the exposure to the sunlight was made at an open window and the apparatus then removed to a dark room before the expansion was made.

Although in these sunlight experiments no nuclei, requiring only slight supersaturation to make condensation take place on them, have been produced, they do not absolutely prove that such nuclei may not be formed by sunlight even in the lower layers of the atmosphere. For it is quite possible that the disappearance of the nuclei produced by weak ultra-violet light, when they are left to themselves, is entirely due to the fact that they very quickly reach the walls of the vessel by diffusion on account of their small size. The time for which they persist would then entirely depend on the size of the vessel containing them. Now the explanation of the fact, that with weak radiation they never grow sufficiently to be caught by slight expansions, may simply be that they reach the walls before any considerable growth has time to take place. In the atmosphere, according to this view they would persist for an almost indefinite time, and might finally become large enough to act like "dust" particles in helping condensation.

In the preliminary note already published* it was pointed out that in the upper regions of the atmosphere sunlight was likely to be rich in ultra-violet rays, and it was suggested that from their action on air and water vapour alone the small particles, to which the blue colour of the sky is due, might arise. TYNDALL† recognised the resemblance between the light of the sky and that scattered by the fogs which he obtained by the action of light on various vapours, pointing out that the light scattered by the fogs is polarised like that of the sky. He concluded that the blue colour of the sky was due to the presence of small particles like those produced in his tubes. The connection between the blue colour of the sky and that of the fogs produced from air and water vapour by the action of ultra-violet light is possibly a still closer one, the small particles to which the colour is due having a similar origin in both cases.

The cloud or nucleus-producing effect of ultra-violet rays has obviously bearings on other meteorological phenomena. The nuclei which enable clouds to form may in many cases arise from this source. The upper clouds especially may owe their formation in this way to the action of sunlight. It is possible, too, that owing to the action of the ultra-violet rays, sunlight may sometimes cause clouds to persist in unsaturated air.

§ 7. NUCLEI PRODUCED BY METALS.

The presence of certain metals in the expansion apparatus was found to give rise to condensation nuclei.

The apparatus used in most of the experiments was of the form shown in fig. 10A. (Only the cloud-vessel is shown, the rest of the expansion apparatus was the same as shown in fig. 1.) The cloud-vessel consisted of a portion of a wide test-tube. This was held in position by means of an indiarubber band. The cloud-vessel was divided into two equal parts by a vertical partition, consisting in most cases of mica on one

* 'Camb. Phil. Soc. Proc.,' vol. 9, p. 392, 1898.

† TYNDALL, 'Roy. Soc. Proc.,' vol. 17, p. 223.

side and the metal under investigation on the other, or simply of metal polished on one side only, or of two metals, such as zinc and copper, in contact.

Another and better form of apparatus which was used in some of the later experiments is that shown in fig. 10B. The top of the tube (2·8 centims. in diameter) was closed by a metal plate, bolted down, with an indiarubber band interposed as a washer. Half of the lower surface of the metal plate was covered with a semicircular sheet of mica, attached to the metal by a little shellac; a vertical mica partition divided the apparatus into two equal parts; the roof of the one compartment being of mica, that of the other of the metal whose effect was to be investigated.

The apparatus first described had this defect—that when a fog had been produced

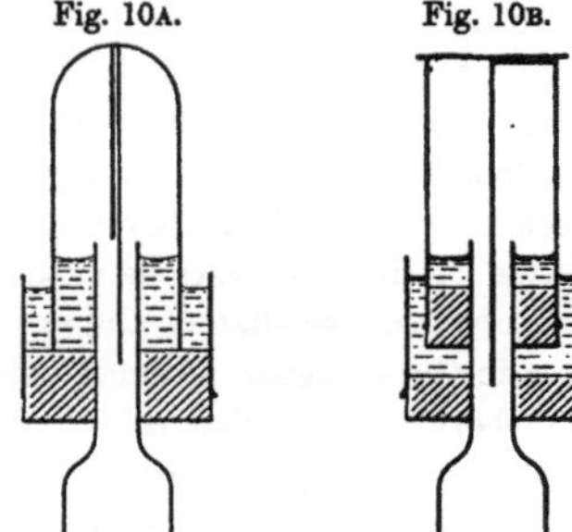

Fig. 10A. Fig. 10B.

by expansion, the metal caused the re-evaporation of the drops in the air near it, before they had time to settle to the bottom of the vessel. This made the process of removing dust particles, by repeatedly forming a cloud by expansion and allowing it to settle, a very slow one in an apparatus of this kind. With the metal at the top there is no difficulty of this kind, for any drops formed near it very quickly fall out of reach of the metal.

In no case were the metals found to produce nuclei requiring only slight expansion to catch them.

This simplified the method of working, making it possible to remove all ordinary dust particles originally present, or any drops remaining in suspension after a cloud had been produced, by the expansion method just referred to, without any arrangement for shielding the air from the action of the metal while this was being done. When this process had been completed, so that with expansions of moderate amount drops were no longer produced, the expansion might be increased without any visible condensation generally to the point ($v_2/v_1 = 1·25$) where a few drops are produced even in the absence of any metal.

When, however, the expansion was such that a rain-like condensation would have resulted in the absence of any metal (v_2/v_1 being between 1·25 and 1·38), the number of drops in the compartment next the metal was generally greater than on the other side of the partition.

With amalgamated zinc comparatively dense fogs are obtained with such expansions; polished zinc and lead also show the effect well; polished copper and tin produce no appreciable effect. A plate of zinc amalgamated on one side and merely polished on the other shows a great difference in the density of the fog on the two sides, the fog next the amalgamated metal being much the denser. If the partition consists of lead with an old surface on one side and a freshly scraped surface on the other, many more drops appear in the half of the tube next the fresh surface. With a zinc copper partition also, the drops are much more numerous on the zinc side.

With amalgamated zinc a few drops may be produced even when v_2/v_1 is rather less than 1·25; they have been observed with a pressure fall of only 13 centims. The effect of the metals is much more marked when the expansion is considerably greater ($v_2/v_1 = 1·30$ or more); in many cases, indeed, the effect of the metal was inappreciable with smaller expansions. When v_2/v_1 exceeds 1·38 no difference can be detected between the comparatively dense fogs which then occupy both sides of the tube.

There can be little doubt that the effect here described is due to the same cause as the influence which these metals have on a photographic plate, studied by RUSSELL* and others. As far as these experiments go they tend to show that the order in which the metals must be arranged to indicate their relative activity in producing nuclei is the same as their order when arranged according to their photographic activity; amalgamated zinc giving the most effect; tin and copper little effect, if any; polished zinc and lead being intermediate in activity.

The experiments described above were all performed with air in the expansion apparatus. In experiments with hydrogen in the apparatus the metals (zinc, amalgamated zinc, and lead) showed only a very slight effect. This, however, does not prove conclusively that more nuclei are produced when the metal is in contact with air than with hydrogen; the difference may be due to the more rapid removal of the nuclei in hydrogen by diffusion to the walls of the vessel.

I have not been able to obtain any effect from metals outside the apparatus, even through celluloid, which RUSSELL found to be penetrated by the photographic action.

§ 8. NUCLEI PRODUCED BY THE ACTION OF ULTRA-VIOLET LIGHT ON A NEGATIVELY ELECTRIFIED ZINC PLATE.

LENARD and WOLFF (*loc. cit.*) were able to show that the condensation of a steam jet becomes dense in the neighbourhood of a negatively electrified zinc plate when this is exposed to the action of ultra-violet light.

The delicacy of the expansion method makes it a matter of some difficulty to

* RUSSELL, 'Roy. Soc. Proc.,' vol. 61, p. 424, 1897; vol. 63, p. 102, 1898. COLSON, 'C. R.,' vol. 123, p. 49, 1896.

investigate by its means this effect in air. For, as has already been shown, not only does ultra-violet light produce nuclei in moist air in the absence of any zinc plate, negatively charged or otherwise, but a zinc plate introduces nuclei into the surrounding air in the absence of both electrification and ultra-violet light. These two effects are apt to disguise the one looked for.

By using very weak ultra-violet light, however, it was found possible to demonstrate the production of condensation nuclei by the action of the light on a negatively charged zinc plate.

Fig. 11.

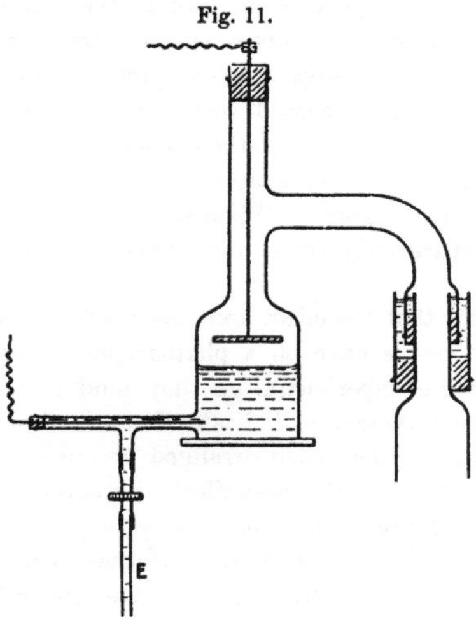

The ultra-violet light (from a zinc spark) entered the apparatus (shown in fig. 11) through a quartz plate covered with 2 or 3 centims. of water, the surface of which served as one plate of a condenser, a zinc plate placed about 1 centim. above the water serving as the other. The condensation from the action of the ultra-violet light alone, or of the zinc alone, was too slight to be detected with a pressure fall below about 18 centims.; the short distance which the few drops that are doubtless produced had to fall, causing them to be overlooked.

On allowing the ultra-violet light to strike the zinc plate, and using a pressure fall of between 15 and 18 centims., a fog was obtained if the zinc plate was made negative (a difference of potential of a few volts being applied between the plates of the condenser); while no condensation could be detected if the zinc was positive or uncharged. The experimental details are given below.

The general construction of the cloud-vessel will be understood by reference to the figure (fig. 11). The zinc plate was soldered to a brass rod, which passed through the small indiarubber stopper closing the upright tube at the top of the apparatus. The lower surface of the zinc was freshly polished, and the upper surface was covered with wet filter paper, to diminish as far as possible the production of cloud nuclei by the zinc. The thickness of the layer of air could be increased or diminished by running water out, or drawing water in, through the tube pointing downwards on the left. Any desired difference of potential could be maintained between the zinc and the water by means of a battery of small secondary cells.

The actual experimental numbers are given in the tables which follow.

Polished Zinc Plate in Air.

ULTRA-VIOLET light from zinc-spark 30 centims. below the quartz plate. Depth of water over the quartz = 2·8 centims. Thickness of air-layer = 0·8 centim.

Gauge reading (in millims.) = p.	E.M.F., in volts.	Result of expansion.	
		Zinc positive.	Zinc negative.
151	9	0	0
159	9	0	Slight fog
167	9	0	Fog
173	9	0	Fog
181	9	Slight fog	Dense fog
161	32	0	Fog
157	32	0	Fog
177	40	0	Fog
177	80	0	Fog
177	240	0	Fog
153	240	0	0
153	80	0	0
183	240	Fog	Fog

No difference could be observed between the results without any E.M.F., and those obtained with the same expansion when the zinc was positive.

ZINC plate freshly polished; zinc-spark 54 centims. below the quartz; depth of water over quartz = 2·3 centims.; thickness of air layer = 1·6 centims.; barometer = 766; temperature = 15° C., π = 13 millims.

Gauge reading (in millims.) = p.	E.M.F. in volts.	Result of expansion.		
		Zinc uncharged.	Zinc positive.	Zinc negative.
163	240	Shower	Shower	Fog
153	240	Very few drops	Very few drops	Very dense shower
151	80	Very few drops	Very few drops	Distinct shower
149	80	Very few drops	Very few drops	Very few drops
149	120	Very few drops	Very few drops	Very few drops
139	120	Very few drops	Very few drops	Very few drops

Gauge reading when the effect of the negative charge is first detected = 151 millims.

Corresponding value of $v_2/v_1 = \frac{B - \pi + m}{B - \pi - p} = 1{\cdot}252$.

In the second series, partly owing to the greater thickness of the air layer, and the consequent greater chance of drops being seen, partly also on account of greater activity of the zinc surface, or greater intensity of the ultra-violet rays, drops were seen, with the expansions used, even when the zinc was positively charged or neutral. The change produced when the charge is negative is however well marked. It will be seen that, although drops are formed even when the expansion is below the limit $v_2/v_1 = 1{\cdot}25$, the effect of the negative electrification first becomes manifest at that point. It will be remembered that the nuclei produced by metals or by ultra-violet light do not show any very definite limit in the least expansion required to catch them.

Experiments were also made in which all conditions were the same as in the experiments just described with this exception, that the zinc plate was covered on the side exposed to the ultra-violet light with wet filter paper. Under these conditions, as is well known,* it ceases to be capable of allowing negative electricity to escape under the influence of ultra-violet light.

The result of a given expansion was now found not to depend on the sign of the charge on the zinc plate; the appearance of the showers or fogs being the same whether the zinc was positive, negative, or uncharged.

Hydrogen.

The phenomena are more easily studied in hydrogen than in air, the effect of ultra-violet light throughout the volume of the gas being so very slight that quite strong radiation may be used.

The hydrogen was obtained by heating palladium which had been charged with the gas. The gas was first allowed to pass through the apparatus at a pressure of a few centimetres of mercury. The tube E was prolonged downwards and passed through the cork of a small wash-bottle containing distilled water, through which the gas had to bubble on its way to the pump. A convenient quantity of this water was finally admitted into the apparatus by closing the tube leading to the pump, and allowing a little air to enter the wash-bottle. The clip between the wash-bottle and the expansion apparatus was then closed and the hydrogen brought to atmospheric pressure by heating the palladium. A parallel beam of ultra-violet light was used, the zinc-spark being placed at the focus of a quartz lens (focal length = 6 centims.) The beam of light was just wide enough to illuminate the whole of the zinc plate.

* STOLETOW, 'Comptes Rendus,' vol. 106, p. 1593, 1888.

ZINC Plate freshly polished. Depth of water over the quartz 1·5 centim. Thickness of hydrogen layer = 2·3 centims.; barometer = 749, temperature = 15° C. = 13 millims.; m = 1 millim.

Gauge reading (in millims.) = p.	E.M.F.	Result of Expansion.		
		Zinc positive.	Zinc negative.	Short circuit.
171	6 Leclanché cells	1 or 2 drops	Fog	
152	6 " "	0	Shower	
161	6 " "	1 or 2 drops	Fog	1 or 2 drops
161	1 " "	1 or 2 drops	Slight fog	
161	3 " "	...	Dense fog	
161	6 " "	...	Fog no denser	
161	20 secondaries	...	Much less dense	
161	120 "	1 or 2 drops	1 or 2 drops	1 or 2 drops
151	6 Leclanché cells	0	Very few drops	
145	6 " "	0	1 or 2 drops	
143	6 " "	0	0	

Gauge reading, when expansion is just sufficient to make condensation take place on the nuclei due to the negative electrification = 145 millims.

Corresponding value of $v_2/v_1 = \frac{B - \pi + m}{B - \pi - h} = 1{\cdot}247$.

It will be noticed that the expansion required to make condensation take place on the nuclei, produced by the action of ultra-violet light on a negatively electrified zinc plate, in air or in hydrogen, is identical with that required in the case of the nuclei produced by X-rays and Uranium rays.

It is only when the zinc plate is negatively electrified that there is any action of this kind.

With the zinc at the given height above the surface of the water the density of the fog produced in hydrogen by a given expansion is a maximum with a comparatively small difference of potential between the zinc and water. The maximum number of drops is obtained with a difference of potential produced by 6 Leclanché cells or less. With the much stronger field, produced by 120 secondary cells (= 240 volts) only a few scattered drops were seen, no more numerous than were obtained with the same expansion when the zinc plate was positively charged, or when the terminals leading to the zinc and water respectively were connected together by a wire.

The diminution of the number of the drops as the electromotive force is increased is easily understood, for when the electric intensity is more than sufficient to remove all the carriers from the zinc as fast as they are produced by the ultra-violet light, then the total number of the carriers which cross from one plate to the other in a given time must remain constant, being equal to the number produced in that time by the ultra-violet light. The velocity of each carrier, however, is proportional to the electric intensity; the number of drops produced on expansion, indicating as it does the number of carriers which at that instant are on their way across between the plates, will therefore be inversely proportional to the difference of potential. The phenomenon is, in fact, closely related to the fact first noticed by STOLETOW,* that the currents produced by the action of ultra-violet light on negatively electrified surfaces approach a saturation value as the electromotive force is increased.

The fact that the source of the rays is discontinuous must also not be forgotten, for with strong fields all the ions produced by the action of one spark may have time to travel across to the other plate, under the influence of the electric field, before the next spark takes place.

The fact that the carriers in hydrogen, as RUTHERFORD† has shown, travel several times as fast as in air, explains why the phenomenon under consideration was only observed in the former gas. With greater differences of potential, or a smaller distance between the plates, it would no doubt be observed in air also.

The fact that the nuclei produced by the action of ultra-violet light on a negatively electrified zinc plate whether in air or hydrogen, require just the same degree of supersaturation in order that they may act as centres of condensation as those produced by Röntgen rays or Uranium rays, is strong evidence that the carriers of the electricity in all these cases are of the same nature. RUTHERFORD‡ has already proved this in quite a different way by his measurements of the velocity with which the carriers move in an electric field.

The very considerable degree of supersaturation necessary to make condensation take place on these nuclei is of itself sufficient to prove that the particles which carry off the negative charge from the zinc-plate are not of the nature of dust particles, but on the contrary must be of almost molecular dimensions (*vide* 'Phil. Trans.,' *loc. cit.*, p. 305).

The conclusion arrived at is therefore opposed to that which LENARD and WOLFF drew from the results of their steam-jet experiments (*loc. cit.*), that the escape of negative electricity from a zinc plate exposed to ultra-violet light is due to the escape of particles arising from the disintegration of the metal. As was pointed out by R. v. HELMHOLTZ and RICHARZ,§ the steam jet is incompetent to distinguish between dust particles and the "ions," to which the latter observers attribute most of the condensation phenomena studied by them.

* STOLETOW, 'Comptes Rendus,' 106, p. 1149, 1888.

† RUTHERFORD, 'Camb. Phil. Soc. Proc.,' vol. 9, p. 401, 1898.

‡ RUTHERFORD, 'Camb. Phil. Soc. Proc.,' *loc. cit.*

§ R. v. HELMHOLTZ und RICHARZ, 'Wied. Ann.,' 40, p. 161, 1890.

§ 9. Nuclei Produced by the Discharge of Electricity from a Pointed Platinum Wire.

The effect of the electric discharge from a pointed wire in altering the appearance of a steam-jet has been studied by many observers.

In expansion experiments, however, no effect appears to have been noticed other than the immediate removal of all dust particles or fog from the air inside the apparatus.* To study the properties of the condensation nuclei arising from this source an expansion apparatus allowing of rapid expansions of large amount is, in fact, necessary.

Fig. 12.

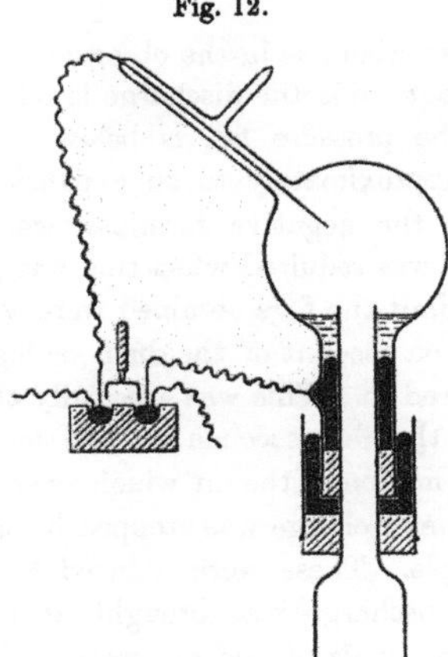

The apparatus which I have used for experiments on this subject is shown in fig. 12. The cloud-vessel consisted of a glass bulb with a side tube through which was sealed a platinum wire, reaching well into the interior of the bulb and ending in a sharp point. This was connected to one terminal of a Wimshurst machine. The other terminal of the machine was connected to earth and to another platinum wire fused through the neck of the bulb, the lower part of which was filled with mercury, covered with distilled water. The side tube was kept dry by warming when necessary with a small flame.

No condensation nuclei were produced, except when the point of the wire was luminous when viewed in the dark.

* First observed by Lodge, 'Nature,' vol. 31, p. 265, 1885.

DISCHARGE from a Pointed Platinum Wire in Air. Expansion made while the Discharge is taking place.

Gauge reading (in millims.) = p.	Result of expansion.	
	Pointed wire positive.	Pointed wire negative.
141	0	0
147	0	Fog
155	Fog	Fog
167	Very dense fog	Dense fog
177	Very dense fog	Very dense fog

The phenomena are simplest when, as in the observations given in the above table, the expansion is brought about while the discharge is taking place. No drops whatever are seen, so long as the pressure fall is below a limit amounting to about 15 centims., corresponding approximately to an expansion, $v_2/v_1 = 1{\cdot}25$. The fogs, when the pointed wire was the negative terminal, were always obtained with a slightly lower expansion than was required when this was positive. With expansions only slightly exceeding this limit the fogs obtained were very dense. The fogs have only a momentary existence, on account of the dust- or fog-removing property of the point discharge already referred to. This was especially the case when the point was positive. The fogs, during the few seconds or less for which they lasted, made manifest the violent eddying motion of the air which accompanies the discharge.

It was found that, when the discharge was stopped before the expansion was made, the results were not so simple. These were reduced to some degree of regularity when the stopping of the discharge was brought about by suddenly connecting together by a wire the two terminals of the apparatus. For this purpose each was connected to one of two mercury cups made near together in a block of paraffin. A short wire was dropped into these cups while the discharge was taking place. This must very quickly have brought the pointed wire and the wet walls of the apparatus to the same potential. Any electrified particles produced by the discharge had, therefore, a greater chance of remaining in suspension on the air than would have been the case if the difference of potential were allowed to exist for any considerable time after it ceased to be sufficient to produce a supply of the particles which act as carriers of the electricity.

DISCHARGE from Pointed Platinum Wire in Air. Terminals joined by a wire before expansion.

Gauge reading (in millims.) = p.	Interval before expansion.	Sign of charge on wire.	Result of expansion.
75	40 secs.	Negative	Fog
75	10 „	„	Slight shower
75	5 mins.	„	Slight fog
26	90 secs.	Negative	0
55	30 secs.	Negative	Fog
55	20 „	„	Slight fog
55	60 „	„	Shower
45	30 secs.	Negative	Very few drops
45	10 „	„	0
45	60 „	„	Very few drops
87	10 secs.	Negative	Very few drops
87	30 „	„	Fog
87	10 „	Positive	Slight fog
87	2 „	„	0
87	30 „		Slight fog
104	2 secs.	Negative	0
104	30 „	„	Fog
104	60 „	„	Fog
104	2 mins.		Slight fog
104	6 „		1 or two drops
170	15 secs.	Negative	Very dense fog
170	2 mins.	„	Slight fog
170	15 secs.	Positive	Dense fog
170	70 „	„	Slight fog

In the second column of the above table is given the time which elapsed from the moment when the discharge was stopped till the expansion was made to take place.

It will be observed that, as this interval is increased, the expansion required to produce a fog diminishes; in other words, the nuclei appear to grow when left to themselves. In no case was fog obtained with a pressure fall of less than 4 or 5 centims.

The nuclei produced by the discharge last for one or two minutes (whether the wire is positive or negative); during this time the number has diminished considerably, and practically none last so long as six minutes.

Hydrogen.

The results obtained with hydrogen when the expansion was made while the discharge was taking place are given in the table which follows :—

Gauge reading (in millims.) = p.	Result of expansion.	
	Pointed wire positive.	Pointed wire negative.
144	0	0
150	0	Shower
156	0	Fog
157	Slight fog	Fog
158	Slight fog	Dense fog

The expansion required is, it will be seen, practically the same as in air. The fogs were again observed with slightly less expansion when the wire was negative than when it was positive.

The results of a second series of experiments made some weeks later are contained in the next table.

Gauge reading (in millims.) = p.	Result of expansion.	
	Pointed wire positive.	Pointed wire negative.
145	0	0
153	Slight shower	Fog
159	Fog	Fog

In a third independent series obtained some months later, the following numbers were obtained :—

Gauge reading (in millims.) = p.	Result of expansion.	
	Pointed wire positive.	Pointed wire negative.
132	0	0
146	0	Few drops
157	Dense fog	Dense fog

The positive fogs thus obtained are much more evanescent than the negative, appearing generally as a momentary flash of brightness in the apparatus, while the negative fogs last for two or three seconds. This is due to the "electric wind" accompanying the positive discharge being the stronger. Possibly the greater expansion required to give visible condensation when the wire is positive may be due to this rapid motion of the contents of the cloud-vessel causing very thin fogs to be overlooked.

Except under certain conditions, to be mentioned immediately, the expansion required to catch the nuclei in hydrogen shows very little diminution (none in the case of the positive discharge) when the discharge is stopped by short-circuiting the terminals before the expansion is made. A slight effect of this kind can be detected when the wire is negative, but the least expansion required to give even a slight shower, whatever interval might be allowed to elapse before making the expansion, was not less than that corresponding to a gauge-reading of 13 centims. The difference between air and hydrogen in this respect is not entirely due to the more rapid diffusion of the nuclei in the latter causing them to reach the sides of the vessel before they have time to grow to any considerable extent, for nuclei, requiring a pressure fall of more than 15 centims., can be detected even 30 seconds after the discharge has been stopped, and when this interval only amounts to 15 seconds they are sufficiently numerous to give quite a dense fog on expansion. Possibly even this slight tendency to become larger exhibited by the nuclei produced in hydrogen when the discharge from the pointed wire is negative, is really a remnant of the effect now to be described.

When the apparatus was first charged with hydrogen, fogs could be obtained under the conditions just described (after the discharge was stopped), even with comparatively slight expansions. The effect was much more marked with the negative than with the positive discharge, and was often absent in the latter case. If, however, the experiments were continued for a day or two the fogs obtained under these conditions became gradually less dense, and finally only a few drops could be obtained with expansions less than that corresponding to a pressure fall of 15 centims.

There can be little doubt that the effect just described is due to some impurity, probably air or oxygen remaining in the apparatus, either mixed with the hydrogen, or absorbed by the platinum wire. This is gradually removed by combination with the hydrogen, the combination being doubtless hastened by the luminous discharge from the point of the wire.

We may conclude from the condensation phenomena attending the discharge of electricity from a pointed platinum wire, that in a discharge of this kind, whether in air or hydrogen, the electricity is carried by "ions" identical with those which are produced in air exposed to Röntgen rays.

The after-effect of the discharge, noticed in air and under certain conditions in hydrogen, is probably a consequence of the chemical combination which can scarcely fail to take place at the glowing point of the wire; where also the ions are, doubtless, liberated. So long as the difference of potential is maintained high enough to produce the discharge, the carriers are driven across to the walls of the vessel, before they have time to grow appreciably. If, however, the electric field be removed suddenly, by short-circuiting the terminals, many of the carriers which have left the point of the wire may not have reached the walls before this process is completed, and their comparatively slow motion when the wire and walls are at the same potential enables them to persist for some time. The growth which then takes place is probably the result of the condensation, upon the nuclei, of some substance produced by the discharge. The substance may be nitric acid or H_2O_2.

§ 10. Behaviour of the various kinds of Nuclei in an Electric Field.

It has already been suggested ('Camb. Phil. Soc. Proc.,' *loc. cit.*) that the nuclei requiring expansions between 1·25 and 1·37 to make condensation take place on them are to be identified with the ions, to which the conductivity of gases exposed to X-rays or Uranium-rays is due. The only evidence there furnished for this view was the fact that in ordinary moist air or other gases such nuclei were found to be present in exceedingly small numbers, while when the gas was made a conductor by being exposed to X-rays or Uranium-rays, immense numbers of these nuclei could be detected.

The experiments with the nuclei produced by the discharge from a pointed platinum wire, as well as with those which are produced by the exposure of a negatively charged zinc plate to ultra-violet light, support this view, at the same time pointing to the conclusion that in all these cases the carriers of the electricity are of the same kind.

A difficulty, however, is introduced by the results obtained with air exposed to weak ultra-violet light or to the action of certain metals, for in both cases nuclei are produced, requiring, in order that water may condense on them, a degree of supersaturation approximately the same as is required in the case of nuclei associated with conducting power in the gas. Now there is no evidence that either the presence of metals or exposure to ultra-violet light causes air to act as a conductor of electricity.

It might be thought that the great delicacy of the condensation method of detecting free ions (each individual carrier being represented by a visible drop on expansion) was the cause of this apparent discrepancy, and that air under the conditions in question really has conducting power, too small to be detected by ordinary methods. The experiments to be described, however, show that the nuclei produced by the presence of metals, as well as those produced by the action of ultra-violet light on moist air, differ from those present in air exposed to X-rays or Uranium-rays in not carrying a charge of electricity, or, to be more exact, in not being affected by an electric field.

To compare the behaviour of the nuclei produced by the action of ultra-violet light on moist air with that of the nuclei produced by Röntgen rays, the apparatus shown in fig. 13 was used. The air is contained between two plates of a condenser, the upper plate consisting of a sheet of aluminium, forming the roof of the cloud-vessel, the lower plate being formed by the upper surface of the water which fills the lower part of this. The aluminium plate was fixed by means of sealing wax. The thickness of the layer of air between the plates was 1·6 centim. By means of a battery of secondary cells, any difference of potential up to 240 volts could be maintained between the plates. The positive terminal of the battery was connected to the aluminium. An aperture at the side, closed by a quartz plate, fixed with sealing-wax, enabled a horizontal beam of ultra-violet light from a zinc spark to enter the apparatus. The light did not impinge on the aluminium plate. The air could be exposed either to the ultra-violet light or to the Röntgen rays from a focus-tube placed above the aluminium plate.

Fig. 13.

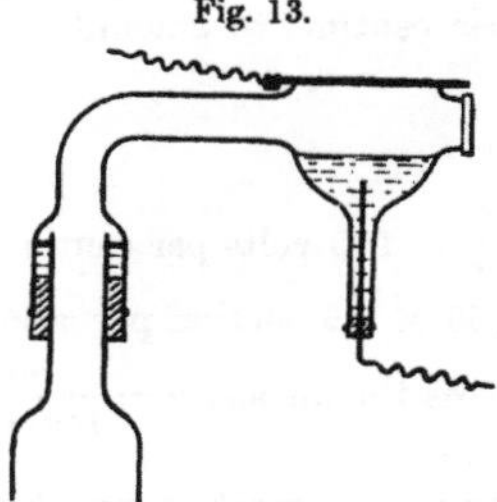

The first experiments were made with ultra-violet light weak enough to give no condensation with pressure fall less than about 15 centims. When a somewhat greater expansion was used ($p = 172$ millims.), an equally dense fog was obtained, whether the difference of potential between the plates was 240 volts or zero. The ultra-violet light in both cases was applied for 30 seconds, and the expansion was made before cutting off the light. In other experiments the expansion was not made till 3 seconds after cutting off the ultra-violet rays; in these experiments the expansion was somewhat greater than before, p being equal to 183 millims. The fogs obtained, when a difference of potential of 240 volts was maintained between the aluminium and the water during the exposure and till after the expansion had been made, were again indistinguishable in appearance from those obtained in the absence of any difference of potential.

On exposing the air to Röntgen rays, instead of ultra-violet light, the expansion being the same as before (gauge reading = 183), very dense fogs were obtained in the absence of electromotive force, while, when a difference of potential of 240 volts was maintained between the metal and water surfaces, only a very slight fog appeared on expansion. An expansion of the same amount, made 3 seconds after the rays were cut off, gave a fog in the absence of any difference of potential, whereas, when the potential difference amounted to 240 volts, no drops at all were produced, even when the expansion was brought about 2 seconds after cutting off the rays. In fact, with the difference of potential just mentioned, no nuclei could be detected if the rays were cut off before the expansion, even if this were effected as quickly as possible after the rays were stopped. On the other hand, when no electromotive force was applied, some of the nuclei lasted for at least 10 seconds after the rays were cut off, a shower being even then obtained on expansion.

There is thus a very marked difference in the behaviour of the nuclei according as they are produced by Röntgen rays or ultra-violet rays, the nuclei produced by the latter being uninfluenced even by a comparatively strong field. The phenomena observed with air exposed to Röntgen rays are easily understood in the light of RUTHERFORD's experiments* on the velocity of the ions in air which has acquired conducting power under the influence of these rays. He finds the velocity, with a potential gradient of 1 volt per centim., to amount to about 1·6 centim. per second in air.

In the present case,

$$\text{Potential gradient} = \frac{240}{1{\cdot}6} = 150 \text{ volts per centim.}$$

$$\text{Velocity of carriers} = 150 \times 1{\cdot}6 \text{ centim. per second.}$$

$$\text{Time taken to travel across the air space} = \frac{1{\cdot}6}{150 \times 1{\cdot}6} \text{ seconds} = \frac{1}{150} \text{ second.}$$

Thus even the carriers which have the greatest possible distance to travel reach one of the plates in less than $\frac{1}{100}$th of a second. This explains how no fogs were obtained when the expansion was made even a very short time after cutting off the Röntgen rays.

Now, when the air was exposed to weak ultra-violet light in place of the Röntgen rays, the difference of potential being, as before, 240 volts, no diminution in the number of the nuclei by the action of the electric field could be detected even 3 seconds after the radiation was cut off. Even in three seconds the distance they have travelled under the influence of the electromotive force is therefore small compared with the thickness of the air layer. These nuclei therefore travel at least 300 times as slowly as those produced by Röntgen rays under the same potential gradient. It is unlikely that this difference is due mainly to a difference in the size of the nuclei, the charge being the same; for with ultra-violet light of the intensity used the two classes of nuclei are indistinguishable from one another, with respect to their power of enabling condensation to take place upon them. There can be little hesitation in concluding that the nuclei produced throughout the volume of the moist air by the action of ultra-violet light differ from those produced by Röntgen rays in being uncharged. If any ions are present in air exposed to ultra-violet light they are exceedingly few in comparison with the uncharged nuclei which are at the same time produced.

* RUTHERFORD, 'Phil. Mag.,' vol. 44, p. 422, 1897.

It still remains possible that the comparatively few nuclei, all requiring large expansions, which can be detected in hydrogen exposed to strong ultra-violet light, may consist of ions produced throughout the volume of the gas. This point could easily be tested by experiments like those just described.

Further experiments were made with air exposed to the much more intense rays which were obtained when a quartz lens was interposed between the zinc points and the quartz plate. The intensity of the ultra-violet rays was then sufficiently great to give fogs with comparatively slight expansion. The apparatus being arranged to give expansion corresponding to a gauge reading $p = 64$ millims., no condensation (in the absence of electromotive force) was obtained with an exposure of 10 seconds, while an exposure of 15 seconds with an expansion of the same amount gave a fog. The nuclei thus took between 10 and 15 seconds to grow large enough to be caught with the degree of expansion used. Yet the application of a difference of potential of 240 volts between the plates did not prevent very dense fogs being obtained with the same expansion with an exposure of 3 minutes. Thus, in spite of the electric field, the nuclei were able to exist for more than 10 seconds; in other words, they took more than 10 seconds to travel across the space between the plates, which were again 1·6 centims. apart. They thus took more than 1000 times as long as the nuclei produced by Röntgen rays to travel the same distance. It is, therefore, very improbable that the growth of the nuclei under the action of strong ultra-violet rays, or the diminution of the expansion required to catch them, is the result of any electrification of the nuclei by the action of the rays. Another explanation, therefore, than the possession of a charge of electricity by the drops, must also be sought for the persistence of the visible fogs, which are the final result of prolonged exposure to strong ultra-violet rays.

The great diminution of the number of drops which are produced on expansion when an electromotive force is applied during the exposure of the air to Röntgen rays is easily understood. For the number of nuclei present at any instant is proportional to the rate at which these are being produced by the rays and to the average length of time for which they persist. Now before the application of the electric field the average life of the nuclei, depending on the rate at which they combine with one another or reach the walls by diffusion, is seen to be something of the order of 1 second, for a large proportion of the nuclei persist for 2 or 3 seconds. Now, when the electric field of the intensity used in the experiments is applied the time for which they persist must, as has been seen, be reduced to something like $\frac{1}{100}$th part of this. The number of drops in the fog will be diminished in the same ratio. The immense difference in the appearance of the fogs with and without the action of the electric field is in complete agreement with this. The complete absence of any such difference in the case of the fogs produced under the action of the ultra-violet light is again a proof that the nuclei on which their production depends do not move under the action of an electromotive force.

Uranium Rays.

For experiments on the influence of an electric field on the nuclei produced by Uranium-rays the apparatus constructed for the experiments on the nuclei arising from a zinc plate exposed to ultra-violet rays was used (fig. 11). A piece of thin sheet tin was substituted for the zinc. A thin float, consisting of a sheet of cork wrapped in tinfoil, formed the lower plate of the condenser. On the upper surface of this was placed a layer of moist uranium oxide. The thickness of the air layer was regulated by allowing some water to escape or drawing a little more into the cloud-vessel, according as an increase or diminution of the thickness was required.

The following results were obtained.

DIFFERENCE of potential used = 240 volts; thickness of air layer = 1 centim.

Gauge reading (in millims.) = p.	Result of expansion.		
	Upper plate positive.	Upper plate negative.	No E.M.F.
162	0	0	Fog
184	0	0	Fog

These experiments were many times repeated with the same results. The effect of the electric field was equally marked when the distance between the plates was increased to 1·3 centims. The drops produced while the electric field was maintained were too few to be detected. All the nuclei produced by the action of the uranium appear therefore to be charged.

Metals.

The same apparatus was used, but the float was omitted and a polished zinc plate was substituted for the tin. The arrangements were in fact exactly the same as in the experiments on the nuclei produced by the action of ultra-violet light on zinc, with the omission of the apparatus necessary for producing these rays. The results obtained were entirely negative.

DIFFERENCE of potential = 240 volts; thickness of air layer = 1·4 centims.

Gauge reading (in millims.) = p.	Result of expansion.		No E.M.F.
	Zinc positive.	Zinc negative.	
165 188	Slight shower Dense shower	Slight shower Dense shower	Slight shower Dense shower
Thickness of air layer = ·8 centims.			
189 217	Dense shower Dense fog	Dense shower Dense fog	Dense shower Dense fog

Exactly similar results were obtained when the zinc was amalgamated, so that a larger number of nuclei might be produced.

The action of zinc in producing nuclei is thus proved to be of quite a different nature to that of uranium oxide. It does not consist in the production of free ions throughout the volume of the air near it by the action of radiation like that from uranium and its compounds.

It might be supposed, however, that the nuclei consisted of ions, not produced throughout the volume of the air, but having their origin at the surface of the zinc. They might in fact be a direct product of the oxidation of the zinc, the oxygen or water molecules being split up, half of the molecule combining with the zinc, the other part escaping into the surrounding gas as a free ion. One would expect the ions, according to this view, to be all charged with electricity of the same sign. There ought therefore to have been a difference in the number of cloud particles produced according as the zinc was made positive or negative. In fact one would expect, as was found to be the case when the zinc was exposed to ultra-violet light, an increase in the number of nuclei when the electromotive force was in one direction, that namely tending to move the ions from the zinc, and a diminution when the field was reversed. The absence of any difference whatever in the appearance of the fogs whether the zinc was the positive or negative terminal or was uncharged, shows that the nuclei do not consist of ions, produced either at the surface of the metal or throughout the volume of the air and in its neighbourhood. They are, like the nuclei produced by the action of ultra-violet light on moist air, uncharged.

Ions are thus not the only nuclei requiring expansions between the limits $v_2/v_1 = 1·25$ and $v_2/v_1 = 1·37$, in order that condensation may take place upon them; both weak ultra-violet light and certain metals produce such nuclei, which experiment shows to be unaffected by an electric field, that is, not to be ions. They have less definite properties as nuclei of condensation than the ions; the minimum expansion required to make condensation take place on them may be less or more than is required for the ions according as their number is great or small, and those produced by ultra-violet light grow, if the rays are strong enough, till they become visible without expansion. To account for the growth of the nuclei under the influence of ultra-violet light, I have already suggested that some compound such as H_2O_2 may be produced by the action of these rays on the nuclei. We may, perhaps, extend this idea somewhat, and regard also the nuclei produced in air by weak ultra-violet light or by metals, as consisting of molecules of H_2O_2 or of aggregates of molecules of H_2O and H_2O_2. SCHÖNBEIN, so long ago as 1866, found that hydrogen peroxide was produced by shaking together amalgamated zinc, oxygen and water ('Journ. für Pr. Chem.,' vol. 98, p. 65). There can be little doubt that the nuclei produced in the neighbourhood of metals have some relation to the active substance, hydrogen peroxide, "active oxygen," or whatever it may be, which is produced in many cases of slow oxidation, and about which there has been so much controversy.

The question now arises in which class of nuclei, the charged or the uncharged, must we place those which always appear to be present in small numbers in moist air, giving rise to the rain-like condensation which takes place with expansions between the limits $v_2/v_1 = 1{\cdot}25$ and $v_2/v_1 = 1{\cdot}38$.

To decide this question the same apparatus as that used for the experiments last described was used, but the zinc plate was replaced by one of sheet tin, because this metal appears to be inactive, that is, it produces no increase as far as can be detected in the number of nuclei present. The plate was, moreover, covered on both surfaces with wet filter paper to prevent direct contact of the metal with the air. The thickness of the air layer was equal to 1·7 centims.; an electromotive force of 225 volts was used.

The expansion used was that corresponding to a gauge reading of 187 millims. This gave a slight shower, and no difference could be detected in its appearance whether the electromotive force was applied or not. This appears to indicate that these nuclei are not charged. It is, however, doubtful whether the tin, even when covered with wet filter paper, is absolutely inactive, and on that account, perhaps, not a great deal of weight can be attached to this experiment. If we assume that the effect of the tin is negligible, there still remains the possibility that although the nuclei requiring expansions considerably exceeding the limit $v_2/v_1 = 1{\cdot}25$ are uncharged, the very few which require an expansion only very slightly exceeding this may be charged and identical with those produced by Röntgen rays. Otherwise we have the somewhat astonishing result that two quite different kinds of nuclei require absolutely the same degree of supersaturation, that, namely, corresponding to the expansion $v_2/v_1 = 1{\cdot}25$, in order that condensation should take place on them. To make experiments of the same kind with expansions only slightly exceeding the limit $v_2/v_1 = 1{\cdot}25$ is difficult on account of the exceedingly small number of the drops. Apparatus on a much larger scale would be better for experiments on this point.

One way of accounting for the fact that an electric field has no effect on the rain-like condensation under consideration would be to suppose that the nuclei are produced at the moment of expansion. They might, for example, be caused by the motion of the piston or plunger through the water. The fact, however, that the condensation is just as easily observed when the cloud-vessel is connected with the rest of the apparatus by a bent tube of considerable length, as for example in the apparatus shown in fig. 2, shows that this is not the source of the nuclei.

The question whether the nuclei which exist in small numbers in moist air are charged or not must, I think, be left an open one for the present. It is manifestly a matter of considerable meteorological interest.

The view which was taken in previous papers concerning the dense fogs which are obtained with expansions exceeding the second limit $v_2/v_1 = 1{\cdot}38$, was that the degree of supersaturation is then great enough to cause condensation to take place independently of all nuclei other than the molecules of gas or vapour themselves. According to this view no effect is to be expected on applying an electric field when expansions so great as this are used. In fact, the same apparatus being used as before, no difference could be detected in the appearance of these fogs, whether they were produced in the absence of any electric field, or with a difference of potential of 225 volts, between the tin and water surfaces, these being 1 centim. apart.

§ 11. On Ions and Condensation.

The experiments described in this paper furnish strong evidence that the passage of electricity through gases is effected by carriers of the same nature, whether the conduction is the result of exposure of the gas to Röntgen rays or Uranium rays, or the action of ultra-violet light on a negatively electrified zinc plate, or consists in the escape of electricity from a pointed platinum wire. It is not only in their efficiency as condensation nuclei that the carriers from the first three of the above-mentioned sources agree, for Rutherford has shown that their velocity in an electric field of the same strength is almost identical.

Further, these carriers are by no means of the nature of dust particles, for unlike the latter, which require only an exceedingly slight supersaturation in order that condensation may take place on them, they do not act as centres of condensation unless the vapour is about 4·2 times as dense as that in equilibrium over a flat surface of water at the same temperature (*vide* 'Phil. Trans.,' *loc. cit.*). In the paper just referred to the number $8{\cdot}6 \times 10^{-8}$ was given as an approximate value of the radius in centims. of water-drops equivalent in their action to these nuclei. The nuclei are therefore not much larger than molecules; the fact that dense condensation takes place with a supersaturation only twice as great when, as far as can be judged, no nuclei are present other than the molecules of vapour and gas, is further evidence that the nuclei with which we are here concerned are not very large compared with molecular dimensions. This again is an agreement with experiments on the velocity of the ions.*

It is, perhaps, of some interest to calculate what charge would be required to keep a drop of the above-mentioned size (radius $= 8 \cdot 6 \times 10^{-8}$ centim.) from evaporating. Making use of the results given by Professor THOMSON ('Applications of Dynamics,' pp. 163, 165), we have, when the charge on the drop just balances the effect of surface tension so that there is no longer any tendency to evaporate in vapour saturated with respect to a flat surface,

$$e^2 = 16\pi Ta^3,$$

where e is the charge on the drop, T is the surface tension, and a is the radius.

This gives us in the present case

$$e = 1 \cdot 5 \times 10^{-9} \text{ electrostatic unit,}$$

which agrees sufficiently nearly with what we have reason to suppose the order of magnitude of ionic charges to be. We must not forget, however, the assumptions made in obtaining the above-mentioned estimate of the size of the nuclei (*vide* 'Phil. Trans.,' *loc. cit.*, p. 305).

TOWNSEND has shown† that freshly prepared gases are often electrified, and that the charge is carried by nuclei on which, even if the gas be not saturated with aqueous vapour, water condenses to form visible drops. He has shown, moreover, that the charge carried by each of these nuclei (in oxygen) amounts to about 3×10^{-10} electrostatic unit, and is presumably the charge carried by one ion. The experiments of H. A. WILSON‡ furnish strong evidence that the growth of the drops in the fogs studied by TOWNSEND is not, however, a direct consequence of the charge which they carry, but is due to the presence of some substance in solution in the drops.

Many of the cases of condensation (apparently with only slight or without any supersaturation) produced by chemical action, which were studied by R. v. HELMHOLTZ and RICHARZ,§ and which were attributed by them to the influence of free ions, are probably also mainly the result of the formation of some substance in solution in incipient drops (of which the original nuclei may be free ions). Professor J. J. THOMSON‖ has suggested that the great influence which the presence of moisture has in facilitating chemical reactions between gases may be due to the presence of minute drops, at the surface of which (or throughout the volume) the combination is able to take place. For example, dry NH_3 and HCl do not combine, but if water vapour be present,

* RUTHERFORD, 'Camb. Phil. Soc. Proc.,' vol. 9, p. 415, 1898.

† TOWNSEND, 'Camb. Phil. Soc. Proc.,' 9, pp. 244 and 345, 1897.

‡ H. A. WILSON, 'Phil. Mag.,' vol. 45, p. 454, 1898.

§ HELMHOLTZ and RICHARZ, 'Wied. Ann.,' vol. 40, p. 161, 1890.

‖ J. J. THOMSON, 'Phil. Mag.,' vol. 36, p. 313, 1893.

combination at once takes place with the formation of a cloud. If the moist gases contain minute water drops, it is evident that combination must take place within the drops, for HCl and NH_3 at once combine when in solution. It appears to me natural to suppose that the fogs produced by this and similar reactions are to be explained by the products of the reaction being formed in solution in incipient drops, in quantities sufficient to counterbalance the effect of the curvature of the surface on the vapour pressure. The drops then grow so long as the products of the reaction continue to accumulate within them. The original droplets may be formed by the action of the ions; but it is quite possible that even in the absence of any ions, minute drops are continually being formed, and on account of surface tension at once evaporating again, unless made permanent by the formation within them of some other substance than water.

AITKEN* found that when proper precautions were taken, no condensation nuclei were produced by the combustion of hydrogen. In this case (if we assume the product of combustion to be pure water only) any growth of the droplets through the lowering of the vapour pressure by a dissolved substance is out of the question. Although these experiments of AITKEN show that in this case there is no production of comparatively large nuclei, such as would be capable of promoting condensation with slight supersaturation, or of travelling a considerable distance along a narrow tube without being removed, they do not prove that no free ions are produced by the combustion, or that these would not act as centres of condensation if the degree of supersaturation were reached, which the experiments described in this paper show to be in general required to cause condensation on the ions.

There is, I think, no evidence that the ions alone, in the absence of other influences, ever act as centres of condensation unless the above-mentioned comparatively great degree of supersaturation (approximately fourfold) be exceeded.

In conclusion, I wish to acknowledge how greatly I am indebted to Professor THOMSON for his suggestions and encouragement during the course of this work.

* AITKEN, 'Trans. Roy. Soc., Edin.,' vol. 39 (1), p. 15. 1897.

INDEX SLIP.

---

st ('Phi
, by the
the plat
he object
ation of t
ned in t
arly appl
cial chen
lish equa
o extend
d physic
d and th
alts in q
or has d
ological a
s, contain
. Soc.,' T
, $6H_2O$ in
Soc.,' T
etals ('J
igations
ical prop
ow the o
(K = 39,
of the at
function

Of all the isomorphous series referred to, the normal sulphates alone

X. *The Thermal Deformation of the Crystallised Normal Sulphates of Potassium, Rubidium, and Cæsium.*

By A. E. TUTTON, *B.Sc.*

Communicated by Captain ABNEY, *C.B., F.R.S.*

Received January 31,—Read February 16, 1899.

IN a communication made to the Royal Society in April last ('Phil. Trans.,' A, vol. 191, 313) the author described an interference dilatometer, by the use of which, owing to the introduction of compensation for the expansion of the platinum-iridium interference apparatus by means of a disc of aluminium laid on the object, the delicate method of FIZEAU is rendered equally sensitive in the determination of the expansion of solid substances, notably crystals, which cannot be obtained in blocks of the relatively large size hitherto required. The method is particularly applicable in the cases of those substances, including the crystals of most artificial chemical salts or other preparations, whose ground surfaces will not take a polish equal to that of glass. The author was led to devise it in order to be able to extend his investigations, concerning the relations between the morphological and physical properties of the crystals of isomorphous series of salts on the one hand and their chemical constitution on the other, to the thermal deformation of the salts in question. In previous communications to the Chemical Society the author has described the results of detailed observations of a large number of morphological and physical properties of the crystals of the series of normal alkali sulphates, containing as metal potassium, rubidium, and cæsium respectively ('Journ. Chem. Soc.,' Trans., 1894, 628); of twenty-two double sulphates of the series $R_2M(SO_4)_2, 6H_2O$ in which R is represented by the same three alkali metals ('Journ. Chem. Soc.,' Trans., 1893, 337 and 1896, 344); and of the normal selenates of these metals ('Journ. Chem. Soc.,' Trans., 1897, 846). The general result of these investigations has been to show that the whole of the investigated morphological and physical properties of the crystals of these salts exhibit progressive variations which follow the order of progression of the atomic weights of the three alkali metals (K = 39, Rb = 85·2, Cs = 132·7), so that the variations may be said to be functions of the atomic weight of the alkali metal, in the broad sense in which the term "function" is usually applied in connection with atomic weight.

Of all the isomorphous series referred to, the normal sulphates alone prove to be

17.6.99

suitable for an investigation of the thermal deformation. The double sulphates are unsuitable on account of the ease with which most of them lose water of crystallisation when their temperature is raised, and a similar remark applies to the double selenates, whose investigation with respect to their morphological and physical properties is now proceeding. The simple selenates offer great difficulties on account of their excessively hygroscopic nature, which is so marked in the case of cæsium selenate, in accordance with the rapidly progressive advance in the solubility of the three salts which has been shown (*loc. cit.* p. 851) to follow the order of the atomic weights of the metals, as to place it in the category of effective desiccating agents. The normal sulphate of potassium is absolutely free from this disadvantage, being one of the least soluble of the salts usually classed as soluble in water, 100 cub. centims. of this liquid at the ordinary temperature only dissolving 10 grams of the salt (*loc. cit.* p. 851 and sulphate memoir *loc. cit.* p. 632). Rubidium sulphate is so slightly hygroscopic, its solubility being only 44 per cent., as to present no difficulty on this ground. Cæsium sulphate is decidedly hygroscopic, the solubility being so relatively great as 163 grams in 100 cub. centims. water. Although this characteristic is by no means so strong as in the analogous selenate, the solubility of cæsium selenate being no less than 245 grams in 100 cub. centims. water, still it is sufficiently marked to render the use of the salt for the purpose in question impossible in damp weather. The difficulty has, however, been successfully overcome in the case of cæsium sulphate, by taking advantage of the driest days of the recent remarkably dry summer, and of a few dry frosty ones of the early winter, together with the expedient of utilising the inner chamber of the air bath of the dilatometer as a desiccator, by placing a vessel containing oil of vitriol therein until the actual moment of commencing the observations.

In the present memoir, therefore, are presented the results of an investigation of the thermal deformation of the orthorhombic normal sulphates of potassium, rubidium, and cæsium. It is scarcely necessary to remark that the series of these particular three metals has been chosen throughout the whole of the author's work on the relations between the chemical composition of salts and the properties of their crystals, because of their well-established close relationship, as being in the strictest sense members of the same family group of the periodic system, the definitely established and relatively large differences between their atomic weights, and the fact that they form the most strongly electro-positive series of elements.

Preparation of the Crystals.

Although the new compensation method does not require crystal blocks of greater thickness than 5 millims., the greatest difficulty has been experienced in obtaining crystals of the commonest of the three salts, potassium sulphate, of adequate thickness in all three of the axial directions along which measurements of expansion or contraction by heat were desired. By the slow evaporation of cold saturated solutions over oil of vitriol *in vacuo* it is possible, given adequate length of time and sufficient amount of solution, to obtain excellent crystals of the more soluble sulphates of rubidium and cæsium, of the requisite size to furnish blocks from 5 to 10 millims. thick in the three axial directions. But in the case of the sparingly soluble potassium salt the crystals are almost invariably small. Moreover, when by exception they are larger, they are either pseudo-hexagonal triplets or other twinned forms, useless for the purpose in view owing to the unequal expansion of the interpenetrating parts due to different axial expansion in accordance with orthorhombic symmetry; or otherwise, they are individuals of an elongated prismatic nature, the elongation being in the direction of the axis *a* and the prisms being too narrow for use along the other two axial directions, particularly that of the axis *b*. It was found exceedingly difficult to induce well-formed individual prisms, deposited from a large quantity of a cooling saturated solution, to grow further to the required transverse dimensions in cold saturated solutions over vitriol *in vacuo*. After attaining a thickness of about 4 millims. it almost always happened that, rather than grow further, fresh crystals began to be deposited. After months of fruitless labour, Messrs. HOPKIN and WILLIAMS kindly undertook to attempt to obtain larger crystals by the use of very much larger quantities of solution, and eventually succeeded in producing seven crystals of exceptional thickness, and which, after a little further growth in a cold saturated solution over vitriol *in vacuo*, have at length yielded transparent blocks 6·5 to 9·6 millims. thick along the direction of the axis *c* and 5·1 to 5·9 millims. along the axis *b*. The author desires to express his great indebtedness to the firm in question for so kindly placing their resources at his disposal, and thus enabling this investigation to be completed.

PREPARATION OF THE PARALLEL-FACED CRYSTAL-BLOCKS.

Improvements on the Cutting and Grinding Goniometer.

The preparation of the crystal-blocks, each provided with a pair of truly plane and truly parallel surfaces accurately perpendicular to the particular crystallographical axis along which the linear expansion or contraction was to be measured, was carried out with the aid of the author's new cutting and grinding goniometer. The instrument in question is similar in principle and general appearance to that which was described to the Royal Society in December 1894, ('Roy. Soc. Proc.,' vol. 57, p. 324), and which is now in the National Collection in the South Kensington Museum. It differs from the latter instrument in including a few slight improvements which prolonged use has shown to be advisable for the sake of greater convenience and ease in manipulation. The perfected instrument affords the highest satisfaction, enabling the most accurately orientated and truly plane surfaces of crystals of any degree of hardness to be obtained with the minimum expenditure of time and trouble. Its cost is necessarily very heavy, and the author wishes it to be clearly understood that it is not intended as an instrument for ordinary laboratory use, but as a means of obtaining, irrespective of cost, plates blocks or prisms of the highest attainable accuracy of orientation and perfection of surface, for the purposes of original investigation. The accompanying illustration represents it.

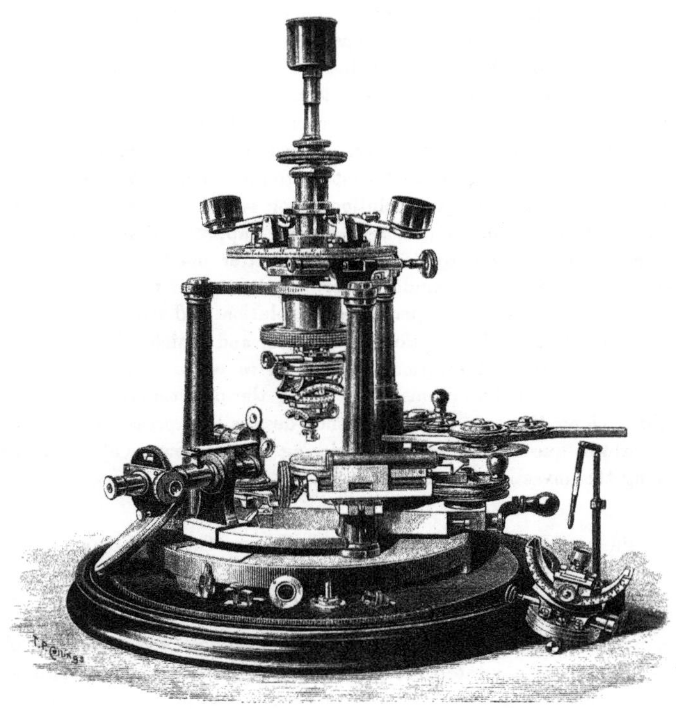

The following description of the improvements on the instrument previously described is appended for the information of other workers, and in response to several enquiries which the author has received both from investigators in this country and abroad.

The most important is a new method of mounting the grinding table, which admits of its movement in its own plane; this is introduced in order to be able to vary the part used in the grinding more considerably than can be achieved merely by use of the centering movements of the crystal-adjusting apparatus, and thus avoid concentric grooving of the grinding surface, besides no longer requiring the centering of the crystal to be disturbed. This is achieved by mounting the grinding gear on a slider, which can be made to traverse a bevelled bed resting on the base of the instrument, by rotation of a long screw gearing with a corresponding thread in the bed and manipulated by a winch handle. The gunmetal bed is 20 centims. long, 6 centims. wide at the top surface, the greater portion of which is planed out into a depression so as to reduce friction by only leaving narrow strips along the long edges to act as guides for the slider, and 4 centims. wide at the base, there being thus a bevel of 1 centim. at each side. The bed is fixed firmly to the main base of the instrument by four large screws. At the outer end is a cup-shaped hollow ending in the tapped horizontal hollow cylinder of steel which gears with the steel screw carried by the slider. The latter is slightly longer than the bed, and 8 centims. broad. It is a solid casting of steel hollowed underneath to the shape of the bed; the thickness above the bed is 8 millims., and the sides, bevelled on their interior, have a depth of 24 millims. In order to ensure close but not inconveniently tight fitting of the slider on its bed, room has been left on one side for the insertion of a thick strip of steel between the bevelled edges of the slider and the bed, and this is attached to the slider by five screws whose heads are on its outside, and which serve as adjusting screws to regulate the fitting of bed and slider. At the outer end the slider narrows off laterally and its termination is deepened to the depth of the bevelled sides, so as to form a stout support for the passage of the thick steel cylinder which, beyond this bearing, where it is flanged to prevent its traversing, is cut with a deep screw-thread corresponding to the hollow one in the bed. Immediately outside this bearing, the steel cylinder passes into the brass winch fitted with ebonite handle. As the other end of the screw is unflanged, the slider can be readily removed from, or re-attached to, the bed; on pushing the slider over the bed it slides unimpeded for a dozen centims., when the screw begins to gear, and further movement is effected by the winch. The grinding gear is similar to that permanently fixed to the base in the previous instrument, the grinding table being provided with an efficient means of adjustment exactly perpendicular to the vertical goniometrical axis, and the friction pulley with a means of adjustment for the tightening of the driving band. The former is necessarily effected differently to the method employed in the previous instrument, as it is no longer possible to penetrate the base for the utilisation of the spiral spring and tripod method. It is here effected by an adjustable stout circular base-plate, which is fixed to the slider by six screws, three of which penetrate both plate and slider, and three pass through the plate only and thus act as adjusting screws; after these are adjusted the table is fixed in the adjusted position by means of the other three screws. The adjustment of the friction pulley is effected by carrying it on a bracket which is fixed at the inner end of the slider by two broad-headed screws, working in slots of the required length in the horizontal claws of the bracket. Both this attachment and the circular-plate-bearing attachment of the driving pulley carried at the outer end of the slider can be slightly tilted if desired, so as to make the pulleys run sufficiently eccentrically to avoid friction of the band at the two crossing points on each side of the grinding table pulley. The slider carries in addition, at the middle of the friction pulley end, a small brass plate supporting a short tube, in which, together with two holes driven into the slider one near each side and on the other side of the grinding table, fit three rod supports for a circular guard to surround the grinding table and prevent projection of the lubricating liquid when grinding. The guard has a window on the side nearest the friction-pulley, for the passage of the crystal and its supporting apparatus as the slider is pushed into position; the window can be closed by a shutter after the passage of the crystal, a strip of metal of similar curvature to the guard itself being fitted round this portion of the guard for this purpose, and made movable with the aid of a little handle along a suitable slot directed by broadheaded guiding pins.

The height of the grinding surface of the lap, when the slider is in position, is exactly the same as that of the cutting disc when the cutting gear is in position, so that no variation of the height of the crystal-carrying axis is required. The cutting gear is exactly as in the former instrument, rotatable about the back pillar, and supported also, when in position, in a traversing apparatus carried by the front pillar. The slider carrying the grinding apparatus is removed when the cutter is in position.

In addition to the laps provided with the previous instrument, two additional ones are furnished. One is a polishing lap for hard crystals, consisting of hard opticians' wax melted into a circular metal tray of the same size as the other laps, and afterwards compressed so as to present a plane surface. This lap, employed with ochre or rouge, enables the opticians' method of polishing glass surfaces to be closely followed in the polishing of hard crystal surfaces. The second is a lap whose grinding surface is formed by a sheet of emery cloth stretched over, and cemented to, a metal base of the size and shape common to all the laps. This lap has been particularly useful for effecting the preliminary grinding down of the relatively large crystals employed in the work whose results are now being communicated, leaving but little for the ground-glass lap to do. The variety of eleven laps now provided, enables any or all of the usual grinding and polishing processes of the optician and lapidary to be followed, besides those described by the author for the grinding and polishing of the softer crystals of artificial salts. One of the laps is shown in position in the illustration, and another to the left leaning against the base of the instrument.

A further small but important addition to the accessories consists of three gripping crystal-holders, which are shown resting on the base. One of them is a triply and widely split tube of a centimetre bore, narrowing at a centimetre from the orifice into a cone which passes into a grooved stem similar to the stems of the ordinary holders used for wax attachment of the crystal. The wide splits are continued down to the stem, and the conical portion is provided with a screw thread, with which gears a milled collar, by the screwing of which down the cone the three portions of the tubular holder are compressed together. The latter are padded inside with broad-cloth, within which, on rotating the collar, the crystal is firmly clamped. This holder is suitable for the gripping of prismatic crystals. The other two, which differ only in the size of their apertures, somewhat resemble miniature tuning forks, the stems being of the same size as those of the other holders and similarly grooved to fit the ribbed socket at the base of the crystal-adjusting apparatus. The two prongs are in each case relatively broad, 1·3 centim., are padded inside with broad-cloth and can be drawn together so as to effect the grip by means of a milled-headed clamping screw, passing loosely through one prong and screwing firmly through the other. One of the forks takes crystals 5 millims. thick and the other takes crystals up to 1·2 centim. thick. If the crystal is not sufficiently tabular for direct gripping by one of these two latter holders, and not sufficiently evenly prismatic for the advantageous use of the split-tube gripper, it is packed in a rectangular block of cork held in the larger forked holder; the cork can be cut with a sharp penknife so as to accommodate the most inconveniently shaped crystal in the position required for grinding. When the prongs are screwed together as much as possible so as to tightly grip the cork setting and the contained crystal, the latter is found to be rigidly held without any danger of cracking, the cork lending itself to an even distribution of pressure. These grip-holders were devised in order to avoid the frequent cracking which large crystals suffer when warm opticians' wax is employed to cement them to the ordinary holders. The importance of this point is obvious, when it is remarked that a cracked crystal is totally unsuitable for use in determinations of thermal expansion by the interference method, the crack being sure to develop further during the observation and derange the interference bands.

Another addition is a special crystal-adjusting apparatus, shown resting on the table to the right in the illustration, intended for use in preparing 60° prisms for refractive index determinations, in cases where it is a difficult matter, by reason of deliquescence or other rapid deterioration of the substance under investigation, to prepare the two inclined surfaces by separate settings on a crystal-holder employed with the ordinary adjusting apparatus. In order to prepare two surfaces inclined at 60° by one setting of the crystal, it is obviously necessary to rotate the crystal for 60° on each side of the particular principal optical plane which has been adjusted vertical to the grinding plane, with the aid of the goniometrical arrangements provided on the instrument, and to which optical plane the two required surfaces are to be symmetrical. The adjusting apparatus provided for ordinary purposes is similar to the second one described in the memoir on the first and smaller pattern grinding goniometer, intended for use in grinding small artificial salt crystals ('Phil. Trans.,' A, 1894, p. 895), but of larger size, corresponding to the larger instrument. It includes two cylindrical adjusting movements provided with divided silver arcs and indicators, and a divided horizontal circle between them to enable the lower movement to be set at any desired angle, usually 90°, to the upper one. The two cylindrical movements, however, only admit of 35° of rotation on either side of the vertical axis in each case; this amount is ample for ordinary purposes, including the preparation of a 60° prism by separately setting the direction of each required prism-face, by rotation of 30° from the plane perpendicular to the bisecting plane, the former of which planes can usually be as readily goniometrically adjusted with reference to the existing crystal faces as the bisecting plane itself. In the exceptional cases referred to, of which the extremely deliquescent cæsium selenate is an excellent instance, it repays to render the adjusting mechanism more cumbrous in order to secure the prime object, and for this reason the new alternative adjusting apparatus is provided.

It is exactly like the one provided for ordinary use as far down as the upper fixed cylindrical segment and its divided silver scale reading 35° on each side, which is suspended by a bracket from the lower disc of the centering arrangement. The latter is given in duplicate, one being always attached to the ordinary and one to the special adjusting apparatus, as it is more readily attached to, or detached from, the inner axis of the goniometer than the adjusting apparatus to or from the centering disc. Sliding in and under the fixed segment, instead of the usual movable segment of the same size, is one of double the size, that is of rather more than 150°. On one face this enlarged movable segment carries a silver index, to indicate the position with respect to the fixed graduated arc above it; and on the other a silver arc graduated to 75° on each side of the centre. In a rabbetted bed on the under side of the large segment slides the carrier of the lower adjusting segments, which are of the same kind as in the ordinary apparatus, arranged permanently at right angles to the two upper ones, the horizontal circle of the ordinary apparatus being omitted in order to avoid complexity. The sliding of the large segment about the upper fixed one is effected, for the 35° of its path on each side, by manipulation of the milled head of a tangent screw arrangement as in the ordinary apparatus. The lowest of the pair of segments arranged at right angles to these upper ones is also manipulated in its segmental bed for 35° each side by a similar tangent screw. But the sliding of the carrier of the two lower segments about the large segment is effected by hand, and fixation at any required position, with reference to the large silver arc as indicated by an index on the carrier, can be brought about by a milled-headed screw-clamp on the opposite side to that on which are situated the manipulating screw of the lower segments and the index just referred to.

The mode of using the apparatus is very simple. The crystal is attached, with the minimum of wax protruding at the sides, to the smallest of the special crystal-holders which are provided with azimuth adjustment, and with the plane which is desired to be the bisectrix of the 60°-prism arranged vertically as nearly parallel to the goniometrical axis as possible and parallel to the lower tangent screw. The latter can be accurately attained by use of the azimuth adjustment of the crystal-holder. The plane referred to is then exactly goniometrically adjusted with the aid of the two tangent screws, that is, in the case of the upper adjusting movement by slight rotation of the large segment about the upper fixed segment, the clamping screw being fixed, with the indicator on the carrier at zero. The screw is then unloosed and the slider, together with the lower segments carried by it, moved round 60° on one side, as indicated by the silver index on the carrier, which travels closely underneath the large divided arc. If the 60° are not conveniently attained by the hand movement of the carrier, the difference can be nicely made up by movement of the large segment about the upper arc by means of its tangent screw. If the original position of the upper segment had been noted on the silver scale it could readily be re-attained after the grinding of the first surface. When the latter has been achieved, the lower part of the apparatus is transferred to a position 60° on the other side of the centre, by a total sliding underneath the large segment of 120°, and the second surface is ground.

Another smaller but very useful addition is a spring-clutch to keep up the counterbalanced gun-metal axis, the apparatus for varying the pressure of the crystal on the grinding lap, when the left hand is removed from that one of the counterpoising levers which it manipulates during grinding. In order to prevent this axis from moving during the adjustment of the crystal, a hooked spring-clutch is arranged about the bearing of the lever, in such a manner that when it is pushed over into position, it is maintained there by the force of a spring. It consists of an arm rotatable about an axis screwed into the lower part of one side of the bearing, and carrying a short horizontal bar attached at its outer end which presses down on the outer arm of the counterpoising lever when in position; a strong spring fixed to the base of the bearing and which has to be overcome on moving over the little arm, presses up against an angle of the latter below the axis in such a manner as to keep the bar of the clutch, after being brought over into position, firmly down on the counterpoising lever. The lever is thus fixed with its elbow resting on the circle plate, and its other arm carrying the knife edge is maintained pressed up against the collar of the gun-metal axis, and thus the axis and the crystal which it carries at its lower extremity is unable to fall out of position during the adjustment.

The remaining improvements are two additions to the telescope of the goniometer, due to the suggestion of the author's friend Professor H. A. MIERS, who had already had such additions made to the telescope of an inverted goniometer constructed for him by the same firm, Messrs. Troughton and Simms, and which was intended for the study of the vicinal faces of crystals while in the act of growth in a cell of mother liquor. As the author's cutting and grinding goniometer forms a most excellent inverted goniometer, it was considered advisable to adopt these additions. A rectangular cell with truly plane glass sides is also included to contain the saturated solution employed in such investigations.

The first of the two consists of a combined goniometer- and micrometer-eyepiece, which provides two fixed spider-lines arranged at 90°, one vertical and adjusted exactly parallel to the vertical axis of the goniometer and the other horizontal, both being diameters of the circular field; and also a third spider-line which is both rotatable and capable of movement perpendicular to itself in the focal plane. The fixed lines are attached in the central aperture of a circle-plate 7 centims. diameter fixed round the optical tube of the eyepiece, and which carries near its periphery a circle divided directly into degrees. The movable spider-line is carried in the aperture of a micrometer box carried in front of a similar circular plate, which latter is fitted closely to the former plate in front of it, and partly enveloping it with a milled flange in such a manner as to be rotatable about it. This front plate is pierced by a window above the micrometer box in such a position that the divided circle of the fixed plate is visible through it. The inner edge of the window is bevelled and carries a vernier, with the aid of which the circle reads to minutes. The movable line is fixed to the front of the traversing frame of the micrometer, at the focus of the double eyepiece which slides in the short portion of the optical tube in front of the box. The fixed lines are brought into the focal plane and almost into contact with the movable one by means of a relatively thick annulus capable of penetrating the traversing frame and attached to the aperture of the fixed circle. The traverse of the frame and its spider line is recorded by a divided drum of the usual kind at the right-hand side of the box; the drum is divided into 100 parts, and the reading is indicated by an index mark fixed alongside. The movable circle and the spider line which it carries can be clamped to the fixed circle and the stationary spider lines when desired, by means of a suitable clamping screw provided with milled head, on the lower part of the periphery opposite to that near which the window is situated. These arrangements enable small movements of the image of the collimator signal, reflected from a crystal surface during growth and due to disturbance of the thermal or other conditions of the solution, to be followed and measured, whether they are lateral, angular, or both.

The other addition to the telescope is that of so arranging the removable lens, usually added to the telescopes of goniometers outside the objective for the purpose of converting the optical system into that of a low power microscope focussing the crystal, as to make it capable of being thrown into position either as usual behind the objective or in front of the eyepiece, and further of making it capable of travelling for some distance along the optical axis. The purpose of this is to enable the image of the signal to be actually followed right up to the image of the crystal itself, in order to be quite certain as to the particular face from which it emanates. This is achieved by supplying two such lenses, mounting the pair on a T-piece, and hinging the stem about a small platform carried above a short tube sliding round the main optical tube and prevented from rotation by a suitable rib and groove. It is only necessary to swing the T-piece over one way or the other for the lens to fall into position either adjoining the objective or the eyepiece, the length of the end cross-piece carrying the lenses being arranged so that either lens falls exactly into the optical axis. The one which falls behind the objective is generally employed close up to the latter in the usual manner. The other one which falls in front of the eyepiece is the one employed to trace an image; it is of such focal length that when it is close up to the eyepiece it permits the image of the signal to be seen almost as well as when it is absent, while as it is drawn more and more in front of and away from the eyepiece it causes the image to pass gradually into that of the particular reflecting face of the crystal itself. The latter is clearly focussed when the sliding tube has been drawn forward to the full extent of its path, and the face affording the signal image is seen brightly illuminated, as well possibly as other vicinal faces, from which it is distinguished by the tracing process just indicated.

The rest of the arrangements of the instrument are precisely as described in the former memoir (*loc. cit.*).

Procedure in Cutting and Grinding the Crystal-blocks.

In selecting crystals from which to prepare a parallel-faced block, those were naturally chosen which were free from traces of turbidity and from cracks and distortions. Crystals of cæsium sulphate are readily obtained perfectly free from turbidity; in the case of rubidium sulphate only very slow growth *in vacuo* yields crystals satisfactory in this respect. The exceptional crystals of the potassium salt, eventually obtained after so much trouble, as has been referred to, were also satisfactory from this point of view.

After removal from the mother liquor, the crystals were carefully dried, and then immediately stored in a desiccator for several days at least before use. With two exceptions each selected crystal was only employed for duplicate determinations, on two successive days, of the linear thermal expansion or contraction along some one particular axial direction. In all, 29 different crystals were employed, 11 of potassium sulphate, 8 of rubidium sulphate, and 10 of cæsium sulphate. The two exceptions were crystals of the rubidium and cæsium salts, the former of which was a particularly fine specimen elongated along one axial direction, and which, when cut in two halves transversely to this direction yielded portions so large that they were separately employed for determinations in two different axial directions; the crystal of cæsium sulphate was cut and ground into a rectangular block for successive determinations in all three axial directions, so as to afford an instance of all three linear values, and from these the value for the cubic expansion, being derived from one and the same crystal, for comparison of the cubic deformation thus obtained with that derived by calculation from measurements of the three linear expansions or contractions exhibited by different crystals. The results were so nearly identical, and the comparison therefore so satisfactory, that there will be no occasion to further refer to this point.

The orientation of the various faces, and the consequential identification of the axial directions of the crystals, was usually an easy matter, as the author was familiar with the salts in question owing to the exhaustive morphological and optical study already made. The identification of the axial directions is such a vital matter, as will be abundantly evident when the results are discussed, that it should be stated that the whole of the work, in common with all the author's previous crystallographical investigations, has been carried out exclusively by the author personally. In every case, the axial directions were actually verified both by goniometrical measurements and by examination of the interference figures in convergent polarised light. Immersion in a cell of benzene, on the inverted goniometrical polariscope, materially facilitates the latter verification, as the refractive index of that liquid is not far removed from the mean of the indices of the three salts, and the interference figures are consequently very clear, and the apparent optic axial angle is very nearly the true angle. Oil cannot be used, as it is apt to penetrate into any minute cavities in the surfaces, and to ooze out during the thermal observations in drops too small to be noticeable without a lens, but which are sufficient to entirely derange the interference bands by lifting the compensator by an amount which is very appreciable in observations of such delicacy.

For the same reason oil cannot be used in grinding the surfaces, and recourse was again made to benzene, which by its volatility rapidly removes itself from cavities. It is naturally unavoidable that greater quantities require to be used than of oil, as it so rapidly dries away. Hence a dropping funnel was arranged above the cutting or grinding disc, to deliver drops sufficiently fast to continually provide adequate lubrication.

The crystal, after verification of the axial direction along which it was desired to determine the linear deformation, was mounted in the grip-holder, in the manner already described, with the axial direction in question approximately vertical, parallel to the goniometrical axis and perpendicular to the cutting disc and grinding table. The approximation was then converted into absolute adjustment, by goniometrical observation and adjustment of the natural zone of faces parallel to the axis in question. If the crystal were so terminated below that much grinding would be necessary to produce the required surface, the cutter was first brought into requisition and the lower end cut off, at such a distance as to afford a surface of the required extent with the least sacrifice of thickness in the axial direction adjusted. The cutting of these crystals of artificial salts, although they are so much more friable than mineral crystals, is nevertheless most successfully performed by the new instrument. No crystal has yet been broken in the process. If no cutting were required the rough grinding of the surface was carried out on the emery-cloth lap, at first with the crystal-holder and lap detached from the instrument and the former held in the hand, and then, after a rough approximation to the desired surface had been attained, with the crystal and its holder and the lap in position. Finally, after verification of the adjustment, which, owing to the mode of fitting of the holder on to the suspended adjusting apparatus, was usually unimpaired, the surface was finely ground with one of the ground-glass laps.

As the method of using the aluminium compensator above the crystal was always adopted in arranging the interference apparatus of the dilatometer, the second of the methods described in the memoir concerning the latter and which is illustrated by a special figure in the German translation of that memoir contributed to the 'Zeitschrift für Krystallographie' (30, 530), there was no necessity to polish the surfaces of the crystal-blocks. For a crystal surface is not required to act as the lower reflecting surface involved in the generation of the interference bands, the upper surface of the compensator performing that function. It was therefore only necessary to complete the block by preparing a similar parallel surface in the same manner, separated from the first one by as much thickness of crystal as the particular specimen admitted of. The crystal-block was then cleaned from crystal dust by washing in benzene, dried with a clean linen cloth free from fluff, and stored in a desiccator until required for the observations. The thickness of the blocks employed varied, as will subsequently be seen from the record of the accurate measurements, from 4·8 to 10·7 millims., the former limit being in the case of the only crystal under 5 millims. in thickness. The great majority were from 7 to 9 millims. thick.

In two or three cases, although only benzene had been used in the treatment of the crystals, the observations of expansion were vitiated by the oozing of minute traces of liquid, which proved to be mother-liquor, between the surfaces of the platinum-iridium tripod table and the crystal, or between the latter and the compensator. For in most cases the three point method of contact was impossible, owing to the prepared crystal surfaces being narrower in one direction than in the other, too narrow to take the third point but not too narrow for stable equilibrium of both crystal and compensator. A comparison of the results for the same direction by the two methods shows, however, no appreciable difference, the surfaces having always been absolutely clean and free from dust. Moreover, the surfaces produced by the author's cutting and grinding goniometer are so absolutely plane, that no rolling, due to slight convexity of surface, has ever been observed with them. Further, the placing of the crystal and compensator in position on the table of the tripod was always effected by sliding, to minimise any intervening compressed air film. In the cases of oozing of mother-liquor referred to, the crystals were subsequently heated slowly to 105° in an air-bath, and maintained at this temperature for twelve hours. On repeating the observations of expansion no further disturbance occurred, successful determinations being obtained, and the results agreed satisfactorily with those obtained for the same direction of the same salt in cases where this treatment had been unnecessary.

The Determinations of Linear Deformation.

Mode of Conducting the Observations.

The determinations of thermal expansion or contraction were made in the manner which is very briefly outlined for crystals at the close of the memoir concerning the dilatometer (*loc. cit.*, lower part of p. 363 and p. 364), after the description and communication of the results of the determinations of the expansion of the platinum-iridium alloy of the interference tripod and of the aluminium of the compensators. The temperatures employed have not been quite so high as in the cases of those metallic substances, the highest limit being in the neighbourhood of 96°, in order that there might be no appreciable deformation due to internal strain, provoked by the attempted vaporisation of the water of mother-liquor contained in the inevitable minute internal cavities. It is impossible to altogether prevent the formation of such cavities, even by slow evaporation *in vacuo*, but the remarkable agreement of the results obtained indicates that any variable deformation due to this cause has been infinitesimal.

Every effort has been made to render the conditions of the determinations as rigidly analogous as possible, so that comparisons of the results can be made with confidence. As far as possible the same aluminium compensator has been used throughout, namely, one 5·25 millims. thick and a centimetre diameter, and unprovided with points as the three-point method of contact was so rarely available; where exceptions have been made, results with the compensator mentioned are available for the same direction of the same salt, and the two series of results agree so well that the change has evidently not introduced any error. This, of course, should be so, for the compensators, including the one 12 millims. thick used for the determination of the expansion of the metal, were all cut from the same casting of pure aluminium. In most of the exceptional cases the other compensators were provided with points, and the three-point method was used, and afforded the results which have already been stated to accord with those where points were not used. The 5·25 compensator gives excellent interference bands, particularly from one of the two surfaces, which was marked and invariably used. The bands afforded by it were slightly curved, due to infinitesimal convexity, an additional advantage as it was always possible, by noting whether they moved outwards from or inwards towards the centre of curvature, to at once ascertain whether the movement of the bands were due to expansion or to contraction. There is a further advantage in employing the compensator above rather than below the crystal, namely, that the polished surface of aluminium reflects light almost equally with the other surface involved in the production of the bands, the lower surface of the large cover-glass which is laid on the platinum-iridium tripod screws and which bears about its centre the miniature silver ring whose centre is the point of reference for the micrometric measurement of the position of the bands.

The air-film between the two reflecting surfaces was in nearly all cases very thin; it was not found advisable to strive so much for exact compensation for the expansion of the screws as to produce the most brilliant bands. For the correction for non-compensation is of course in all cases accurately determined from the known expansion of the tripod alloy and aluminium. The screw-length corresponding to 5·25 millims.

of aluminium is about 13·77 millims. which leaves room for a crystal 8·37 millims. thick and an air film of 0·15 millim. As the crystals only usually varied 1 or 2 millims. each side of the thickness mentioned, the amount of under or over compensation was never very large.

The same thermometers have been employed as were fully described in the previous memoir. Their fixed points were carefully redetermined after the completion of the determinations. The inner bent thermometer whose bulb was in contact with the tripod and whose indications were those accepted, was found to have altered only to the extent of 0°·1, the indications at 0° and 100° in ice and steam, after applying the pressure correction for the latter, being 0°·1 and 100°·1 respectively. Hence the interval had remained unchanged, and as only differences of the temperatures are employed in calculating the coefficients of expansion, no correction of these latter is required for change of interval.

The usual *modus operandi* was to expend the greater part of three days in carrying out a duplicate pair of determinations, of the linear thermal deformation of any one crystal along the direction perpendicular to the two prepared parallel surfaces. The afternoon of the first day was employed in adjusting the crystal and the whole apparatus so as to afford a suitable field of interference bands. Each of the two succeeding days was utilised for the carrying out of a complete series of observations of the position and transit of bands for two intervals of temperature, the operations on each day occupying 5 to 7 hours, during the whole of which time the author followed the bands without intermission. Naturally, the carrying out of sixty-four such observations has proved very trying and fatiguing, the observer being continually afraid of such highly delicate measurements being vitiated by earth tremors due to street traffic or other disturbance, in spite of the rigid mounting of the apparatus on a slate table. Fortunately, this fear has not proved to have had much foundation, as the author's laboratory is happily situated in an exceptionally quiet part of Oxford well removed from the city and the railway. But the experience has shown that the observations would have been far more difficult, if not impossible, in a large city with a network of underground railways such as London. Although this source of disturbance has been minimised, several observations have been lost, generally after spending hours upon them, by the cracking of the crystal under the influence of the rise of temperature, slow as it always was in order to avoid this catastrophe.

The further experience gained during this work indicates that in the case of crystals the Abbe method, of calculating the number of bands which pass the point of reference between two temperatures from initial and final observations of the positions of the bands nearest the reference point, for two wave-lengths, is generally inapplicable. The only guarantee that the observation has been a trustworthy one, that no disturbance due to any of the causes already referred to has occurred, is obtained by carefully following the bands for the whole of the temperature-interval, and observing that they maintain their regular distances and exhibit no appreciable twisting round the centre or any other irregularity, throughout the whole of the interval of time. Frequently slight cracking of the crystal is accompanied by widening or narrowing of the bands during one part of the observation and movement in the inverse direction during the other, or possibly by twisting for a whole revolution, and very frequently by merely jumping several bands, the appearance at the end being much the same as at the beginning. Such an observation is, of course, valueless, but the Abbe method would not detect this fact. Undoubtedly, the author's method, although very fatiguing, is the only one which is trustworthy when fragile substances are under investigation.

The counting of the bands was achieved precisely as described in the dilatometer memoir (p. 348) with the aid of the tape-puncturing recorder, the induction coil which illuminated the hydrogen Geissler tube being actuated at sufficiently rapidly succeeding intervals to enable the author to observe the passage of at least every quarter of a band. Timing the transit with the watch is an excellent aid, as, if the observation is trustworthy, there should be no sudden changes of rapidity in the movement of the bands. When the Fletcher ring-gas-burner below the double air-bath is first ignited, the bands move very slowly, the rapidity then growing with a regular increment until it reaches, in the case of large expansions where at least forty bands pass during the interval of 45° of temperature, a maximum of two bands per minute; the rapidity then as gradually diminishes until, with the attainment of constancy at the higher limit for that particular interval, the bands cease to move altogether. Moreover, if the temperature recorded by the inner bent thermometer, whose bulb is tied to and in contact with the platinum-iridium tripod, shows any slight tendency to descend a fraction of a degree, the bands should immediately begin to retrace their steps to a corresponding extent. No observation has been accepted during which these conditions were not fulfilled.

The temperature limits employed were respectively the ordinary temperature, obtained as low as possible by commencing work about 7 A.M., the neighbourhood of 56°, and that of 96°. The determinations of the positions at these temperatures, of the two bands nearest to the reference point, were made precisely as described in the previous memoir (p. 346). The monochromatic light employed throughout was red hydrogen light, corresponding to the C line of the solar and hydrogen spectrum, separated from all other radiations by a train of prisms in the manner described in the dilatometer memoir (pp. 322 and 342). The wave-length of this radiation employed in the calculations was 0·0006562 millim.

When adequate time for complete cooling had elapsed, after the second series of observations, the measurement of the thickness of the crystal and the length of the tripod screws was made, by means of the thickness measurer described on p. 337 of the former memoir. For this purpose the interference chamber was carefully raised out of the bath by means of the rackwork on the pedestal, and the tripod, together with the supported crystal and compensator, after cutting the thread binding the thermometer to it and gently drawing the latter aside, was removed from the chamber and transferred to the thickness measurer. The greatest care was taken not to disturb the positions of the crystal and compensator on the table of the tripod. The large cover-glass (cover-wedge of the previous memoir) was too large to be taken through one of the windows of the chamber along with the tripod, so was left inside, being raised with the left hand while the tripod was removed with the right to the nearest resting place, the top of the air-bath ; the cover-wedge was then turned over so as not to injure the silver reference ring and laid on the floor of the chamber. The tripod was then removed with both hands to such a position on the glass base of the thickness measurer that the agate pointed end of the measuring bar would fall exactly on the centre of the compensator, over which the silver reference ring of the cover-wedge had been situated during the observations. The height of the plane of the tops of the three tripod screws at this point was then first determined by laying on the screws a large circular disc of glass similar to the cover-wedge, and whose surfaces were truly plane and the thickness of which had previously been repeatedly determined at a position near the centre which was conveniently indicated by a small internal bubble. The disc was laid so that the bubble was over the centre of the compensator. The measurement was then made by lowering the counterpoised bar into gentle contact with the top of the disc and reading the scale with the aid of the micrometer. This height, *minus* the known thickness of the disc, gave the height of the plane of the tops of the screws. The disc was then removed and the bar lowered down upon the compensator, and the height again noted. The difference between this and the height of the screws gave the thickness of the air film. The bar was again raised and the compensator next removed, without disturbing the crystal, a matter requiring some nicety of manipulation with a pair of small ivory-tipped forceps ; the bar was then allowed to fall gently on the crystal, when another measurement was taken. The difference of this and the last was of course the thickness of the compensator, as nearly as possible 5·250. It then only remained to once more raise the bar, remove the crystal, allow the bar to fall on to the table of the tripod, and take a final measurement of the height of this. The difference between this reading and the previous one afforded the measure of the thickness of the crystal. In cases where the three-point method was employed the only difference was to determine the mean height of the three particular table points used, with the aid of a small disc of glass, of known thickness at the centre, and placed on the same points, instead of determining the height of the surface of the table itself. The length of the screws was evidently afforded by subtracting from the height of the screws the height either of the table itself or of the points, according to the method of supporting the crystal employed.

The four desired basal quantities, L_{t_1} the thickness of the crystal, l_a the thickness of the compensator, l the length of the screws, and d the thickness of the air-film, were thus determined exactly along the vertical line passing through the centre of reference of the interference band observations, and so any error due to minute lack of parallelism of the surfaces involved was obviated.

An example taken at random from the actual measurements will render the process quite clear. It refers to the fourth crystal of cæsium sulphate along the direction of the morphological axis *b*.

	millims.		millims.
Height of top of glass disc	40·857		
Known thickness of glass disc . . .	6·117		
Height of screws	34·740	$d =$	0·145
„ top of compensator . . .	34·595	$l_a =$	5·253
„ „ crystal	29·342	$L_{t_1} =$	8·379
„ „ tripod table. . . .	20·963	$l =$	13·777

The Nature of the Problem with reference to the Crystallographical Symmetry.

The symmetry of the three salts under investigation being orthorhombic, the three axes of the thermal ellipsoid coincide in direction in each case with the crystallographical axes, just as do the axes of the optical ellipsoid already fully elucidated in a previous memoir. The amounts of thermal deformation along these three axial directions should not, from general considerations, be equal, as in crystals belonging to the cubic system, nor even would any two of them be likely to exhibit the same amount of expansion, as in the case of crystals exhibiting tetragonal or hexagonal symmetry. Orthorhombic symmetry requires that if a sphere of the substance of any one of these crystallised salts could be procured at any specific temperature, at any other temperature such sphere would have become converted into an ellipsoid with three unequal axes, and that these axes would coincide in direction with the three rectangular crystallographical axes. One of these morphological axes would thus be the direction of maximum expansion or contraction, another that of minimum and the remaining one that of intermediate deformation. The problem of the determination of the nature and amount of this thermal deformation consequently resolves itself into the determination of the amount of linear thermal expansion or contraction along the respective directions of the three morphological axes. From these fundamental data can be calculated the cubical expansion, in other words, the difference in volume between the sphere of unit radius and the deformation ellipsoid produced therefrom as the effect of change of temperature.

The Determinations and Computations.

The work has thus consisted in the determination of nine quantities, namely, the linear coefficients of thermal expansion or contraction along each of the three crystallographical axes of each of the three salts. It may be at once stated that in no case has contraction been observed, expansion in every direction having been found to be the invariable rule with regard to all three sulphates. Every one of the nine quantities has been determined at least six independent times on three different crystals, and with respect to five of the quantities eight determinations have been made on four separate crystals. In all sixty-four independent determinations have been carried out, on different days, and using the twenty-nine different crystals of which the details have already been given (p. 465).

Each determination afforded, as already fully explained in connection with the determinations of the expansion of platinum-iridium and aluminium in the dilatometer memoir (*loc. cit.* p. 352), the two constants required for a complete statement of the thermal behaviour, namely, the constant a, the coefficient of expansion at 0°, and b, half the increment of the coefficient *per* degree of temperature, the coefficient not being a fixed quantity for all temperatures but varying regularly with the temperature. The coefficient of thermal expansion is signified by α, and the expression for the actual coefficient at any temperature t, as also for the mean coefficient between any two temperatures whose mean is t, is:

$$\alpha = a + 2bt.$$

The mean coefficient of expansion between 0° and t° is, however

$$a + bt.$$

The data afforded by observations of the positions of the interference bands at three adequately separated temperatures, and of the number of bands passing the reference point during the intervals between these temperatures, together with a knowledge of the original thicknesses of the block of crystal and of the aluminium compensator, and the length of the platinum-iridium screws projecting above the tripod table or its raised points, are ample to enable the two constants a and b to be calculated. For it is only necessary to insert respectively in three equations of the form

$$L_t = L_0 (1 + at + bt^2)$$

the known values of the three temperatures and the lengths (thicknesses) of the crystal block at those temperatures, and to solve the three equations thus provided, for the three unknown quantities L_0, a, and b.

The solution of these equations furnishes expressions for the three required quantities of the forms

$$a = \frac{\theta}{L_0}, \quad b = \frac{\phi}{L_0}, \quad \text{and} \quad L_0 = L_{t_1} - \theta t_1 - \phi t_1^2,$$

in which θ and ϕ are terms involving the differences of the lengths, L_{t_1}, L_{t_2}, L_{t_3}, at the three temperatures t_1, t_2, and t_3, and the sums and differences of those temperatures.

The actual expressions for θ and ϕ employed throughout the observations were:

$$\theta = \frac{(t_1 + t_3)(L_{t_2} - L_{t_1})}{(t_2 - t_1)(t_3 - t_2)} - \frac{(t_1 + t_2)(L_{t_3} - L_{t_1})}{(t_3 - t_1)(t_3 - t_2)},$$

$$\phi = \frac{L_{t_3} - L_{t_1}}{(t_3 - t_1)(t_3 - t_2)} - \frac{L_{t_2} - L_{t_1}}{(t_2 - t_1)(t_3 - t_2)}.$$

The results of the determinations of linear thermal expansion are presented in the next section in tabular form. Each table represents the results for one axial direction of a particular salt, and is divided into three portions. In the first portion is given the essential experimental data afforded by the observations and measurements. L_{t_1}, is the measured thickness of the crystal block, l_a that of the compensator, l the length of the platinum-iridium screws, and d the thickness of the air-film, each measured in the manner described on p. 471. Next come the temperatures, and subsequently the corresponding barometric pressures. The next column contains f_2, the number of interference bands which effected their transit past the reference point during the interval between t_1 and t_2. In the succeeding column is given the small correction to be applied to the number of bands, rendered necessary by the alteration in the wave-length of the monochromatic light employed, which accompanies the change in the refraction of air consequent on the considerable rise of temperature and possible alteration of pressure. The nature and amount of this correction were fully discussed in the previous memoir (*loc. cit.* p. 350), and the formula for it there given was invariably followed. The barometric pressures and d are essential terms of that formula. The corrected number of bands, f_2', is given in the next column, and the three remaining columns contain f_3, the number of bands for the temperature interval between t_1 and t_3, its correction, and f_3' the corrected number for that interval.

In the second portion are given, in the first two columns, the calculated values of the apparent expansion, obtained by multiplying the corrected number of bands by half the wave-length of the red C hydrogen light employed, 0·0003281 millim., according to the fundamental principle of the method; in the next six columns the calculated quantities involved in the correction to be applied to the apparent expansion for lack of compensation are recorded; and in the last two columns the actual expansion of the crystal obtained by use of the correction. For a fuller discussion of the principle of the method as touching the first two columns, the memoir concerning the dilatometer may be referred to; it need only be remarked here that the transit of each band past the reference spot corresponds to an alteration in the thickness of the air-film d, between the compensator and the cover-wedge at the position of the reference spot, equal to half a wave-length of the monochromatic light employed. The determination of the correction for non-compensation involves the calculation of the actual expansion of the platinum-iridium screws, which is given in the third and fourth columns for the two respective temperature-intervals, and of the aluminium compensator, which is given in the fifth and sixth columns. These values were calculated with the aid of the coefficients of linear expansion of the two metals, as previously determined with the greatest care by the author and published in the memoir concerning the dilatometer (pp. 356 and 360). The following were the actual expressions used, l and l_a being the values given in the fourth and third columns respectively of the first portion of the table

For the screws: $l\left[10^{-9}\left(8600 + 4\cdot56\frac{t_1 + t_2}{2}\right)\right](t_2 - t_1)$ for the first interval, and similarly for the second interval, substituting t_3 for t_2.

For the compensator: $l_a\left[10^{-8}\left(2204 + 2\cdot12\frac{t_1 + t_2}{2}\right)(t_2 - t_1)\right.$ for the first interval, and a like expression for the second interval with t_3 substituted for t_2.

In each case the actual expansion of the metal is thus calculated by multiplying the length (thickness) of the metal by the mean coefficient of the linear expansion between the two temperatures, that is by $a + 2bt$ for that metal where t is the mean of the limiting temperatures of the interval, namely $\frac{1}{2}(t_1 + t_2)$ or $\frac{1}{2}(t_1 + t_3)$; and also by the amount of the temperature interval, that is, $t_2 - t_1$ or $t_3 - t_1$. Actually, of course, one uses $b(t_1 + t_2)$ instead of $2b \cdot \frac{1}{2}(t_1 + t_2)$.

The differences between the amounts of expansion of the screws and the compensator are given in the next two columns headed "correction for non-compensation." The correction is obviously positive, given an expanding crystal, when the screws expand most, and negative when the compensator expands to the greater extent. For in the former case the effect is to increase the thickness of the air-film, and consequently the amount of diminution of the thickness of the air-film due to the expansion of the crystal is not fully evident, the actually observed amount being less than that really effected by the expanding crystal by the amount of this excess of expansion on the part of the screws. This latter amount should, therefore, be added. The inverse is the case when the excess is on the part of the compensator; causing, as it does, additional diminution of the thickness of the air-film, it should be subtracted. The values given in the last two columns, representing the actual expansions of the crystal during the two intervals of temperature, $L_{t_2} - L_{t_1}$ and $L_{t_3} - L_{t_1}$, were obtained by applying the correction for non-compensation, in the sense just indicated, to the apparent expansions $f_2'\lambda/2$ and $f_3'\lambda/2$.

In the last portion of the table are given the calculated values of θ, ϕ, and L_0, and of the two required constants of the coefficient of linear expansion, a, the coefficient at 0° and b, half the increment of the coefficient *per* degree of temperature. In the last column are given the values of the coefficient of linear expansion, α, for 50°, calculated by means of the formula $\alpha = a + 2bt$. FIZEAU invariably gave the coefficient at 40°, a specific temperature in the neighbourhood of the mean of the extreme limits employed by him, in addition to a and b. As 50° is nearer the mean of the author's limiting temperatures, this specific temperature has been chosen in preference, for which to record a particular calculated value of α.

THE RESULTS.

In the following tables are presented the results of the determinations and calculations.

Potassium Sulphate, Direction of Axis a.

Experimental Data.

Crystal.	L_{t_1}.	l_a.	l.	d.	t_1.	t_2.	t_3.	b_1.	b_2.	b_3.	f_2.	Corrn.	f_2'.	f_3.	Corrn.	f_3'.
	millims.	millims.	millims.	millims.	°	°	°	millims.	millims.	millims.						
1	10·355	5·237	15·786	0·194	17·2	56·4	96·6	766·5	765·0	763·5	43·90	– 0·02	43·88	90·42	– 0·04	90·38
					16·8	56·3	96·9	762·0	762·5	763·0	44·10	– ·02	44·08	91·26	– ·04	91·22
2	7·834	5·250	13·337	0·253	11·6	56·5	96·6	752·4	752·6	752·9	39·98	– ·03	39·95	77·32	– ·05	77·27
					11·6	56·5	96·4	757·3	758·0	758·8	39·94	– ·03	39·91	76·87	– ·05	76·82
3	8·814	5·254	14·223	0·155	13·2	56·7	96·8	753·0	754·0	755·0	43·36	– ·02	43·34	84·36	– ·03	84·33
					12·7	58·4	96·2	756·0	756·5	757·0	45·40	– ·02	45·38	84·37	– ·03	84·34
4	6·568	5·248	12·055	0·239	15·8	58·1	97·5	760·2	760·1	760·0	33·10	– ·03	33·07	65·10	– ·05	65·05
					12·9	56·8	96·3	759·5	759·4	759·3	34·61	– ·03	34·58	66·70	– ·05	66·65

Calculated Actual Expansions.

Diminution of thickness of air-layer.		Expansion of tripod screws.		Expansion of aluminium compensator.		Correction for non-compensation.		Expansion of crystal.	
$f_2' \lambda/2$.	$f_3' \lambda/2$.	1st interval.	2nd interval.	1st interval.	2nd interval.	1st interval.	2nd interval.	$L_{t_2} - L_{t_1}$.	$L_{t_3} - L_{t_1}$.
0·0143970	0·0296540	0·0054256	0·0110887	0·0046848	0·0096523	+0·0007408	+0·0014364	0·0151378	0·0310904
·0144630	·0299300	·0054665	·0112019	·0047195	·0097508	+ ·0007470	+ ·0014511	·0152100	·0313811
·0131078	·0253525	·0052430	·0100291	·0053656	·0103471	– ·0001226	– ·0003180	·0129852	·0250345
·0130945	·0252046	·0052430	·0100051	·0053656	·0103221	– ·0001226	– ·0003170	·0129719	·0248876
·0142200	·0276688	·0054194	·0105240	·0052065	·0101930	+ ·0002129	+ ·0003310	·0144329	·0279998
·0148892	·0276720	·0056953	·0105085	·0054731	·0101756	+ ·0002222	+ ·0003329	·0151114	·0280049
·0108503	·0213432	·0044714	·0087245	·0050664	·0099646	– ·0005950	– ·0012401	·0102553	·0201031
·0113459	·0218680	·0046353	·0088967	·0052478	·0101534	– ·0006125	– ·0012567	·0107334	·0206113

Calculated Coefficients of Linear Expansion.

θ.	ϕ.	L_0.	a.	b.	$\alpha_{50°}$.
0·000 375 23	0·000 000 148 5	10·3485	0·000 036 26	0·000 000 014 4	0·000 037 70
·000 365 51	·000 000 230 7	10·3488	·000 035 32	·000 000 022 3	·000 037 55
·000 280 17	·000 000 132 7	7·8307	·000 035 78	·000 000 016 9	·000 037 47
·000 281 08	·000 000 114 8	7·8307	·000 035 90	·000 000 014 7	·000 037 37
·000 326 32	·000 000 078 1	8·8097	·000 037 04	·000 000 008 9	·000 037 93
·000 321 78	·000 000 125 0	8·8099	·000 036 52	·000 000 014 2	·000 037 94
·000 235 66	·000 000 091 8	6·5643	·000 035 90	·000 000 014 0	·000 037 30
·000 239 85	·000 000 066 8	6·5649	·000 036 53	·000 000 010 2	·000 037 55
	Mean values		0·000 036 16	0·000 000 014 4	0·000 037 60

Potassium Sulphate, Direction of Axis b.

Experimental Data.

Crystal.	L_{t_1}.	l_a.	l.	d.	t_1.	t_2.	t_3.	b_1.	b_2.	b_3.	f_2.	Corrn.	f_2'.	f_3.	Corrn.	f_3'.
	millims.	millims.	millims.	millims.	°	°	°	millims.	millims.	millims.						
1	5·392	5·250	10·753	0·111	12·3	56·9	95·7	753·8	753·9	754·0	27·81	−0·01	27·80	53·02	−0·02	53·00
					13·1	56·9	95·9	754·7	754·8	754·9	27·28	−0·01	27·27	52·70	−0·02	52·68
2	5·931	5·253	11·322	0·138	8·5	57·0	96·2	764·8	764·7	764·6	31·69	−0·02	31·67	58·74	−0·03	58·71
					8·7	57·3	96·4	763·9	763·7	763·5	31·94	−0·02	31·92	58·77	−0·03	58·74
3	5·103	5·255	10·514	0·156	12·4	57·8	95·9	757·2	756·6	756·0	27·17	−0·02	27·15	50·87	−0·03	50·84
					12·7	57·1	95·9	747·0	746·0	745·0	26·52	−0·02	26·50	50·72	−0·03	50·69
4	5·210	5·250	10·603	0·143	10·0	57·6	96·0	735·0	734·5	734·0	29·57	−0·02	29·55	54·50	−0·03	54·47
					9·8	56·3	95·3	731·0	730·5	730·0	28·62	−0·02	28·60	53·53	−0·03	53·50

Calculated Actual Expansions.

Diminution of thickness of air-layer.		Expansion of tripod screws.		Expansion of aluminium compensator.		Correction for non-compensation.		Expansion of crystal.	
$f_2'\ \lambda/2$.	$f_3'\ \lambda/2$.	1st interval.	2nd interval.	1st interval.	2nd interval.	1st interval.	2nd interval.	$L_{t_2}-L_{t_1}$.	$L_{t_3}-L_{t_1}$.
0·0091212	0·0173896	0·0042001	0·0079334	0·0053325	0·0101517	−0·0011324	−0·0022183	0·0079888	0·0151713
·0089475	·0172846	·0041255	·0078783	·0052387	·0100828	− 0032	− ·0022045	·0078343	·0150801
·0103911	·0192628	·0048043	·0087763	·0057920	·0106650	− 697	− ·0018887	·0094034	·0173741
·0104730	·0192726	·0048149	·0087771	·0058055	·0106669	− ·0009906	− ·0018898	·0094824	·0173828
·0089080	·0166809	·0041816	·0077669	·0054358	·0101747	− ·0012542	− ·0024078	·0076538	·0142731
·0086948	·0166314	·0040889	·0077396	·0053150	·0101393	− ·0012261	− ·0023997	·0074687	·0142317
·0096955	·0178719	·0044182	·0080625	·0056871	·0104585	− ·0012689	− ·0023960	·0084266	·0154759
·0093838	·0175533	·0043144	·0080136	·0055516	·0103935	− ·0012372	− ·0023799	·0081466	·0151734

Calculated Coefficients of Linear Expansion.

θ.	ϕ.	L_0.	a.	b.	$\alpha_{50°}$.
0·000 174 15	0·000 000 071 7	5·3898	0·000 032 31	0·000 000 013 3	0·000 033 64
·000 173 02	·000 000 083 5	5·3897	·000 032 10	·000 000 015 5	·000 033 65
·000 186 83	·000 000 107 7	5·9294	·000 031 51	·000 000 018 2	·000 033 33
·000 189 89	·000 000 079 2	5·9293	·000 032 03	·000 000 013 4	·000 033 37
·000 164 25	·000 000 061 6	5·1010	·000 032 20	·000 000 012 1	·000 033 41
·000 163 11	·000 000 073 2	5·1009	·000 031 98	·000 000 014 4	·000 033 42
·000 171 87	·000 000 076 2	5·2083	·000 032 99	·000 000 014 6	·000 034 45
·000 171 36	·000 000 058 1	5·2083	·000 032 90	·000 000 011 2	·000 034 02
Mean values			0·000 032 25	0·000 000 014 1	0·000 033 66

Potassium Sulphate, Direction of Axis c.

Experimental Data.

Crystal.	L_{t_1}.	l_a.	l.	d.	t_1.	t_2.	t_3.	b_1.	b_2.	b_3.	f_2.	Corrn.	f_2'.	f_3.	Corrn.	f_3'.
	millims.	millims.	millims.	millims.	°	°	°	millims.	millims.	millims.						
1	6·871	5·251	12·304	0·182	12·8	57·1	96·6	735·1	736·3	737·6	38·00	−0·02	37·98	75·24	−0·04	75·20
					9·0	56·1	96·6	740·4	740·3	740·2	40·28	−0·02	40·26	78·28	−0·04	78·24
2	6·537	5·254	11·837	0·046	14·3	57·7	96·8	766·0	766·5	767·0	36·29	0·00	36·29	71·84	−0·01	71·83
					13·5	56·6	96·8	767·4	767·7	768·0	35·97	0·00	35·97	72·45	−0·01	72·44
3	9·642	5·250	15·023	0·131	13·3	57·7	97·3	761·9	762·5	763·1	49·16	−0·02	49·14	97·31	−0·03	97·28
					13·0	57·1	96·6	764·7	765·3	765·9	48·96	−0·02	48·94	96·96	−0·03	96·93

Calculated Actual Expansions.

Diminution of thickness of air-layer.		Expansion of tripod screws.		Expansion of aluminium compensator.		Correction for non-compensation.		Expansion of crystal.	
$f_2'\lambda/2$.	$f_3'\lambda/2$.	1st interval.	2nd interval.	1st interval.	2nd interval.	1st interval.	2nd interval.	$L_{t_2}-L_{t_1}$.	$L_{t_3}-L_{t_1}$.
0·0124615	0·0246734	0·0047744	0·0091243	0·0052992	0·0102087	−0·0005248	−0·0010844	0·0119367	0·0235890
·0132093	·0256708	·0050699	·0095288	·0056216	·0106527	− ·0005517	− ·0011239	·0126576	·0245469
·0119069	·0235679	·0045024	·0086457	·0051996	·0100638	− ·0006972	− ·0014181	·0112097	·0221498
·0118019	·0237679	·0044690	·0087279	·0051592	·0101578	− ·0006902	− ·0014299	·0111117	·0223380
·0161232	·0319176	·0058444	·0111709	·0053130	·0102365	+ ·0005314	+ ·0009344	·0166546	·0328520
·0160572	·0318031	·0058035	·0111150	·0052748	·0101836	+ ·0005287	+ ·0009314	·0165859	·0327345

Calculated Coefficients of Linear Expansion.

θ.	ϕ.	L_0.	a.	b.	$\alpha_{50°}$.
0·000 248 14	0·000 000 304 9	6·8678	0·000 036 13	0·000 000 044 4	0·000 040 57
·000 250 28	·000 000 283 5	6·8687	·000 036 44	·000 000 041 3	·000 040 57
·000 239 50	·000 000 260 9	6·5335	·000 036 66	·000 000 039 9	·000 040 65
·000 239 77	·000 000 257 4	6·5337	·000 036 70	·000 000 039 4	·000 040 64
·000 346 44	·000 000 403 8	9·6373	·000 035 95	·000 000 041 9	·000 040 14
·000 348 64	·000 000 391 5	9·6374	·000 036 18	·000 000 040 6	·000 040 24
Mean values			0·000 036 34	0·000 000 041 3	0·000 040 47

Rubidium Sulphate, Direction of Axis a.

Experimental Data.

Crystal.	L_{t_1}.	l_a.	l.	d.	t_1.	t_2.	t_3.	b_1.	b_2.	b_3.	f_2.	Corrn.	f_2'.	f_3.	Corrn.	f_3'.
	millims.	millims.	millims.	millims.	°	°	°	millims.	millims.	millims.						
1	7·824	5·253	13·751	0·674	13·0	56·3	97·3	749·0	749·4	749·8	39·24	−0·09	39·15	78·24	−0·16	78·08
					12·9	56·2	97·4	752·2	752·7	753·1	39·15	−0·09	39·06	78·33	−0·16	78·17
2	9·217	5·252	14·722	0·253	17·7	56·9	96·4	758·6	758·9	759·2	40·48	−0·03	40·45	83·06	−0·05	83·01
					17·9	56·6	96·0	757·4	757·6	757·8	40·55	−0·03	40·52	84·13	−0·05	84·08
3	9·918	5·253	15·306	0·135	13·1	56·8	96·1	759·0	758·9	758·8	47·58	−0·02	47·56	92·26	−0·03	92·23
					12·3	57·0	95·7	757·9	757·5	757·3	48·66	−0·02	48·64	92·68	−0·03	92·65

Calculated Actual Expansions.

Diminution of thickness of air-layer.		Expansion of tripod screws.		Expansion of aluminium compensator.		Correction for non-compensation.		Expansion of crystal.	
$f_2' \lambda/2$.	$f_3' \lambda/2$.	1st interval.	2nd interval.	1st interval.	2nd interval.	1st interval.	2nd interval.	$L_{t_2} - L_{t_1}$.	$L_{t_3} - L_{t_1}$.
0·0128450	0·0256180	0·0052147	0·0102610	0·0051804	0·0102780	+0·0000343	−0·0000170	0·0128793	0·0256010
·0128160	·0257000	·0052264	·0103010	·0051916	·0103140	+ ·0000348	− ·0000130	·0128508	·0256870
·0132718	·0272358	·0050613	·0102655	·0047003	·0096093	+ ·0003610	+ ·0006562	·0136328	·0278920
·0132947	·0275867	·0049966	·0101869	·0046402	·0095354	+ ·0003564	+ ·0006515	·0136511	·0282382
·0156045	·0302608	·0058615	·0112417	·0052319	·0101144	+ ·0006296	+ ·0011273	·0162341	·0313881
·0159588	·0304089	·0059920	·0112925	·0053479	·0101137	+ ·0006441	+ ·0011788	·0166029	·0315877

Calculated Coefficients of Linear Expansion.

θ.	ϕ.	L_0.	a.	b.	$a_{50°}$.
0·000 286 84	0·000 000 152 9	7·8202	0·000 036 68	0·000 000 019 6	0·000 038 64
·000 283 48	·000 000 182 6	7·8203	·000 036 25	·000 000 023 4	·000 038 59
·000 335 23	·000 000 168 2	9·2110	·000 036 39	·000 000 018 3	·000 038 22
·000 336 04	·000 000 224 0	9·2109	·000 036 48	·000 000 024 3	·000 038 91
·000 359 60	·000 000 170 0	9·9133	·000 036 27	·000 000 017 1	·000 037 98
·000 358 31	·000 000 189 3	9·9136	·000 036 14	·000 000 019 1	·000 038 05
Mean values			0·000 036 37	0·000 000 020 3	0·000 038 40

Rubidium Sulphate, Direction of Axis b.

Experimental Data.

Crystal.	L_{t_1}.	l_a.	l.	d.	t_1.	t_2.	t_3.	b_1.	b_2.	b_3.	f_2.	Corrn.	f_2'.	f_3.	Corrn.	f_3'.
	millims.	millims.	millims.	millims.	°	°	°	millims.	millims.	millims.						
1	6·324	5·255	11·756	0·177	13·0	56·7	96·6	763·1	763·4	763·6	30·62	−0·02	30·60	60·10	−0·04	60·06
					12·5	56·6	96·4	764·2	764·4	764·6	30·18	−0·02	30·16	58·94	−0·04	58·90
2	7·697	5·250	13·050	0·103	14·9	58·4	96·4	753·5	753·3	753·1	35·33	−0·01	35·32	68·51	−0·02	68·49
					15·1	57·2	96·3	757·7	758·5	759·3	33·85	−0·01	33·84	66·66	−0·02	66·64
3	8·027	5·255	13·398	0·116	13·9	57·4	97·9	762·0	761·7	761·3	35·54	−0·01	35·53	70·13	−0·02	70·11
					13·7	57·3	96·8	760·0	759·7	759·3	35·97	−0·01	35·96	69·88	−0·02	69·86

Calculated Actual Expansions.

Diminution of thickness of air layer.		Expansion of tripod screws.		Expansion of aluminium compensator.		Correction for non-compensation.		Expansion of crystal.	
$f_2' \lambda/2$.	$f_3' \lambda/2$.	1st interval.	2nd interval.	1st interval.	2nd interval.	1st interval.	2nd interval.	$L_{t_2} - L_{t_1}$.	$L_{t_3} - L_{t_1}$.
0·0100399	0·0197061	0·0044998	0·0086978	0·0052309	0·0101932	−0·0007311	−·0014954	0·0093088	0·0182107
·0098956	·0193255	·0045403	·0087273	·0052773	·0102261	−·0007370	−·0014988	·0091586	·0178267
·0115886	·0224719	·0049746	·0094167	·0052109	·0099353	−·0002363	−·0005186	·0113523	·0219533
·0111030	·0218650	·0048154	·0093823	·0050407	·0098992	−·0002253	−·0005169	·0108777	·0213481
·0116576	·0230033	·0051069	·0099655	·0052110	·0102520	−·0001041	−·0002865	·0115535	·0227168
·0117986	·0229213	·0051184	·0098555	·0052223	·0101358	−·0001039	−·0002803	·0116947	·0226410

Calculated Coefficients of Linear Expansion.

θ.	ϕ.	L_0.	a.	b.	$a_{50°}$.
0·000 204 61	0·000 000 120 7	6·3213	0·000 032 37	0·000 000 019 1	0·000 034 28
·000 199 35	·000 000 120 5	6·3215	·000 031 54	·000 000 019 1	·000 033 45
·000 244 79	·000 000 220 9	7·6933	·000 031 82	·000 000 028 7	·000 034 69
·000 250 02	·000 000 115 7	7·6932	·000 032 50	·000 000 015 0	·000 034 00
·000 257 06	·000 000 120 3	8·0234	·000 032 04	·000 000 015 0	·000 033 54
·000 260 61	·000 000 107 1	8·0234	·000 032 56	·000 000 013 3	·000 033 89
Mean values			0·000 032 14	0·000 000 018 4	0·000 033 98

Rubidium Sulphate, Direction of Axis c.

Experimental Data.

Crystal.	L_{t_1}.	l_a.	l.	d.	t_1.	t_2.	t_3.	b_1.	b_2.	b_3.	f_2.	Corrn.	f_2'.	f_3.	Corrn.	f_3'.
	millims.	millims.	millims.	millims.	°	°	°	millims.	millims.	millims.						
1	8·638	5·250	14·153	0·265	12·9	56·6	96·2	755·8	755·9	756·0	42·25	−0·03	42·22	84·35	−0·05	84·30
					12·7	56·4	95·6	757·0	757·2	757·4	42·34	−0·03	42·31	83·70	−0·05	83·65
2	6·797	5·247	12·169	0·125	14·9	57·2	96·1	755·4	755·5	755·6	34·38	−0·02	34·36	68·90	−0·03	68·87
					13·4	55·7	96·4	756·1	756·3	756·5	34·26	−0·02	34·24	70·19	−0·03	70·16
3	10·490	5·250	15·877	0·137	11·1	56·7	96·0	737·5	735·9	734·3	51·68	−0·02	51·66	100·18	−0·03	100·15
					11·0	56·7	95·8	732·1	731·7	731·4	51·65	−0·02	51·63	99·77	−0·03	99·74

Calculated Actual Expansions.

Diminution of thickness of air-layer.		Expansion of tripod screws.		Expansion of aluminium compensator.		Correction for non-compensation.		Expansion of crystal.	
$f_2' \lambda/2$.	$f_3' \lambda/2$.	1st interval.	2nd interval.	1st interval.	2nd interval.	1st interval.	2nd interval.	$L_{t_2} - L_{t_1}$.	$L_{t_3} - L_{t_1}$.
0·0138526	0·0276592	0·0054170	0·0104323	0·0052256	0·0101443	+0·0001914	+0·0002880	0·0140440	0·0279472
·0138819	·0274461	·0054164	·0103798	·0052244	·0100918	+ ·0001920	+ ·0002880	·0140739	·0277341
·0112735	·0225964	·0045114	·0087480	·0050613	·0098919	− ·0005499	− ·0011439	·0107236	·0214525
·0112342	·0230197	·0045079	·0089390	·0050542	·0101053	− ·0005463	− ·0011663	·0106879	·0218534
·0169496	·0328594	·0063383	·0119217	·0054484	·0103298	+ ·0008899	+ ·0015919	·0178395	·0344513
·0169399	·0327250	·0063521	·0119067	·0054602	·0103162	+ ·0008919	+ ·0015905	·0178318	·0343155

Calculated Coefficients of Linear Expansion.

θ.	ϕ.	L_0.	a.	b.	$\alpha_{60°}$.
0·000 296 58	0·000 000 356 7	8·6341	0·000 034 35	0·000 000 041 3	0·000 038 48
·000 300 02	·000 000 318 7	8·6341	·000 034 75	·000 000 036 9	·000 038 44
·000 233 71	·000 000 274 6	6·7935	·000 034 40	·000 000 040 4	·000 038 44
·000 234 63	·000 000 261 1	6·7938	·000 034 54	·000 000 038 4	·000 038 38
·000 366 08	·000 000 370 8	10·4859	·000 034 91	·000 000 035 4	·000 038 45
·000 365 14	·000 000 370 0	10·4859	·000 034 82	·000 000 035 3	·000 038 35
Mean values			0·000 034 63	0·000 000 038 0	0·000 038 43

Cæsium Sulphate, Direction of Axis a.

Experimental Data.

Crystal.	L_{t_1}.	l_a.	l.	d.	t_1.	t_2.	t_3.	b_1.	b_2.	b_3.	f_2.	Corrn.	f_2'.	f_3.	Corrn.	f_3'.
	millims.	millims.	millims.	millims.	°	°	°	millims.	millims.	millims.						
1	5·864	3·879	10·271	0·528	128	56·3	95·1	756·0	754·0	752·6	26·69	− 0·07	26·62	52·38	− 0·12	52·26
	5·792*	3·907†	10·779	1·080	14·6	55·9	94·2	754·8	754·6	754·5	24·52	− 0·14	24·38	48·88	− 0·24	48·64
2	10·737	5·249	16·230	0·244	19·6	56·5	95·6	752·0	751·8	751·7	40·50	− 0·03	40·47	85·35	− 0·05	85·30
					18·3	57·0	96·1	755·0	755·6	756·2	41·65	− 0·03	41·62	85·65	− 0·05	85·60
3	8·277	5·254	13·851	0·320	18·0	56·3	96·0	759·3	759·2	759·1	34·55	− 0·04	34·51	72·00	− 0·07	71·93
					18·5	57·1	96·3	761·6	761·8	762·0	34·64	− 0·04	34·60	71·58	− 0·07	71·51
4	8·324	5·252	13·783	0·207	12·4	57·0	95·9	747·2	746·0	745·0	40·66	− 0·02	40·64	77·99	− 0·04	77·95
					10·9	56·1	95·6	736·5	735·0	733·5	40·98	− 0·02	40·96	78·65	− 0·04	78·61

Calculated Actual Expansions.

Diminution of thickness of air layer.		Expansion of tripod screws.		Expansion of aluminium compensator.		Correction for non-compensation.		Expansion of crystal.	
$f_2' \lambda/2$.	$f_3' \lambda/2$.	1st interval.	2nd interval.	1st interval.	2nd interval.	1st interval.	2nd interval.	$L_{t_2} - L_{t_1}$.	$L_{t_3} - L_{t_1}$.
0·0087341	0·0171470	0·0039128	0·0074776	0·0038425	0·0074013	+ 0·0000703	+ 0·0000763	0·0088044	0·0172233
·0079991	·0159590	·0039000	·0075917	·0036769	·0072129	+ ·0002231	+ ·0003788	·0082222	·0163378
·0132780	·0279870	·0052544	·0109320	·0044253	·0092794	+ ·0008291	+ ·0016526	·0141071	·0296396
·0136555	·0280854	·0055095	·0111885	·0046393	·0094960	+ ·0008702	+ ·0016925	·0145257	·0297779
·0113230	·0236004	·0046521	·0095719	·0045937	·0095273	+ ·0000584	+ ·0000446	·0113814	·0236450
·0113525	·0234628	·0046902	·0095495	·0046323	·0095065	+ ·0000579	+ ·0000430	·0114104	·0235058
·0133340	·0255759	·0053837	·0101817	·0053349	·0101688	+ ·0000488	+ ·0000129	·0133828	·0255888
·0134389	·0257923	·0054529	·0103231	·0054006	·0103065	+ ·0000523	+ ·0000166	·0134912	·0258089

Calculated Coefficients of Linear Expansion.

θ.	ϕ.	L_0.	a.	b.	$\alpha_{50°}$.
0·000 190 15	0·000 000 177 2	5·8615	0·000 032 44	0·000 000 030 2	0·000 035 46
·000 187 74	·000 000 161 0	5·7892	·000 032 43	·000 000 027 8	·000 035 21
·000 367 34	·000 000 196 6	10·7297	·000 034 24	·000 000 018 3	·000 036 07
·000 361 09	·000 000 189 3	10·7303	·000 033 65	·000 000 017 6	·000 035 41
·000 285 97	·000 000 150 6	8·2718	·000 034 57	·000 000 018 2	·000 036 39
·000 283 02	·000 000 166 5	8·2717	·000 034 22	·000 000 020 1	·000 036 23
·000 288 68	·000 000 164 1	8·3204	·000 034 69	·000 000 019 7	·000 036 66
·000 287 91	·000 000 157 9	8·3208	·000 034 60	·000 000 019 0	·000 036 50
Mean values			0·000 033 85	0·000 000 021 4	0·000 035 99

* Crystal surfaces were ground for second observation ; natural faces were used in first. † Different compensator.

Cæsium Sulphate, Direction of Axis b.

Experimental Data.

Crystal.	L_{t_1}.	l_a.	l.	d.	t_1.	t_2.	t_3.	b_1.	b_2.	b_3.	f_2.	Corrn.	f_2'.	f_3.	Corrn.	f_3'.
	millims.	millims.	millims.	millims.	°	°	°	millims.	millims.	millims.						
1	8·218	5·255	13·672	0·199	21·0	56·6	95·9	762·0	761·0	760·0	30·00	−0·02	29·98	64·27	−0·04	64·23
					22·0	57·2	95·6	760·2	760·1	760·0	29·81	−0·02	29·79	63·30	−0·04	63·26
2	8·003	5·252	13·489	0·234	20·9	57·3	97·5	758·6	757·9	757·2	29·73	−0·03	29·70	64·53	−0·05	64·48
					20·4	56·7	95·5	755·2	754·8	754·4	30·14	−0·03	30·11	63·64	−0·05	63·59
3	8·348	5·261	13·903	0·294	20·4	56·0	95·4	754·8	754·6	754·4	29·53	−0·04	29·49	64·73	−0·06	64·67
					19·5	56·6	94·8	756·0	755·8	755·5	30·81	−0·04	30·77	64·20	−0·06	64·14
4	8·379	5·253	13·777	0·145	8·8	56·3	96·3	745·0	744·0	743·0	40·49	−0·02	40·47	76·17	−0·03	76·14
					12·4	56·7	96·4	741·0	740·0	739·0	37·10	−0·02	37·08	72·10	−0·03	72·07

Calculated Actual Expansions.

Diminution of thickness of air-layer.		Expansion of tripod screws.		Expansion of aluminium compensator.		Correction for non-compensation.		Expansion of crystal.	
$f_2'\lambda/2$.	$f_3'\lambda/2$.	1st interval.	2nd interval.	1st interval.	2nd interval.	1st interval.	2nd interval.	$Lt_2 - Lt_1$.	$Lt_3 - Lt_1$.
0·0098365	0·0210740	0·0042719	0·0090795	0·0042771	0·0091625	−0·0000052	−0·0000830	0·0098313	0·0209910
·0097742	·0207560	·0042256	·0089236	·0042320	·0090066	− ·0000064	− ·0000830	·0097678	·0206730
·0097447	·0211563	·0043101	·0091649	·0043719	·0093715	− ·0000618	− ·0002066	·0096829	·0209497
·0098792	·0208641	·0042971	·0089797	·0043576	·0091778	− ·0000605	− ·0001981	·0098187	·0206660
·0096756	·0212183	·0043428	·0092427	·0042797	·0091806	+ ·0000631	+ ·0000621	·0097387	·0212804
·0100958	·0210446	·0045253	·0092762	·0044593	·0092115	+ ·0000660	+ ·0000647	·0101618	·0211093
·0132782	·0249816	·0057249	·0106559	·0056715	·0106427	+ ·0000534	+ ·0000132	·0133316	·0249948
·0121661	·0236461	·0053448	·0102395	·0052992	·0102341	+ ·0000456	+ ·0000054	·0122117	·0236515

Calculated Coefficients of Linear Expansion.

θ.	ϕ.	L_0.	a.	b.	$\alpha_{50°}$.
0·000 268 07	0·000 000 104 2	8·2123	0·000 032 64	0·000 000 012 7	0·000 033 91
·000 270 53	·000 000 088 1	8·2120	·000 032 94	·000 000 010 7	·000 034 01
·000 251 46	·000 000 186 0	7·9977	·000 031 44	·000 000 023 3	·000 033 77
·000 261 15	·000 000 121 1	7·9976	·000 032 65	·000 000 015 1	·000 034 16
·000 253 82	·000 000 258 5	8·3427	·000 030 42	·000 000 031 0	·000 033 52
·000 261 12	·000 000 168 2	8·3428	·000 031 30	·000 000 020 2	·000 033 32
·000 272 55	·000 000 124 6	8·3765	·000 032 54	·000 000 014 9	·000 034 03
·000 265 39	·000 000 148 6	8·3757	·000 031 69	·000 000 017 7	·000 033 46
Mean values			0·000 031 95	0·000 000 018 2	0·000 033 77

Cæsium Sulphate, Direction of Axis c.

Experimental Data.

Crystal.	L_{t_1}.	l_a.	l.	d.	t_1.	t_2.	t_3.	b_1.	b_2.	b_3.	f_2.	Corrn.	f_2'.	f_3.	Corrn.	f_3'.
	millims.	millims.	millims.	millims.	°	°	°	millims.	mil ims.	millims.						
1	5·883	3·879	10·604	0·842	13·0	56·4	96·0	751·4	752·5	753·0	30·07	−0·11	29·96	60·63	−0·19	60·44
					12·9	56·5	96·0	758·5	57·5	757·0	30·21	−0·12	30·09	60·73	−0·20	60·53
2	5·122	3·879	10·165	1·164	9·5	56·0	95·4	760·5	759·0	758·0	28·10	−0·17	27·93	54·10	−0·28	53·82
					10·3	56·5	95·9	762·0	762·4	762·8	28·01	−0·17	27·84	53·90	−0·28	53·62
3	4·816	3·879	10·566	1·870	14·8	56·0	95·3	753·1	753·1	753·0	23·33	−0·24	23·09	47·42	−0·42	47·00
					15·9	56·2	96·0	752·9	752·4	751·9	22·57	−0·23	22·34	47·22	−0·41	46·81
4	6·712	5·256	12·085	0·117	9·5	56·8	95·9	768·0	768·1	768·1	38·93	−0·01	38·92	73·81	−0·02	73·79
					9·6	56·1	96·4	768·3	768·4	768·5	38·15	−0·01	38·14	74·09	−0·02	74·07

Calculated Actual Expansions.

Diminution of thickness of air-layer.		Expansion of tripod screws.		Expansion of aluminium compensator.		Correction for non-compensation.		Expansion of crystal.	
$f_2'\lambda/2$.	$f_3'\lambda/2$.	1st interval.	2nd interval.	1st interval.	2nd interval.	1st interval.	2nd interval.	$L_{t_2}-L_{t_1}$.	$L_{t_3}-L_{t_1}$.
0·0098299	0·0198300	0·0040306	0·0077879	0·0038343	0·0074678	+0·0001963	+0·0003201	0·0100262	0·0201501
·0098726	·0198600	·0040492	·0077970	·0038520	·0074765	+ ·0001972	+ ·0003205	·0100698	·0201805
·0091639	·0176580	·0041355	·0077181	·0041006	·0077144	+ ·0000349	+ ·0000037	·0091988	·0176617
·0091344	·0175930	·0041103	·0076938	·0040766	·0076920	+ ·0000337	+ ·0000018	·0091681	·0175948
·0075758	·0154210	·0038140	·0075284	·0036422	·0072467	+ ·0001718	+ ·0002817	·0077476	·0157027
·0073303	·0153590	·0037320	·0074943	·0035648	·0072164	+ ·0001672	+ ·0002779	·0074975	·0156369
·0127697	·0242109	·0050024	·0092306	·0056542	·0105160	− ·0006518	− ·0012854	·0121179	·0229255
·0125138	·0243024	·0049170	·0092749	·0055567	·0105679	− ·0006397	− ·0012930	·0118741	·0230094

Calculated Coefficients of Linear Expansion.

θ.	ϕ.	L_0.	a.	b.	$\alpha_{50°}$.
0·000 210 43	0·000 000 296 8	5·8802	0·000 035 79	0·000 000 050 5	0·000 040 84
·000 210 07	·000 000 300 9	5·8802	·000 035 73	·000 000 051 2	·000 040 85
·000 184 89	·000 000 197 5	5·1202	·000 036 11	·000 000 038 6	·000 039 97
·000 186 39	·000 000 180 3	5·1201	·000 036 40	·000 000 035 2	·000 039 92
·000 175 42	·000 000 178 4	4·8134	·000 036 44	·000 000 037 1	·000 040 15
·000 169 41	·000 000 230 6	4·8132	·000 035 20	·000 000 047 9	·000 039 99
·000 240 69	·000 000 234 1	6·7097	·000 035 87	·000 000 034 9	·000 039 36
·000 239 49	·000 000 241 4	6·7097	·000 035 69	·000 000 036 0	·000 039 29
	Mean values		0·000 035 90	0·000 000 041 4	0·000 040 04

SUMMARY OF THE RESULTS FOR THE LINEAR COEFFICIENTS OF EXPANSION.

Collating the mean values given at the foot of each of the foregoing tables, the following is a statement of the essential results of the work as regards the linear coefficients.

MEAN Coefficients of Linear Expansion, $a + bt$, between 0° and t°.

Potassium Sulphate.

For the direction of the axis a . . .	0·000 036 16 + 0·000 000 014 4t.
" " b . . .	0·000 032 25 + 0·000 000 014 1t.
c . . .	0·000 036 34 + 0·000 000 041 3t.

Rubidium Sulphate.

For the direction of the axis a . . .	0·000 036 37 + 0·000 000 020 3t.
" " b . . .	0·000 032 14 + 0·000 000 018 4t.
c . . .	0·000 034 63 + 0·000 000 038 0t.

Cæsium Sulphate.

For the direction of the axis a . . .	0·000 033 85 + 0·000 000 021 4t.
" " b . . .	0·000 031 95 + 0·000 000 018 2t.
c . . .	0·000 035 90 + 0·000 000 041 4t.

In abbreviated notation will next be given a list of the true coefficients. The suffix attached to α indicates the axial direction.

TRUE Coefficients α of Linear Expansion at t°, or Mean Coefficients between any Two Temperatures whose Mean is t. $\alpha = a + 2bt$.

Potassium Sulphate.

$$\alpha_a = 10^{-8} (3616 + 2{\cdot}88t).$$
$$\alpha_b = 10^{-8} (3225 + 2{\cdot}82t).$$
$$\alpha_c = 10^{-8} (3634 + 8{\cdot}26t).$$

Rubidium Sulphate.

$$\alpha_a = 10^{-8} (3637 + 4{\cdot}06t).$$
$$\alpha_b = 10^{-8} (3214 + 3{\cdot}68t).$$
$$\alpha_c = 10^{-8} (3463 + 7{\cdot}60t).$$

Cæsium Sulphate.

$$\alpha_a = 10^{-8} (3385 + 4{\cdot}28t).$$
$$\alpha_b = 10^{-8} (3195 + 3{\cdot}64t).$$
$$\alpha_c = 10^{-8} (3590 + 8{\cdot}28t).$$

A comparison of the coefficients of expansion along analogous directions in the three salts is presented in the following table, which also includes a comparison of the coefficients for the particular temperature of 50°.

Comparative Table of the Linear Coefficients of Expansion for the Three Salts.

THE Constant a, the Coefficient of Expansion at 0°.

Crystallographical axial direction.	K_2SO_4.	Rb_2SO_4.	Cs_2SO_4.
a	0·000 036 16	0·000 036 37	0·000 033 85
b	·000 032 25	·000 032 14	·000 031 95
c	·000 036 34	·000 034 63	·000 035 90
Sums of values for all three directions:	0·000 104 75	0·000 103 14	0·000 101 70

THE Constant b, Half the Increment of the Coefficient per Degree.

Axial direction.	K_2SO_4.	Rb_2SO_4.	Cs_2SO_4.
a	0·000 000 014 4	0·000 000 020 3	0·000 000 021 4
b	·000 000 014 1	·000 000 018 4	·000 000 018 2
c	·000 000 041 3	·000 000 038 0	·000 000 041 4
Sums of values for all three directions:	0·000 000 069 8	0·000 000 076 7	0·000 000 081 0

$\alpha_{50°}$, the Coefficient of Expansion, $\alpha = a + 2bt$, for 50°.

Axial direction.	K_2SO_4.	Rb_2SO_4.	Cs_2SO_4.
a	0·000 037 60	0·000 038 40	0·000 035 99
b	·000 033 66	·000 033 98	·000 033 77
c	·000 040 47	·000 038 43	·000 040 04
Sums of values for all three directions:	0·000 111 73	0·000 110 81	0·000 109 80

It will be observed that in the preceding table the sums of the values of each constant for the three axial directions of each particular salt are taken. These sums represent the constants of the cubical coefficients of expansion. For when the expression for the product of the expansions in the three rectangular axial directions, which naturally gives the expansion of the solid, is examined, it is found to consist of a large number of terms of which the only ones that affect the fourth and last place of significant figures in the coefficient of expansion for any temperature are the sums of the constants a and b respectively.

In the next table is presented a summary of the constants of the cubical coefficients of expansion, and of the cubical coefficients for 50°, in a form which readily admits of a comparison of the values for the three salts.

COEFFICIENTS of the Cubical Expansion of the three Sulphates.

	a.		b.		$\alpha_{50°}$.	
K_2SO_4 . . .	0·000 104 75		0·000 000 069 8		0·000 111 73	
		Diff. 161		Diff. 69		Diff. 92
Rb_2SO_4 . . .	0·000 103 14		0·000 000 076 7		0·000 110 81	
		Diff. 144		Diff. 43		Diff. 101
Cs_2SO_4 . . .	0·000 101 70		0·000 000 081 0		0·000 109 80	

The mean coefficients of the cubical expansion of the three salts between 0° and $t°$ are therefore as follows :

For potassium sulphate . 0·000 104 75 + 0·000 000 069 8t, or 10^{-8} (10475 + 6·98t).
,, rubidium sulphate . 0·000 103 14 + 0·000 000 076 7t, or 10^{-8} (10314 + 7·67t).
,, cæsium sulphate . 0·000 101 70 + 0·000 000 081 0t, or 10^{-8} (10170 + 8·10t).

The actual coefficients of cubical expansion, α, at any temperature t, and also the mean coefficients of cubical expansion between any two temperatures whose mean is t, are the following, in which $\alpha = a + 2bt$:

For potassium sulphate . 0·000 104 75 + 0·000 000 139 6t, or 10^{-8} (10475 + 13·96t).
,, rubidium sulphate . 0·000 103 14 + 0·000 000 153 4t, or 10^{-8} (10314 + 15·34t).
,, cæsium sulphate . 0·000 101 70 + 0·000 000 162 0t, or 10^{-8} (10170 + 16·20t).

DISCUSSION OF THE RESULTS, AND CONCLUSIONS THEREFROM.

The Cubical Expansion.

The most striking result of the investigation is apparent from an inspection of the comparative table of the cubical coefficients of expansion. It may be stated in the following words :

The coefficients of cubical expansion of the normal sulphates of potassium, rubidium, and cæsium exhibit a progression, corresponding to the progression of the atomic weights of the three respective metals. This is true of both the constants a and b in the general expression for the coefficient of cubical expansion, the values of each constant for the rubidium salt being intermediate between the corresponding values for the potassium and cæsium salts.

It may be further stated that:

The differences between the values of the constant a, which represents the coefficient of cubical expansion for 0°, for the three salts, are small, amounting to only one and a half per cent.; this is an amount, however, which is five times as great as the possible experimental error in the determinations.

Also that:

The order of progression of the two constants of the cubical coefficient of expansion is inverted; a, the coefficient at 0°, diminishes with increasing atomic weight of the metal contained in the salt, while b, half the increment of the coefficient per degree, increases.

This latter fact leads to an interesting result, namely, that the coefficients, in increasing with rise of temperature, approach each other in value, until for three certain temperatures between 110° and 170° they become identical in pairs; moreover, in the neighbourhood of the second of these temperatures the three values approximate so closely to each other that their difference comes within the limits of experimental error. For temperatures higher than those of coincidence, the values diverge and exhibit an inverted order of progression. This will be rendered clear by a table showing the true coefficients of cubical expansion, $a + 2bt$, for intervals of 50° up to 200°, and for the three temperatures of coincidence. These latter are 114° for the identity of cubical expansion of potassium and rubidium sulphates, 136° for potassium and cæsium sulphates, and 168° for the coincidence of expansion of the rubidium and cæsium salts.

COEFFICIENTS of Cubical Expansion for Various Temperatures from 0° to 200°

Salt.	0°.	50°.	100°.	114°.	136°.	150°.	168°.	200°.
K_2SO_4	10^{-8} 10475	11173	11871	12066	12373	12569	12820	13267
Rb_2SO_4	10^{-8} 10314	11081	11848	12065	12400	12615	12891	13382
Cs_2SO_4	10^{-8} 10170	10980	11790	12017	12373	12600	12891	13410

The point may be graphically demonstrated by plotting out the values on curve paper, taking temperatures as abscissæ and coefficients of cubical expansion as ordinates. The three straight lines thus obtained, shown in the reproduction given, will be observed to converge from 0° towards the three temperatures of coincidence, where crossing of the lines in pairs occurs, beyond which they diverge. The relative nearness of the lines to each other in the middle of the part where crossing occurs, together with the fact that the increment $2b$ is not determinable with the accuracy of the constant a, the coefficient at 0°, suggests the probability that the three lines should all cross at one point, somewhere near 136°. At this temperature the value for rubidium sulphate only exhibits a difference of one in five hundred from the two identical values for the two other salts, so that the three values are identical within the limits of experimental error.

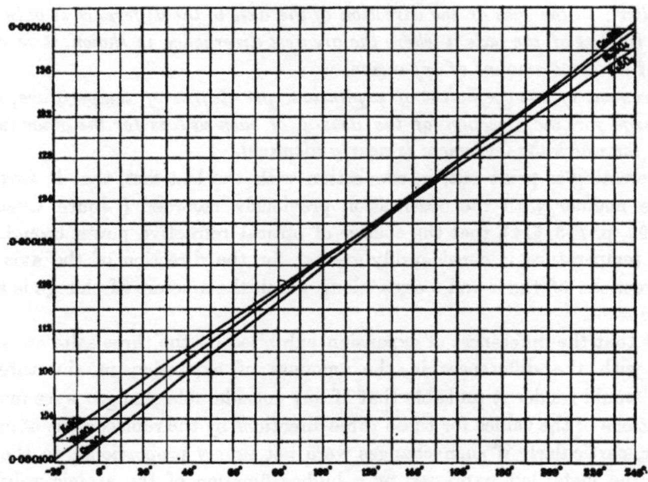

Graphical Expression of Cubical Expansions.

These considerations may be summarised in the following addition to the last italicised statement.

In consequence of this fact the coefficients of cubical expansion of the three salts converge with rise of temperature towards equality, which, within the limits of experimental error, they reach at 136°. *Beyond the temperature at which identity of expansion occurs the coefficients of expansion exhibit increasing divergence, the order of progression being inverted, an increase in the atomic weight of the metal being now accompanied by an increase in the coefficient of cubical expansion.*

The Linear Expansion.

The first conclusions to be drawn from an inspection of the coefficients of linear deformation are the following:

The thermal deformation is of the nature of an expansion along all directions in the crystals of the three sulphates, no contraction occurring in any direction.

The amount of the expansion is relatively large, compared with the expansions of metals, being four times that of platinum and one and a half times as great as the large expansion of aluminium.

The differences between the amounts of linear expansion along the three axial directions of any one salt are small, the difference between the maximum and minimum being about twelve per cent. of the total expansion in the case of each salt. But the differences between the values for the same direction of the three salts are much smaller; in the case of the direction of the axis b, the difference is only one per cent., and in that of the axis a where the greatest divergence is shown, it is only six per cent. of the total amount of expansion.

The increment of the coefficient of expansion, per degree of temperature, is about twice as large for the direction of the axis c of each salt as for the other two axial directions, for which the increment is nearly identical.

It is interesting to point out, in connection with the last fact, that it agrees in a remarkable manner with the observation previously recorded ('Journ. Chem. Soc.,' Trans., 1894, p. 715, §14), that the change of optical refractive power brought about by rise of temperature is considerably greater for the direction of the axis *c*, than for the directions of the *a* and *b* axes, along which the amount of change is approximately the same.

The fact that the differences of expansion exhibited by the three salts are so small, compared with the differences in the amounts of expansion in the three axial directions, would render it probable that if any considerable changes were introduced in the relations of the values for these three directions by the replacement of one metal by another, particularly if such changes were not simply proportional to the atomic weight of the metal but expressed by a higher function of the atomic weight, such change would suffice to negative the possibility of a direct progression of the linear coefficients of expansion for each axial direction of the three salts, corresponding to the atomic weights of the metals present. That simple proportionality to the atomic weight was not to be expected directionally, has been indicated by the whole of the morphological and physical work on both sulphates and selenates. In the case of the refractive indices, however, the directional changes were small in comparison with the differences exhibited by different salts; hence in the case of these, as of other, optical constants, the interesting progression according to the atomic weight of the metal was not interfered with. But although such perturbations of a directional character are able, in the case of the thermal constants, to obliterate such a progression of the linear coefficient of expansion, they would mutually compensate each other when the total solid change was considered. Hence it must be apparent that if the influence of atomic weight were indeed a progressive one it would only be clearly revealed in the case of the coefficients of cubical expansion. These latter constants have been shown to exhibit such a progression in the clearest possible manner.

In accordance with the above considerations, a progression of the linear coefficients of expansion is found to be prevented by the slight directional perturbations due to the different natures of the molecules of the three salts. The replacement of the atoms of one metal by those of another of higher atomic weight is possibly, and even probably, accompanied by movement of the relative positions of the constituents of the molecules of their spheres of motion, as well as by the purely chemical change; for not only is the substitution accompanied by an increase of mass, but also by an increase in the electro-positive energy of the metallic atoms, which may very reasonably be expected to result in a closer approximation to the negative atoms, probably of oxygen, to which they are attracted. There are, however, several interesting facts exhibited by the linear coefficients, which connect their relations very intimately with those of the optical constants, for which a true progression in the order of atomic weight has been clearly demonstrated. Before passing to the consideration of these indications of parallelism between the thermal and optical behaviour of the crystals of the three salts, attention must be drawn to the two following salient facts which are apparent from an inspection of the linear coefficients. It is that:

The amount of expansion along the direction of the crystallographical axis b is practically identical for all three sulphates, indicating that the interchange of the three metals is without influence on the thermal behaviour along the macrodiagonal axis of the crystals. Moreover, the crystals of all three salts expand least along this direction, which is therefore that of the minimum axis of the thermal ellipsoid.

These two facts are doubtless of significance with respect to the structure of the molecule, apparently indicating absence of the metallic atoms or their spheres of motion from the immediate proximity of the axis *b*. The significance becomes enhanced in view of the fact that the author has shown ('Journ. Chem. Soc. Trans.,' 1896, p. 507) that the whole of the work on the sulphates and double sulphates points to the conclusion that the structural unit of the crystals of the simple sulphates is the simple chemical molecule, a conclusion which is supported by the work of FOCK (referred to *loc. cit.*) on the solubility of mixed crystals.

The relations between the amounts of expansion along the directions of the other two crystallographical axes, *a* and *c*, are much more complicated, and are evidently influenced by the replacement of one metal by another. Considering the coefficients of expansion for 0°, the amount of expansion along the direction of the axis *c* is the greater in the case of both the potassium and cæsium salts; and for all three salts the increment is, as already indicated, greater for this than for any other direction. But in the case of the rubidium salt a remarkable excess of expansion is observed to occur along the direction of the *a* axis, which is the maximum thermal axis at 0°, at the expense of that along the *c* axis, which becomes reduced to the intermediate thermal axis. The increments per degree, however, for these two directions in the rubidium salt, remain of the same order as for the other two salts. Now it is an interesting fact, and doubtless not without significance, that the directions of maximum thermal effect coincide with those of the first median line of the optic axial angles of all three salts, which is the axis c in both the potassium and cæsium salts and the axis a in the case of the rubidium salt. In order to follow the parallelism further, it will be necessary to compare the linear coefficients of expansion for higher temperatures, the values for 50° and 100° sufficing for the purpose. In the following table are given the values of $\alpha = a + 2bt$ for 0°, 50° and 100°. The directions are also indicated of the axes of the optical indicatrix, namely, the first median line, the second median line, and the intermediate axis of the optical ellipsoid. The sign of the double refraction of the crystals is also given, as this determines whether the first median line is the maximum or the minimum axis of the optical indicatrix, the former being the case for positive double refraction and the latter for negative.

COMPARISON of Linear Expansions at Different Temperatures with the Optical Indicatrix.

Salt.	Sign of double refraction.	Crystallographical axis.	Direction in optical ellipsoid.	Linear coefficients of expansion—		
				At 0°.	At 50°.	At 100°.
K_2SO_4	Positive . .	a	Intermediate axis	10^{-8} 3616	3760	3904
		b	Second median line	3225	3366	3507
		c	First median line	3634	4047	4460
Rb_2SO_4	Very feebly positive	a	First median line	3637	3840	4043
		b	Second median line below 50°	3214	3398	3582
		c	Intermediate axis below 50° .	3463	3843	4223
Cs_2SO_4	Negative .	a	Second median line	3385	3599	3813
		b	Intermediate axis	3195	3377	3559
		c	First median line	3590	4004	4418

It will be apparent from the table that the relations of the linear expansions at 0° still hold good, as the temperature is raised, in the cases of the potassium and cæsium salts. A similar observation has been shown to be valid with respect to the optical properties. But in the case of the rubidium salt a remarkable change occurs. Owing to the greater increment of the expansion along the direction of the axis c, the preponderance of the expansion along the axis a diminishes, until at 50° the amounts of expansion along these two axial directions become equal. That is to say, in the neighbourhood of 50° the crystals of rubidium sulphate simulate uniaxial symmetry as regards their thermal behaviour. This is rendered the more interesting by the fact that about this temperature the crystals of this salt also exhibit uniaxial optical properties. Owing to the extremely feeble double refraction, it was shown (*loc. cit.*, p. 693) that the slight alterations of the relations of the refractive power in the three

axial directions, brought about by rise of temperature according to a
all three salts, result in bringing two of the refractive indices to equ
temperature indicated, and the optic axial angle diminishes until
temperature the circular rings and rectangular cross of an uniax
exhibited in convergent polarised light. The exact temperatures
uniaxial interference figure is produced differ slightly according to t
of the light, owing to there being a large amount of dispersion of
They are respectively 42° for red lithium light, 44° for red C. hydroge
sodium light, 52° for green thallium light and 58° for greenish blue F. h
the average is thus exactly 50°, the temperature at which the crysta
sulphate are thermally uniaxial. But the two ellipsoids of revolutio
properties are not similarly orientated, the principal axis for the op
being the crystallographical axis *a*, while for the thermal property it is

This close parallelism between the thermal and optical properties is c
importance, inasmuch as the optical constants were shown to exhibi
gression corresponding to the progression of the atomic weights of the
the differences between the optical constants for the three salts being
relatively to their variations in the three axial directions of any one sal
case of the thermal properties.

On following the growth of the coefficients of linear expansion furthe
observed that the continued gain on the part of the *c* value has now
expansion along this axis clearly the maximum, thus reversing the orde
expansions which had obtained below 50°. The maximum axis of the the
for 100°, and temperatures superior to this, is thus the crystallograph
all three sulphates; moreover the axis *b* is the minimum thermal axis, a
the axis of intermediate thermal expansion for all three salts. To compl
lelism of the thermal and optical properties, it may be mentioned that at
superior to the neighbourhood of 50°, where the uniaxial optical interf
is produced, the figure again breaks up into a biaxial one, but with th
separated in the plane perpendicular to that which formerly contained t
optic axes separate more and more in this new plane until about 18
becomes the first median line instead of the axis *a*. At this temperatu
the crystallographical axis *c* is the first median line for all three salts.
shown that at temperatures superior to 100° the axis *c* is also the
maximum thermal expansion. Hence, at higher temperatures the rule
the lower ones, that the direction of maximum thermal expansion is
optical first median line, is equally valid.

The foregoing considerations, concerning the relations of the linear c
expansion, may be summarised as follows:—

The smallness of the difference in the coefficient of expansion along an direction in the crystals, which is introduced by the replacement of one by another, compared with the larger differences of expansion exhibited in the three axial directions of any one salt, together with the fact that the change of metal is accompanied by considerable modifications of these latter relative expansions for two of the axial directions, a and c, prevent the coefficients of expansion for any one direction of the three salts from exhibiting any progression corresponding to that of the atomic weights of the three metals. These directional perturbations are, however, mutually compensative, the increase of expansion in one of the two directions referred to being more or less balanced by the diminution in the other; consequently the effect of interchange of the metals is clearly exhibited by the solid deformation, the cubical expansion, the coefficients and increments of which have been shown to exhibit a well-defined progression following the order of the atomic weights of the three metals.

Before proceeding to summarise the interesting analogy between the thermal and the optical properties, it may be of advantage to consider what the dimensions of the linear change, on replacing one metal by another, would probably be, provided no directional perturbations occurred. The difference between the cubical coefficients of expansion of potassium and rubidium sulphates is 0·00000161, and of the rubidium and cæsium salts 0·00000144. The mean is 0·0000015,* and the linear differences might be reasonably expected to be about one-third of this, namely 0·0000005. Even if the linear expansions along the axis *b* are accepted as free from perturbation and unaffected by change of metal, the linear differences for the other two directions could not exceed 0·0000008. Now the directional perturbation in which the rubidium salt exhibits a reversal of the relative directions of the maximum and intermediate thermal axis compared with the potassium salt, amounts to more than twice this amount, namely 0·0000017. Hence it is clearly apparent that a progressive change, of the maximum possible amount, would be completely masked by the larger directional perturbation. A brief summary of the nature of the perturbation and its relation to the optical changes may next be given.

The chief directional perturbation consists of a reversal, for temperatures below 50°, of the directions of the maximum and intermediate axes of the thermal ellipsoid in the rubidium salt, compared with their directions in the potassium and cæsium salts. The maximum thermal axis is the crystallographical axis c for the two latter salts, but the a axis for the rubidium salt. A similar reversal of the direction of the first median line, the maximum axis of the optical ellipsoid (the indicatrix), from the direction c to the direction a, occurs for similar temperatures, in the case of the rubidium salt. Hence, the maximum thermal axis is identical in all three salts with the first median line.

At higher temperatures the same relations still hold for the potassium and cæsium salts, both thermally and optically. But owing to the increment of expansion along the axis c being so much greater than for other directions, the exceptional intermediate expansion along the axis c of rubidium sulphate is rapidly brought up to equality, at 50°, with the expansion along the axis a; and beyond this temperature c becomes the maximum thermal axis for the rubidium salt, as it is for the other two sulphates. Hence, at 50° the crystals of rubidium sulphate are apparently thermally uniaxial. At temperatures varying 10° each side of 50° for different wave-lengths of light, they are also apparently optically uniaxial. The thermal and optical ellipsoids of revolution, however, are not identically orientated, the axis of the former being the crystallographical axis b, and of the latter a. Further, the change of direction of the maximum thermal axis of rubidium sulphate, from a to c, is followed optically at 180° by the change of the first median line from a to c. Thus the first optical median line corresponds, as at lower temperatures, to the maximum thermal axis, for all three sulphates.

This parallelism between the linear thermal expansions and the optical constants is of significance, inasmuch as the latter constants, which, unlike the former ones, exhibit differences between the three salts of much greater magnitude than the directional differences for any one salt, show a clear progression, in the order of the atomic weights of the metals contained in the three sulphates.

It will be interesting in conclusion, to compare the results for the thermal deformation thus obtained by the refined interference method, with those previously obtained from the much cruder method of combining determinations of specific gravity at the ordinary and higher temperatures with measurements of the morphological angles at those temperatures. Such an attempt to determine the coefficients of expansion was described in the previous memoir on the sulphates ('Journ. Chem. Soc., Trans.,' 1894, p. 653). It naturally depended for success on the possibility of employing a liquid in the pyknometer which was absolutely without action on the salts, as well as upon the degree of accuracy with which such determinations and angular measurements, the latter involving total deviations of less than two minutes of arc, can be carried out, even with the aid of the extremely delicate instruments employed.

The actual values found for the total cubical expansion for 40° (between 20° and 60°) were :—

For potassium sulphate . . .	0·0053
„ rubidium „ . . .	0·0052
„ cæsium „ . . .	0·0051.

Thus a diminution of expansion was found to occur as the atomic weight of the metal increased, a result which is fully borne out by the more accurate determinations now presented. The figures for the three salts were so near, however, that they were taken as identical, having reference to the method by which they were obtained, for the purpose of calculating the coefficients of linear expansion with the aid of the angular deviations for the same temperatures. For the linear coefficients of expansion λ for 100° the following numbers were given :—

$$\lambda_a = 0{\cdot}00437, \quad \lambda_b = 0{\cdot}00385, \quad \lambda_c = 0{\cdot}00479,$$

It was therefore concluded that "the crystals of the three salts, on heating, expand most in the direction of the vertical axis c, and least along the macrodiagonal axis b." We have seen that this is indeed the case, except for temperatures below 50° in the case of λ_a and λ_c of the rubidium salt, a fact which the method could not possibly have indicated, as at 60°, the temperature of the higher density determinations, the rule found is really true.

The actual values now published for the total expansion of potassium sulphate, taking this salt as an example, for the 100° between 20° and 120°, calculated by the formula $\alpha = a + 2b\left(\frac{20 + 120}{2}\right)$, are as follows:

$$\alpha_a = 0{\cdot}003818, \qquad \alpha_b = 0{\cdot}003422, \qquad \alpha_c = 0{\cdot}004212.$$

The difference between these highly accurate values and the approximate ones obtained by the rougher method is not great considering the nature of the latter, and the order is the same.

An attempt to determine the expansion of crystals of potassium and rubidium sulphates, by means of the weight-thermometer method, has been described by SPRING ('Bull. de l'Acad. de Belgique,' 1882, 197, and 'Ber. Deut. Chem. Ges.,' 15, 1940). Olive oil was employed as the liquid, and the density determinations were carried up to 100°. The value obtained for the cubical expansion of potassium sulphate for 100° was 0·0126, and for rubidium sulphate 0·0111. The latter value is extraordinarily near the truth according to the results now presented, the value for 0° to 100° being actually 0·01108. But the impossibility of trusting this method, equally with all the relatively coarser density methods, to afford correct differences between the values for different salts, is clearly demonstrated by the fact that the difference shown between the values for the potassium and rubidium salts, namely 0·00150, is seventeen times as great as the real difference (·01117–0·01108 = 0·00009), which is now shown to exist.

SUMMARY OF CONCLUSIONS.

The principal results of the investigation are presented in the following summary.

1. The coefficients of cubical expansion of the orthorhombic crystals of the normal sulphates of potassium, rubidium and cæsium exhibit a progression, corresponding to the progression of the atomic weights of the three respective metals. This is true of both the constants a and b in the general expression for the coefficient of cubical expansion, $\alpha = a + 2bt$, for any temperature t.

2. The order of progression of the two constants is inverted; a, the coefficient for 0°, diminishes with increasing atomic weight of the metal, while b, half the increment of the coefficient per degree of temperature, increases.

3. In consequence of rule 2 the coefficients of cubical expansion of the three salts converge, with rise of temperature, and attain equality, within the limits of

experimental error, at 136°. Beyond the temperature of identity divergence occurs and an increase of atomic weight is now accompanied by an increase in the coefficient of cubical expansion.

4. The thermal deformation is of the nature of an expansion in all directions in the crystals of all three sulphates.

5. The differences between the coefficients of linear expansion along the three crystallographical axial directions of any one salt, although only amounting to one-eighth of the total coefficient, are large compared with the differences between the values for the same direction of the three salts.

6. The operation of rule 5, together with the fact that the replacement of one metal by another is accompanied by considerable modifications of the relations of two of the three values for the original salt, those corresponding to the axes *a* and *c*, prevent the coefficients of linear expansion for any one direction of the three salts from exhibiting any progression corresponding to that of the atomic weights of the three metals. These directional perturbations are, however, mutually compensative, so that the effect of interchange of the metals is clearly exhibited by the solid deformation, the cubical expansion, the coefficients of which and their increments have been shown to exhibit a progression according to the atomic weight of the metal, as stated in rule 1.

7. The increment of the linear coefficient of expansion for the direction of the vertical axis *c* of each salt, is about twice as large as the increments for the other two directions *a* and *b*, for which latter the increments are nearly equal. This thermal property is analogous to the optical behaviour, the refractive power being altered (diminished) by rise of temperature much more in the direction of the axis *c* than in the other two directions, in which the lesser amounts of change are nearly equal.

8. The amount of expansion along the direction of the crystallographical axis *b* is approximately identical for all three sulphates, indicating that interchange of the metals is without influence on the thermal behaviour along the macrodiagonal axis of the crystals. The crystals of all three salts also expand least in this direction, which is therefore the common minimum axis of the thermal ellipsoid.

9. The chief of the directional perturbations, referred to under 6, consists of a reversal, for temperatures below 50°, of the directions of the maximum and intermediate axes of the thermal ellipsoid for rubidium sulphate, compared with their directions in the potassium and cæsium salts. The maximum thermal axis is the crystallographical axis *c* for the two latter salts, but *a* for the rubidium salt. A similar reversal of the direction of the first median line, the maximum axis of the optical ellipsoid (the indicatrix), from *c* to *a* occurs for the same temperatures, in the case of rubidium sulphate. The maximum thermal axis is identical in all three salts with the optical first median line.

10. At higher temperatures the same relations still obtain for the potassium and cæsium salts, both thermally and optically. But owing to the increment of expansion

along the axis *c* being so much greater than for the other directions, the intermediate expansion along *c* for rubidium sulphate attains equality at 50° with the expansion along *a*, and beyond this temperature *c* becomes the maximum thermal axis for this salt, as it is for the other two sulphates. Consequently, at 50° the crystals of rubidium sulphate are apparently thermally uniaxial. At temperatures varying 10° each side of 50° for different wave-lengths of light, they have previously been shown to simulate uniaxial optical properties. The thermal and optical ellipsoids of revolution are not, however, identically orientated, the axis of the former being the axis *b* and of the latter *a*. Further, the change of direction of the maximum thermal axis of rubidium sulphate from *a* to *c* is followed optically at 180° by the change of the first median line from *a* to *c*, rendering the last sentence of rule 9 again valid.

11. A close parallelism between the linear thermal expansion and the directional optical behaviour is thus shown to exist, and is indicative that the same progressive effect of variation of the atomic weight of the metal is in operation with regard to the former, as was clearly demonstrated in a former memoir with respect to the latter, and that this effect would be manifest in the former were it not masked by the larger effect indicated under 6.

12. The thermal deformation constants best capable of indicating the effect of the replacement of one alkali metal by another, in the crystals of the normal alkali sulphates, have thus been shown to be the cubical coefficients of expansion and their increments; and these have been further demonstrated to exhibit a regular progression, which follows the order of progression of the atomic weights of the metals in question. Moreover, the linear coefficients and their increments have been shown to exhibit variations which present a remarkable analogy to those of the optical constants, for which, the values for the three salts being very much more widely separated and consequently undisturbed by the modification of the directional differences for the same salt which are relatively so much more important in the case of the linear thermal constants, a clear progression according to the atomic weight of the alkali metals has been proved.

The final conclusion of this investigation, therefore, is that:

The thermal deformation constants of the crystals of the normal sulphates of potassium, rubidium, and cæsium exhibit variations, which, in common with the morphological, optical, and other physical properties previously investigated, follow the order of progression of the atomic weights of the alkali metals which the salts contain.

This result is, therefore, in perfect agreement with the principle enunciated at the conclusion of the memoir concerning the alkaline selenates ('Journ. Chem. Soc., Trans.,' 1897, p. 920), which reads as follows:

The difference in the nature of the elements of the same family group which is manifested in their regularly varying atomic weights, is also expressed in the similarly regular variation of the characters of the crystals of an isomorphous series of salts of which these elements are the interchangeable constituents.

XI. *On the Electrical Conductivity of Flames Containing Salt Vapours.*

By HAROLD A. WILSON, *B.Sc.* (*Lond. and Vic.*), 1851 *Exhibition Scholar, Cavendish Laboratory, Cambridge.*

Communicated by Professor J. J. THOMSON, *F.R.S.*

Received March 10,—Read April 27, 1899.

IN a recent paper* on the electrical conductivity and luminosity of flames containing salt vapours, by Professor A. SMITHELLS, Mr. H. M. DAWSON, and the writer, the similarity between the conductivity of flames and that of gases exposed to Röntgen rays was pointed out, and it was shown that the relation of the current between two electrodes in the flame to the potential difference between them could be represented by the formulæ

$$I - i = k_2 \frac{i^2}{E^2},$$

$$C = i + k_1 E,$$

where C = the current,

E = the P.D. between the electrodes.

I, k_1, k_2 are constants, and i is defined by the second equation.

When E is large, these equations become

$$C = I + k_1 E,$$

and if $k_1 = 0$, then they reduce to

$$I - C = k_2 \frac{C^2}{E^2},$$

which represents the relation between the current and P.D. for the conductivity of Röntgenised gases. (See a paper by J. J. THOMSON and E. RUTHERFORD, 'Phil. Mag.,' Nov., 1896.)

The experiments described in the present paper were undertaken with the object of following up the analogy between the conductivity of salt vapours and that of Röntgenised gases, and especially of getting some information about the velocities of the ions in the flame itself.

The paper is divided into the following sections:—

* Abstract at 'Roy. Soc. Proc.,' vol. 64, p. 142.

26.6.99

A summary of the earlier work done on this subject is given in WIEDEMANN'S 'Lehre von der Elektricität,' vol. 4 B. ARRHENIUS'S paper ('Wied. Ann.,' vol. 42, 1891) is referred to in our paper mentioned above.

(1.) *Description of the Apparatus used for Producing the Flame.*

The apparatus used for producing the flame was similar in principle to that used in the investigation referred to above. Carefully regulated supplies of coal gas and air were mixed together, along with spray of salt solution, and the mixture burnt from a brass tube, 0·7 centim. in diameter. The apparatus is shown in fig. 1.

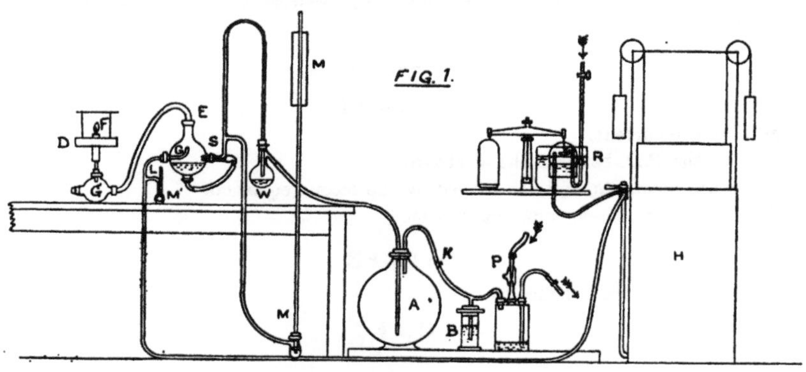

P, Water pump.	M M', Water manometers.	E, Exit tube from G.
B, Mercury blow-off.	R, Gas regulator.	G', Second globe.
A, Carboy.	H, Gasometer.	F, Flame.
W, Water flask.	L, Constriction.	D, Wood block.
S, Gouy sprayer.	G, Globe containing salt solution.	

The air supplied by the water pump, P, partly escapes by bubbling through mercury in B, and then passes into a carboy, A. From A the air passes through a flask, W, containing water, to the Gouy sprayer, S, and its pressure is measured by the water manometer, M. The air supply is regulated by means of a pinch-cock, K, and by altering the water supply to the pump. A very considerable change in the water supply was necessary to appreciably alter the pressure indicated by the manometer, M. The air pressure used was 180 centims. of water, and it was very easily kept constant within 2 or 3 millims., only very occasional adjustments being required for this.

The coal gas was passed through a regulator, R, consisting of a bell-jar suspended in water from one arm of a balance and arranged as shown in fig. 1, so that, when the gas in the jar attained the required pressure, the jar rose and cut off the gas supply. After passing through this regulator the gas supply was connected with a gasometer, H, which served to maintain the pressure steady. The weights on the gasometer were adjusted so as to produce a pressure equal to that at which the regulator partly cut off the gas. The gas was passed through a constriction, L, and then allowed to mix with the air and spray from the sprayer, S, in the globe, G.

The pressure of the gas supply was measured on the water manometer, M′, by means of a cathetometer reading easily to 0·01 centim. The gas pressure used was 3·62 centims. of water, and it was easily kept constant within 0·01 centim. by occasionally altering the weights on the gasometer and in the pan of the balance.

The mixture of gas and air passed along the tube, E, into a globe, G′, and from this into the flame tube, T. The tube, T, was supported by a wider brass tube, provided at its upper end with three screws for centreing the flame tube, and fixed into an octagonal wooden block, D, 3·5 centims. thick and 20 centims. across (see fig. 2). A cylindrical glass shade, 15 centims. in diameter and 16 centims. high, rested in a circular groove in the block, D, and on this a flat tin plate was placed, having a circular hole at its centre, 3 centims. in diameter, for the escape of the products of combustion. Three holes, each 1 centim. in diameter, in the block admitted air to the flame, F.

The flame thus obtained was steady, and measurements of its conductivity, when a particular salt solution was sprayed, did not differ more than 1 or 2 per cent. on different days.

The gas consumed by the flame amounted to 43 litres per hour. The height of the inner sharply-defined green cone was 1·5 centim., and that of the outer cone 7·5 centims.

(2.) *The Relation between the Current and E.M.F.*

Some experiments were first done on the relation between the current and E.M.F. in the flame. The electrodes used each consisted of a brass disk, 14 centims. in diameter and 0·2 centim. thick, having a circular hole 5·6 centims. in diameter at its centre, covered with a grating of platinum wires (see fig. 2).

These two disks were each supported by three glass rods horizontally one above the other symmetrically about the axis of the flame. Two parallel slots were cut on each side of the hole in the upper disk, and through these the platinum gauze was stretched across and kept in position by two brass strips screwed down over the slots. The gauze was thus stretched across the under side of the upper disk. The gauze had a mesh 0·06 centim. square, with wire 0·02 centim. thick. It was found necessary to use a wider mesh for the lower electrode to allow the flame to pass easily. The grating on the lower disk was, therefore, made by winding platinum wire between

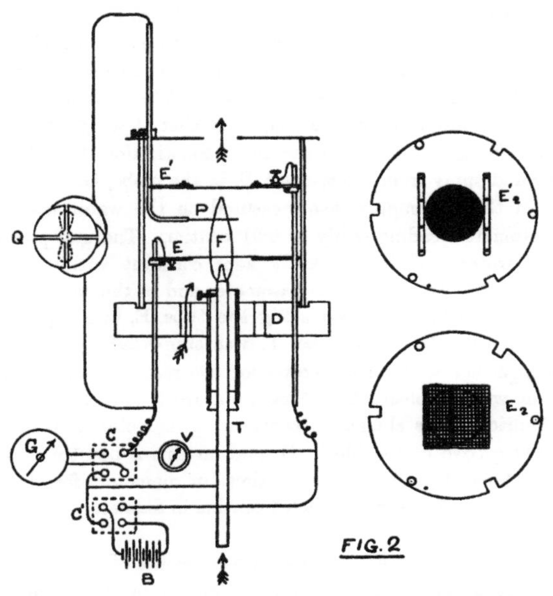

F,	Flame.	Q,	Quadrant electrometer.
D,	Wood block.	C C',	Commutators.
T,	Flame tube.	G,	Galvanometer.
E E',	Electrodes.	V,	Voltmeter.
E'_2,	Upper surface of upper electrode.	B,	Battery.
E_2,	Upper surface of lower electrode.		

small brass pegs fixed in the upper surface of the disk. The wires were 0·3 centim. apart.

The lower electrode was connected through a galvanometer to one pole of a battery of small secondary cells, and the upper electrode was connected to the other pole. The flame tube was also connected to the same pole of the battery as the lower electrode. The potential difference between the two electrodes could be measured by means of a multicellular electrostatic voltmeter when it was above 250 volts. A circular patch at the centre of the grating on each electrode was heated by the flame. The lower electrode was nearly white hot usually, and the upper red hot, more or less, according to its position.

Diagram No. 1 shows the relation between the current and E.M.F. when a $\frac{1}{20}$ normal solution of potassium carbonate was sprayed for four positions of the upper electrode, which was charged positively, the lower electrode being in the same position in each case, viz., 5·6 centims. above the flame tube.

Diagram No. 1.

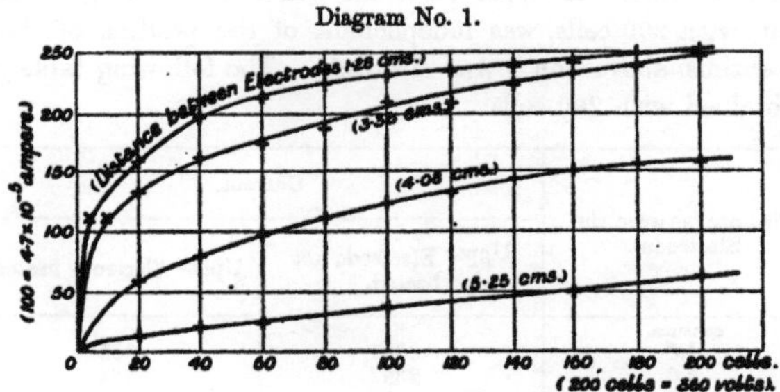

It will be observed that as the distance between the electrodes is increased, the E.M.F. necessary to produce approximate saturation increases very rapidly, but the saturation value of the current, where it is actually reached, is independent of the position of the upper or positive electrode. Diagram No. 2 shows the relation between the current with 200 cells and the distance between the electrodes, the lower electrode being kept fixed and negatively charged as before.

Diagram No. 2.

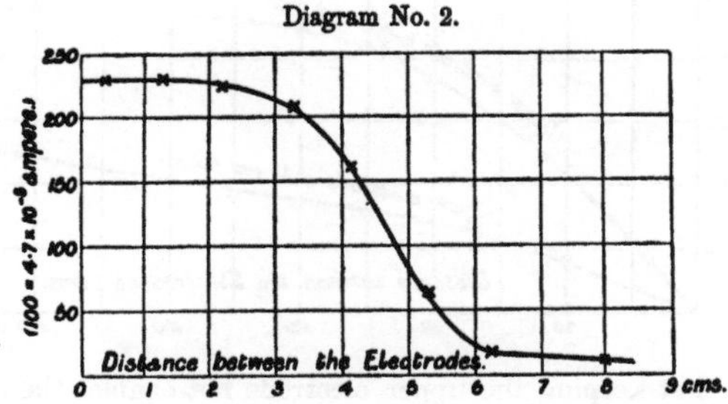

The falling off in the current as the upper electrode is raised is very rapid at from 3 to 6 centims. distance. Up to 3 centims. distance the temperature of the upper electrode remained sensibly constant, but above this distance it became cooler as the electrode was raised higher in the flame, and at 6 centims. distance it was only red at the centre, but it remained visibly hot even at 9 or 10 centims. above the lower electrode. It appeared probable that the cooling of the upper electrode might affect the observed current, and to test this point an electrode was constructed similar to those already described, but so arranged that the grating of platinum wires could be heated by passing an electric current through it. The grating was insulated by strips of mica, and the battery used to heat up the grating was insulated, and one end of the grating wire connected to the brass disk supporting it.

It was found that when the upper electrode was kept at a bright red heat in this way the current, with 200 cells, was independent of the position of the upper electrode up to 8 centims. above the lower electrode. The following table gives some of the numbers obtained with 200 cells.

Distance between the Electrodes.	Current.	
	Upper Electrode not heated.	Upper Electrode heated.
centims.		
1·3	235	235
3·0	236	234
4·2	180	230
6·2	18	227
8·0	10	235

Diagram No. 3 shows the change in the relation of the current to the E.M.F. when the upper electrode is heated in this way.

Diagram No. 3.

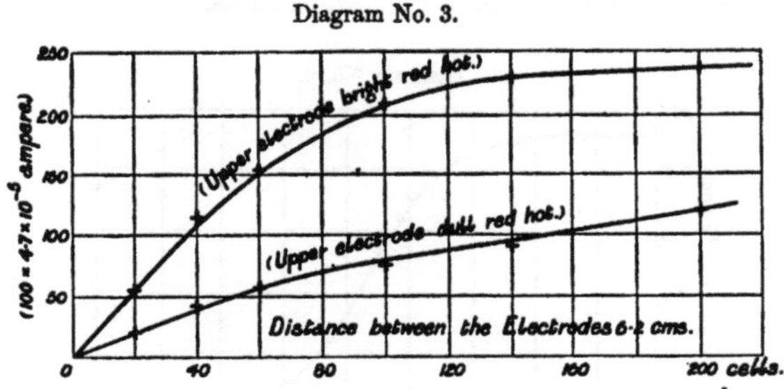

Thus it appears that keeping the upper electrode hot enables the current to attain its saturation value with a much smaller E.M.F. than is necessary when the electrode is not specially heated.

Diagram No. 4 shows the relation between the current and E.M.F. for several positions of the upper electrode when this is negatively charged, the lower electrode being in the same position in each case as before.

Diagram No. 4.

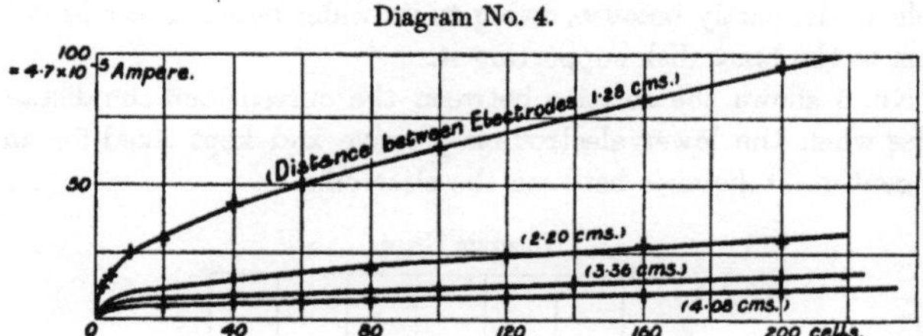

In this case, in which the positive or lower electrode is hotter than the negative electrode, the current does not show much sign of arriving at a saturation value as the E.M.F. is increased. To see whether this depends on the relative positions of the electrodes with reference to the direction of motion of the flame gases, the lower electrode was fixed very near the base of the flame, so that it was less heated than the upper electrode. It was then found that when the lower electrode was positive, the current became nearly saturated with about 100 cells, whereas when the lower electrode was negative, the current showed no sign of attaining a saturation value.

Thus the saturation of the current depends on the temperatures of the electrodes, and not on the motion of the flame gases.

Diagram No. 5 shows the relation between the current and E.M.F. when the distance between the electrodes was 0·3 centim., the lower electrode being slightly hotter than the upper electrode.

Diagram No. 5.

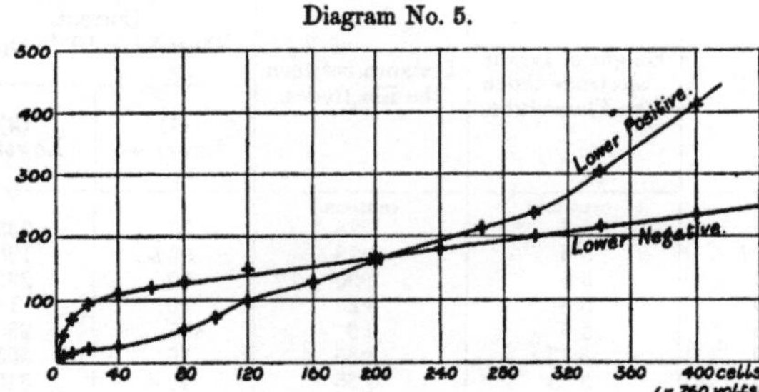

When the E.M.F. is less than 150 volts the two curves are similar to those obtained when the electrodes were at greater distances apart, but at higher E.M.F.'s the current with the lower electrode positive increases rapidly with the E.M.F., and becomes greater than that with the lower electrode negative. In all these experiments except those in which the upper electrode was heated by an electric current, the heated surface on the upper electrode was considerably greater than that on the lower electrode, owing to the upper electrode being of finer gauze than the lower electrode, but the lower electrode was usually heated to a higher temperature than the upper electrode, partly because, owing to its wider mesh, it lost heat less readily by conduction to the brass disk supporting it.

Diagram No. 6 shows the relation between the current and the distance between the electrodes when the lower electrode is positive and kept fixed for an E.M.F. of 70 volts per centim. of distance between the electrodes.

Diagram No. 6.

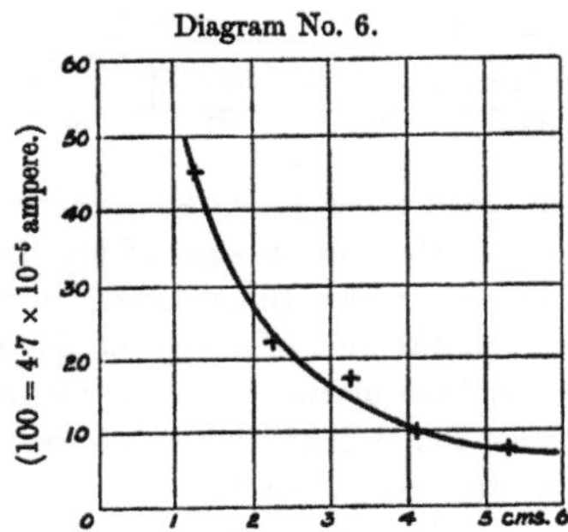

The current was always greater when the hotter electrode was negative than when it was positive, except when the electrodes were very near together, so that a very great electromotive intensity could be applied.

The following table gives some of the currents observed showing this :—

E.M.F.	Height of Lower Electrode above the Flame tube.	Distance between the Electrodes.	Current. (100 = $4\cdot7 \times 10^{-5}$ ampere.)	
			(1) Lower +	(2) Lower −
volts.	centims.	centims.		
335	5·6	1·28	75	233
100·5	5·6	1·28	36·5	198
330	5·6	2·20	30	233
370	5·6	6·2	10	18
725	5·6	0·3	407	235
350	3·65	3·35	16	355
175	3·65	3·35	10·5	315

The theoretical bearing of the results described in this section of the paper is discussed in Section 7.

(3.) *The Fall of Potential between the Electrodes.*

To examine the fall of potential along the flame between the electrodes, a horizontal insulated platinum wire was put in the flame, and its potential measured either by means of a quadrant electrometer or by connecting it through a galvanometer to a point on the battery used to charge the electrodes and adjusting the position of the wire until no current passed through the galvanometer. The wire and connections with the electrometer are shown in fig. 2.

The wire took up the potential of the flame very quickly, so that even if it was connected to earth through a high resistance its potential was not affected appreciably. The wire was always kept as nearly as possible so as to pass through the axis of the flame; if this was not done the potential curves obtained were considerably altered, although their general character remained the same.

Diagram No. 7 shows some of the results obtained when the electrodes were 3·8 centims. apart, the upper electrode being positively charged and the salt solution sprayed a $\frac{1}{50}$ normal rubidium chloride solution.

Diagram No. 7.

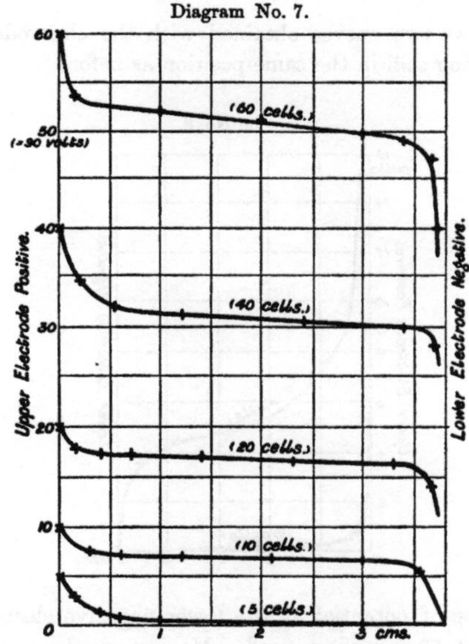

In this case, in which both of the electrodes were bright red hot, the fall of potential between the electrodes is very similar to that observed in gases at low pressures. Near each electrode there is a rapid fall of potential and in the intervening space an approximately uniform small potential gradient. This potential gradient is approximately proportional to the potential difference between the electrodes. The variation of the drop of potential at the negative electrode with the potential difference between the electrodes can be represented by the formula

$$0{\cdot}873E - 4{\cdot}4,$$

where E is the P.D. between the electrodes expressed in terms of the E.M.F. per cell used (1·80 volts). When E is less than about 5 the drop at the negative electrode is zero.

The following table shows this :—

E.	0·873E − 4·4.	Negative drop observed.
60	48·0	48·5
40	30·5	30·0
20	13·1	16·0
10	4·3	6·0
5	0·0	0·0

Diagram No. 8 shows two curves obtained with the electrodes 5 centims. apart, the lower electrode being still in the same position as before.

Diagram No. 8.

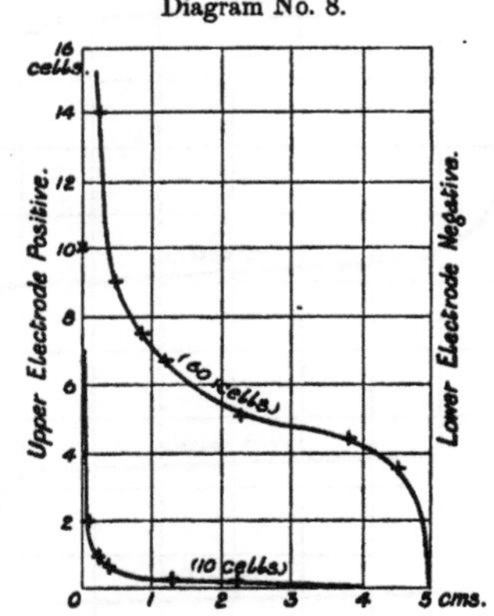

In this case the drop of potential at the lower negative electrode did not become appreciable until about 60 cells were used. Moving up the upper electrode to the colder parts of the flame rapidly increased the P.D. at which the negative drop appeared. Thus at 8·8 centims. above the lower electrode there was no negative drop even when 400 cells were used.

When the upper electrode was charged negatively the character of the potential curves was completely changed. In this case nearly all the fall of potential occurred near to the upper negative electrode. Diagram No. 9 shows two curves got with the electrodes 5 centims. apart. Diagrams 10 and 11 each show two curves, one with the upper electrode positively and the other with it negatively charged.

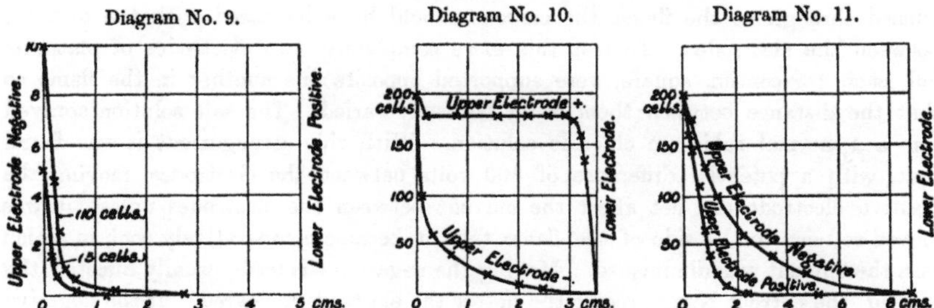

(4.) *The Ionisation of the Salt Vapour.*

ARRHENIUS ('Wied. Ann.,' 42, p. 18, 1891) concluded from the results of his experiments that the conductivity of salt vapours in flames is due to partial ionisation of the salt by the high temperature of the flame, and that the conductivity of a flame containing a salt vapour is very analogous to the conductivity of an aqueous solution of the salt. In our paper referred to above we have not seen any reason to doubt the general accuracy of ARRHENIUS' conclusions.

There are, however, a number of important facts which do not readily lend themselves to explanation by the hypothesis just mentioned. The phenomena of unipolar conduction are among these. HITTORF (see WIEDEMANN'S 'Elektricität,' vol. 4 B) showed that the current depends very greatly on the negative electrode, and that it is greater when a bead of salt in the flame is near the negative electrode than when it is near the positive electrode, and he concluded that nearly all the resistance to the passage of the current is at or near to the surface of the negative electrode, at any rate in the case of flames free from salt vapour.

The experiments of ARRHENIUS and those described in our paper referred to above, in which the current between two electrodes very near together in the flame was measured, were not adapted for the examination of unipolar conduction and allied phenomena, and it was sufficient, in considering the results obtained, to suppose the conductivity due to ionisation of the salt vapour, without making any further hypothesis as to exactly how and where the ionisation occurs. At the same time, it was more or less tacitly assumed that the salt vapour is ionised throughout the volume of the flame, just as a salt is ionised in an aqueous solution.

I have concluded, from the results described in this paper, that the ionisation of the salt takes place entirely, or very nearly so, at the surface of the glowing platinum electrodes, and not throughout the volume of the flame. The experiments described above on the variation of the current with the distance between the electrodes, show that when the two electrodes are both kept hot, then the saturation current is independent of the distance between the electrodes, whereas if the salt vapour were ionised throughout the flame, the current should have increased with the distance between the electrodes. To test this more completely, two electrodes of platinum foil, each 1·5 centim. square, were supported opposite one another in the flame so that the distance between them could be easily varied. The salt solution sprayed was a $\frac{1}{50}$ normal rubidium chloride solution. With this arrangement it was found that, with a potential difference of 800 volts between the electrodes, moving the positive electrode did not affect the current between the electrodes unless it was moved so near to the side of the flame that it became comparatively cool, in which case the current was diminished. Moving the negative electrode usually affected the current, the current being greater the hotter the electrode appeared. If the negative electrode was placed at about the axis of the flame, then it could be moved several millimetres either way without appreciably affecting the current, but the effect of moving it to at all near the sides of the flame was to diminish the current, this effect being much more marked than in the case of the positive electrode.

The amount of salt vapour passing between the electrodes is roughly proportional to the distance between them, so that since the potential difference used was fully enough to approximately saturate the gas, the current should have increased with the distance between the electrodes if the ionisation took place throughout the volume of the flame.

If the two platinum electrodes just described were placed one on each side of the flame, just far enough from it not to be visibly heated, and about half-way up the flame, only a very small current could be passed between them, even when an E.M.F. of 400 volts was applied. This current, moreover, was only slightly increased when the flame was filled with a salt vapour. The following are the currents observed in one case with 45 volts E.M.F. :—

(1.) Both electrodes just outside the flame and not visibly hot. Distance between the electrodes 2·5 centims.—

Current without salt . . . . 3 divisions.
Current with $\frac{1}{50}$ Rb_2CO_3 . . . . . 12 ,,

(2.) Both electrodes just inside the flame and red hot. Distance between the electrodes 2 centims.—

Current without salt . . . . 18 divisions.
Current with $\frac{1}{50}$ Rb_2CO_3 . . . . . 610 ,,

It is clear that the heating of the electrodes enormously increases the available conductivity, exactly as though the ionisation did not take place unless the electrodes were red hot. There is, nevertheless, a very small amount of conductivity even when cold electrodes are used, as has been known for a long time. This is no doubt due to a small amount of ionisation really taking place throughout the volume of the flame, both in the case of the flame gases and of the salt vapour.

This conductivity is, however, only a minute fraction of that which is observed when the electrodes are both hot enough to glow. If one electrode only is hot, then the current is much greater than when both are cold, but is still small compared with that obtained when both are hot. The explanation of this is considered in Section 7.

If a piece of platinum foil is put in the flame midway between the two electrodes when both are just outside the flame and not visibly hot, then the current is greatly increased, showing that the presence of the glowing platinum enables ionisation to take place. The following currents were observed with an E.M.F. of 45 volts and $\frac{1}{80}$ Rb_2CO_3 :—

(1.) Both electrodes not visibly hot. Current 12 divisions.
(2.) With a piece of platinum foil in the flame between the electrodes and insulated. Current 40 divisions.
(3.) Foil connected to earth. Current 400 divisions.

The great increase in the current on connecting the foil to earth appears to be due to the much greater ease with which an electrode in the flame loses negative electrification than positive. This causes the foil when insulated to be positively charged to nearly the same potential as the positive electrode, which diminishes the current. The explanation of this will be considered in Section 7.

The effect of putting a small bead of salt near the electrodes was also tried, the flame being otherwise free from salt. It was found that if the salt vapour only came in contact with the positive electrode, then the increase in the current due to the salt was very small, whereas if the salt vapour came in contact with the negative electrode the current was greatly increased. If the salt vapour passed between the electrodes without coming in contact with either, then the current was not increased at all. The following numbers were obtained with a bead of potassium carbonate :—

(1.) Flame without salt. Current 20 divisions.
(2.) Salt vapour on positive electrode. Current 60 divisions.
(3.) Salt vapour on negative electrode. Current 720 divisions.

It is clear from this that unless the salt vapour actually comes into contact with the glowing electrodes, the conductivity of the flame is not affected by its presence.

GIESE ('Wied. Ann.,' vol. 17, p. 517, 1882) showed that when two pairs of electrodes were placed one above the other in a flame free from salt, or rather in the gases immediately above the flame, then applying an E.M.F. to the lower pair diminishes the conductivity between the upper pair. This effect was evidently due to the removal of the ions from the stream of gas by the lower pair of electrodes, and should, therefore, not happen in the case of the conductivity of the salt vapour if there is no ionisation of the vapour throughout the volume of the flame. Two pairs of electrodes were arranged as shown in fig. 3. The upper pair was supported by a glass tube passing through the tin plate above the flame, and could be fixed at any height above the lower pair. The battery and galvanometer connected with the upper pair were insulated so that the electrodes took up the potential of the flame. It was found that applying an E.M.F. to the lower electrodes affected the current between the upper pair very little, if at all, in the case of the conductivity due

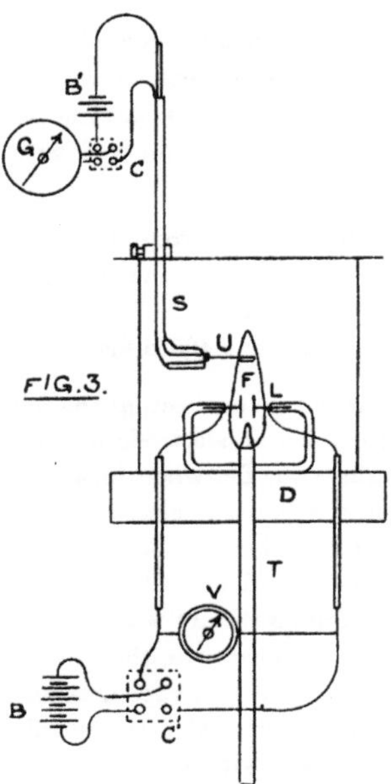

U, Upper electrodes. L, Lower electrodes. S, Tube supporting upper electrodes.

to a salt vapour. With a $\frac{1}{10}$ normal solution of potassium carbonate the following currents were obtained; the upper electrodes being about 3 centims. above the lower :—

(1.) No E.M.F. on the lower pair—
Current between the upper pair . . . . 66 divisions.

(2.) 360 volts on the lower pair—
Current between the upper pair . . . . 65 „

With 700 volts on the lower pair the current between the upper pair was slightly increased or diminished, according to the direction of the current between the lower pair, showing that the upper pair took up some of the current from the lower electrodes.

This effect became less as the distance between the two pairs was increased.

The results of this experiment are therefore entirely in accord with the view that the ionisation of the salt vapour takes place only at the surface of the glowing electrodes.

Another experiment very strikingly in favour of this view was also tried. The two horizontal electrodes of platinum gauze were fixed in position about 4 centims. apart in the flame. A bead of potassium carbonate or other salt was then held in the flame between the two electrodes so that the salt vapour only came in contact with the upper electrode. It was found that the current was almost independent of the height of the bead above the lower electrode, unless it was brought so near to the lower electrode that the salt vapour from the bead came into contact with the glowing platinum, when an increase in the current between the electrodes occurred, which was very great if the lower electrode was negatively charged.

The following currents were obtained with a bead of lithium carbonate and potential difference of 380 volts :—

(1.) Bead very near the lower electrode—

Current with lower electrode negative . .	130·0 divisions.
,, ,, ,, positive . .	4·3 ,,

(2.) Bead 0·2 centim. above lower electrode—

Current with lower electrode negative . .	84·0 divisions.
,, ,, ,, positive . . .	2·8 ,,

(3.) Bead 1 centim. above lower electrode—

Current with lower electrode negative . .	3·5 divisions.
,, ,, ,, positive . . .	2·6 ,,

To further test the view that the salt vapour is not ionised in the flame, except at the surface of the glowing electrodes, the conductivity of the flame alone for very rapidly-alternating currents was compared with that of the flame containing salt vapour by the method described by Professor J. J. Thomson ('Cambridge Phil. Soc. Proc.,' vol. 8, Part V.).

The outer coatings of two Leyden jars were connected through two coils of wire, each consisting of five or six turns of well-insulated wire, and the inner coatings were charged by means of a Wimshurst machine; so that when the charges on the inner coatings were allowed to discharge to each other rapid electrical oscillations passed through the two coils. An electrodeless discharge bulb containing bromine vapour placed in one of the coils served to indicate, by the intensity of the light from its discharge, the absorption of energy when a conductor was placed in the other coil. With this arrangement the conductivity of a large Bunsen flame could be distinctly detected, but the introduction of salt into the flame produced little or no effect, although enough salt was introduced to have increased the current between two electrodes in the flame by several hundred times.

It thus appears that nearly all the ionisation of the salt vapour, to which the conductivity of the flame is due, takes place at the surfaces of the glowing electrodes, although the electrodes are certainly colder than the flame gases. If we regard a molecule of the salt as consisting of oppositely-charged ions or electrons held together by the attractions between their charges, then, when the molecule is very near to a conductor like the electrode in the flame, the induced charges on the conductor diminish the attraction between the ions composing the molecule which may enable it to be ionised, even while the molecules not near the electrodes are quite stable, though at a higher temperature.

(5.) *The Relative Velocities of the Ions in the Flame.*

The way in which the current through the flame depends on the temperature of the negative electrode indicates that the part played by the negative ions in carrying the current is more important than that played by the positive ions. This fact, and the results obtained in investigating the fall of potential between the electrodes, suggested the idea that the velocity of the negative ions, due to a given potential gradient, is much larger than the corresponding velocity of the positive ions.

To test this, experiments were made in which the potential difference between the upper and lower electrodes necessary to cause the positive or negative ions to move down the flame against the upward stream of gases was determined. The apparatus used for this purpose is shown in fig. 4.

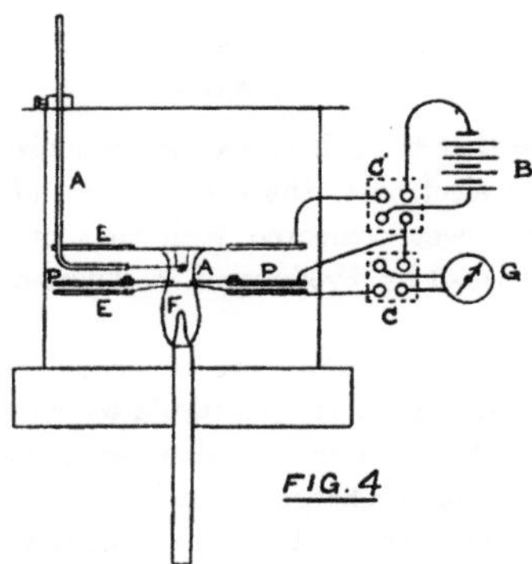

AA, bead of salt and support. P, screen above lower electrode.

A bead of salt was put in the flame between the electrodes, and the current from the lower electrode measured. It was found that when the upper electrode was positive, introducing the bead caused no increase in the current unless the potential difference between the electrodes was greater than 100 volts, when the electrodes were 5 centims. apart. When the upper electrode was negative, other conditions being the same, the current increased on introducing the bead, even with a potential difference of 1 or 2 volts.

To prevent any ions reaching the lower electrode by passing down the sides of the flame where the velocity of the blast is small, a screen, PP (fig. 4), was placed above the lower electrode. The hole in the platinum gauze of the screen at D was 2 centims. in diameter, and it was completely filled by the flame, which also passed through the gauze round the hole. The platinum grating on the lower electrode was bent up so that it was only 2 or 3 millims. below the gauze screen. Diagram No. 12 shows the results obtained with a bead of potassium carbonate.

Diagram No. 12.

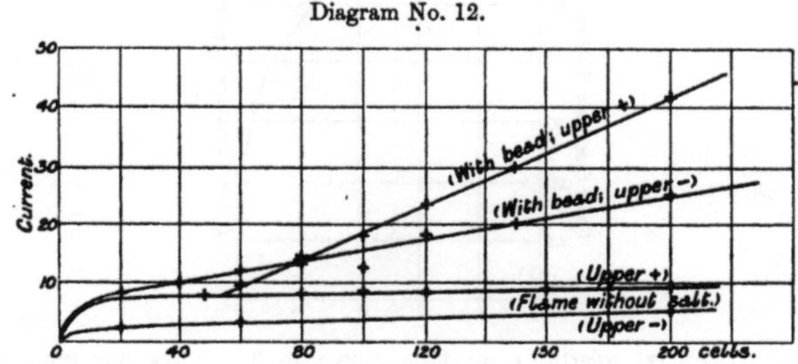

It will be observed that the introduction of the bead produced no increase in the current when the upper electrode was positive until a definite E.M.F. was applied. It was possible to determine this E.M.F. within about 5 volts with certainty with salts of Cs, Rb, and K. With salts of sodium and lithium the amount of current obtained was small compared with the current through the flame without salt, which made it impossible to determine the necessary E.M.F. with any approach to accuracy. The following table gives the results obtained with carbonates of the alkali metals for the positive ions:—

Cæsium	104 volts.
Rubidium	100 ,,
Potassium	107 ,,
Sodium	90 to 110 ,,
Lithium	90 to 100 ,,

The E.M.F. between the electrodes 5 centims. apart required to cause the positive ions of the carbonates of K, Rb, Cs, Na, and Li to move down against the blast of the flame is about 100 volts, so that the positive ions of these salts must all have nearly the same velocity due to the slope of potential in the flame. With beads of the chlorides of the same metals the same result appeared to hold, but the much greater volatility of the chlorides made it difficult to obtain satisfactory results.

To compare the velocity of the positive ion due to a potential gradient of 1 volt per centim. with the velocity of the flame gases, it is necessary to know the potential gradient between the electrodes when the P.D. is just enough to make the ions go down the flame. Diagram No. 13 shows the fall of potential in the flame without salt when this is the case with a P.D. of 107 volts between the electrodes 5 centims. apart, the upper electrode being positive.

Diagram No. 13.

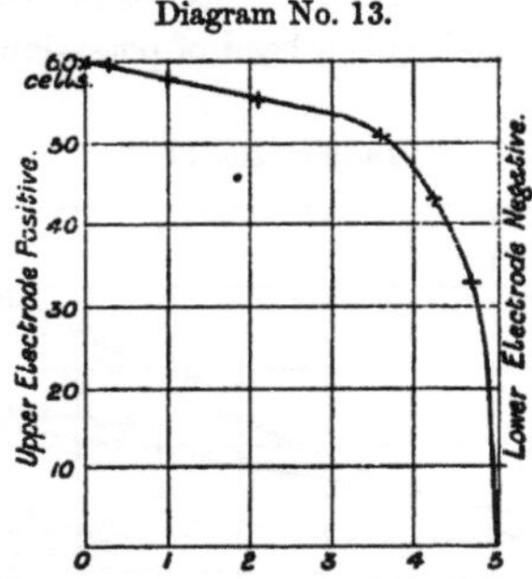

The smallest potential gradient on this curve is 3·3 volts per centim., which is, therefore, the gradient necessary to make the positive ions of the alkali metal salts move against the gases of the flame.

The mean velocity of the mixture of gas and air in the tube leading to the flame was approximately 206 centims. per second, as determined from the volumes of gas and air supplied to the apparatus. On entering the flame the gases of course expand very greatly, but the effect of this expansion is probably not great on the upward velocity of the gases, since the flame is free to expand laterally. Taking the velocity of the gases in the flame as 206 centims. per second, the velocity of the positive ions in the flame due to a potential gradient of 1 volt per centim. is therefore

$$\frac{206}{3 \cdot 3} = 62 \frac{\text{centims.}}{\text{sec.}}.$$

In determining the P.D. necessary to make the negative ions move down against the flame gases, a long wire through which a current was passing was used to supply the potential differences required. Diagram No. 14 shows the results obtained with a bead of Na_2CO_3.

When the E.M.F. is less than 1 volt, there is a small current opposed to the applied E.M.F. At about 1 volt the current with the bead suddenly begins to increase rapidly, with the E.M.F. indicating that the negative ions have begun to go down against the blast of the flame.

The explanation of the small inverse current when the E.M.F. is small is not easy. It was increased by the introduction of the bead, which may have been due to some salt particles getting carried round to the lower electrode by the circulation of the air between the electrodes. The presence of this inverse current prevented the E.M.F. necessary to bring down the negative ions being determined very exactly, as it varied with the size of the bead and with the volatility of the salt used.

The P.D. required by the negative ions was about 1 volt for both oxy and haloid salts of any of the alkali metals. Taking the potential gradient corresponding to this as 0·2 volt per centim. gives for the velocity of the negative ion due to 1 volt per centim.

$$1030 \frac{\text{centims.}}{\text{sec.}},$$

which is 17 times the corresponding velocity of the positive ions.

Diagram No. 14.

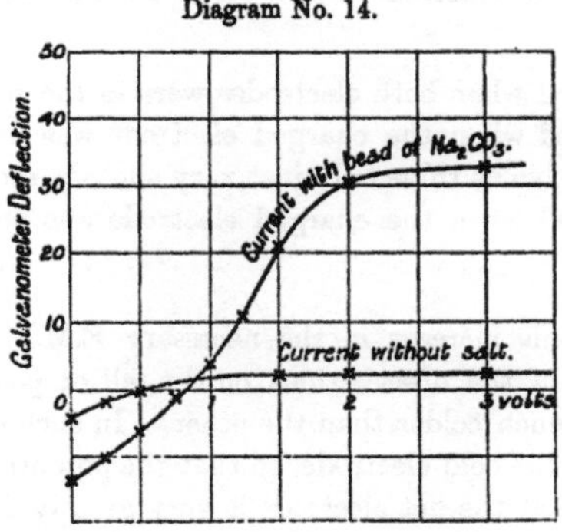

Another method of estimating the velocities of the ions was also tried. Two electrodes of platinum foil were fixed opposite each other in the flame, about 1·5 centim. apart, each electrode being 1·5 centim. square. A bead of salt was held just below one of these, so that the salt vapour only came into contact with this electrode which was connected to earth. The other electrode was charged, and any leak from it measured by the galvanometer. It was found that the current was not increased by the presence of the bead of salt unless the E.M.F. used was greater than a definite amount, which was much greater when the electrode was negatively charged than when it was positively charged. This E.M.F. necessary for the current to increase when the salt is introduced is evidently that required to drag the ions from one electrode to the other, across the blast of the flame gases. The salt vapour from the bead rapidly spreads out in the flame, so that unless the charged electrode was placed near the edge of the flame, the vapour came in contact with both electrodes, in which case the current was increased by the presence of the salt even with very small E.M.F.'s.

It is difficult to form any estimate of the velocities from this experiment, because the ions must spread out from the electrode at which they start by mere diffusion, which will make the velocity appear too great. The fall of potential between the electrodes also is not uniform, and cannot be determined conveniently when the electrodes are so near together. Nevertheless, the results obtained were of the same order of magnitude as those obtained by the method described above. When the electrode was positively charged an E.M.F. of 0·25 volt was enough to increase the current when the salt vapour was introduced. This gives for the velocity of the negative ions 1200 $\frac{\text{centims.}}{\text{sec.}}$. If the positively-charged electrode was placed outside the flame, then the E.M.F. necessary for the current to increase when the salt vapour was introduced at the other electrode was much greater than before. The following numbers show this. The electrodes were kept at the same distance apart, 1·5 centim. in each case.

(1.) E.M.F. required when both electrodes were in the flame . . 0·25 volt.
(2.) E.M.F. required when the charged electrode was at the side of the flame, so as to be only just very slightly red hot . . 1·2 volt.
(3.) E.M.F. required when the charged electrode was outside the flame . . . . 2·3 volts.

The explanation of this increase in the necessary E.M.F. can be readily seen by considering the results of the observations on the fall of potential between the two electrodes when one is much colder than the other. In such a case nearly all the fall of potential occurs near the cold electrode, so that the potential gradient available for dragging across the ions at the hot electrode is very greatly diminished.

A third estimate of the velocity of the negative ions was obtained by diverting the current between two electrodes, one above the other, in the flame by means of a pair of charged electrodes placed outside the flame. Two small platinum electrodes were fixed one above the other in the flame, and 3 centims. apart. The two platinum foil electrodes already described were placed one on each side of the flame, and on a level with the lower of the small electrodes. The lower small electrode was put to earth and the upper one charged positively to $\frac{1}{10}$ of a volt, and the current from it measured. The two large electrodes were then charged positively, when it was found that the current observed was diminished and reversed if they were charged to more than 0·4 volt. Since the observed current was probably mainly carried by negative ions moving upwards with the blast, because the E.M.F. used was not enough to drag down any positive ions, the stoppage of the current on charging the large electrodes must have been due to the negative ions being dragged across from the lower small electrode to the large electrodes. These latter were 2 centims. apart, so that the potential gradient between them and the lower small electrode was anyhow less than 0·4 volt. per centim. This gives for the velocity of the negative ions, supposing that they travelled from the lower small electrode to the tops of the larger electrodes, which were 0·7 centim. higher up and 1 centim. distant horizontally,

$$\frac{206}{0\cdot4 \times 0\cdot7} = 740 \frac{\text{centims.}}{\text{sec.}}.$$

This result agrees quite as well as can be expected with the other two estimates made of the velocity of the negative ions due to a potential gradient of 1 volt per centim.

To compare the velocities of the negative ions of various salts some measurements were made of the E.M.F. necessary to drag them from an electrode in the flame to one placed opposite to it, but outside the flame. As I have already mentioned, this E.M.F. increases rapidly as the distance of the outside electrode from the flame is increased. By placing it about 0·5 centim. from the edge of the flame the E.M.F. necessary was increased to about 12·5 volts. This necessary E.M.F. was determined for the following salts, KCL, NaCL, Li_2CO_3, KF, KI, KBr, K_2CO_3, and was found in every case to lie between 12 and 13 volts. Thus it appears that the velocity of the negative ions is independent of the nature of the salt from which they are derived in the flame.

(6.) *The Relative Velocities of the Ions in Hot Air.*

The experimental difficulties in the determinations of the velocities of the ions in the flame described above prevented anything more than rough estimations of the respective relative velocities being obtained. I, therefore, endeavoured to devise a method in which the conditions of the experiments should be simpler and more under control. It is clear that the gases of a flame, even when this is kept in as steady a state as possible, are not a very suitable medium for accurate work. I, therefore, tried treating the electrodes by means of electric currents passing through them, and put salt on one or both electrodes. After several trials, with various forms of apparatus, this method was abandoned, owing to the difficulty of getting the air between the electrodes sufficiently hot to keep the salt vapour from condensing.

A new apparatus was now constructed, in which a current of very hot air practically replaced the flame in the experiments already described. The water pump and pressure-regulating apparatus, already described, were used to supply the air. The air was heated by passing it through a platinum tube, 50 centims. long and 1·3 centim. in diameter, maintained at a bright red heat in a Fletcher's gas-tube furnace. The stream of hot air so obtained would heat a platinum wire red hot 5 or 6 centims. from the end of the tube. To prevent the furnace gases getting into the air at the end of the tube, a plane flange of platinum, 6 centims. in diameter, was fitted to the end of the tube.

Fig. 5 shows the essential parts of the apparatus for getting the velocity of the ions in the stream of hot air.

F is the flange on the platinum tube. An electrode, A, consisting of a grating of platinum wires, 0·25 millim. in diameter and 2 millims. apart, wound on pegs in a brass disk, 5·5 centims. in diameter, was placed opposite to the flange, and the stream of hot air kept the grating red hot. This electrode could be charged up by means of battery of small accumulators, and so an electric field maintained between A and F. A small bead of the salt to be investigated was put midway between A and F in the current of hot air, and was supported by a platinum wire held by a glass tube, B.

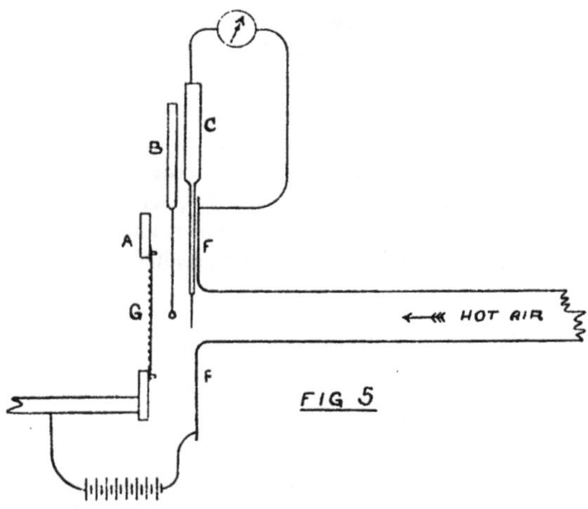

FIG 5

An electrode, C, was also introduced into the stream of hot air as near the mouth of the tube as possible. This consisted simply of a ring of fine platinum wire supported in a glass tube, which insulated the wire leading to the ring.

The ring electrode was connected to a galvanometer of 11,000 ohms resistance, and through this to the platinum tube, which was also connected to "earth." The bead of salt was slowly volatilised in the current of hot air and its vapour passed through the grating. On the hot grating some of it was ionised, and if the electric field between the grating and the ring electrode and flange was strong enough, some of the ions moved against the blast of air to the ring electrode, so producing a current which was indicated by the galvanometer.

It was thus possible to determine the P.D. necessary to make the ions move against the blast of hot air exactly in the same way as was done in the flame. If the velocity of the blast and the potential gradient corresponding to this necessary P.D. were known, the absolute velocity of the ions could be at once calculated, but, unfortunately, it is not possible to get anything more than rough estimates of these quantities. All that was attempted, therefore, was to determine the necessary P.D.'s for various ions under the same conditions. This gives the relative values of their velocities in the hot air.

The following diagram shows the variation of the current with the P.D. when the platinum flange and the grating electrode were 1 centim. apart, the grating being positively charged with beads of KI and KOH :—

Diagram No. 15.

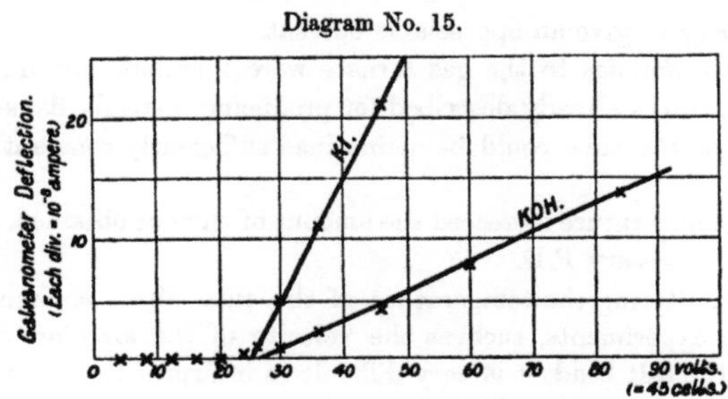

It will be observed that in each case the current is zero until a P.D. of about 25 volts. is reached, when it increases nearly uniformly with the P.D.

Other conditions remaining the same, the necessary P.D. should vary directly as the distance between the electrodes. Actually it was found that the necessary P.D. increased more rapidly than the distance, except when this was less than about 1·5 centim. Below 1·5 centim. distance, the necessary P.D. varied as the distance between the electrodes. Thus the following results were obtained with a Rb_2CO_3 bead :—

Distance between Electrodes.	Necessary P.D.
centims.	cells
1·5	17·5
1·0	12·5
0·8	9·5

With other salts similar results were obtained. In most of the measurements made the grating and the flange were kept 1 centim. apart.

The effect of varying the velocity of the air blast was also tried. It was found that for small variations of the velocity not greater than 20 per cent. of its usual value, the necessary P.D. varied directly as the velocity of the blast. The blast could not be increased more than about 20 per cent. with the apparatus available. For diminutions of the velocity of the blast greater than 20 per cent., the necessary P.D. diminished more rapidly than the velocity of the blast. This was probably due to the effects of accidental air currents and diffusion, which, of course, become more important when the blast of air is made less rapid. If the velocity of the blast was reduced to much less than one-half its usual value, the grating did not become hot at all, and practically no current at all could be obtained.

Any change in the temperature of the platinum tube produced a change in the amount of current obtained with any given P.D., and also changed the value of the least P.D. necessary to give an appreciable current.

The gas and air supplies to the gas furnace were therefore carefully regulated by means of the apparatus already described for producing a steady flame. In this way the temperature of the tube could be maintained sufficiently constant for any length of time.

Increasing the temperature increased the amount of current obtained, and diminished slightly the least necessary P.D.

Since, however, altering the temperature of the tube alters so many of the other conditions of the experiments, such as the velocity of the air blast and the rate of volatilisation of the salt bead, it is very difficult to interpret the meaning of this last result. Consequently in all the experiments the temperature of the tube was kept as nearly as possible the same.

The accuracy with which the necessary P.D. could be determined varied with the distance between the electrodes and with the temperature of the tube. If the electrodes were too near together, a current was obtained with any P.D., however small.

Increasing the temperature diminished the least distance at which satisfactory results could be obtained.

Finally, a distance of 1 centim. between the electrodes and a definite supply of gas and air to the furnace and to the tube were adopted, as giving the best results. The P.D. necessary to produce an appreciable current could be determined within 1 volt in any one experiment, and the results of different experiments with the same salt usually agreed within the same limit.

Occasionally larger discrepancies than this occurred, which, however, could generally be traced to some accidental circumstance affecting the supply of gas or air to the furnace.

The amount of current obtained with any one salt depended much less than might be expected on the size of the bead used. With some very volatile salts such as KI, a bead about 3 millims. in diameter only lasted two or three minutes, yet during nearly all this time the current with a given P.D. remained very nearly constant and only began to fall off when the bead had almost disappeared. This, of course, greatly facilitated the measurements, since it was not necessary to keep the bead very constant in size during a series of determinations of the current with different P.D.'s.

The rate at which the current increased with the P.D. after the least necessary P.D. was reached varied greatly with different salts, and with the sign of the charge on the grating. In general, the salt which was the more volatile gave the greater rate of increase of current with P.D. The rate of increase of current with P.D. was also usually greater when the grating was negatively charged than when it was positively charged.

When the grating was negatively charged it was found that the P.D. at which the current began to increase with the P.D. between the electrodes was, other conditions remaining unchanged, approximately 7·0 volts for all the salts tried, viz. :—

Na_2CO_3, NaOH, NaCL, K_2CO_3, KOH, KCL, KBr, KI, KF, LiCL,
Li_2CO_3, RbCL, Rb_2CO_3, CsCL, Cs_2CO_3, $CaCL_2$, $BaCL_2$, $SrCL_2$, $Ba(NO_3)_2$.

The following salts were also tried, but although they gave a large amount of current, it was so unsteady and lasted such a short time, owing to the decomposition of the salt, that no definite results could be obtained : $FeCL_3$, $AlCL_3$, $ZnCL_2$, $MgCL_2$.

When the grating was positively charged, the least necessary P.D. was very approximately 25 volts with each of the salts of Li, Na, K, Rb, and Cs mentioned above, and very approximately 48 volts for each of the salts of Ca, Sr, and Ba.

The mean velocity of the air blast was estimated to be about 180 $\frac{\text{centims.}}{\text{sec.}}$, which gives for the velocities of the ions due to one volt per centim. the following values :—

(1.) Negative ions . . . . 26 $\frac{\text{centims.}}{\text{sec.}}$.

(2.) Positive ions of salts of Li, Na, K, Rb, and Cs . 7·2 $\frac{\text{centims.}}{\text{sec.}}$.

(3.) Positive ions of salts of Ba, Sr, and Ca . . . . 3·8 $\frac{\text{centims.}}{\text{sec.}}$.

Only the relative value of these velocities has any pretensions to accuracy. As already explained, it was not found possible to obtain more than the order of magnitude of the absolute velocities.

The result obtained for the velocities of the ions in the flame, viz., that all the negative have the same velocity, and likewise the positive ions of salts of Li, Na, K, Rb, and Cs, was thus found to hold good also in hot air.

The velocities in the hot air are much smaller than those in the flame. The values given for the velocities in the hot air were obtained by assuming the fall of potential between the grating and the ring electrodes to be uniform. Since there was practically no current between them up to the least necessary P.D., this assumption cannot be very far from correct; still, any departure from a uniform fall would cause the values given to be too small.*

* Since the salt vapour is not ionised till it reaches the hot gauze, there are no ions between the gauze and flange unless the ions go against the blast.

Taking this source of error into account, it is still clear that the velocities in the flame are very much greater than those in the hot air. The average temperature of the flame may be taken as say 2000° C., whilst that of the hot air was nearly 1000° C.; consequently, if the size of an ion is supposed to be independent of the temperature, the velocity of the ions in the flame should have been about double that in the hot air. Actually the negative ions were found to move forty times quicker in the flame than in the hot air.

The results obtained by McClelland* for the velocities of the ions in the hot gases coming from flames appear to bear on this point. He found that the velocity of the ions rapidly diminished as the distance from the flame was increased, and this pointed to "a rapid condensation on the charged carrier of some uncharged body greatly increasing its mass."

The theoretical velocity of an ion supposed to be one atom, carrying the same charge that an ion does in solutions, can be calculated by means of the formula

$$u = \frac{Xe}{mk} D$$

where u is the velocity of the ion,

D the coefficient of interdiffusion of the ions and the gas,

m the mass of an ion,

e the charge on it,

X the electromotive intensity,

k the quotient of pressure by density for the ions.†

This gives for the velocity of an ion of molecular weight 32 in air at the ordinary temperature about 40 centims. per second. At a temperature of 2000° C., its velocity would be 300 centims. per second, and at 3000° C., about 400 centims. per second.

The velocity of the negative ions in the flame, 1000 centims. per second, is, therefore, of the same order of magnitude as the theoretical velocity of an ion consisting of one atom.

All the other velocities are less than this, which indicates that the ions consist of clusters of atoms, that is, assuming that they carry the same charge as the ions in electrolysis of solutions. This result is in agreement with those of Rutherford on the velocities of the ions in Röntgen ray and uranium ray conductivity, and of McClelland on the velocities of the ions in the hot gases coming from flames.

The size of this cluster appears to be much greater in the case of positively-charged ions than in the case of negatively-charged ions. Zeleny ('Phil. Mag.,' July, 1898) has shown that the velocity of the negative ions produced in gases by

* 'Phil. Mag.,' July, 1898.

† See J. J. Thomson, 'Brit. Assoc. Report,' 1894, Art., "Diffusion," 'Encyclopædia Britannica,' and E. Rutherford, 'Phil. Mag.,' Nov., 1897.

Röntgen rays is greater than that of the positive ions in air by about 25 per cent. McCLELLAND has found the same thing for the ions in the gases coming from flames, and RUTHERFORD has shown that the same result holds good for the ions produced by uranium radiation.

In the flame the negative ions of alkali salt vapours move 17 times as fast as the positive ions, and in air at 1000° C. 3·6 times faster with alkali salts, and 7 times faster with salts of Ba, Sr, and Ca.

It seems reasonable to suppose that since the ions in each of the three classes, viz.:

1 Negative ions,
2 Positive ions of alkali metal salts,
3 Positive ions of Ba, Sr and Ca salts,

have equal velocities, they are equal clusters of atoms. It thus appears that ions which in solutions have equal charges, have equal velocities in the gaseous state. This points to the conclusion that the size of the cluster of atoms forming a gaseous ion, depends, at a given temperature, only on the charge on the ion. Those ions, therefore, which have equal charges, have also equal velocities in the same medium.

(7.) *Conclusion.*

Since the ionisation on which the conductivity of the salt vapour depends takes place entirely at the surfaces of the glowing electrodes, there is therefore at the surfaces a thin layer in which very rapid ionisation and recombination are going on. The number of ions dragged out from the surface of an electrode will depend on the slope of potential at the surface, and if this is great enough to drag out all the ions of one sort before they can recombine the current will be as great as possible. Owing to the much greater velocity of the negative ions they will be far more easily dragged out than the positive ions, so that unless the slope of potential is great enough to drag out all of either kind of ions, the current from an electrode with a given slope of potential at it will be greater when the electrode is negatively charged than when it is positively charged.

Consider the case in which one electrode is white hot and the other comparatively cool, so that little or no ionisation occurs at it. In this case only one kind of ions will be present in the space between the electrodes, viz., those of the same sign as the hot electrode, so that there will be a charge in the gas which will diminish the slope of potential near the hot electrode and increase it near the cool electrode. The experimentally-determined slopes of potential with one electrode cool show this effect very clearly.

In this case, in which the fall of potential is nearly all at the cold electrode, the smallness of the potential gradient at the hot electrode is not favourable to the attainment of the saturation value of the current, and the current E.M.F. curves consequently continue to slope up, even with the greatest E.M.F.'s that could be applied.

If the electrodes are connected to the galvanometer simply without any battery, there is a small deflection indicating a current from the colder electrode to the hotter through the flame. This is easily explained by supposing that some of the negative ions at the electrode diffuse out, owing to their high velocity leaving an excess of positive behind. The blast of the flame will assist this action by blowing away the negative ions as soon as they get away from the surface of the electrode. According to this the hotter electrode loses more negative ions than the colder, since there are more ions formed at it. In the same way a wire immersed in the flame becomes positively charged. This action is almost exactly analogous to the charging of a polished zinc plate, when ultra-violet light falls on it, described by HALLWACHS, RIGHI, and others.

If both the electrodes are hot enough to produce ionisation, and if the fall of potential at the upper electrode is great enough to make the ions from it come down against the blast, then there will be both sorts of ions present in the space between the electrodes, which will modify the fall of potential. Since the positive ions move so much more slowly than the negative ions, they remain in the gas much longer and so have a greater effect on the fall of potential; consequently the fall of potential is much greater at the negative electrode than at the positive electrode. The great velocity of the negative ions enables a much smaller slope of potential to drag them all out than is required to drag out the positive ions, so that unless the E.M.F. applied is very great the current is mainly carried by the negative ions. These points are very well illustrated by the experiments described above, on the effect of specially heating the upper electrode when it is positive. When it is heated the positive ions coming from it transfer the fall of potential from the positive electrode to the negative electrode, so that all the negative ions there are dragged out and the current attains its saturation value as far as the negative ions are concerned.

If the positive electrode only is hot then the positive ions moving across will cause all the fall of potential nearly to be at the negative electrode, and owing to their small velocity this effect will be more complete than when the negative ions alone are present. Owing to this, and to the greater difficulty of dragging out the slow positive ions from the electrode, the current, when the positive electrode is hot, will be much smaller than when the negative electrode is hot. In this way it is easy to explain all the phenomena of unipolar conduction.

In the case where both electrodes are hot, the fall of potential being mostly at the negative electrode, the current due to the negative ions attains its saturation value at a comparatively small E.M.F., but the current carried by the positive ions continues to increase nearly uniformly with the E.M.F.

According to this then, in the equations

$$C = I + k_1 E,$$

$$I - i = k_2 \frac{i^2}{E^2},$$

I is the maximum current carried by the negative ions coming from the negative electrode, and k_1E is the current due to the positive ions, which is still far from its saturation value.

If E is made sufficiently great, the current due to the positive ions should also become saturated, and with very high E.M.F'.s it is found that the current increases less rapidly with the E.M.F. than the formula $C = I + k_1E$ represents.

The equation $I - i = k_2 \frac{i^2}{E^2}$ was originally obtained for the conductivity of Röntgenised gases in which most of the ionisation takes place throughout the volume of the gas. In this case $k_2 = \frac{\alpha l^3}{\epsilon U^2}$, where α is the constant in the equation $\frac{dN}{dt} = \alpha N^2$ representating the rate of recombination of the ions, l is the distance between the electrodes, ϵ the charge carried by each ion, and U the sum of the velocities of the positive and negative ions under an electromotive intensity of one volt per centim.

When all the ionisation takes place at the surfaces of the electrodes, the above signification of k_2 no longer holds, but the fact that the equation can still represent the observed relation between the current and E.M.F. shows that the general nature of the conductivity is the same in both cases.

In our paper referred to above, we showed that the conductivity of the halogen salts is approximately proportional to the square root of the concentration of the salt vapour in the flame. The conductivity of the oxysalts followed the same law when the concentration was very small, but with greater concentrations was nearly proportional to the concentration of the salt vapour. Now when the concentration of the salt vapour in the flame is very small, the amount of salt in an extremely thin layer at the surface of an electrode in the flame will be proportional to $C^{2/3}$, where C is the concentration of the salt vapour, that is, supposing that the mean distances between the molecules of salt is larger than the thickness of the layer in which ionisation takes place. Consequently the conductivity should vary as $C^{2/3}$ if it is supposed that the amount of ionisation is proportional to the amount of salt in this layer. If the concentration of the salt vapour in the flame is so great that the mean distance between the molecules of salt is small compared with the thickness of this layer, then the amount of salt in the layer will be proportional to the concentration of the salt in the flame, so that at such concentrations the conductivity should be proportional to the concentration.

In the following table the observed conductivity of KCL is compared with the calculated by assuming it proportional to (1) $C^{2/3}$, (2) $C^{1/2}$:—

Concentration of solution, sprayed.	Current calculated.		Current observed.*
	(1.)	(2.)	
0·2	(31·9)	(31·9)	31·9
0·1	22·6	20·1	21·0
0·05	15·9	12·7	14·1
0·01	7·14	4·33	6·0

It will be observed that in each case the observed current lies between the two calculated currents.

The results of the observations described above on the fall of potential between the electrodes show that there is a very close analogy between the conductivity of salt vapours in flames and the conductivity of gases at low pressures. In both cases there is a greater fall of potential near the negative electrode than near the positive electrode, with a small slope of potential in the intermediate space, and it seems very likely, therefore, that the peculiar form of the fall of potential in gases at low pressures is due, like that in the flame, to a great difference between the velocities of the positive and negative ions.

In conclusion, I desire to express my best thanks to Professor J. J. Thomson for many valuable suggestions during the course of these experiments, and also to Professor A. Smithells for his kindness in allowing me to continue the investigation of this subject begun in his laboratory.

* These numbers are taken from the previous paper on this subject.

TO THE

PHILOSOPHICAL TRANSACTIONS,

SERIES A, VOL. 192.

B.

D.

E.

F.

G.

H.

L.

M.

N.

P.

ERRATA.

P. 123, line 9. *For* $f'(\alpha', \beta', \gamma', \ldots)$, *read* $f(\alpha', \beta', \gamma', \ldots)$.

P. 128. The expression at the end of § 20 should be multiplied by k^2.

P. 131, line 8. *For* $2a^2/n$, *read* $2a^4/n$.

LONDON:
HARRISON AND SONS, PRINTERS IN ORDINARY TO HER MAJESTY,
ST. MARTIN'S LANE, W.C.

THE PHILOSOPHICAL TRANSACTIONS OF THE ROYAL SOCIET .

	£	s.	d.
1801. Part II...	0	17	6
1802. Part I...	0	11	0
—— Part II...	0	17	6
1803. Part I...	0	12	6
—— Part II...	0	13	6
1804. Part I...	0	10	6
—— Part II...	0	12	6
1805. Part I...	0	10	0
—— Part II...	0	11	6
1806. Part I...	0	13	6
—— Part II...	0	17	6
1807. Part I...	0	10	0
—— Part II...	0	15	6
1808. Part I...	0	9	6
—— Part II...	0	14	6
1809. Part I...	0	14	6
—— Part II...	1	0	6
1810. Part I...	0	10	6
—— Part II...	0	12	0
1811. Part I...	0	15	0
—— Part II...	0	15	0
1812. Part I...	0	17	6
—— Part II...	0	17	6
1813. Part I...	0	14	0
—— Part II...	0	18	0
1814. Part I...	0	18	0
1815. Part II...	1	2	0
1816. Part II...	0	17	6
1819. Part III...	0	10	0
1823. Part II...	1	8	0
1824. Part I...	0	12	6
—— Part II...	1	0	0
—— Part III...	1	4	0
1825. Part I...	1	4	0
—— Part II...	2	0	0
1826. Part I...	1	2	6
—— Part II...	0	12	6
—— Part III...	2	0	0
—— Part IV...	1	2	6
1827. Part I...	0	18	0
—— Part II...	0	18	0
1828. Part I...	1	1	0
—— Part II...	0	10	0
1829. Part I...	0	16	0
—— Part II...	0	14	0
1830. Part I...	1	10	0
—— Part II...	1	1	0
1831. Part I...	1	10	0
—— Part II...	1	12	0
1832. Part I...	1	1	0
—— Part II...	2	0	0

	£	s.	d.
1833. Part I...	1	1	0
—— Part II...	2	18	0
1834. Part I...	0	17	0
—— Part II...	2	2	0
1835. Part I...	1	2	0
—— Part II...	0	14	0
1836. Part I...	1	10	0
—— Part II...	2	0	0
1837. Part I...	1	8	0
—— Part II...	1	8	0
1838. Part I...	0	13	0
—— Part II...	1	8	0
1839. Part I...	0	18	0
—— Part II...	1	1	6
1840. Part I...	0	18	0
—— Part II...	2	5	0
1841. Part I...	0	10	0
—— Part II...	1	10	0
1842. Part I...	0	16	0
—— Part II...	1	2	0
1843. Part I...	0	10	0
—— Part II...	1	10	0
1844. Part I...	0	10	0
—— Part II...	1	10	0
1845. Part I..	0	16	0
—— Part II...	1	0	0
1846. Part I...	0	7	6
—— Part II...	1	12	0
—— Part III...	1	12	0
—— Part IV...	1	12	0
1847. Part I...	0	14	0
—— Part II...	0	16	0
1848. Part I...	1	0	0
—— Part II...	0	14	0
1849. Part I...	1	0	0
—— Part II...	2	5	0
1850. Part I...	1	10	0
—— Part II...	3	5	0
1851. Part I...	2	10	0
—— Part II...	2	10	0
1852. Part I...	1	0	0
—— Part II...	2	5	0
1853. Part I...	0	18	0
—— Part II...	0	12	0
—— Part III...	1	2	0
1854. Part I...	0	12	0
—— Part II...	0	16	0
1855. Part I...	0	16	0
—— Part II...	1	6	0
1856. Part I...	2	0	0
—— Part II...	1	4	0

	£	s.	d.
1856. Part III...	1	4	0
1857. Part I...	1	8	0
—— Part II...	1	4	0
—— Part III...	1	2	0
1858. Part I...	1	8	0
—— Part II...	3	0	0
1859. Part I...	2	10	0
—— Part II...	2	5	0
1860. Part I...	0	16	0
—— Part II...	2	1	6
1861. Part I...	1	3	0
—— Part II...	1	5	0
—— Part III...	1	7	6
1862. Part I...	2	14	0
—— Part II...	3	0	0
1863. Part I...	1	14	0
—— Part II...	1	7	6
1864. Part I...	0	11	0
—— Part II...	1	7	6
—— Part III...	1	10	0
1865. Part I...	2	2	0
—— Part II...	1	5	0
1866. Part I...	1	14	0
—— Part II...	2	7	6
1867. Part I...	1	3	0
—— Part II...	1	15	0
1868. Part I...	2	5	0
—— Part II ..	2	0	0
1869. Part I...	2	10	0
—— Part II...	3	3	0
1870. Part I...	1	10	0
—— Part II...	1	18	0
1871. Part I...	1	10	0
—— Part II...	2	5	0
1872. Part I...	1	12	0
—— Part II...	2	8	0
1873. Part I...	2	10	0
—— Part II...	1	5	0
1874. Part I...	2	8	0
—— Part II...	3	0	0
1875. Part I...	3	0	0
—— Part II. .	3	0	0
1876. Part I...	2	8	0
—— Part II...	2	8	0
1877. Part I...	1	16	
—— Part II...	2		
Vol. 168 (extra)			
1878. Part I.			
—— Part			
1879. P			
——			

	£	s.	d.
1880. Part I...	2	5	0
—— Part II...	2	0	0
—— Part III...	1	1	0
1881. Part I...	2	10	0
—— Part II...	1	10	0
—— Part III...	2	2	0
1882. Part I...	1	14	0
—— Part II...	2	0	0
—— Part III...	2	10	0
—— Part IV...	1	0	0
1883. Part I...	1	10	0
—— Part II...	2	10	0
—— Part III...	1	12	0
1884. Part I...	1	8	0
—— Part II...	1	16	0
1885. Part I...	2	10	0
—— Part II...	2	5	0
1886. Part I...	1	8	0
—— Part II...	1	15	0
1887. (A.)	1	3	0
—— (B.)	1	16	0
1888. (A.)	1	10	0
—— (B.)	2	17	6
1889. (A.)	1	18	0
—— (B.)	1	14	0
1890. (A.)	1	16	6
—— (B.)	1	5	0
1891. (A.)	2	2	0
—— (B.)	3	3	0
1892. (A.)	2	1	0
—— (B.)	2	2	0
1893. (A.)	3	14	0
—— (B.)	2	13	0
1894. (A.) Pt. I.	1	5	0
—— (A.) Pt. II.	1	10	0
—— (B.) Pt. I.	3	10	
—— (B.) Pt. II.	1		
1895. (A.) Pt. I.			
—— (A.) Pt.			
—— (B.)			
—— (P			
1896			
——			

*** When the Stock on hand exceeds One Hund
Years may be purchased by Fellows at One-Third

SOLD BY HARRISON

Lightning Source UK Ltd.
Milton Keynes UK
UKHW020840201118
332647UK00011B/971/P